Plant Pathology and Diseases

Plant Pathology and Diseases

Editor: Edward Parker

STATES
ACADEMIC PRESS
www.statesacademicpress.com

States Academic Press,
109 South 5th Street,
Brooklyn, NY 11249, USA

Visit us on the World Wide Web at:
www.statesacademicpress.com

ISBN: 978-1-63989-422-2 (Hardback)

Cataloging-in-Publication Data

Plant pathology and diseases / edited by Edward Parker.
p. cm.
Includes bibliographical references and index.
ISBN 978-1-63989-422-2
1. Plant diseases. 2. Diseased plants. 3. Plants--Wounds and injuries.
4. Plant pathologists. I. Parker, Edward.
SB601 .P53 2022
632--dc23

Table of Contents

Preface

Every book is a source of knowledge and this one is no exception. The idea that led to the conceptualization of this book was the fact that the world is advancing rapidly; which makes it crucial to document the progress in every field. I am aware that a lot of data is already available, yet, there is a lot more to learn. Hence, I accepted the responsibility of editing this book and contributing my knowledge to the community.

Plant pathology is the scientific study of diseases in plants caused by infectious organisms and physiological factors. Plant disease is an impairment of the normal state of a plant that interrupts or remodels its primary functions. All species of plants are prone to diseases and their occurrence and prevalence vary from season to season. They also depend on the presence of the pathogen, environmental conditions, and the crops and varieties grown. Plant pathology deals with the study of pathogen identification, disease etiology, disease cycles, economic impact and plant disease epidemiology. The topics covered in this extensive book deal with the core subjects of this discipline. It will also provide interesting topics for research that interested readers can take up. Students, researchers, experts, and all associated with plant pathology will benefit alike from this book.

While editing this book, I had multiple visions for it. Then I finally narrowed down to make every chapter a sole standing text explaining a particular topic, so that they can be used independently. However, the umbrella subject sinews them into a common theme. This makes the book a unique platform of knowledge.

I would like to give the major credit of this book to the experts from every corner of the world, who took the time to share their expertise with us. Also, I owe the completion of this book to the never-ending support of my family, who supported me throughout the project.

Editor

1

Wheat rust evolution in Spain: a historical review

Fernando MARTÍNEZ-MORENO*, Ignacio SOLÍS

ETSIA, Dep. Ciencias Agroforestales, University of Seville, Ctra de Utrera km1, 41013 Seville (Spain)

*Corresponding author: fernan@us.es.

Competing Interests: The Author(s) declare(s) no conflict of interest.

Editor: Andy Tekauz, AAFC Cereal Research Centre,Winnipeg MB, Canada (retired).

Abstract. Rusts are important wheat diseases worldwide. The three rust diseases of wheat are yellow rust, leaf rust and stem rust, and each has characteristic features. The Guadalquivir valley in the south-west and Girona in the north-east are the areas Spain most affected by these diseases. Key factors for understanding the history of wheat rust epidemics in Spain are historical weather records in the rust-prone areas and characterization of rust resistance in historical varieties currently preserved in seed banks. These diseases in Spain have been of moderate importance, with stem rust being the most significant disease. During the second half of the 20th century several major epidemics occurred. In 1958 and 1978 severe outbreaks of yellow rust occurred in the Guadalquivir valley. These were probably associated with crop intensification, especially a large number of host landraces replaced by just a few cultivars, as well as immigration of external pathogen inoculum. From the early 1970s, CIMMYT elite cultivars arrived in Spain. These possessed good resistance to leaf and stem rust (*Sr2*), and had early heading dates. Subsequently, stem rust severity rapidly decreased in Spanish fields, but leaf rust epidemics became frequent during 1998-2008 on durum wheat in south-west Spain. In 2013, races virulent on *Lr14a* gene were first reported in Spain, but they did not result in disease epidemics. In 2012–16 yellow rust epidemics were recorded at many locations due to incursion of the 'Warrior' race. Despite the availability of effective fungicides and resistant cultivars to manage the three rust diseases, these diseases continue to threaten wheat production in Spain. In 2016, stem rust caused epidemics on durum wheat in Sicily (Italy), which has similar climatic conditions to those found in the south and east of Spain. Alert systems and international co-operation are needed to characterize the resistance of cultivars, and to monitor the movement and virulence of the wheat rust fungi.

Keywords. *Triticum aestivum, Triticum turgidum, Puccinia triticina, Puccinia striiformis, Puccinia graminis,* rust diseases.

INTRODUCTION

Wheat rusts are diseases caused by obligate biotrophic pathogenic fungi that belong to the division *Basidiomycota*, order *Pucciniales*, and genus *Puccinia*. The diseases are caused by three distinct species producing three different diseases. Yellow rust is caused by *Puccinia striiformis* West. f. sp.

tritici, leaf rust is caused by *P. triticina* Eriks., and stem rust is caused by *P. graminis* Pers. f. sp. *tritici* Eriks & E. Henn. Wheat rusts have complex life cycles (macrocyclic) that involve five types of spores, i.e., urediniospores, teliospores, basidiospores, pycniospores, and aeciospores. These pathogens are heteroecious, requiring alternate hosts to complete their life cycles (Kolmer, 2013). On wheat, uredioniospores are produced asexually from uredinia ("pustules") that develop on the leaves, and each pustule may produce up to 3,000 spores per day over a period of 20 days (Roelfs *et al.*, 1992). Polycyclic infections occur in wheat field within growing seasons, when the newly formed urediniospores re-infect the same populations of wheat plants, and these infections have potential to develop into epidemics. Urediniospores are airborne and may travel as far as several thousand kilometres from their initial sources (Ordoñez *et al.*, 2010).

CHARACTERISTICS OF WHEAT RUSTS

Yellow (or stripe) rust

Typical symptoms of *Puccinia striiformis* f. sp. *tritici* infections are yellow–orange pustules forming stripes on the leaves, and awns can also be affected. Since the discovery that barberry (*Berberis* spp.) are the alternate hosts, *P. striiformis* is now considered as a macrocyclic and heteroecious rust (Jin *et al.*, 2010). Temperatures for infection and disease development are low (2–15°C), with an optimum temperature of approx. 10°C (Roelfs *et al.*, 1992). Yellow rust is the prevailing rust at high altitudes or in places with moderate or cold winter, although new races with tolerance to warmer temperatures have been recently reported (Hovmøller *et al.*, 2011). Yield losses can reach 24 to 39% in winter wheat cultivars in Central Asia (Sharma *et al.*, 2016). According to McIntosh *et al.* (2013), 67 yellow rust resistance genes (*Yr1* to *Yr67*), and 42 with temporary designations, have been discovered.

Leaf (or brown) rust

Leaf rust is the most common wheat rust worldwide. Pustules of *P. triticina* are reddish–brown in colour. Leaf rust is heteroecious, with *Thalictrum speciosissimum, Isopyrum fumaroides* and *Anchusa azurea* reported as alternate hosts (Anikster *et al.*, 1997; Bolton *et al.*, 2008). The disease progresses at temperatures from 10 to 30°C, with an optimum temperature of 20°C (Roelfs *et al.*, 1992), and causes yield losses attributable mainly to reduced kernel weights (Saari and Prescott, 1985; Huerta–Espino *et al.*, 2011). However, in environments conducive to leaf rust, losses can surpass 30% of potential yield (Cátedra and Solís, 2003). Cultivars resistant to leaf rust carry *Lr* genes. More than 100 of these genes and their alleles have been identified in wheat, and 71 of these are officially named (Singh *et al.*, 2013).

Stem (or black) rust

Stem rust is characterized by dark reddish–brown pustules. *Puccinia graminis* f. sp. *tritici* can cause disease at temperatures ranging from 15 to 35°C, with an optimum of approx. 25°C (Roelfs *et al.*, 1992). In the past this was considered to be the most harmful wheat rust, causing yield losses of up to 50% (Leonard and Szabo, 2005). Stem rust is heteroecious with several species of barberry and *Mahonia* spp. as alternate hosts (Leonard and Szabo, 2005). Cultivars resistant to stem rust carry *Sr* genes, and at least 60 *Sr* genes are known (Chen *et al.*, 2018).

CLIMATIC CONDITIONS AND HISTORIC WHEAT AREAS OF SPAIN

Spain has a prevailing Mediterranean climate, that is characterized by dry, hot summers and mild, wet winters, in the *Csa* 'Köppen–Geiger' classification (Kottek *et al.*, 2006). In the north of Spain, where temperatures are generally less than in the south, most areas have *Csb* (Mediterranean climate with a warm summer) or *Cfb* classifications (temperate oceanic climate). It is also important to discriminate between the 'standard Mediterranean' climate of lowland regions with moderately–cool winters, and 'continentalized Mediterranean' climate of the interior with cold winters. The standard Mediterranean climate covers coastal areas (excluding the northern Atlantic coast), the Guadalquivir river basin, and the lower reaches of the Tagus and Guadiana basins to the west of the country (Figure 1). The continentalized Mediterranean climate predominates in Spain's plateau (*Meseta*) of the interior, at more than 600 m above sea level (AEMET, 2011).

Water deficits are the main constraints to high yields in Spanish field crops. Two thirds of the country receives less than 400 mm of annual rainfall, and the amount of solar radiation is high, especially in the south (Rivero, 2013). Winds coming from Morocco blowing from SW to NE are important for transport of rust inoculum from early to late maturing wheat varieties (Bout–Roumazeilles *et al.*, 2007).

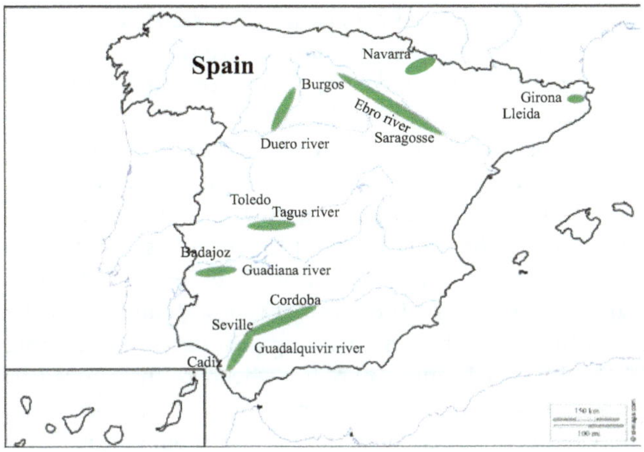

Figure 1. Main wheat growing regions (green) and provinces (red font) in Spain, that are prone to rust outbreaks.

Wheat has been, and continues to be, cultivated in the best dryland soils of Spain. These are located near the basins of the main Spanish rivers that are, from north to south, the Ebro, Duero, Tagus, Guadiana, and Guadalquivir (Figure 1). The deep clay soils of these areas are suitable for wheat production (Rivero, 2013). Sowing of wheat takes place during October through to December each year, while harvest dates range from late May (Guadalquivir basin) to mid-July (Ebro and Duero basin). Extreme heat in summer (especially in the south) prevents extended growing cycles. Winter and facultative wheat cultivars are planted in the Ebro, Duero, Tagus, and Guadiana basins. Spring wheat cultivars are winter-sown in southern Spain (Guadalquivir basin). Bread wheat is the main wheat class, with a current area of approx. 1.8 million ha in north and central Spain, while durum wheat is grown on approx. 0.3 million ha and is the prevailing class in southern Spain (AETC, 2018).

ORIGIN OF WHEAT RUSTS AND THEIR PRESENCE IN SPAIN

Wheat was domesticated as cereal grain approx. 8,000 years BCE, in the Fertile Crescent (Zohary *et al.*, 2012). The most important species of cultivated wheat are *Triticum aestivum* L. subsp. *aestivum* (bread wheat) and *Triticum turgidum* L. subsp. *durum* (Desf.) Husnot (durum wheat). Leaf rust first originated from rust infected *Aegilops speltoides* in Israel while rust isolates from durum wheat derive from virulent genotypes on bread wheat (Liu *et al.*, 2014). The putative centre of origin of yellow rust is reported to be near the Himalaya region (Ali *et al.*, 2014), while that of stem rust is Central Asia, which is also the origin of barberry (Leonard and Szabo, 2005).

Three rust diseases probably adapted quickly to domesticated wheat in a 'host tracking' manner (Stukenbrock and McDonald, 2008). Cultivated wheat spread eastward and westward at approx. 1 km per year (Zohary *et al.*, 2012). In the west, wheat reached Turkey and later spread to the entire Mediterranean Basin, first arriving in Spain approx. 5,600–5,700 years BCE when the 'Neolithic culture' entered Spain. The first two introductions were likely via maritime travel on the Mediterranean Sea from Italy to the Valencia region (east of Spain), and by land through the Pyrenees to northern Spain. Another route is recorded via Morocco to southern Spain approx. 200 years later (García-Martínez de Lagrán, 2015). The expansion of the Neolithic culture was rapid in the Iberian Peninsula (during less than 300 years), and was likely facilitated by the network of Mesolithic peoples. The rust diseases presumably spread along with the cultivation of wheat.

Wheat rusts were important diseases in ancient times. Aristotle (384–322 BCE) mentioned that humidity produced the 'rust disease', and Theophrastus (371–287 BCE) recorded that rust was caused by sunshine and dew (Arthur, 1929). Stem rust uredospores have been retrieved from excavations in Israel from sites established in the late Bronze Age (1,300 BCE) (Kislev, 1982).

In Roman times (8th BCE to 5th ACE centuries) a divinity dedicated to wheat rusts was created, i.e. the god *Robigus*. The celebration *Robigalia* was established on April 25, and was one of the many agricultural festivals celebrated in the month of April. A prayer was recited and a red dog puppy was sacrificed on an altar in a forest outside Rome, Italy (current Via Cassia–Via Sesto Miglio cross, coordinates; 41.967156 N, 12.439740 W). There is also evidence of the Robigalia celebration taking place in *Hispania* (Roman Spain). At the archaeological sites of Mas Castellar de Pontós (Girona) and La Huelga (Palencia), remains of sacrificed dogs have been found that could be connected to Robigalia ceremonies (Adroher *et al.*, 1993; Lettow-Vorbeck *et al.*, 2014). The Spanish word for rust, *roya* derives from the Latin *Robigus*.

Pope Gregory I declared in 590 ACE that April 25 was a day to pray for a good harvest, i.e. the (major) Rogation Day. In western Christianity, a rogation is a public supplication consisting of blessing the fields, and asking for God's mercy through prayers like the Litany of the Saints (Gozalo de Andrés, 2003; Wikipedia, Rogation days, 2018). The coincidence of Rogation Day with Saint Mark's day is chance. Saint Mark died on that day in 68 ACE as a martyr in Alexandria (Egypt). Saint Mark's day (*día de San Marcos*) is still celebrated as a feast day in many villages and small towns throughout Spain, and it is common that people gather and eat together in the countryside on April 25 (Table 1).

DESCRIPTIONS OF WHEAT RUSTS IN THE MIDDLE AND MODERN AGES (5TH TO 19TH CENTURIES ACE)

It is not known whether the occurrence of wheat rust increased in Spain in the Middle Ages compared to Roman times. The 'honeymoon' hypothesis suggests that the severity of grain, especially wheat, diseases was low in Roman times and increased in the Middle Ages in North Europe, as bread wheat became the prevalent cereal (Dark and Gent, 2001). It is possible that Spain also underwent this 'honeymoon' effect, but the scanty records of rust attacks does not allow corroboration of this hypothesis. In the Book of Agriculture of G. Alonso de Herrera (16th century) there is a mention of 'rust', but on barley, and it is not clear if this referred to rust or to any foliar disease favoured by fog (Alonso de Herrera, 1818).

Barberry (*Berberis vulgaris*) is present in Spain, and three subspecies are reported to occur, subsp. *vulgaris*, *seroi* and *australis*. Moors may have planted barberry bushes as thorny hedges to border fields and gardens of *Al-Andalus* (Muslim Spain), and they also produced good quality jam from the berries (Roelfs *et al.*, 1992). Andalusian botanist Abu'l-Khayr cited the presence of barberry bushes in his 'Book of Agriculture' in the 11th century (Harvey, 1993). The expansion of barberry plantings may have contributed to increases in rust outbreaks in Spain, since this these plants are alternate hosts for stem rust and, as recently reported by Jin *et al.* (2010), also for yellow rust. As barberry only grows at altitudes above 1,000 m in southern Spain, but in the north can grow at lower elevations (300 m and above; López, 2007), it is likely that rust outbreaks due to barberry proximity first occurred in wheat cultivated in northern Spain.

Drought, locusts, birds, and ants were frequently cited as the main constraints to Spanish wheat production in the Late Middle and Modern Ages (12th to 19th century) (Alberola-Romá, 2012; Zadoks, 2013; Páscoa *et al.*, 2017), but rusts were rarely mentioned. Throughout the Middle Ages there were periods of climatic instability. From 1645 to 1715, the 'minimum of Maunder' (a reduction of solar activity) led to global cooling resulting in the so-called 'Little Ice Age' (Barriendos, 1997; Alberola-Romá, 2014). In the 40 year period of 1760 to 1800 the 'Maldà Oscillation' produced a 'moody' climate throughout the Spanish Mediterranean area, with years of severe drought and others of heavy rains (Barriendos and Llasat, 2009). Rust epidemics were probably not frequent, but they may have occurred with some regularity, especially in years with heavy rains in winter and spring.

A possible consequence of increased rust severity during this period was the retrieval of ancient wheat hulled species, einkorn, emmer and spelt, which are considered to be more resistant to wheat rusts than non-hulled varieties (Campbell, 1997; Hussien *et al.*, 1998; Zapata *et al.*, 2004). These cereal types, dominant in the Neolithic Period but in declining production at the time of the Roman Empire, may have been re-introduced to some extent to many places in Spain and Europe in the Middle Ages, especially in humid and mountainous land as found in many places in northern Spain (e.g. Asturias). Their cultivation continued until recently in that region (Dugan, 2008).

In S. Clemente's supplement (*adición*) to the book *Libro de Agricultura General* from Alonso de Herrera, it is stated that "*chamorro* [awnless] wheat is more resistant to rust because dew droplets do not build up in their spikes", but *candeal velloso* is more infected by rust due to its awned heads. To quote Alonso de Herrera (1818), "*Chapado velloso* and *moruno velloso* wheat [two kinds of durum wheat] resist well the rust".

DESCRIPTIONS OF WHEAT RUSTS IN THE 19TH AND EARLY 20TH CENTURIES

Throughout the 19th century European scientists reported the presence of rusts in wheat fields, but severe epidemics were rarely noted. Epidemics that occurred were of only local or regional importance, probably because of low crop intensification, long rotation periods between wheat crops, and diversity of wheat landraces. Yellow rust was the first to appear in February–March each growing season, followed by leaf rust in March–April and stem rust, the most damaging, in May. In Castille, straw and other materials were commonly burned to avoid dew formation on the leaves, and dew was removed by ropes each held by two people walking along the crop furrows. In 1877 stem rust epidemics were reported in Navarra (Ruiz de Casaviella, 1878), and leaf rust was also recorded in Teruel, Castellón and Guadalajara (González-Fragoso, 1918).

In 1923, during a trip to northern Spain, American plant pathologist E.C. Stakman observed greater presence of stem rust in fields close to barberry bushes than in other areas (Stakman, 1923). A brochure by the Spanish Ministry of Agricuture (*Ministerio de Agricultura de España*) in 1932 described the state of wheat rusts in

Table 1. Selected locations in Spain where Saint Mark's day is a popular feast.

Location	Province	Population[1] (2017)	Characteristics
Adra	*Almería*	24,697	Celebration
El Ejido	*Almería*	88,096	Local pilgrimage
Noreña	*Asturias*	5,210	Gastronomic fair
El Barraco	*Ávila*	1,886	Local pilgrimage
Almendralejo	*Badajoz*	33,540	Local pilgrimage
Talayuela	*Cáceres*	7,338	Celebration
Puente Genil	*Córdoba*	30,173	Local pilgrimage
Arroyo del Ojanco	*Jaén*	2,353	Local pilgrimage
Beas de Segura	*Jaén*	5,275	Local pilgrimage, bullfighting festival
Noia	*La Coruña*	14,295	Horse, livestock and machinery fair
Oleiros	*La Coruña*	35,198	Celebration
San Martín de la Vega	*Madrid*	18,824	Celebration, procession, bullfighting festival
Valdemoro	*Madrid*	73,976	Meal on park, typical sweet (*hornazo*)
Benaoján	*Málaga*	1,497	Festival, local pilgrimage
Cuevas de San Marcos	*Málaga*	3,722	Local pilgrimage, own name village, typical sweet (*hornazos de San Marcos*)
Alfarnate	*Málaga*	1,113	Local pilgrimage
Bullas	*Murcia*	11,546	Local pilgrimage
Yecla	*Murcia*	34,092	Local pilgrimage
Palencia	*Palencia*	78,892	Local pilgrimage
Agulo	*S.C. Tenerife*	1,066	Bonfire jumping
Tegueste	*S.C. Tenerife*	11,108	Local pilgrimage
Numancia de la Sagra	*Toledo*	4,755	Celebration
Gatika	*Vizcaya*	1,672	Celebration

[1] Locations shown are those with populations of more than 1,000 inhabitants.

Spain. The most common rust was listed to be stem rust, and the main control measures suggested were removal of barberry bushes near wheat fields, hot water treatment of wheat seed to kill the spores, and use of resistant wheat varieties. Praying to God was still considered as a disease control measure at the time (*Ministerio de Agricultura*, 1932).

Unamuno Irigoyen (1943), in his reception speech to the *Real Academia de Ciencias Exactas, Físicas y Naturales*, stated that, "from all known rust fungi there is no doubt that wheat rusts are the most important from an agricultural economy point of view, because of the great losses they cause on harvest". He also confirmed the absence of published data about the importance of rust fungi in Spain.

DESCRIPTIONS OF WHEAT RUST EPIDEMICS, RACES, AND RESISTANCE BREEDING FROM THE SECOND HALF OF THE 20TH CENTURY

The most severe rust attacks from 1950 onwards were in areas where wheat crop areas were largest and most intensified, with the Guadalquivir basin region (western Andalusia, south–west Spain) having the greatest rust severity. There are also reports of rust attacks in Girona (Catalonia, north–east). Until the 1950s, there was no information about rust races in Spain, and on many occasions wheat rust diseases and their causal species were confounded, especially leaf and stem rust. The first scientific studies were carried out by M.J. Urríes after his stay in 1947 with W.Q. Loegering at the Minnesota Agricultural Experiment Station (Co-operative Rust Laboratory at St. Paul, Minnesota, USA) to study stem rust. He later collaborated with agronomists R. Cañamas and J. Salazar from INIA in Madrid (Fernández, 1964). J. Salazar continued his research at the *Centro de Cerealicultura* (Center for Cereal Research) in Madrid in the 1950s, 1960s, and 1970s. He surveyed the virulence of the main stem and leaf rust races in Spain. The main highlights relating to epidemics, race surveys, and breeding for resistance to the three wheat rust diseases are presented here.

YELLOW RUST

Yellow rust epidemics in Andalusia

Yellow rust rarely produced epidemics in Spain, but in 1957 and 1960 severe yellow rust outbreaks were reported on bread wheat in the Guadalquivir basin (Zadoks, 1960). Only a few years earlier, French (Florence Aurore) and Italian (Mara, Impeto) cultivars had been imported in an attempt to increase the low yields of typical older Spanish landrace varieties. Concurrently, changes in cultural practices such as increased seed rates and enhanced N fertilization were implemented for wheat crops. In 1978 another more severe yellow rust epidemic occurred in the Guadalquivir basin. According to Nagarajan *et al.* (1984), these epidemics were caused by a 'perfect storm' of conditions favourable to yellow rust, including:

1. Replacement of varieties. The cultivars Mahissa 1 and Siete Cerros (of CIMMYT origin) had replaced a multitude of landraces and varieties, but the replacements were highly susceptible to yellow rust. However, other new cultivars such as Cajeme, Yecora, Cocorit and Mexicale were resistant (Alvarado and Morillo, 1978).

2. Presence of yellow rust inoculum. In 1976 and 1977, severe yellow rust epidemics were recorded in many parts of the south-west Mediterranean Basin, such as Tunisia and Algeria. Those countries were also replacing old landraces with high yielding, semi-dwarf CIMMYT cultivars, and airborne spores could easily have reached Andalusia from these countries.

3. Favourable weather for the development of the disease. Heavy rains in the Guadalquivir basin were reported in the spring of 1978, with many cloudy days and mild temperatures.

However, Nagarajan *et al.* (1984) made *P. striiformis* isolate surveys, and ruled out the presence of a new virulent race. The main races were 40E008, 40E136 and 41E136 (Johnson *et al.*, 1972), which belonged to the 'French group' and the 'Levantine group'. At this time wheat cultural practices also intensified (increased plant density and fertilizer applications). The epidemics prompted the former *Instituto Nacional de Semillas y Plantas de Vivero* (National Institute of Seed and Nursery Plants) to demand resistance to yellow rust as a prerequisite to register a variety (Palmero *et al.*, 2008). Yellow rust severity rapidly decreased thereafter.

The 'Warrior' yellow rust race

From 1980 yellow rust did not cause significant damage in Spain, but in 2012 yellow rust epidemics were again reported on bread wheat at several Spanish locations. The newly created Global Rust Reference Center (GRRC) in Denmark (GRRC, 2018) surveyed Spanish yellow rust races, and all matched the 'Warrior' race. This race was first detected in 2011 in Denmark, France, Germany, Sweden, United Kingdom, and Spain, often at high frequencies. The race replaced previous populations, from which it was clearly different, since it originated from a population close to the Himalayas (Hovmøller *et al.*, 2016). The rapid spread of the 'Warrior' race from northern to southern Europe was due to its virulence on most bread wheat cultivars, including the British cultivar 'Warrior', from which the race takes its name. The race displays a high number of virulence alleles, including those for *Yr1, Yr2, Yr3, Yr4, Yr6, Yr7, Yr9, Yr17, Yr25, Yr32*, and *YrSp* (Table 2).

Previously, in 2008 and 2009, yellow rust was found on some bread wheat cultivars in Navarra (northern Spain). In 2011 yellow rust severity increased in Navarra, and in 2012 spread to other locations in northern Spain (Aragon, Castilla y León). In 2013 yellow rust spread south (Castilla La Mancha, Andalusia) and east-

Table 2. Races of *Puccinia striiformis* f. sp. *tritici* (yellow or stripe rust) in Spain during 2010–2016, according to the Global Rust Reference Center.

Pu. striiformis f. sp. *tritici*	Year						
	2010	2011	2012	2013	2014	2015	2016
Race name	PstS2v17v27	PstS2v27, PstS2v27v17, PstS3v2v25v27, Warrior	Warrior	Warrior	Warrior	Warrior, Warrior(-), Triticale 2006	Warrior(-)
Frequency (%)	100	29/14/43/14	100	100	100	44/44/12	100
No. isolates	2	8	6	8	6	9	5

[1] Virulence of race PstS2v27 (*Yr2,6,7,8,9,25,27*), PstS2v17v27 (*Yr2,6,7,8,9,17,25,27*), PstS3v2v25v27 (*Yr2,6,7,8,25,27*), Warrior (*Yr1,2,3,4,6,7,9,17,25,32,Sp,Amb*), Warrior (-) (*Yr1,2,3,4,6,7,9,17,25,32,Sp*), and Triticale2006 (*Yr2,6,7,8,10*).

wards (Catalonia). Farmers had to treat bread wheat crops with fungicides in many locations. The GENVCE (Spanish Group for Screening Field Crops New Varieties) field tests in 2013/14 found that the cultivars Artur Nick, Gazul, and Valbona were resistant (Gómez-Caño, 2016).

In 2014 yellow rust epidemics intensified. As an example, in 2013 variety Altamira was susceptible only in Navarra, but in 2014 it was also susceptible in Castilla y León, Aragón, and Madrid. As well, important cultivars in Spain such as CCB Ingenio, Artur Nick and Nogal became susceptible to yellow rust. In 2015, race Triticale 2015, first detected in Scandinavia and virulent on many triticale cultivars, was found in Spain by GRRC researchers (GRRC, 2018).

For durum wheat, in 2013 the cultivars Gigadur, Lecitur, Clovis and Massimo Meridio showed yellow rust symptoms in Andalusia and Extremadura. In 2014 yellow rust was also recorded in Castilla y León in durum wheat crops. Yellow rust severity on durum wheat increased and was similar to that on bread wheat in many regions in 2016. In research on yield losses in durum wheat due to yellow rust, performed at a field trial in Aranjuez (Madrid), Vergara-Diaz *et al.* (2015) reported a mean yield loss of 18%, while two cultivars suffered 57% losses. The numbers and weights of kernels decreased, but head numbers were not affected by the disease.

A new yellow rust race was detected in 2016 in Morocco and Sicily (Italy) by the GRRC. This race, named PstS14 and virulent on *Yr2, Yr3, Yr6, Yr7, Yr8, Yr9, Yr17, Yr25, Yr32,* and *YrSp* was detected in Spain in 2018 (GRRC, 2018). Yellow rust severity on the Avocet differential host set at three locations of Andalusia (2016–2017) is presented in Table 3. Resistance genes *Yr5, Yr10, Yr15* were effective in all locations, and Avocet lines with *Yr1* and *Yr8* genes showed resistance to most tested isolates. Yellow rust infections decreased slightly by 2017 in durum and bread wheat crops throughout Spain.

Table 3. Severity of yellow rust (*Puccinia striiformis*) on the Avocet bread wheat near-isogenic differential, set at three locations in the Guadalquivir basin of Andalusia during 2016 and 2017.

Differential genotype	Yellow rust severity		
	Escacena del campo (Huelva)	Écija (Sevilla)	Jerez de la Frontera (Cádiz)
Year	2016	2017	2017
Avocet S	70	-	15
Avocet (*YrA*)	30	20	30
Yr1/6*Avocet S	30	0	0
Yr5/6*Avocet S	0	0	0
Yr6/6*Avocet S	60	40	15
Yr7/6*Avocet S	60	40	20
Yr8/6*Avocet S	5	8	10
Yr9/6*Avocet S	50	80	10
Yr10/6*Avocet S	0	0	0
Yr15/6*Avocet S	0	0	0
Yr17/6*Avocet S	40	40	5
Yr18/6*Avocet S	40	80	20
Yr24/6*Avocet S	15	30	15
Yr26/6*Avocet S	15	20	5
Yr27/6*Avocet S	20	7	2
Yr32/6*Avocet S	50	40	15
YrSp/6*Avocet S	30	3	0

LEAF RUST

First race surveys (1968–1977)

J. Salazar initiated the first leaf rust surveys in Spain. In 1973 he published research on 159 rust samples collected from 1968 to 1971 (151 from bread wheat, seven from durum, and one from triticale) that were characterized for their virulence. He identified 31 races, using a differential set based on five varieties. The most common race was found in 23 isolates, and the five main races comprised 52% of total isolates (Salazar and Brañas, 1973b). In another study performed during 1972 to 1975, 178 samples were collected and 33 races identified. Rust samples were taken from Tortosa (Tarragona), Borjas Blancas (Lleida), La Alberca (Murcia), and La Coruña. None of the races were virulent on variety Malakoff (*Lr1*), and the prevailing races were similar to those found previously (Salazar and Brañas, 1977). These results were published as reports without details regarding methods, which makes it difficult to fully understand the study.

Leaf rust epidemics on durum wheat (1997–2004)

Durum wheat is a traditional crop in southern Spain with average crop areas of approx. 200,000 ha since the 19th century. In 1992, however, area of durum wheat increased when the European Union implemented a supplement subsidy (in the Common Agricultural Policy) for cultivation of this crop (Royo, 2005). Durum acrea reached a peak of 940,000 ha in 2003/04. Although durum wheat was generally considered more resistant to leaf rust than bread wheat, from 2001/02 leaf rust epidemics on durum wheat were recorded at many locations in the Guadalquivir basin. Most durum cultivars were susceptible. Only the Italian variety Colosseo (with *Lr14a* gene) was resistant. Other resistant Italian cultivars such as Italo, Vinci, and Virgilio were brought to Spain in the 2004 growing season (RAEA, 2004). However, most bread wheat cultivars were resistant to leaf rust during this period.

In 2001 a severe leaf rust epidemic was reported in Mexico where durum wheat variety Altar C84, resistant to leaf rust for almost 20 years, became susceptible. The new virulent race, designated as BBG/BN, caused losses valued at $US 32 million in Mexico over the course of 3 years (Singh *et al.*, 2004). Many durum cultivars from CIMMYT, including Altar, carried the *Lr72* gene that protected from many races, including leaf rust races from bread wheat (Herrera *et al.*, 2014). A mutation at the virulence locus corresponding to *Lr72* resulted in a race virulent to most durum wheat cultivars. The variety Gallareta, an Altar sibling, was one of the most cultivated cultivars at that time in Spain, and this displayed susceptibility. As most durum wheat cultivars sown in Spain had CIMMYT origins, many probably carried the *Lr72* gene.

In field trials on yield losses due to foliar diseases in durum wheat in the Guadalquivir basin during the 2000–2001 and 2001–2002 growing seasons, yield losses reached 28 to 30% in susceptible cultivars. Both seasons were characterized by high levels of leaf rust infections (Cátedra and Solís, 2003).

Martínez *et al.* (2005) studied the virulence of 56 single pustule leaf rust isolates from durum and bread wheat, collected from several Andalusian locations during 1998–2000 (Table 4). Thirty-five races were identified using the Thatcher near isogenic lines as differentials. No race was virulent to genes *Lr9* and *Lr24*. Bread wheat races were different from durum wheat races, and the bread wheat races were more variable than those from durum wheat (Kolmer *et al.*, 2013). None of the durum races were virulent to *Lr1, Lr3, Lr15, Lr16* and *Lr17* genes, whereas several bread wheat isolates were virulent to these genes.

A worldwide study of durum wheat leaf rust virulence performed by the Cereal Disease Laboratory (CDL) (Minnesota, USA) determined that Spanish and Italian races grouped together (Kolmer and Liu, 2000). The races were classified as DBBDML, having virulence/avirulence spectra of *Lr2c, 10, 14b, 20, 23 / Lr1, 2a, 2c, 3, 9, 16, 24, 26, 3ka, 11, 17, 30, B, 3bg, 14a, 15, 18, 28*. In a later study, seven isolates taken from durum wheat in Andalusia during the period 2000 to 2003 were found to be similar to isolates from Spain, France, and those from America, suggesting a common origin (Ordoñez and Kolmer, 2007).

Table 4. Frequency of *Puccinia triticina* (leaf or brown rust) isolates virulent on Thatcher near-isogenic lines, collected in Spain and Andalusia in different years.

Genes / locations[1]	Salazar 1972–75	Andalusia 1997	Andalusia 1998–2000	Andalusia 1998–2000 (durum)	Andalusia 2000–03 (durum)	Andalusia 2009–13 (durum)
Lr1	3	0	30	0	0	1.2
Lr2a	17	0	2	0	0	0
Lr2b	52	0	34	47		
Lr2c	88	100	73	100	0	100
Lr3	53	40	41	0	0	3.6
Lr3bg		40	41	0	0	3.6
Lr3ka		10	25	0	0	0
Lr9		0	0	0	0	0
Lr10		100	89	87	100	97.6
Lr11	73	0	96	100	0	0
Lr12			100	100		
Lr13			56	20		
Lr14a		20			0	13.2
Lr14b		100	98	93	14.3	98.8
Lr15		0	30	0	0	
Lr16		30	30	0	0	1.2
Lr17		20	32	0	0	0
Lr18		50	96	100	0	57.8
Lr19		0	0	0	0	
Lr20		60	71	93	100	100
Lr21		0	98	100		
Lr22			100	100	14.3	
Lr23		100	79	87	100	
Lr24		0	0	0	0	0
Lr25		0	0	0	0	
Lr26		0	7	7	0	0
Lr28		0	9	0	0	0
Lr30		10	13	0	0	0
Lr32		0	9	0	0	
Lr34			100	100		
Lr35			82	73		
Lr37			100	100		
LrB		60	64	47		
Thatcher		100	100	100	100	100
No. isolates	178	10	56	15	8	83

[1] Modified and here reproduced with the permission of the copyright holder from Martínez *et al.* (2005).

From 2004 onwards, rust attacks in Spain began to decrease. Farmers regularly applied fungicides to their crops, new durum cultivars from breeders were released in the Guadalquivir basin with resistance to leaf rust, and the environmental conditions of subsequent seasons (mainly low rainfall) were unfavourable to rust epidemic development.

Many of the recently released durum wheat resistant cultivars carried *Lr14a* located on chromosome 7BL. This gene, from the emmer cultivar Yaroslav, was transferred to bread wheat and then to durum wheat. The Italian cultivar Creso (released in 1974) carried this gene, and for more than 30 years was resistant to all leaf rust races. Many durum wheat cultivars worldwide inherited the *Lr14a* gene, including Italian variety Colosseo and Spanish variety Don Jaime (Martínez *et al.*, 2005).

New races virulent to Lr14a in Spain

Pustules of high infection type were observed on the cultivars Colosseo and Don Jaime in the Spring of 2013, at widely separated field trials at Conil de la Frontera (Cádiz, southern Spain) and La Tallada d'Empordà (Girona, north-east Spain). This confirmed that virulence to *Lr14a* gene was present in Spain (Table 5). Additionally, this race was different from the reported French races virulent to *Lr14a* that were avirulent to *Lr72*, while the Spanish races were also virulent to *Lr72* (Soleiman *et al.*, 2016).

Durum wheat breeding programmes are also using other genes for resistance to leaf rust. Herrera-Foessel *et al.* (2005) reported the presence of new genes in durum cultivars, such as the complementary genes *Lr27* + *Lr31* (present in CIMMYT variety Jupare or in Spanish Don Ricardo), *Lr61* (present in CIMMYT variety Guayacán), *LrCam* (present in Camayo) and *Lr3* (present in Storlom). At the same time a high level of partial resistance was found in some durum wheat cultivars (Herrera-Foessel *et al.*, 2008). Further research has identified novel resistance genes to be deployed in the near future in durum wheat from wild relatives but already in the bread wheat background, i.e. *Lr19* (from *Lophopyrum ponticum*), *Lr47* (from *Triticum speltoides*), and *Lr37* (from *Triticum ventricosum*). Other research has focused on selecting durum landraces, lines or old varieties with leaf rust resistance. Loladze *et al.* (2016) found usable resistance in the durum varieties Gaza, Amria, Geromtel_3, Geruftel_1, Tunsyr_2, and Biblos, of varied origins.

In a study of gene postulation in bread wheat, the presence of resistance genes *Lr1*, *Lr10*, *Lr13*, *Lr20*, *Lr26* and *Lr28* was confirmed (Martínez *et al.*, 2007). In durum wheat, the resistance responses of most cultivars when inoculated with several leaf rust isolates did not match with any of the *Lr* genes from the Thatcher differential series. Some cultivars also had some levels of partial resistance (Martínez *et al.*, 2007). Another study was performed by the same group to characterize leaf rust resistance in a collection of 917 accessions from the Spanish Center for Genetic Resources (CRF-INIA). Susceptible reactions were normal, and only 6% of total entries (4.8% in durum, 11.9% bread wheat) had severities less than 20% compared to the susceptible check (Figure 2). Seven susceptible accessions (six bread, one rivet wheat) were also identified with fair levels of partial resistance (Martínez *et al.*, 2001).

Table 5. Resistance responses of 20 near-isogenic lines and eight durum genotypes, each at the 5th leaf growth stage, to two races of *Puccinia triticina* (leaf or brown rust), from 2009-11 and the race of 2013.

Genotype	Race and infection type		
	Isolates 2009-11		Isolate 2013
	DBB/BN	DBB/CN	DBB/BS (Conil, Girona)
Thatcher	S	S	S
Lr1	R	R	R
Lr2a	R	R	R
Lr2c	S	S	S
Lr3	R	R	R
Lr3bg	R	R	R
Lr3ka	R	R	R
Lr9	R	R	R
Lr10	S	S	S
Lr11	R	R	R
Lr13	R	R	R
Lr14a	R	R	S
Lr15	R	R	R
Lr16	R	R	R
Lr17	R	R	R
Lr18	R	S	R
Lr23	S	S	S
Lr24	R	R	R
Lr26	R	R	R
Lr30	R	R	R
Jupare (*Lr27+31,Lr72*)	R	R	R
Gallareta (*Lr72*)	S	S	S
Somateria (*Lr14a+*)	R	R	S
Colosseo (*Lr14a+*)	R	R	S
Don Jaime (*Lr14a+*)	R	R	S
Storlom (*Lr3,Lr72*)	R	R	R
Guayacán (*Lr61*)	R	R	R
Camayo (*LrCam*)	R	R	R

[1] R indicates a resistance response and S a susceptible response. Modified and here reproduced with the permission of the copyright holder from Soleiman *et al.* (2016).

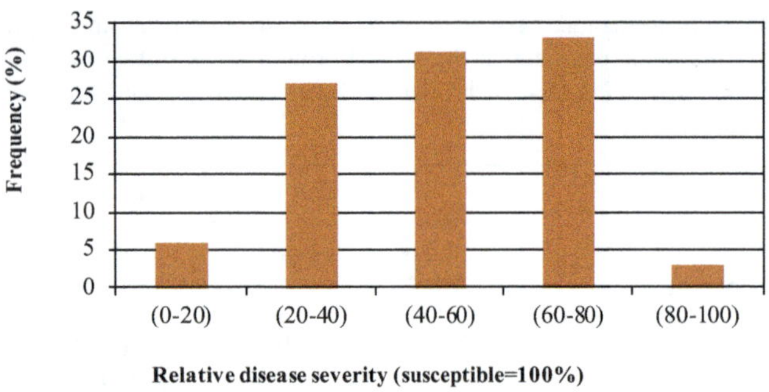

Figure 2. Relative leaf rust (*Puccinia triticina*) severity in a collection of 917 Spanish wheat landraces, planted at Cordoba, Spain in 1996/97. Here reproduced with the permission of the copyright holder from Martínez *et al.* (2001).

STEM RUST

Initial studies (1948–1971)

M.J. Urríes collected stem rust samples during 1948 to 1961 when this rust predominated, for identification of races using the Stakman differential set (Urríes and Arzoz, 1961). Subsequently, J. Salazar and M. Brañas monitored stem rust in the period 1968 to 1971, and reported only mild attacks in the field. From the 120 rust samples they collected, the majority were taken from bread wheat, although some were from barley and *Aegilops* spp. A total of 15 races were found, and their virulence profiles in samples from 1948, 1961 and 1968 to 1971 are displayed in Table 6. Samples collected near barberry bushes belonged to a group of races different from the isolate group collected at distance from barberry. The majority of races tested were avirulent to varieties Reliance (*Sr5*), Vernal (*Sr9e*) and Khapli (*Sr13*) (Salazar and Brañas, 1973a).

Table 6. Virulence of isolates of *Puccinia graminis* f. sp. *tritici* (stem rust) collected in Spain over several time periods and tested on a set of differentials wheat genotypes.

Differential	*Sr* gene	Year[1]		
		1948	1961	1968–71
Little Club	Susc.	100	100	100
Marquis	*Sr7b*	46.6	37.2	55.8
Reliance	*Sr5*	6.6	0.5	9.2
Kota	*Sr28*	22.6	33.3	32.5
Arnautka	*Sr9d*	86.6	47.6	51.7
Mindum	*Sr9d*	86.6	47.6	52.5
Spelmar	*Sr9d*	86.6	47.6	53.3
Kubanka	*Sr9g*	100	100	81.7
Acme	*Sr9g*	100	100	79.2
Einkorn CI2433	*Sr21*	65.3	21.9	40.0
Vernal	*Sr9e*	2.6	0.2	3.3
Khapli	*Sr13*	0	0	0
No. isolates		–	510	120

[1] Analyses performed in 1948 and 1961 by Urríes and Arzoz (1961). The 1968–71 analyses were performed by Salazar and Brañas (1973a).

Stem rust vanishes by the mid–1970s

The first CIMMYT wheat cultivars were introduced to Spain in the early 1970s and these had earlier heading dates than most cultivars cultivated in Spain at that time. In Andalusia, harvest date was advanced by about a month, to May 25 to June 15. Stem rust did not appear on wheat leaves until the end of April each year, by which time it was too late to cause epidemics and the disease gradually disappeared from wheat fields. Furthermore, these new CIMMYT cultivars carried the *Sr2* gene that provided partial resistance and was effective worldwide (Singh and Rajaram, 2002). However, at present in off-season summer field trials under irrigation performed by some wheat breeding companies, pustules of stem rust can be seen at some locations, e.g. Conil de la Frontera (Cadiz) in southern Spain.

The 'Sicily threat' of 2016

In 1999 a new virulent stem rust race, named Ug99, appeared in Uganda, and subsequently spread throughout East Africa (Wanyera *et al.*, 2006). In 2003 this race was found in Ethiopia and then crossed the Red Sea. In 2007 it was detected in Iran, but was controlled by releasing resistant cultivars led by international co-operation. No sign of stem rust was seen in Spain during this period.

In 2016 a severe stem rust epidemic was recorded in Sicily affecting 20,000 ha of durum wheat. The race involved was different from Ug99 and was virulent to genes *Sr9e* and *Sr13*. Both genes are common in durum cultivars (Bhattacharya, 2017). A risk analysis by the GRRC predicted the possibility that spores of the Sicilian stem rust race (TTRTF) were already in the eastern regions of the Iberian Peninsula (GRRC, 2018). However, to our knowledge stem rust race TTRFT has not yet been detected in Spain.

CONCLUSIONS

Wheat rust outbreaks occur from time to time but are difficult to predict. The main factors promoting epidemics are: 1) the presence of airborne inoculum that can be dispersed over very long distances, 2) novel pathogen races with virulence to the main *R*-genes, 3) deployment of susceptible host cultivars, 4) particular cultural practices, such as a high crop seed rates, monoculture, and irrigation, and 5) changing weather conditions. These factors are difficult to control, as continents and many countries are involved. International co-ordination is needed to monitor the movement of inoculum, identify prevailing races and their virulence, and gain knowledge of the rust resistance genes present in the cultivated wheat cultivars, in Spain and worldwide. As such, and emphatically, the monitoring and research carried out by the Borlaug Global Rust Initiative (BGRI, 2018), and the Global Rust Reference Center (GRRC, 2018) are crucial. Additionally, at the Spanish national level, it is important that disease alert systems such as RAIF in Andalusia (RAIF, 2018) are adequately maintained into the future.

The use of rust resistant cultivars is important, preferably deploying effective *R*-genes against the prevailing pathogen races prevailing in a region. Many plant breeding companies and institutions are attempting to combine the common type of resistance (monogenic, vertical and hypersensitive) with more difficult to manage partial resistance (polygenic, and horizontal), to enhance the durability of the resistance.

Diversification of crop genetics is also relevant. Rust epidemics in the second half of the 20th century were partly the result of crop intensification and reductions in the number of wheat cultivars grown, which were often genetically related and shared the same *R*-genes. Diversity solutions such as extended crop rotations, cultivar mixtures, segregating host populations, regional control to deploy cultivars with different *R*-genes, and genes for partial resistance, are all advisable rust management strategies (McDonald and Linde, 2002). The consequences of new cultural practices for wheat (including irrigation, new fertilizers, increased seeding rates, hybrid cultivars, and high-yielding susceptible cultivars) with respect to rust epidemics are yet to be seen. This may be controversial, as cultivars with *R*-genes normally have small yield penalties, and many farmers prefer to use high-yielding but susceptible cultivars, and apply fungicides to maximize their incomes. It is not always possible, however, to apply fungicides or do so in a timely manner. In addition, susceptible cultivars may act as inoculum reservoirs for any of the three wheat rust pathogens.

An array of fungicide classes (triazoles, strobilurins, carboxamides) are available for farmers. Fungicides are relatively affordable, effective, and they can control other foliar diseases as well as rusts. However, their application should be discouraged when resistant cultivars are available, as fungicide use will reduce farmer profitability and contribute to increased environmental footprints. Fungicides are certainly warranted when resistance is not available, or is not fully effective, particularly if rust outbreaks are forecasted to cause significant yield losses.

What may be expected of the three rusts in future

Yellow rust is the most unpredictable rust. The epidemics of 1958, 1978, and 2012 in Spain appeared suddenly and caused significant yield losses. Potential damage from epidemics decreased in importance when breeders released new resistant wheat cultivars and farmers applied fungicides. New races adapted to warm temperatures may become more prevalent and infect wheat for longer periods than expected.

Leaf rust is the most common rust in Spain, although it rarely causes major problems. Most of the durum and bread wheats grown contain effective *Lr* genes. However, virulence analyses of the prevailing races, and severity assessments in rust-prone regions, must continue to be carried out.

Stem rust has not been detected in Spain since the early 1970s although it may be present at low frequencies. It is important to carefully monitor this rust in view of the durum wheat stem rust outbreak of 2016 in Sicily.

ACKNOWLEDGMENTS

The authors thank Dr Julián Rodríguez-Algaba (GRRC, Aarhus University, Denmark) for his constructive comments on the manuscript of this paper. F. Martínez-Moreno acknowledges funding from the FIUS (Research Foundation of the University of Seville) projects PRJ201702960 and PRJ201602711.

LITERATURE CITED

Adroher A.M., Pons i Brun E., Ruiz de Arbulo J., 1993. El yacimiento de Mas Castellar de Pontós y el comercio del cereal ibérico en la zona de *Emporion* y *Rhode* (SS. IV–II A.C). *Archivo Español de Arqueología* 66: 31–70.

AETC (Spanish Asociation of Cereal Experts), 2018. http://www.aetc.es/.

AEMET (Spanish Agency of Meteorology), 2011. *Iberian Climate Atlas.* Agencia estatal de Meteorología, Ministerio de Medio Ambiente y Medio Rural y Marino (Spain) - Instituto de Meteorología de Portugal (Portugal), 80 pp. http://www.aemet.es/es/conocermas/recursos_en_linea/publicaciones_y_estudios/publicaciones/detalles/Atlas-climatologico.

Alberola-Romá A., 2012. Plagas de langosta y clima en la España del siglo XVII. *Relaciones* 129: 21–50.

Alberola-Romá A., 2014. *Los Cambios Climáticos. La Pequeña Edad del Hielo en España.* Editorial Cátedra, Madrid, Spain, 344 pp.

Ali S., Gladieux P., Leconte M., Gautier A., Justesen AF, ... Vallavieille-Pope C., 2014. Origin, migration routes and worldwide population genetic structure of the wheat yellow rust pathogen *Puccinia striiformis* f. sp. *tritici. PLoS Pathogens* 10(1), e1003903. doi:10.1371/journal.ppat.1003903.

Alonso de Herrera G., 1818. *Agricultura general de Gabriel Alonso de Herrera* (revised versión, 1st publ. 1513). Pub. Real Sociedad Económica Matritense, Madrid, Spain, 188 pp.

Alvarado M., Morillo F., 1978. *Enfermedades del trigo,* Hojas divulgadoras. Ministerio de Agricultura, Spain. 15 pp. http://www.mapama.gob.es/ministerio/pags/biblioteca/hojas/hd_1978_08.pdf.

Anikster Y., Bushnell W.R., Eilam T., Manisterski J., Roelfs A.P., 1997. *Puccinia recondita* causing leaf rust on cultivated wheat, wild wheat, and rye. *Canadian Journal of Botany* 75: 2082–2096.

Arthur, J.C., 1929. *The Plant Rust (Uredinales).* John Wiley & Sons, New York, USA, 446 pp.

Barriendos M., 1997. Climatic variations in the Iberian Peninsula during later Maunder Minimum (ad 1675–1715): an analysis of date from rogation ceremonies. *The Holocene* 7(1), 105–111.

Barriendos M., Llasat C., 2009. El caso de la anomalía "Maldà" en la cuenca mediterránea occidental (1760–1800). Un ejemplo de fuerte variabilidad climática. In: *Desastre Natural, Vida Cotidiana y Religiosidad Popular en la España Moderna y Contemporánea,* (A. Alberola, J. Olcina, ed.). Publicaciones de la Universidad de Alicante, Spain, 253–286.

Bhattacharya S., 2017. Deadly new wheat disease threatens Europe's crops. Nature News. http://www.nature.com/news/deadly-new-wheat-disease-threatens-europe-s-crops-1.21424.

BGRI (Borlaug Global Rust Initiative), 2018. https://www.globalrust.org/.

Bolton M.D., Kolmer J.A., Garvin D.F., 2008. Wheat leaf rust caused by *Puccinia triticina. Molecular Plant Pathology* 9(5): 563–575.

Bout-Roumazeilles V., Combourieu Nebout N., Peyron O., Cortijo E., Landais A., Masson-Delmotted V., 2007. Connection between South Mediterranean climate and North African atmospheric circulation during the last 50,000 yr BP North Atlantic cold events. *Quaternary Science Reviews* 26: 3197–3215.

Cátedra M.M., Solís I., 2003. Effect of a fungicide treatment on yield and quality parameters of new varieties of durum wheat (*Triticum turgidum* L. ssp. *durum*) and bread wheat (*Triticum aestivum* L.) in western Andalusia. *Spanish Journal of Agricultural Research* 1(3): 19–26.

Campbell E.A., 1997. Spelt: Agronomy, genetics, and breeding. In: *Plant Breeding Reviews* (J. Janick ed.), John Wiley, New York, USA, 187–213.

Chen S., Guo Y., Briggs J., Dubach F., Chao S., ... Dubcovsky J., 2018. Mapping and characterization of wheat stem rust resistance genes *SrTm5* and *Sr60* from *Triticum monococcum. Theoretical and Applied Genetics* 131: 625–635.

Dark P., Gent H., 2001. Pests and diseases of prehistoric crops: a yield "honeymoon" for early grain crops in Europe? *Oxford Journal of Archeology* 20: 59–79.

Dugan F.M., 2008. *Fungi in the Ancient World.* American Phytopathological Society, St. Paul, MN, USA, 152 pp.

Fernández E., 1964. Don Manuel Jordán de Urríes y Azara (1909–1962) (nota necrológica). *Anales del Instituto Botánico A.J. Cavanilles* 21(2): 331–336.

García-Martínez de Lagrán I., 2015. Recent data and approaches on the neolithization of the Iberian Peninsula. *European Journal of Archeology* 18(3): 429–453.

Gómez-Caño A., 2016. *Estudio sobre la evolución de la resistencia a la roya amarilla (Puccinia striiformis) en las nuevas variedades comerciales de trigos duros, harineros y triticales en campo (2008-2015).* MsC thesis, ETSIA, University of Seville, Spain, 127 pp.

González-Fragoso R., 1918. La roya de los vegetales. Enumeración y distribución geográfica de los uredales conocidos hasta hoy en la Península Ibérica e Islas Baleares. *Trabajos del Museo Nacional de Ciencias Naturales, serie botánica núm. 15.* Madrid, Spain.

Gozalo de Andrés C., 2003. Meteorología popular. Las rogativas. *Revista del Aficionado a la Meteorología* 15 (nov. 2003).

GRRC (Global Rust Reference Center), 2018. Aarhus University, Denmark. www.wheatrust.org.

Harvey J.H., 1993. The Book of Abu'l-Khayr. *Garden History* 21: 118–120.

Herrera-Foessel S.A., Singh R.P., Huerta-Espino J., Yuen J., Djurle A., 2005. New genes for leaf rust resistance in CIMMYT durum wheats. *Plant Disease* 89: 809–814.

Herrera-Foessel S.A., Singh R.P., Huerta-Espino J., Crossa J., Djurle A., Yuen J., 2008. Genetic analysis of slow-rusting resistance to leaf rust in durum wheat. *Crop Science* 48 (6): 2132–2140.

Herrera-Foessel S.A., Huerta-Espino J., Calvo-Salazar V., Lan C.X., Singh R.P., 2014. *Lr72* confers resistance to leaf rust in durum wheat cultivar Atil C2000. *Plant Disease* 98: 631–635.

Hovmøller M.S., Sørensen C.K., Walter S., Justesen A.F., 2011. Diversity of *Puccinia striiformis* on cereals and grasses. *Annual Review of Phytopathology* 49: 197–217.

Hovmøller M.S., Walter S., Bayles R.A., Hubbard A., Flath K., ... Vallavieille-Pope, C. 2016. Replacement of the European wheat yellow rust population by new races from the centre of diversity in the near-Himalayan region. *Plant Pathology* 65: 402–411.

Huerta-Espino J., Singh R.P., Germán S., McCallum B.D., Park R.F., … Goyeau H., 2011. Global status of wheat leaf rust caused by *Puccinia triticina*. *Euphytica* 179: 143–160.

Hussien T., Bowden R.L., Gill B.S., Cox T.S., 1998. Chromosomal locations in common wheat of three new leaf rust resistance genes from *Triticum monococcum*. *Euphytica* 101: 127–131.

Jin Y., Szabo L.J., Carson M., 2010. Century-old mystery of *Puccinia striiformis* life history solved with the identification of *Berberis* spp. as an alternate host. *Phytopathology* 100: 432–435.

Johnson R., Stubbs R.W., Fuchs E., Chamberlain N.H., 1972. Nomenclature for physiologic races of *Puccinia striiformis* infecting wheat. *Transactions of the British Mycological Society* 58 (3): 475–480.

Kislev M.E., 1982. Stem rust of wheat 3300 years old found in Israel. *Science* 216: 993–994.

Kolmer J.A., Liu J.Q., 2000. Virulence and molecular polymorphism in international collections of the wheat leaf rust fungus *Puccinia triticina*. *Phytopathology* 90: 427–436.

Kolmer J.A., 2013. Leaf rust of wheat: pathogen biology, variation and host resistance. *Forests* 4: 70–84.

Kottek M., Grieser J., Beck C., Rudolf B., Rubel F., 2006. World map of the Köppen-Geiger climate classification updated. *Meteorologische Zeitschrift* 15(3): 259–263.

Lettow-Vorbeck C.L., Esparza A., Sánchez A., 2014. Bones as rubbish or a ritualized deposit? Dog butchering in La Huelga (Dueñas, Palencia). *Zephyrus* 74: 89–115.

Leonard K.J., Szabo L.J., 2005. Stem rust of small grains and grasses caused by *Puccinia graminis*. *Molecular Plant Pathology* 6(2): 99–111.

Liu M., Rodrigue N., Kolmer J., 2014. Population divergence in the wheat leaf rust fungus *Puccinia triticina* is correlated with wheat evolution. *Heredity* 112: 443-453.

Loladze A., Kthiri D., Pozniak C., Ammar K., 2016. Genetic analysis of leaf rust resistance in six durum wheat genotypes. *Phytopathology* 104: 1322–1328.

López G., 2007. *Guía de los árboles y arbustos de la Península Ibérica y Baleares*. Mundi-Prensa, Madrid, Spain, 896 pp.

Martínez F., Niks R.E., Moral A., Urbano J.M., Rubiales D., 2001. Search for partial resistance to leaf rust in a collection of ancient Spanish wheats. *Hereditas* 135: 193–197.

Martínez F., Sillero J.C., Rubiales D., 2005. Pathogenic specialization of *Puccinia triticina* in Andalusia from 1998 to 2000. *Journal of Phytopathology* 153: 344-349.

Martínez F., J.C. Sillero, Rubiales D., 2007. Resistance to leaf rust in cultivars of bread wheat and durum wheat grown in Spain. *Plant Breeding* 126: 13–18.

McIntosh R.A., Yamazaki Y., Dubcovsky J., Rogers J., Morris C., ... Xia X.C., 2013. Catalogue of gene symbols for wheat. *12th International Wheat Genetics Symposium*, 8–13 September 2013, Yokohama, Japan.

McDonald B.A., Linde C., 2002. The population genetics of plant pathogens and breeding strategies for durable resistance. *Euphytica* 124: 163–180.

Ministerio de Agricultura, Industria y Comercio, 1932. *Las royas de los cereales*. Dirección General de Agricultura. http://www.mapama.gob.es/ministerio/pags/biblioteca/hojas/hd_1932_18.pdf.

Nagarajan S., Kranz J., Saari E.E., Seiboldt G., Stubbs R.W., Zadoks J.C., 1984. An analysis of the 1978 epidemic of yellow rust on wheat in Andalusia, Spain. *Zeitschrift Für Pflanzenkrankheiten Und Pflanzenschutz / Journal of Plant Diseases and Protection* 91 (2): 159–70.

Ordoñez M.E., Kolmer J.A., 2007. Virulence phenotypes of a worldwide collection of *Puccinia triticina* from durum wheat. *Phytopathology* 97: 344–351.

Ordoñez M.E., Germán S.E., Kolmer J.A., 2010. Genetic differentiation within the *Puccinia triticina* population in South America and comparison with the North American population suggests common ancestry and intercontinental migration. *Phytopathology* 100: 376–383.

Palmero D., García M.C., Santos M., Iglesias C., Tello J.C., 2008. Evaluación de la incidencia de enfermedades criptogámicas foliares en cereales de invierno y primavera en España. Periodo 1993–1996. *Agroecología* 3: 41–50

Páscoa P., Gouveia C.M., Russo A., Trigo R.M., 2017. The role of drought on wheat yield interannual variability in the Iberian Peninsula from 1929 to 2012. *International Journal of Biometeorology* 61: 439–451.

RAEA, 2004. *Trigos blandos y duros, cebadas y triticales, Ensayo de variedades, campaña 2003–2004.* Consejería de Agricultura y Pesca, Junta de Andalucía. 63 pp. https://www.juntadeandalucia.es/agriculturaypesca/ifapa/servifapa/contenidoAlf?id=c3cfdb6d-eea1-445b-a6fa-9cd074c4d08c&l=informes.

RAIF, 2018, https://www.juntadeandalucia.es/agriculturaypesca/raif/.

Rivero J., 2013. *Los cambios técnicos del cultivo de cereal en España (1800–1930).* MAGRAMA, Madrid, Spain, 850 pp.

Roelfs A.P., Singh R.P., Saari E.E., 1992. *Rust Diseases of Wheat: Concepts and Methods of Disease Management.* CIMMYT, Mexico D.F., Mexico, 81 pp.

Royo C. 2005. Durum wheat improvement in Spain. In: *Durum Wheat Breeding: Current Approaches and Future Strategies* (C. Royo, M. Nachit, N. Di Fonzo, J.L. Araus, W.H. Pfeiffer, G.A. Slafer, ed.). Haworth Press, New York, USA, 883–906 (2nd vol.).

Ruiz de Casaviella J. 1878. Ligeras observaciones sobre la epifitia observada en Navarra el año 1877, y vulgarmente llamada la royada.. *Anales de Historia Natural* 7: 269–276.

Saari, EE, Prescott J.M., 1985. World distribution in relation to economic losses. In: *The Cereal Rusts Vol. II; Diseases, Distribution, Epidemiology, and Control* (A.P. Roelfs, W.R. Bushnell, ed.). Academic Press, Orlando, USA, 259–298.

Salazar J., Brañas M., 1973a. Physiologic races of wheat black rust (*Puccinia graminis* Pers. var. *tritici* Eriks. et Henn.) detected in Spain in the years 1968–1971. *Cereal Rusts Bulletin* 1: 21–23.

Salazar J., Brañas M., 1973b. Physiologic races of *Puccinia recondita* f. sp. *tritici* detected in Spain, years 1968–1971. *Cereal Rusts Bulletin* 1: 28–29.

Salazar J., Brañas M., Martínez M., 1977. Physiological specialization of wheat leaf rust in Spain, 1972–1975. *Cereal Rusts Conferences*, Interlaken, Switzerland.

Sharma R.C., Nazari K., Amanov A., Ziyaev Z., Jalilov A.U., 2016. Reduction of winter wheat yield losses caused by stripe rust through fungicide management. *Journal of Phytopathology* 164: 671–677.

Singh R.P., Rajaram S., 2002. Breeding for disease resistance in wheat. In: *Bread Wheat Improvement and Production* (B.C. Curtis, S. Rajaram, H. Gómez-Macpherson, ed.), FAO, Rome, Italy, 141–156.

Singh R.P., Huerta-Espino J., Pfeiffer W., Figueroa-Lopez P., 2004. Occurrence and impact of a new leaf rust race on durum wheat in northwestern Mexico from 2001 to 2003. *Plant Disease* 88: 703–708.

Singh D., Mohler V., Park R.F., 2013. Discovery, characterization and mapping of wheat leaf rust resistance gene *Lr71*. *Euphytica* 190: 131–136.

Soleiman N.H., Solis I., Soliman M.H., Sillero J.C., Villegas D., … Martinez-Moreno F., 2016. Emergence of a new race of leaf rust with combined virulence to *Lr14a* and *Lr72* genes on durum wheat. *Spanish Journal of Agricultural Research* 14(3), e10SC02.

Stakman E.C., 1923. European travels of a cereal pathologist. In: *Transactions of Wisconsin State Horticultural Society. Fifty-third annual report of the Wisconsin Horticultural Society*. Madison, Wisconsin, USA, 113–119.

Stukenbrock E.H., McDonald B.A., 2008. The origins of plant pathogens in agro-ecosystems. *Annual Review of Phytopathology* 46: 75–100.

Unamuno Irigoyen L.M., 1943. *Algunas aplicaciones de la micología a diversos ramos de la ciencia y de la industria.* Discurso de recepción. Real Academia de Ciencias Exactas, Físicas y Naturales, 67.

Urríes M.J., Arzoz M., 1961. Razas fisiológicas de *Puccinia graminis tritici* en España, en 1961. Anales del Instituto Botánico Cavanilles 20: 93–96.

Vergara-Diaz O., Kefauvera S.C., Elazaba A., Nieto-Taladriz M.T., Araus J.L., 2015. Grain yield losses in yellow-rusted durum wheat estimated using digital and conventional parameters under field conditions. *The Crop Journal* 3: 200–210.

Wanyera R., Kinyua M.G., Jin Y., Singh R., 2006. The spread of stem rust caused by *Puccinia graminis* f. sp. *tritici*, with virulence on *Sr31* in wheat in Eastern Africa. *Plant Disease* 90(1), 113.

Wikipedia, 2018. Rogation days. https://en.wikipedia.org/wiki/Rogation_days.

Zadoks J.C., 1960. *Verslag van een reis naar Spanje en Portugal ter bestudering van de epidemiologie van gele roest (maart, april, 1958).* Stichting Nederlands Graancentrum, Wageningen, Techn. Ber. Nr. 3.

Zadoks J.C., 2013. *Crop Protection in Medieval Agriculture.* Pub. Sidestone Press, NL, 333 pp.

Zapata L., Peña-Chocarro L., Pérez-Jordá G., Stika H.P., 2004. Early Neolithic agriculture in the Iberian Peninsula. *Journal of World Prehistory* 18(4): 283–325.

Zohary D., Hopf M., Weiss E., 2012. *Domestication of Plants in the Old World: The Origin and Spread of Cultivated Plants in West Asia, Europe, and the Nile Valley* (4th ed.). Pub. Oxford University Press, UK, 264 pp.

2

Characterization of *Phytophthora capsici* isolates from lima bean grown in Delaware, United States of America

NILWALA S. ABEYSEKARA[1], HEATHER HICKMAN[2], SARA WESTHAFER[2], GORDON C. JOHNSON[3], THOMAS A. EVANS[2,*], NANCY F. GREGORY[2,*], NICOLE M. DONOFRIO[2,*]

[1] *Current address: Department of Microbiology and Plant Pathology, University of California-Riverside, Riverside, CA, 92507, USA*
[2] *Department of Plant and Soil Sciences, University of Delaware, Newark, DE, 19716, USA*
[3] *Department of Plant and Soil Sciences, University of Delaware, Georgetown, DE, 19947, USA*
*Co-corresponding authors:
Nicole Donofrio: ndonof@udel.edu
Nancy Gregory: ngregory@udel.edu
Thomas Evans: tomevans@udel.edu

Competing Interests: The Author(s) declare(s) no conflict of interest.

Editor: Jenny Davidson, South Australian Research and Development Institute, Adelaide, Australia.

Summary. Pod rot of lima bean (*Phaseolus lunatus* L.), caused by the broad host range oomycete *Phytophthora capsici,* is an emerging threat to lima bean production in the mid-Atlantic region of the United States of America (USA). There is little known about survival and spread of this pathogen in the State of Delaware, an area of major lima bean production. Irrigation water was sampled in 2014 and 2015 for the presence of *P. capsici* using baiting methods. Over three seasons, isolations from water sources, weeds, and soil samples did not yield *P. capsici*. However, field samples from symptomatic lima bean, watermelon, muskmelon, pepper, pickling cucumber, and pumpkin yielded 64 *P. capsici* isolates. Characterization of the isolates showed that 42 were of the A2 mating type, 31 were sensitive to mefenoxam, 18 were intermediately sensitive, and four were insensitive to this fungicide. All isolates were pathogenic on the eight lima bean and two snap bean cultivars tested. Three EST-SSR markers, PCSSR19, PCN3, and PCN7, used in combinations of PCSSR19/PCN3 or PCSSR19/PCN7 were significantly associated with mefenoxam sensitivity. This study is the first of its kind in Delaware, providing key information as a basis for effective management of *P. capsici,* including mating type, mefenoxam insensitivity, host range, and survival.

Keywords. Fungicide resistance, oomycete, plant pathogen, mefenoxam.

INTRODUCTION

The causal agent of pod rot of lima bean, *Phytophthora capsici* (Leonian 1922), infects members of at least 27 plant families, including vegetable crops in the Cucurbitaceae, Solanaceae, and Fabaceae, as well as conifers, weeds, and tropical crops around the world (Erwin and Ribeiro, 1996; Davidson *et al.* 2002; Gevens *et al.*, 2008; Roberts *et al.*, 2008; Quesada-Ocampo *et al.*, 2009; Granke *et al.*, 2012). This heterothallic, hemi-biotrophic oomycete, belonging to the Peronosporales and Pythiaceae, causes root, stem, fruit and crown rot, foliar blight, and stunting on various hosts (Gevens *et al.*, 2008; Quesada-Ocampo *et al.*, 2016). The pathogen has a broad host range and can cause 50% crop losses in agro-ecosystems (reviewed in Sanogo and Ji, 2012). Lima bean (*Phaseolus lunatus* L.) and snap bean (*Phaseolus vulgaris* L.), however, are the only reported legumes affected by *P. capsici* (Davidson *et al.*, 2002; Tian and Babadoost, 2003; Gevens *et al.*, 2004). Lima bean is the cornerstone of the Delaware vegetable processing industry and a greater area is grown in the mid-Atlantic region (MAR) than elsewhere in the United States of America (USA). Approximately 5,600 ha of lima bean crops are planted in Delaware annually (https: //www.nass.usda.gov/Statistics_by_State/Delaware/index.php). Lima bean fields are often planted after an early season vegetable crop, such as peas or cucumbers. Planting susceptible vegetable crops in rotation or as a double crop ahead of lima bean in fields with previous histories of *P. capsici* may increase the risk of lima bean pod rot (Hausbeck and Lamour, 2004). Along with lima bean, snap bean crops are also grown in and around Delaware, often as warm season vegetable crops for autumn harvest. In 2017, 890 ha of snap bean crops were grown in Delaware (https: //www.nass.usda.gov/Quick_Stats/Ag_Overview/stateOverview.php?state=DELAWARE). Though there are reports of *P. capsici* affecting snap beans in other states of the USA (Gevens *et al.*, 2008; McGrath and Dillard, 2011), occurrence of snap bean pod rot in Delaware has not yet been reported.

The asexual sporangia of *P. capsici* produce motile zoospores that may spread in irrigation water or rain (Ristaino *et al.*, 1992; Granke *et al.*, 2012). Zoospores may remain viable for hours or days in water (Roberts *et al.*, 2005). Dispersal of *P. capsici* sporangia by wind is not frequent and the dispersal to other fields solely by wind is unlikely (Granke *et al.*, 2009). Sexual oospores are produced in the presence of A1 and A2 mating types (MT), and oospores survive in soil for variable times (Babadoost and Pavon, 2013). oospores may survive in infested soil and plant debris for more than 5 years, resulting in infective propagules after crop rotations with non-host crops (Lamour and Hausbeck, 2001). This could result in significant genetic variation of outcrossing populations in the field (Lamour *et al.*, 2012). Weeds may be alternative hosts in the absence of host crops, and host weeds found and reported in the eastern USA for *P. capsici* include black nightshade (*Solanum americanum*, *S. nigrum*), common purslane (*Portulaca oleracea)*, velvet leaf (*Abutilon theophrasti)*, and Carolina geranium (*Geranium carolinianum)* (Tian and Babadoost, 2003).

Studies on *Phytophthora capsici* have demonstrated broad genetic diversity, demonstrated with genetic fingerprinting and molecular markers (Hu *et al.*, 2013; Lamour *et al.*, 2012), with genetic clustering (Granke *et al.*, 2012), and with physiological race testing in pepper in New Mexico (Glosier *et al.*, 2008; Monroy-Barbosa and Bosland, 2011). These results indicate that physiological races exist in *P. capsici*. A physiological race is defined as "a subdivision of a pathogen species, particularly fungi, distinguished from other members of the species by specialization for pathogenicity in different host cultivars" (Kirk *et al.*, 2001). Cultivar differentials are generally used to identify physiological races.

Management strategies for *P. capsici* include applications of the fungicide mefenoxam, which has been used widely for *Pythium* and *Phytophthora*. However, prolonged use of this compound has contributed to the emergence of fungicide insensitivity in *P. capsici* populations and some *Pythium* species (Brent and Hollomond 1998; Parra and Ristaino 1998; Weiland *et al.*, 2014). Fungicide insensitivity in *P. capsici* to mefenoxam, hymexazol, cyazofamid, pyrimorph, and flumorph has been reported (Jackson *et al.*, 2012; Pang *et al.*, 2013, 2016; Jones *et al.*, 2014). In the Mid-Atlantic region of the USA, mefenoxam insensitive isolates of *P. capsici* from lima bean were first reported in 2008 (Davey *et al.*, 2008). Traditionally, mefenoxam sensitivity of an isolate is tested using *in vitro* assays with fungicide-amended media (Parra and Ristaino, 2001; Hausbeck and Lamour, 2004; Keinath, 2007). While these assays are still performed with *P. capsici* isolates (Qi *et al.*, 2012; Ma *et al.*, 2018), we wished to develop a molecular marker-based method to rapidly identify mefenoxam sensitivity.

In the last two decades, molecular markers have been used to identify specific traits or changing populations in *Phytophthora* species. Lamour and Hausbeck (2001) used amplified fragment length polymorphism (AFLP) markers to resolve population dynamics of a recombinant field population of *P. capsici*. Additionally, Pei-Qing *et al.*, (2013) identified four expressed sequence tag, simple sequence repeat (EST-SSR) markers to unravel diversity in *P. capsici* populations in China. While Hu *et al.*, (2014) reported a sequence characterized amplified region (SCAR) marker that can distinguish mefenoxam insensitive populations and sensitive populations of *P. nicotianae*, to date there are no reported markers capable of distinguishing between insensitive and sensitive *P. capsici* isolates.

Surface water sources used for irrigation in Michigan, Georgia, New York, and other states in the USA have been shown to carry *P. capsici* (Bush *et al.*, 2003; Roberts *et al.*, 2005; Wang *et al.*, 2009; Gevens *et al.*, 2007; Jones *et al.*, 2014). Though the pathogen is not known to overwinter in irrigation water sources, water may aid in its spread. The objectives of the present study were to characterize *P. capsici* isolates collected from irrigation water sources, weed and crop hosts and soil samples from the states of Delaware and Maryland (USA) to (1) determine *P. capsici* mating types, mefenoxam insensitivity, host range, and dispersal; (2) develop molecular markers to distinguish mefenoxam sensitive isolates from insensitive isolates; and (3) detect the presence of other *Phytophthora* species. This information would aid development of effective disease management strategies. We undertook these studies in order to better understand this economically important, broad host range pathogen on an important crop in our region; if lima bean cannot be produced profitably in the MAR, other processing vegetables such as peas, snap beans, sweet corn and spinach, would not be produced, resulting in severe economic losses to the region (Evans *et al.*, 2007).

MATERIALS AND METHODS

Sampling: baiting, infected field samples, water, weeds, and soil samples

Thirteen surface water sources including lakes, streams, and naturally-fed ponds were sampled in Kent and Sussex counties in Delaware during the summers of 2014, and 2015. Soil proximal to water sources, known potential *P. capsici* weed hosts, and infected fruits from grower fields were also sampled in the summers of 2014, 2015 and 2016. Baiting traps were each constructed by attaching a polyethylene foam cylinder (5.7 cm in diameter) to a mesh laundry bag (30 cm × 30 cm) with a zipper. Two unripe pears, one whole eggplant, and two cucumbers were placed in each trap as bait along with two rhododendron leaves. Fruits and leaves were surface-sterilized using 0.825% sodium hypochlorite (NaOCl) with two ≅10 μL drops of Tween 20 (Agdia Inc.) for 100 mL of the solution, for 2 min, and washed with sterile distilled water before adding to the bait bag. Bait bags were kept in the water for 4–6 d.

Numerous infected plant samples from 18 crop fields were collected or obtained from the University of Delaware Plant Diagnostic Clinic, from locations in Delaware and Maryland over the 3 year study. Extensive field sampling was carried out in one location in Bridgeville, Delaware (field 17, Table S1) in the summer of 2016. This field was planted with pickling cucumber (*Cucumis sativus*) early in the growing season, then double-cropped with lima beans later in the same growing season. At least 20-25 samples from different locations of this field were sampled to avoid sampling bias that could lead to non-recovery of isolates belonging to both mating types.

Twenty-six water samples (two 1 L bottles per source) were collected from 13 water sources. Two rhododendron leaves were added to one water bottle (1 L) and incubated in the dark. After 3 d, water was removed, and the leaves were washed with sterile distilled water and kept under moist conditions at 25°C for 3 d until lesions developed. The remaining water was vacuum filtered in 100 mL batches using 3.5 cm diam. (P5 Fisher Brand, 3 μm) filter papers. Filter papers were then placed face down on PARP-V8 selective medium (Ferguson and Jeffers, 1999) and incubated at 25°C in the dark for 3–5 d for colony development.

Two 15 g soil samples were collected near the edge of each surface water source. For each 15 g of soil, 200 mL of distilled water was added, mixed to uniformity, and then filtered through several layers of cheesecloth to remove soil and other debris. Filtered water samples were then processed using the vacuum filtration technique described above.

Roots of weeds near water banks were collected if the plants were previously reported as hosts for *P. capsici*. Carolina geranium (*Geranium carolinianum*) and common purslane (*Portulaca oleracea*) were the only weeds known to be hosts for *P. capsici* that were present near the water sources sampled. Roots were washed thoroughly with running tap water and surface sterilized as described previously.

All samples collected (except the water and soil samples) were rinsed in sterile distilled water and dried under ambient laboratory conditions in a biosafety hood. Tissues from margins of lesions from each of these samples were placed on PARP-V8 medium and incubated at 25°C in the dark for 3–5 d.

Morphological identification and generation of single zoospore cultures

Hyphal tips of potential *P. capsici* isolates growing on PARP-V8 medium were transferred to 60% strength potato dextrose agar (PDA; BD Difco) and unclarified V8 medium (600 mL distilled water, 163 mL unfiltered V8 juice, 12 g agar, 1.7 g $CaCO_3$) to examine isolate growth patterns. Sporangium formation was induced in 7–10 d-old cultures grown on unclarified V8 medium using Chen-Zentmeyer salt solution (Chen and Zentmeyer, 1970), and the isolates were identified based on morphometric characteristics (Erwin and Ribeiro 1996; Gallegly and Hong 2008). Single zoospore cultures (SZC) were obtained from field isolates positively identified as *P. capsici* (morphologically and molecularly). Sporangium formation was induced on 10 d-old *P. capsici* isolates with Chen-Zentmeyer salt solution. After 18–20 h, the solution was decanted and 10 mL of sterile distilled water added. Plates were maintained at 4°C for 1 h followed by 15–30 min at room temperature, to induce release of zoospores. Twenty to 50 μL of each zoospore suspension was spread on a water agar plate, incubated at 25°C for 2–3 d, and then examined under a dissecting microscope. Three germinating zoospores per each hyphal-tipped field isolate were each transferred to a low strength PDA plate, and then to a V8 agar plate. Single zoospore cultures were stored long term in sterile distilled water within sterile screw-capped tubes each containing two hemp seeds and three cucumber seeds, and were maintained at 20°C.

DNA extraction

For each isolate tested, a 7 mm diam. plug of actively growing 7-d-old culture was added to 25 mL of lima bean broth (Calvert *et al.*, 1960) and incubated at 25°C for 3 d on an orbital shaker. Mycelial mats growing on broth were harvested using vacuum filtration through Whatman #1 filter paper. After removing the original colonized agar plugs, the mycelial mats were washed with sterile distilled water, lyophilized, and then kept in -80°C until processed. DNA was extracted with a Wizard Genomic DNA purification kit (Promega). DNA was dissolved in 100 μL of DNA rehydration solution and diluted to a final concentration of 25 ng μL^{-1}.

Molecular identification and primers

DNA samples were amplified using two sets of *P. capsici* specific primers (Zhang *et al.*, 2006, Lan *et al.*, 2013) to confirm identification. PCR reactions were carried out with a nested PCR protocol with universal ITS1/ITS4 for the first PCR round, and the primers PC-1 (5'-GTCTTGTACCCTATCATGGCG-3') and PC-2 (5'-CGCCACAGCAGGAAAAGCATT-3') for the second PCR round, as described by Zhang *et al.* (2006). The expected amplified product size was 560 bp. Isolates were also characterized using the *P. capsici* specific primers Pc1F (5'-GTATAGCAGAGGTTTAGTGAA-3') and Pc1R (5'-ACTGAAGTTCTGCGTGCGTT-3'), as described in Lan *et al.* (2013). The expected product size was 364 bp. A 25 μL PCR reaction mixture containing 2 μL of template DNA (25 ng μL^{-1}), 5 μL 5× LongAmp *Taq* Master Mix (New England Biolabs), 300 μM dNTPs, 0.4 μM primer, and 2.5 U LongAmp *Taq* DNA Polymerase (New England Biolabs) was used for marker amplification. The PCR amplification protocol used was as follows: initial denaturation at 94 °C for 10 min, 35 cycles of 94°C for 30 s, 56–60 °C for 30 s, 72°C for 30 s, followed by a final extension at 72°C for 5 min. PCR products were separated on a 2% agarose gel containing ethidium bromide in 10× Tris-Borate-EDTA buffer. Electrophoresis was carried out at a constant 58 V for 2 h and DNA bands were visualized under a UV trans-illuminator. A 100 bp DNA ladder (New England Biolabs) was used to determine amplicon sizes. DNA of isolates confirmed as *P. capsici* were then tested with one Inter-Simple Sequence Repeats (ISSR) marker and five SSR markers, selected from published *P. capsici* population markers, to identify polymorphic molecular markers using a subset of five isolates (Wang *et al.*, 2009, Pei-Qing *et al.*, 2013). One polymorphic marker reported in Pei-Qing *et al.* (2013), PCSSR19 (F-5'-GTCTTCGCTAAAGCCTCCG- 3', R- 5'-AGATGGCCAACAGCGGTTA-3'), showed co-segregation with mefenoxam sensitivity in our population, so this was utilized for further study. Polymorphism of a marker was initially tested with five isolates with varying mefenoxam sensitivity (two mefenoxam sensitive and three mefenoxam insensitive isolates). Markers showing possible co-segregation with mefenoxam sensitivity were further tested with a small subset of 18 isolates before screening all 64 isolates. PCSSR19 primers were searched using BLAST in Fungidb (http: // fungidb.org/fungidb/), and matched one predicated gene in *P. capsici* (PHYCA_548602T0), which when searched using BLAST in NCBI, was shown to be most similar (84% similarity at nucleotide level) to an RNA helicase from *P. infestans*. This coding sequence was then utilized to design additional primers pairs (designated PCN3 and PCN7), using the Primer3 program (Koressaar *et al.*, 2007; Untergasser *et al.*, 2012). The three primer pair sets PCSSR19, PCN3, and PCN7 were then tested on a small set of 18 isolates, and the amplification products were found to be polymorphic between sensitive and insensitive isolates. These three primer sets, PCN3 (F-5'-CGTGGCTTAACCAGTGTTCT-3', R-5'- GACGGTCATAACCACCGTAG-3'), PCN7 (F-5'-CGTTTTCCTACCGATTCCAA-3', R-5'-GACGCGGTACGTATGCAGAT-3'), and PCSSR19, were therefore used to screen all isolates. Conditions for PCR reactions were as follows: initial denaturation at 94°C for 10 min, 35 cycles of 94°C for 30 s, 56–60°C for 30 s, 72°C for 30 s, followed by a final extension at 72°C for 5 min. Products of PCR were separated on a 4% agarose gel (low range ultra agarose; Biorad) containing ethidium bromide in 10× Tris-Borate-EDTA buffer (Fisher Scientific). Electrophoresis was conducted at a constant 58 V for 5 h, and DNA bands were visualized under a UV trans-illuminator. Isolates were screened with markers two additional times to confirm the correct banding pattern.

Mating type determination

Mating type of each *P. capsici* isolate was determined by pairing single-zoospore cultures with known isolates of *P. capsici*, [A1 (10193) or A2 (SP98) MT], as described by Gevens *et al.* (2007). A 7 mm plug of 7-d-old SZC of *P. capsici*, grown on a V-8 plate at 25°C, was placed 3 cm from a 7 mm plug of either the A1 or A2 MT tester isolate on a V8 plate, and incubated at 25°C in the dark. Plates were examined under a compound microscope for the presence of oospores after 14 to 21 d.

Mefenoxam sensitivity

Mefenoxam sensitivity was determined as outlined in Gevens *et al.* (2007). A 7 mm plug from an actively growing plate of SZC of *P. capsici* was placed on three, 100 mm diam. plates of V8 agar amended with 100 ppm of mefenoxam (Technical grade, or Ridomil Gold SL, 45.3%, Syngenta) and two V8 non-amended controls. The active ingredient in Ridomil Gold SL mefenoxam at 45.3%, with the remainder of the formulation being proprietary additives. Technical grade mefenoxam was dissolved in acetone and Ridomil Gold SL was dissolved in sterile distilled water prior to adding to the media cooled to the touch (approx. 55°C). Three SZC from each of the 64 *P. capsici* field isolates were tested with Ridomil Gold SL. After incubating plates at 23°C in the dark for 3 d, perpendicular colony diameters were measured, and percent average growth of an isolate was obtained by comparing growth on V8 plates with growth on V8 amended with mefenoxam. Isolates were rated as follows: sensitive isolates had <30% growth compared to the controls, intermediately insensitive isolates had 30-90% growth compared to the controls, and insensitive isolates had >90% growth compared to the controls, as described in Gevens *et al.* (2007). The experiment was repeated. All three of the SZC's from each isolate were observed and recorded for mefenoxam phenotype, because in several instances, one SZC differed from the other two. To account for these discrepancies, homogeneity of variance among SZC's from the two experiments were assessed with Bartlett tests. Statistical significance of the mefenoxam reactions between SZC was also tested using a Student's t test at $\alpha = 0.05$, using JMP Pro 13 (JMP®, Version *13*. SAS Institute Inc., 1989-2019). The correlations between mefenoxam sensitivity and MT were tested with the Pearson coefficient at $\alpha = 0.05$ also using JMP Pro 13.

Pathogenicity of Phytophthora capsici *to lima bean and snap bean*

Plants of the commercial lima bean cultivars 'Cypress', 'C- Elite Select', 'Fordhook (FH) 242', 'Bridgeton', 'Eastland', 'Maffei (M) 15', '8-78', and '184-85' were each inoculated with isolates of *P. capsici* collected during 2014-2016. A lima bean plant introduction PI477041 from Arizona, determined to have some resistance to *P. capsici* in field and greenhouse inoculations (N. Gregory, personal communication) was also included. Plants were grown under greenhouse conditions at 25°C with a photoperiod of 10 h and light intensity of 2500-3500 μEm^{-2} sec^{-1} until pod set. Accession PI477041 was grown with a photoperiod of 12 h to induce flowers and pods. Young, flat pods were collected and challenged, as described by McGrath (2009) for snap bean pods, with slight modifications. Pods were surface sterilized using a 0.825% NaOCl solution (as above) and completely dried under a laminar flow hood. Two pods were placed on each Petri plate (100 × 15 mm Fisherbrand™) lined with a filter paper moistened with sterile distilled water. A 7 mm diam. plug of an actively growing *P. capsici* culture, grown on V8 plates for 7 d at 25°C, was placed on each pod. Plugs of V8 culture medium without the pathogen were placed on pods as experimental controls. Plates were sealed with Parafilm and incubated at 25°C and a 10 h photoperiod. Pods were assessed as susceptible or resistant 3 to 5 d post inoculation. *Phytophthora capsici* was re-isolated from symptomatic tissue placed on PARP-V8 selective media. The experiment was repeated twice. Two isolates identified as a *Pythium/Phytopythium*-like spp., isolated from water baits, were also tested on the differential group of lima bean cultivars, using the method described above, to determine whether they were pathogenic on lima bean.

Snap bean cultivars 'Provider' and 'Caprice' were inoculated with isolates of *P. capsici* under laboratory and greenhouse conditions. Thirteen *P. capsici* isolates out of the 64 total were tested on young snap bean pods in the laboratory, using the protocol adopted from McGrath (2009). Six pods were tested per isolate and the experiment was replicated once. Greenhouse testing was done for three of the 13 isolates used in the laboratory testing to observe *in planta* reactions. Lima bean cultivars 'M15' and '8-78' were used as positive controls, and greenhouse assays consisted of four plants of each cultivar per isolate. Isolates were grown on V8 agar plates for 14 d, which were subsequently flooded with sterile distilled water, with sporangia being dislodged with a glass rod. Pods were spray-inoculated with a spore suspensions (3×10^4 mL^{-1}) until run off and kept under moist conditions (plastic chamber on a greenhouse bench at 24°C with misting for 1 min every hour during daylight hours from 7 am to 7 pm) until symptom development. Control plants were sprayed with sterile distilled water. Symptoms were evaluated 4 an 12 d post-inoculation. Pods from a total of four replicated plants were examined, and rated as susceptible if they showed typical symptoms of *P. capsici* infections. This experiment was repeated once. Infections were confirmed by re-isolating *P. capsici* from inoculated and symptomatic tissues.

RESULTS

Isolate collection and identification

More than 200 microbial isolates were collected from irrigation water, soil samples, plant hosts, and weed hosts. However, none of the isolates recovered from direct sampling of irrigation water, baited fruits, leaf baits, soil samples, or roots of weed hosts were *P. capsici*. *Pythium/Phytopythium*-like isolates were mainly recovered from surface water sources in late August and early September. *Phytophthora capsici* isolates were recovered only from the infected plant material from crop fields. The numbers of *P. capsici* isolates collected from infected plant material from these fields were 22 in 2014, 16 in 2015, and 26 in 2016, for a total of 64 isolates. All the isolates were positively identified morphologically as *P. capsici*, and yielded the expected 560 and 364 bp bands, when amplified, respectively, with *P. capsici* specific primers PC1/PC2 (used in the nested PCR with ITS1/ITS4 primers) and Pc1F/Pc1R primers. Isolate information is presented in Table 1.

Mating type (MT) determination testing

Isolates belonging to both MT of *P. capsici* were recovered in all three years. Of the 64 *P. capsici* isolates recovered, 42 out of 63 field isolates were A2 MT, with 17 A2 MT isolates recovered from lima bean (Tables 1 and S1). Mating types A1 and A2 were recovered from four field locations in Delaware (fields 8, 12, 17, and 18), including the intensively sampled field in Bridgeville Delaware in the summer of 2016 (Table S1; field 17, 20 samples).

Pathogenicity on commercial lima bean and snap bean cultivars

Pathogenicity testing of *P. capsici* isolates showed all 64 isolates were virulent on all the commercial lima bean cultivars tested ('Cypress', 'C- Elite Select', 'Ford-

Table 1. Numbers of *Phytophthora capsici* isolates, their host sources, mating types, mefenoxam sensitivities, and locations, for isolates collected in the 2014, 2015 and 2016 field seasons.

Host	Number of isolates	Mating types		Mefenoxam sensitivity[a]					Location	
		A1	A2	S[b]	S-IS[c]	IS[d]	IS-I[e]	I[f]	DE[g]	MD[h]
Lima bean	27	10	17	12	2	9	2	2	20	7
Pumpkin	5	2	3	3	1	1	0	0	4	1
Pepper	4	0	4	2	1	1	0	0	3	1
Muskmelon	1	1	0	0	0	0	0	1	1	0
Watermelon	15	2	13	7	2	5	0	1	15	0
Pickling cucumber	12	7	5	7	3	2	0	0	12	0
Totals	64	22	42	31	9	18	2	4	55	9

[a] Mefenoxam sensitivity of field isolates are categorized based on the reaction of the three single zoospore cultures tested for each field isolate.
[b] Sensitive.
[c] Sensitive to Intermediately sensitive.
[d] Intermediately sensitive.
[e] Intermediately sensitive to Insensitive.
[f] Insensitive.
[g] Delaware.
[h] Maryland.

hook (FH) 242', 'Bridgeton', 'Eastland', 'Maffei (M) 15', '8-78', and '184-85'), evidenced by clear signs of white sporulation. Isolate PC67 is an accurate representation of susceptible interactions on these cultivars (Figure. 1). Controls showed no signs of infection or discolouration (Figure 1). Most isolates were pathogenic on PI477041, with the exceptions of isolate PC33 (no sporulation), PC61 and PC62 (reduced sporulation and browning), and PC51 and PC37 (reduced sporulation) (Figure S1). Two of the *Pythium/Phytopythium*-like isolates identified in this study were tested on pods of the lima bean cultivars, and were non-pathogenic, with the exception of isolate Phy4 on 'Eastland' (Figure S2). The two snap bean cultivars, 'Caprice' and 'Provider', were susceptible to the 13 isolates tested in the laboratory, and to a subset of three representative isolates tested in the greenhouse, showing similar signs of sporulation. Control pods inoculated with sterile distilled water showed no symptoms. *Phytophthora capsici* was re-isolated from the infected pods.

Mefenoxam sensitivity

Since the replicates were homogeneous between experiments (Bartlett test - F ratio = 0.0046, $P = 0.9954$), the results from both experiments were averaged to obtain the mefenoxam reaction for the isolates. Of the 64 field isolates tested, 31 were sensitive (S), 18 intermediately sensitive (IS), and four insensitive (I) to Ridomil Gold SL (Table 1). Single zoospore cultures derived from eleven isolates showed variability in their reaction to mefenoxam (Tables 1 and S1). Nine isolates showed S-IS phenotype (of these, six were statistically significant at $\alpha = 0.05$) while two isolates showed IS-I phenotype (neither were statistically significant at $\alpha = 0.05$) (Table S1). There was no significant correlation between the mefenoxam sensitivity and the mating type at $\alpha = 0.05$ (Pearson coefficient = -0.23 $P = 0.063$, Table S1). We wished to determine whether SZCs showed different responses to Ridomil Gold SL (45.3% mefenoxam) versus technical grade mefenoxam (97% ai). This assay showed that SZCs from 14 field isolates exhibited the same reactions while SZCs from several other field isolates gave variable reactions to both mefenoxam formulations (Table S2). For example, all three SZCs derived from two field isolates (PC33 and PC43) gave different reactions, while two out of three SZCs from one field isolate (PC34), and one of three SZCs from four field isolates (PC36, 44, 46, and 55) showed different reactions to both formulations (Table S2). However, only the reactions observed for SZCs of PC33 and PC34 were statistically significant at $\alpha = 0.05$, suggesting that reaction of SZCs

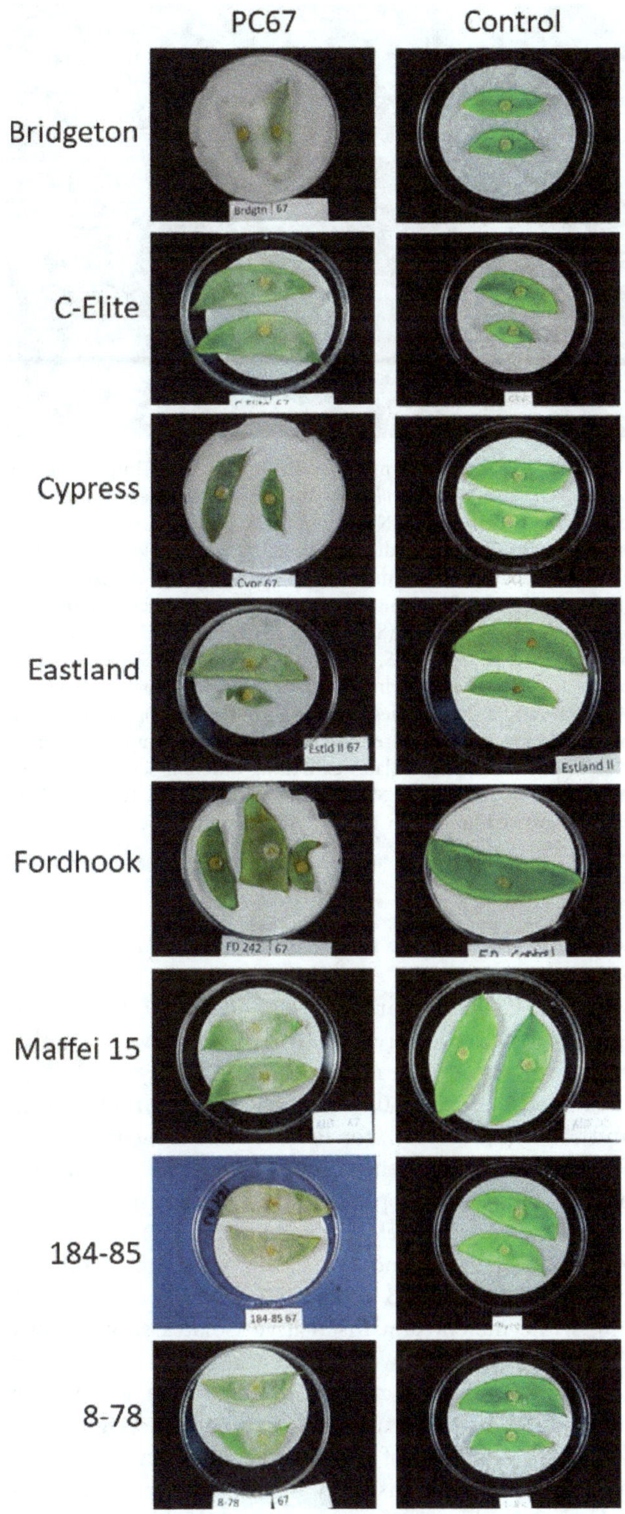

Figure 1. Lima bean pod assay. Pathogenicity testing of *Phytophthora capsici* isolate 67 on lima bean pods from eight different cultivars. Isolate 67 inoculations resulted in sporulation on all eight cultivars, representative of all of other isolate reactions with the lima bean cultivars. Pods were photographed 5 d post inoculation. The experiment was repeated once, with similar results.

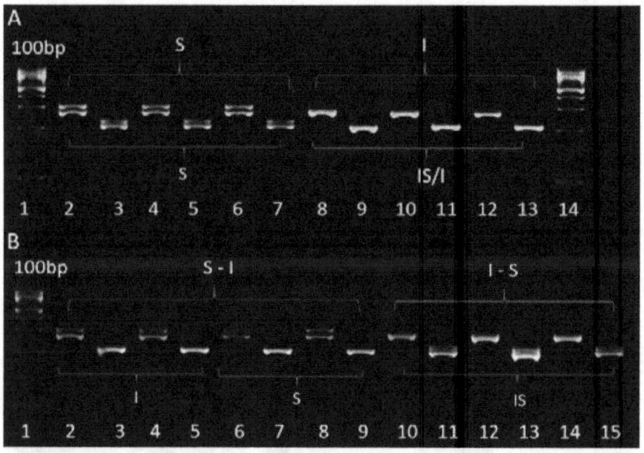

Figure 2. Segregation of mefenoxam sensitive and intermediately sensitive/insensitive *Phytophthora capsici* isolates, based on the SSR markers PCSSR19 and PCN3. S = sensitive, IS = intermediately sensitive, I = insensitive. Predictions based on the markers are shown above the fragments, and mefenoxam sensitivities of the isolates are shown below the fragments. Fragment sizes for each primer pair are as follows: PCSSR19 (S, 263 bp, 281 bp and 325 bp; IS/I, 252 bp and 270 bp) , PCN3 (S, 240-250 bp and 205-215 bp; IS/I, 205-215 bp). (A) Isolates showing the same segregation pattern for both PCSSR19 and PCN3. Lanes 1 and 14 are the 100 bp ladder. Even numbered lanes 2 to 12 are marker PCSSR19, odd numbered lanes 3 to 13 are PCN3. (B) Isolates exhibiting different segregation patterns with the markers PCSSR19 and PCN3. Lane 1, 100 bp ladder, even numbered lanes 2 to 14, PCSSR19, odd numbered lanes 3 to 15, PCN3.

to the technical grade mefenoxam and to Ridomil Gold SL were very similar.

Three EST-SSR markers tested, PCSSR19, PCN3, and PCN7 (which co-segregates with PCN3), were significantly correlated with mefenoxam sensitivity at $\alpha = 0.05$ (Pearson coefficients 0.507 (P = <0.001) for PCSSR 19 and 0.402 ($P = 0.001$) , for PCSSR 19 and for PCN3 and PCN7). Mefenoxam sensitive isolates yielded three bands of 263, 281, and 325 bp, while the isolates assessed as IS or I to mefenoxam yielded two bands of 252 and 270 bp with PCSSR19. Mefenoxam sensitive isolates produced two bands with PCN3 and PCN7 (240-250 and 205-215 bp), while mefenoxam insensitive isolates showed only the 205-215 bp band with PCN3 and PCN7. Mefenoxam sensitivity was predicted more accurately when both markers yielded the same banding pattern (Figure 2A, Table S3). Accuracy of sensitivity prediction when both markers predicted sensitivity was 96%, or for insensitivity was 85%. Twenty-five out of the 26 field isolates predicted to be mefenoxam sensitive by both markers were sensitive to Ridomil Gold SL in plate-based assays, and 11 out of the 13 field isolates predicted to be insensitive to mefenoxam exhibited insensitivity to Ridomil Gold SL. When PCSSR19 predicts insensitivity and PCN3 predicts sensitivity to mefenoxam, the plate-based reaction was IS (Figure 2B, Table S3). Isolates IS to mefenoxam could at times yield the bands for sensitivity with the PCN3 marker. Prediction of the actual sensitivity reaction was less accurate when PCSSR19 predicted sensitivity and PCN3 predicted insensitivity (Figure 2B). Based on these there was a 56% probability for the actual reaction to be IS (Table S3). Correlations between the marker combinations PCSSR19/PCN3 or PCSSR19/PCN7 with mefenoxam reaction, not considering instances where PCSSR19 predicted sensitivity and PCN3 predicted insensitivity, were significantly greater (Person coefficient 0.745 ($P < 0.001$).

DISCUSSION

Phytophthora capsici is an emerging threat to lima bean production in the main USA growing regions of Delaware, Maryland and Virginia. The goal of the present study was to isolate, identify and characterize many lima bean-infecting pathogen isolates in Delaware, and to determine their sensitivity to the widely-applied fungicide mefenoxam. In the course of 3 years, we recovered 64 *P. capsici* field isolates out of 200 samples collected. To identify a potential source of this important pathogen in grower fields, samples were collected from irrigation water sources, weed hosts, soils, and infected plant material. Of all these sources, *P. capsici* isolates were only obtained from infected plant material in crop fields. We were unable to identify any *P. capsici* isolates from irrigation sources, although they have been identified from irrigation sources in Florida, Michigan, and Georgia (Roberts *et al.*, 2005; Gevens *et al.*, 2007; Wang *et al.*, 2009). Our baiting method was similar to those of Wang *et al.*, (2009), although they used infected bait fruits to then inoculate pepper stems, which resulted in increased numbers of *P. capsici* isolates recovered. They observed that plating directly from bait fruits can favour organisms which outgrow *P. capsici*. Thirty seven isolates representative of the isolates recovered from irrigation sources were assessed for molecular identification. Details of this identification are listed in Table S4. None of these species were known pathogens of lima bean, though some where known to cause diseases on other hosts. Many of the isolates recovered from irrigation sources were morphologically similar to *P. capsici*. Based on the sequence data for the maternally inherited Cox1 gene and the ITS, however, several of the isolates recovered from water sources in Delaware were identified as the putative maternal parents of *Phytophthora* x *stag-*

num (Yang *et al.*, 2014; Table S4). The hybrid species *P.* x *stagnum* was recently described (Yang *et al.*, 2014). The paternal parent of *P.* x *stagnum* is reported as *P. chlamydospora*, previously known as *P.* taxon Pgchlamydo, (Hansen *et al.*, 2015), and the maternal parent is thought to be an unknown species close to *P. mississippiae*. The present study is the first report of the identification of the putative maternal parent of *P.* x *stagnum* (C. Hong, personal communication). Of the other species recovered from the water sources, isolates of *P. irrigata* have been reported from water sources in other studies (Kong *et al.*, 2003; Hong *et al.*, 2008). *Phytopythium* is a novel genus in the Pythiaceae, exhibiting morphological characteristics in between *Pythium* and *Phytophthora* (de Cock *et al.*, 2015). However, both isolates of the *Pythium*/*Phytopythium*-like species tested on the lima bean cultivars in the present study did not cause disease on pods except for isolate 4 on 'Eastland'. This indicates that these isolates were non-pathogenic on lima bean, but further greenhouse testing is needed, in particular, further testing on the 'Eastland' cultivar.

The presence of both MTs of *P. capsici* in the same field has been reported previously in New Jersey and North Carolina (Papavizas *et al.*, 1981; Ristaino, 1990; Dunn *et al.*, 2010). Mating type A2 appears to be dominant in Delaware, as approx. 66% of the field isolates tested were of this MT. Our results also demonstrate that both MTs can be present in the same field. This has significant ramifications for management of *P. capsici*, as new strains could quickly emerge due to sexual reproduction. Oospores can be viable in fields for more than 5 years, even after rotations with non-host crops (Lamour and Hausbeck, 2001). Hence, rotations of lima bean with non-host crops could become a less effective management strategy in grower fields in Delaware, if both MTs were present in the same field. *Phytophthora capsici* isolates belonging to both MTs were recovered from a field in Bridgeville, Delaware (field 17), which was planted with pickling cucumber early in the 2016 field season and then with lima bean later in the same growing season. Field 12 also tested positive for *P. capsici* in both 2015 and 2016, with both MTs found in 2016 (Table S1). This indicates that the pathogen survived over time, which could have been due to the presence of compatible MTs leading to formation of oospores are able to survive adverse conditions. Only one isolate was collected in 2015 from this field, and this could possibly be the reason for not detecting both MTs in this field in 2015.

Sampling a field multiple times could affect the accuracy of a study. While large numbers of samples could increase chances of detecting isolates of both MTs and different mefenoxam sensitivities within a field, these could also result in the possibility of sampling clones of individual isolates multiple times. This could affect the ratios for MTs and the accuracy of markers used to predict mefenoxam sensitivity. We sampled fields in multiple locations to offset this issue. Assuming the isolates of one MT and the same mefenoxam sensitivity to be clones of a single isolate within a field (disregarding the year collected and the source), we found 34 isolates of which 23 (68%) were of A2 MT. Though slightly greater, this proportion was is not significantly different from the 66% reported for the 64 isolates ($\chi^2_{df=1} = 0.184$, $P = 0.668$). At the same time, despite having the same MT and mefenoxam sensitivity, isolates collected from different years or different sources from the same field could also be different. Therefore, further analyses using sequence data are required to clearly identify the clones of individual isolates.

Out of 64 isolates collected, 31 were sensitive to mefenoxam, four were insensitive, and 18 were of intermediate sensitivity to the compound. There was variability among the SZCs of the remaining field isolates, indicating that each field isolate was a population of individuals. For example, SZCs of 11 field isolates had S to IS phenotypes, and two SZCs had IS to I phenotypes. Six out of these 11 reactions were significantly different between the individual zoospore isolates. These results indicate that *P. capsici* isolates insensitive to mefenoxam have increased in field populations since 2008, when mefenoxam insensitive isolates were first reported from MAR (Davey *et al.*, 2008). This emphasizes the need for alternative fungicides for management of *P. capsici* diseases. We tested a subset of 21 field isolates for reaction to technical grade mefenoxam to determine whether this gave the same result as the commercial formulation Ridomil Gold SL, which contains only 45.3% mefenoxam. Enough technical grade fungicide was available to test three SZCs each for 21 field isolates with both forms of mefenoxam. Results indicated that 91% of the isolates give the same, or very similar, reaction (reactions not statistically different), regardless of whether they were exposed on technical grade mefenoxam or Ridomil Gold SL (Table S2). The two field isolates (PC33 and PC34) that gave significantly different phenotypes ($\alpha = 0.05$) could have resulted from effects of proprietary additives in the commercial Ridomil Gold SL formulation. However, this also suggests that each *P. capsici* field isolate is likely to be a population of individuals; each SZC cultured from the original field isolate, is not necessarily a clone. This assay provided the same phenotypes in two separate replicates. Further experiments should include at least 25 SZC from a single field isolate in plate-based and molecular-based mefenoxam sensitivity assays, to provide insights into the nature of each field isolate.

Molecular markers were identified as associated with mefenoxam sensitivity, which could be important tools for initial screening of isolates. Mefenoxam sensitivity prediction was more accurate when both markers, PCSSR19 and PCN3, predicted the same reaction to mefenoxam. In a rare case, the two markers were inconsistent with each other, when PCSSR19 predicted sensitivity and PCN3 predicted insensitivity. In these instances, a plate-based assay should be used to confirm the reactions. Isolates PC34 and PC36, which showed S-IS reaction to Ridomil Gold SL, were S to technical grade mefenoxam (Table S2). Both PCSSR19 and PCN3 predicted sensitivity to mefenoxam for these two isolates. These markers are potentially robust tools for rapidly identifying fungicide-insensitive isolates. Future studies will include testing on a larger sample of *P. capsici* isolates than reported here, as well as generation and testing of a segregating population of the pathogen.

Results of the pod assays indicated that all commercial lima bean cultivars used in this study were susceptible to all *P. capsici* isolates tested, including isolates recovered from cucurbits and crops other than lima bean. Susceptibility of all commercial lima bean cultivars highlights one of the biggest disease management challenges. Accession PI477041 exhibited limited sporulation to some *P. capsici* isolates. It is hoped that this line and other resistant landraces can be used in breeding for resistance to *P. capsici* in the future. The detached pod assay could be used as an initial screening system for identifying resistance to this pathogen in future breeding efforts. Extensive greenhouse and field-testing could then be carried out on selected germplasm lines, with reduced cost and resources. Snap bean inoculations in this study indicated that *P. capsici* isolates from Delaware can cause disease on snap beans under optimal conditions for infection. Snap beans are widely grown in Delaware with no reports of *P. capsici* infections in field sites. However, there are reports of *P. capsici* infecting snap bean in other states (Gevens *et al.*, 2008; McGrath and Dillard, 2011). Examining the host range of *P. capsici* in Delaware is important information for farmers, as well as for developing effective management strategies to control *P. capsici* in the MAR.

As a pathogen, *P. capsici* has a broad host range and can attack different host tissues (Hausbeck and Lamours 2004). The presence in a field of both MTs of this pathogen, and isolates which are insensitive to mefenoxam, could result in oospore production and long-term survival of the pathogen, and also in reduced fungicide efficacy against the diseases it causes, posing severe management implications for lima bean production.

ACKNOWLEDGEMENTS

The authors thank Dr. C. Hong at Virginia Polytechnic Institute and State University, VA, for identification of cultures recovered from water samples, and Dr. M. Hausbeck at Michigan State University, MI, for providing the tester isolates used in the mating type analyses. This project was funded by the USDA-SCRI award No. 2012-51181-19776 and DDA grant No. 12-25-B-1661.

LITERATURE CITED

Babadoost, M., Pavon C., 2013. Survival of oospores of *Phytophthora capsici* in soil. *Plant Disease* 97: 1478-1483.

Brent K. J., Hollomond D. W., 1998. *Fungicide resistance: The assessment of risk*. FRAC Monograph No. 2., Global Crop Protection Federation, Brussels.

Bush E.A., Hong C.X., Stromberg E.L., 2003. Fluctuations of *Phytophthora* and *Pythium* spp. in components of a recycling irrigation system. *Plant Disease* 87: 1500–1506.

Calvert O. H., Williams L. F.,Whitehead M. D., 1960. Frozen lima bean agar for culture and storage of *Phytophthora sojae*. *Phytopathology* 50: 136-137.

Chen D. W.,Zentmyer G. A., 1970. Production of sporangia by *Phytophthora cinnamomi* in axenic culture. *Mycologia* 62: 397-402.

Davey J. F., Gregory N. F., Mulrooney R. P., Evans T. A., Carroll R. B., 2008. First Report of Mefenoxam-Resistant Isolates of *Phytophthora capsici* from Lima Bean Pods in the Mid-Atlantic Region. *Disease notes* 656 http: //dx.doi.org/10.1094/PDIS-92-4-0656A.

Davidson C. R., Mulrooney R. P., Carroll R. B., Evans T.A., 2002. First report of *Phytophthora capsici* on lima bean in Delaware. *Plant Disease* 85: 886.

de Cock A.W.A.M., Lodhi A.M., Rintoul T.L., Bala K., Robideau G.P., Gloria Abad Z., Coffey M.D., Shahzad S., Lévesque C.A., 2015. *Phytopythium:* molecular phylogeny and systematics. *Persoonia* 34: 25–39.

Dunn A. R., Milgroom M. G., Meitz J. C., McLeod A., Fry W. E., McGrath M. T. *et al.*,2010. Population structure and resistance to mefenoxam of *Phytophthora capsici* in New York State. *Plant Disease* 94: 1461–1468.

Erwin D.C., Ribeiro O.K., 1996. Phytophthora disease worldwide. American Phytopathological Society Press, St. Paul, MN. ISBN 0-89054-212-0.

Evans T. E., Mulrooney R., Gregory N. F., Kee E., 2007. Lima bean downy mildew: impact, etiology and management strategies for the mid-atlantic US. *Plant Disease* 91: 128-135.

Ferguson A. J., Jeffers S. N. 1999. Detecting multiple species of *Phytophthora* in container mixes from ornamental crop nurseries. *Plant Disease* 83: 1129-1136.

Gallegly M.E., Hong C. 2008. Phytophthora: Identifying Species by Morphology and DNA. American Phytopathological Society Press. St. Paul, MN.

Gevens A. J., Hausbeck M. K., 2004. Characterization and distribution of *Phytophthora capsici* from irrigation water near Michigan cucurbit fields: A first report of *Phytophthora capsici* in irrigation water in Michigan. (Abstr.). *Phytopathology* 94(suppl.): S157.

Gevens A. J., Donahoo R. S., Lamour K. H., Hausbeck M. K., 2007. Characterization of *Phytophthora capsici* from Michigan surface irrigation water. *Phytopathology* 97: 421-428.

Gevens A. J., Donahoo R. S., Lamour K. H., Hausbeck M. K., 2008. Characterization of *Phytophthora capsici* causing foliar and pod blight of snap bean in Michigan. *Plant Disease* 92: 201- 209.

Glosier B. R., Ogundiwin E. A., Sidhu G. S., Sischo D. R., Prince J. P., 2008. A differential series of pepper (*Capsicum annuum*) lines delineates fourteen physiological races of *Phytophthora capsici. Euphytica* 162: 23–30.

Granke L. L., Quesada-Ocampo L., Lamour K., Hausbeck M. K., 2012. Advances in research on *Phytophthora capsici* on vegetable crops in the United States. *Plant Disease* 95: 1588–1600.

Granke L.L., Windstam S.T., Hoch H.C, Smart C.D., Hausbeck, M.K., 2009. Dispersal and movement mechanisms of *Phytophthora capsici* sporangia. *Phytopathology* 99: 1258-1264.

Hansen E. M., Reeser P., Sutton W., Brasier, C. M., 2015. Redesignation of *Phytophthora* taxon Pgchlamydo as *Phytophthora chlamydospora* sp. nov. *North American Fungi* 10(2): 1-14. http: //dx.doi: 10.2509/naf2015.010.002.

Hausbeck M.K., Lamour, K.H., 2004. *Phytophthora capsici* on Vegetable Crops: Research Progress and Management Challenges. *Plant Disease* 88: 12, 1292-1303.

Hong C.X., Gallegly M.E., Richardson P.A., Kong P., Moorman, G.W., 2008. *Phytophthora irrigata*, a new species isolated from irrigation reservoirs and rivers in eastern United States of America. *FEMS Microbiology Letters* 285: 203–211.

Hu J., Diao Y., Zhou Y., Lin D., Bi Y., Pang Z. *et al.*. 2013. Loss of Heterozygosity Drives Clonal Diversity of *Phytophthora capsici* in China. *PLoS ONE*: 8(12), e82691. http: //doi.org/10.1371/journal.pone.0082691.

Hu J. Li Y., 2014. Inheritance of mefenoxam resistance in *Phytophthora nicotianae* populations from a plant nursery. *European Journal of Plant Pathology* 139: 545–555.

Jackson K. L., Yin Y. F., Ji P. S., 2012. Sensitivity of *Phytophthora capsici* on vegetable crops in Georgia to mandipropamid, dimethomorph, and cyazofamid. *Plant Disease* 96: 1337–1342.

Jones L.A., Worobo R.W., Smart C.D., 2014. Plant-pathogenic oomycetes, *Escherichia coli* strains, and *Salmonella* spp. frequently found in surface water used for irrigation of fruit and vegetable crops in New York State. *Applied and Environmental Microbiology* 80: 4814. doi. 10.1128/AEM.01012-14.

Keinath A. P., 2007. Sensitivity of populations of *Phytophthora capsici* from South Carolina to mefenoxam, dimethomorph, zoxamide, and cymoxanil. *Plant Disease* 91: 743-748.

Kirk P. M., Cannon P. F., David J. C., Stapler J. A., 2001. Dictionary of the Fungi. 9th ed. CABI Publishers, Wallingford, Oxon, UK.

Kong P., Hong C. X., Richardson P. A. Gallegly M. E., 2003. Single-strand-conformation polymorphism of ribosomal DNA for rapid species differentiation in genus *Phytophthora. Fungal Genetics and Biology* 39: 238–249.

Koressaar T, Remm M.,2007. Enhancements and modifications of primer design program Primer3. *Bioinformatics* 23(10): 1289-91

Lan C.Z., Liu P.Q., Li B.J., Chen Q.H., Weng Q.Y., 2013. Development of a specific PCR assay for the rapids and sensitive detection of *Phytophthora capsici. Australasian Plant Pathology* 42: 379-384.

Lamour K. H., Hausbeck, M.K., 2001. Dynamics of mefenoxam insensitivity in a recombinant population of *Phytophthora capsici* characterized with amplified fragment length polymorphism markers. *Phytopathology* 91: 553-557.

Lamour K., Mudge J., Gobena D., Hurtado-Gonzáles O., Shmutz J., Kuo A. *et al.*, 2012. Genome sequencing and mapping reveal loss of heterozygosity as a mechanism for rapid adaptation in the vegetable pathogen *Phytophthora capsici. Molecular Plant-Microbe Interactions* 25: 1350- 1360.

Leonian L. H., 1922. Stem and fruit blight of peppers caused by *Phytophthora capsici* sp. nov. *Phytopathology* 12: 401-408.

Ma D., Zhu J., He L. Cui K., Mu W. *et al.*, 2018.Baseline sensitivity of *Phytophthora capsici* to the strobilurin fungicide benzothiostrobin and the efficacy of this fungicide. *European Journal of Plant Pathology.* https: //doi.org/10.1007/s10658-018-1514-8.

McGrath M.T. Dillard H.R., 2011. First report of Phytophthora blight caused by *Phytophthora capsici* on snap bean in New York. *Plant Disease* 95: 1028.

McGrath M.T., 2009. Phytophthora blight affects snap

bean on Long Island, NY. Vegetable MD online. http: //vegetablemdonline.ppath.cornell.edu/NewsArticles/Bean_phytoJune09.html.

Monroy-Barbosa A. Bosland P.W., 2011. Identification of novel physiological races of *Phytophthora capsici* causing foliar blight using the New Mexico recombinant inbred pepper lines set as a host differential. *Journal ofthe AmericanSociety for Horticultural Science*136: 1–6.

Papavizas G.C., Bowers J,H., Johnson S.A., 1981. Selective isolation of *Phytophthora capsici* from soils. *Phytopahtology* 71: 129-133.

Parra G. Ristaino J., 1998. Insensitivity to Ridomil Gold (Mefenoxam) Found Among Field Isolates o f *Phytophthora capsici* Causing Phytophthora Blight on Bell Pepper in North Carolina and New Jersey. *Plant Disease* 82(6): 711-711.

Parra G., Ristaino J.B., 2001. Resistance to mefenoxam and metalaxyl among field field isolates of *Phytophthora capsici* causing Phytophthora blight of bell pepper. *Plant Disease* 85: 1069-1075.

Pang Z., Shao J., Chen L., Lu X., Hu J., ...Liu X., 2013. Resistance to the Novel Fungicide Pyrimorph in *Phytophthora capsici*: Risk Assessment and Detection of Point Mutations in CesA3 That Confer Resistance. *PLoS ONE* 8(2): e56513. doi: 10.1371/journal.pone.0056513.

Pang Z., Chen L., Mu W., Liu L., Liu X., 2016. Insights into the adaptive response of the plant-pathogenic oomycete *Phytophthora capsici* to the fungicide flumorph. *Scientific Reports* 6: 24103.

Pei-Qing L., Min-Liang W., Ben-Jin L., Cheng-Zhong L., Qi-Yong, W. *et al.*, 2013. Development of expressed sequence tag-derived simple sequence repeat markers and diversity analysis of *phytophthora capsici* in china. *ESci Journal of Plant Pathology* 02 (03): 137-146.

Qi R., Wang T., Zhao W., Li P., Ding J. *et al.*, 2012. Activity of ten fungicides against *Phytophthora capsici* isolates resistant to Metalaxyl. *Journal of Phytopathology* 160: 717–722.

Quesada-Ocampo L. M., Fulbright D. W., Hausbeck M. K., 2009. Susceptibility of Fraser fir to *Phytophthora capsici. Plant Disease* 93: 135–141.

Quesada-Ocampo L. M., Vergas A. M., Naegele, R.P., Francis, D.M., and Hausbeck, M.K., 2016. Resistance to crown and root rot caused by *Phytophthora capsici* in a tomato advanced backcross of *Solanum habrochaites* and *Solanum lycopersicum*. Plant Disease 100: 829-835.

Roberts P.D., Urs R.R., French-Monar R.D., Hoffine M.S., Seijo T.E. *et al.*, 2005. Survival and recovery of *Phytophthora capsici* and oomycetes in tailwater and soil from vegetable fields in Florida. *Annals of Applied Biology* 146: 351–35.

Roberts P. D., Gevens A. J., McGovern R. J., Kucharek T. A., 2008. Vegetable diseases caused by *Phytophthora capsici* in Florida. Florida Cooperative Extension Service, Institute of Food and Agricultural Sciences, University of Florida. https: //edis.ifas.ufl.edu/pdffiles/VH/VH04500.pdf.

Ristaino J. B., 1990. Intraspecific variation among isolates of *Phytophthora capsici* from pepper and cucurbit fields in North Carolina. *Phytophatholoy* 80: 1253-1259.

Ristaino J. B., Hord M. J., Gumpertz M. L., 1992. Population densities of *Phytophthora capsici* in field soils in relation to drip irrigation, rainfall, and disease incidence. *Plant Disease* 76: 1017-1024.

Sanogo S., Ji P., 2012. Integrated management of *Phytophthora capsici* on solanaceous and cucurbitaceous crops: current status, gaps in knowledge and research needs. *Canadian Journal of Plant Pathology* 34: 479-492.

Tian D. Babadoost M., 2003. Genetic variation among isolates of *Phytophthora capsici* from Illinois. *Phytopathology* 93: S84. Publication no. P-2003-0613-AMA.

Untergasser A, Cutcutache I, Koressaar T, Ye J, Faircloth B.C., *et al.*, 2012. Primer3 - new capabilities and interfaces. *Nucleic Acids Research* 40(15): e115.

Wang Z.Y., Langston D.B., Csinos A.S., Gitaitis R.D., *et al.*, 2009. Development of an improved isolation approach and simple sequence repeat markers to characterize *Phytophthora capsici* populations in irrigation ponds in southern Georgia. *Applied and Environmental Microbiology* 75: 5467-5473.

Weiland J. E., Santamaria L., Grunwald N. J., 2014. Sensitivity of *Pythium irregulare*, *P. sylvaticum*, and *P. ultimum* from forest nurseries to mefenoxam and fosetyl-Al, and control of Pythium damping-off with fungicide. *Plant Disease* 98: 937-942. 10.1094/PDIS-09-13-0998-RE.

Yang X., Richardson P.A., Hong C., 2014. *Phytophthora* × *stagnum* nothosp. nov., a New Hybrid from Irrigation Reservoirs at Ornamental Plant Nurseries in Virginia. *PLoS ONE* 9(7): e103450. doi: 10.1371/journal.pone.0103450.

Zhang Z.G., Li Y.Q., Fan H., Wang Y.C., *et al.*, 2006. Molecular detection of *Phytophthora capsici* in infected plant tissues, soil, and water. *Plant Pathology* 55: 770-775.

3

Virulence factors of *Fusarium* spp., causing wheat crown and root rot in Iran

SAEEDEH DEHGHANPOUR-FARASHAH, PARISSA TAHERI*, MAHROKH FALAHATI-RASTEGAR

Department of Plant Protection, Faculty of Agriculture, Ferdowsi University of Mashhad, P.O.Box: 91775-1163, Mashhad, Iran

*Corresponding author: p-taheri@um.ac.ir

Abstract. Crown and root rot of wheat, caused by *Fusarium* spp., limit crop yields worldwide, from rotting of seeds, seedlings, crowns, roots and basal plant stems. Virulence factors and virulence and aggressiveness of *Fusarium* spp. were investigated for isolates from Iran, obtained from wheat plants with crown and root rot symptoms. Forty isolates of *Fusarium* were used in this research. Among the isolates, nivalenol (NIV) was detected as the dominant trichothecene chemotype produced. Production of trichothecenes and zearalenone (ZEA) in autoclaved rice cultures of *Fusarium* isolates was analyzed using high performance liquid chromatography. The levels of NIV ranged from 258 to 1246 μg kg^{-1}, of deoxynivalenol (DON) from 45 to 1411 μg kg^{-1}, and of ZEA from 53 to 3220 μg kg^{-1}. All *Fusarium* isolates produced cellulase and pectinase enzymes. Positive correlation was observed between activity of cell wall degrading enzymes (CWDEs) produced by the isolates and their pathogenicity on wheat leaf segments. Virulence of trichothecene-producing isolates was greater than that of non-trichothecene-producing isolates. Considerable association was detected between the capability of *Fusarium* spp. isolates to produce virulence factors (such as mycotoxins and CWDEs) and their pathogenicity on wheat.

Keywords. Cell wall degrading enzymes, mycotoxins.

Competing Interests: The Author(s) declare(s) no conflict of interest.

Editor: Dimitrios Tsitsigiannis, Agricultural University of Athens, Greece.

INTRODUCTION

Crown and root rot of wheat, caused by several *Fsarium* species, occur in most cereal producing regions of the world, including Europe, Australia, North America, South America, West Asia, South Africa and North Africa (Smiley *et al.*, 1996; Paulitz *et al.*, 2002; Smiley *et al.*, 2005). In Iran, several *Fusarium* spp. have been isolated from crown and root rot symptoms in wheat growing regions (Besharati Fard *et al.*, 2017). Fusaria produce a diverse array of toxic secondary metabolites (mycotoxins), which are involved in pathogenicity of these fungi to host plants. The most important *Fusarium* mycotoxins are trichothecenes and zearalenone, which can contaminate agricultural products, making them unsuitable for food or feed (Ma *et al.*, 2013).

Trichothecenes are sesquiterpenoid molecules of which many variants are known. These have been chategorized as type A, including T-2 toxin, HT-2 toxin, diacetoxyscirpenol (DAS), monoacetoxyscirpenol (MAS), and type B, including deoxynivalenol (DON, vomitoxin), nivalenol (NIV) and their mono- and di-acetylated derivatives (Yli-Mattila, 2010).

The ability of *Fusarium* spp. to produce particular mycotoxins can be investigated using biochemical and molecular techniques. Molecular techniques are based on detection of different gene clusters involved in production of mycotoxins. Chemotype determination can be performed with primers for several genes involved in trichothecene biosynthesis such as *Tri3, Tri5, Tri7, Tri12* and *Tri13* (Mahmoud and Shehata, 2017). The type of trichothecene produced by an isolate can be predicted based on genetic markers derived from the *Tri* gene cluster, containing the genes involved in tricothecene biosynthesis. Different mycotoxins have different toxicological properties. NIV is more toxic than DON to humans and domestic animals. Trichothecenes such as DON are potent inhibitors of eukaryotic protein biosynthesis (Van der Lee *et al.*, 2015). Trichothecenes also play important roles as virulence factors in fungal pathogenesis.

Based on several previous reports, different types of trichothecene chemotypes are produced by *F. graminearum* and *F. culmorum*, and possibly by other *Fusarium* species (Khaledi et al. 2017, Desjardins et al. 1993; Sarver et al. 2011; Li et al. 2016). DON is one of the factors influencing virulence and aggressiveness of *Fusarium* spp. Some studies have reported that chemotype diversity depends on geographical distribution. Both DON and NIV chemotypes are reported from several countries in Asia, Africa, Europe, South and North America. The DON chemotype is reported as the major trichothecene chemotype present in North America, while the NIV chemotype has not been detected in this region. The NIV chemotype was most frequently isolated from some Asian and European countries (Gilbert *et al.*, 2002; Zeller *et al.*, 2003; Panthi *et al.*, 2014).

Among the *Fusarium* isolates, NIV, 3-ADON and 15-ADON chemotypes were detected from different fields of Mazandaran and Golestan provinces in the northern region of Iran (Haratian *et al.*, 2008; Malihipour *et al.*, 2012), while 15-ADON was the only chemotype detected among the isolates collected from fields of Ardabil province in the North West of Iran (Malihipour *et al.*, 2012; Davari *et al.*, 2013). Among *Fusarium* isolates from fields of Golestan province in the North of Iran, the NIV genotype occurred more frequently, followed by 3-acetyl deoxynivalenol (3-ADON) and 15-acetyl deoxynivalenol (15-ADON) genotypes (Khaledi *et al.*, 2016).

DON is the most predominant mycotoxin associated with progress of disease in host plants (Mirocha et al. 2003; Taheri 2018). DON is more phytotoxic on cereals than NIV (Desjardins 2006), disrupting normal cell function by inhibiting protein biosynthesis, and is a major virulence factor of *Fusarium* spp. pathogenic on cereals (Khaledi et al. 2017; Yu et al. 2008; Zhang et al. 2010).

Zearalenone (ZEA), a polyketide mycotoxin, has chronic estrogenic effects on mammals, causing reproductive problems in farm-raised pigs, experimental animals, livestock and humans (Gaffoor and Trail, 2006). ZEA can be produced pre- or post-harvest in maize and other cereals. Contaminations to ZEA by *Fusarium* spp. in maize, wheat and barley were reported from North, West and South of Iran (Karami-Osboo and Mirabolfathy, 2008; Ehsani *et al.*, 2014). Production of ZEA by *F. acuminatum, F. crookwellense, F. culmorum, F. equiseti, F. graminearum, F. oxysporum, F. sporotrichioides* and *F. semitectum* have been reported (Jiménez *et al.*, 1996; Ezekiel *et al.*, 2008).

The other mechanism involved in aggressiveness and virulence of *Fusarium* spp. is production of extracellular enzymes, which degrade host plant cell walls. These cell wall degrading enzymes (CWDEs), such as cellulases and pectinases, are crucial in the processes of colonization and disease establishment (Wanyoike *et al.*, 2002; Kikot *et al.*, 2010). The CWDEs are involved in softening the cell walls, increasing accessibility of cell wall components for degradation by other enzymes, which enables success of further infection steps and spread of fungal mycelia into the inner host plant tissues (Roncero *et al.*, 2003; Ortega *et al.*, 2013).

The objectives of the present study were: (i) to investigate the capability of *Fusarium* spp. isolates (obtained from wheat plants with crown and root rot symptoms) for producing various virulence factors such as trichothecenes, ZEA and CWDEs such as cellulase and pectinase; and (ii) to examine pathogenicity and aggressiveness of *Fusarium* isolates on wheat, and determine relationships between ability of the isolates to produce virulence factors and their pathogenicity.

MATERIALS AND METHODS

Fungal inoculum preparation

Forty isolates, belonging to eight species of *Fusarium*, were obtained from wheat plants showing symptoms of crown and root rot in Yazd province of Iran. The isolates were deposited in the fungal collection in Ferdowsi University of Mashhad, Iran (Table 1). Fun-

Table 1. Capability of *Fusarium* spp. isolates obtained from wheat plants for production of pectinase and cellulase, based on quantitative (µg mL^{-1}) and qualitative (mm) analyses, together with virulence and aggressiveness traits of *Fusarium* spp. isolates on inoculated wheat leaf segments. Virulence was evaluated on the leaf segments at 7 d post inoculation.

Isolate code	Maximum of enzyme activity (µg mL^{-1})		Zone diameter (mm)		Virulence on leaf segments (LL[a], mm)	Aggressiveness on leaf segments (hpi[b])
	Pectinase±SE	Cellulase±SE	Pectinase on pectin agar media±SE	Cellulase on CMC agar media±SE		
Ab1	4028±891 mn [c]	1129.5±20.5 abc	6.6±0.8 efg	7.3±0.3 fg	11±2.3 def	24
Ab2	3865±1 no	674±28 kl	5±0.1 ij	3.6±0.3 lm	6±0.5 j	96
Ab3	3727±10.5 rst	735.5±23.5 ij	5±1.1 ij	4.6±0.8 jkl	7.6±0.6 gh	96
Ar1	4309±86 gh	1087.5±27.5 abcde	8.3±0.3 bcde	8.3±0.3 cd	12.6±2.4 de	36
Ar2	4264±49 ghijkl	1010±10 bc	7.3±0.8 de	6±0.5 hi	10±0.1 def	48
As1	4221±207.5 hi	954±31 efg	7.6±0.8 cd	7±0.5 gh	9±2 ef	36
As2	5218±28 bc	1139.5±4.5 abc	12.3±1.3 a	10.3±0.3 ab	25±1.7 a	12
As3	4135±42.5 jk	965±15 efg	6.6±0.8 efg	6±1.5 hi	9.6±1.4 ef	72
Ba1	3770±182.5 qr	787.5±7.5 hi	6±0.1 gh	4.3±0.3 jkl	8.3±2 gh	96
Ba2	4008±3 lmn	690.5±3.5 jkl	7±0.5 ef	5±0.5 jk	8.6±0.8 fg	72
Ba3	3982±17 lmn	967.5±7.5 defg	7±0.5 ef	5.6±0.6 ij	9.3±1.2 ef	48
Ba4	4460±182 ghij	1068±61.5 abcde	10±0.5 abcd	10±0.1 abc	17±1.7 bc	24
Ba5	3938±66.5 mn	514.5±31.5 no	4.6±0.3 jk	3±0.5 m	6±0.1 j	120
Ba6	3725±30 rst	845±25 ghi	6.3±0.6 fg	5.6±0.8 ij	9±1.5 ef	72
Ba7	5031±28.5 cd	1082±22 abcde	10.6±0.6 ab	9.6±0.3 abcd	20.6±1.8 ab	24
Ba8	4021±84 lmn	750±23 hi	6±1.5 gh	6±0.1 hi	8.3±1.6 gh	72
De1	5445±130 ab	1145±12 abc	12.3±1.4 a	11.3±0.8 a	25.6±2 a	12
De2	4089±11 lm	956±4 efg	8±0.1 bcde	4.6±0.3 jkl	8.3±1.6 gh	96
Kh1	3733±108.5 rst	543±133 mno	4.6±1.2 jk	4±0.1 kl	6.6±0.3 ij	120
Kh3	3790±60 pq	700±10 jkl	6.3±0.8 fg	5±0.1 jk	8.3±0.8 gh	96
Kh4	4864±112 def	971±36 defg	9±0.5 bc	9±0.5 bcd	15.6±1.6 bcde	12
Kh5	4171±58.5 ij	810±20 hi	5.6±1.2 hi	4±0.5 kl	7±1.1 hij	120
Kh6	4487±160 ghi	1122.5±12.5 abc	9.3±0.3 b	8.6±0.3 bcde	12±2.6 def	24
Meh1	3831±54 op	476±32 no	4.6±0.3 jk	3±0.1 m	5.6±0.3 j	120
Meh2	4392±75.5 ghijk	1085±50 abcde	6±1.5 gh	7.3±0.3 fg	10±0.5 def	48
Meh3	4080.5±30.5 kl	1009±36 bc	7.3±0.3 de	6±1 hi	9.3±2.3 ef	48
Mey1	3872±62.5 no	1167± 7.5 a	10.3±0.6 abc	9.3±0.3 bc	16±3.7 bcd	24
Mey2	3942±47 mn	599±14 lm	6±1 gh	4.6±0.3 jkl	7.6±0.6 gh	96
Mey3	3679±26 st	498±127 no	4.3±0.8 kl	3.3±0.8 lm	6±0.5 j	120
Mey4	5017±7.5 cd	1102.5 ±2.5 abcde	7.3±1.2 de	8±0.1 de	11±2.8 def	24
Ta1	3671±81 st	443.5±94.5 o	4±0.5 l	3±0.5 m	5±0.1 j	120
Ta2	3763±77 qr	1117.5±2.5 abcd	7±0.1 ef	5±1.1 jk	10.6±0.6 def	72
Ta4	3777±99 pq	1045±95 abcde	7.6±0.3 cd	7±0.5 gh	11.3±0.3 def	36
Ta5	5174±39 bcd	1159.5±15.5 ab	10.3±0.8 abc	10±0.5 abc	20.3±1.6 abc	24
Ta6	3546.5±106.5 t	829±6 ghij	5.6±1.4 hi	5±0.5 jk	8±2 gh	72
Ya1	4098.5±91.5 kl	890±20 fgh	6.6±1.3 ef	5.6±0.3 ij	9.3±0.3 ef	72
Ya2	4903±27 cde	1057 ±67.5 abcdef	8.3±0.3 bcde	7.6±0.3 ef	11.6±1.6 def	36
Ya3	4604±209.5 efg	1003±2 cd	8.6±0.3 bcd	9±0.1 bcd	14.3±3.9 bcdef	24
Ya4	5641.5±195.5 a	1151±8 abc	9±0.1 bc	9.3±0.3 bc	14±5.6 cd	24
Ya5	4552±289.5 fgh	1062.5±52.5 abcde	7.6±0.6 cd	8±0.5 de	12±3.5 def	36

[a] LL, Lesion length.

[b] hpi, hours post inoculation

[c] Different letters indicate significant differences ($P = 0.05$), according to Duncan analysis. Each experiment was repeated twice, with similar results.

gal inocula were produced in Mung Bean Broth (MBB) using the method of Zhang *et al.* (2013). Conidium suspensions were diluted in water containing 0.05% (v/v) Tween 20, to final concentration of 1×10^5 conidia mL^{-1}.

Virulence assays

Spring wheat cultivar (cv. Falat), obtained from Agricultural Research Center of Khorassan Razavi province in Iran, was grown in a greenhouse, with 12 h photoperiod, RH of 75%, and a day:night temperature regime of 18°C:12 °C. After 14 d, 7 cm segments from the mid-section of the first leaf of plants were harvested, and placed adaxial surface up on the surfaces of 0.5 % water agar in Petri plates, as described by Browne and Cooke (2004). Leaf segments were each inoculated at the center of the adaxial surface with 5 μL of conidium suspension. Sterile distilled water containing 0.05% (v/v) Tween 20 was applied on the control (non-inoculated) leaf segments. Petri dishes were incubated at 25°C with a 12 h:12 h light:dark cycle. After 7 d, the lesion length at the point of inoculation on each leaf was determined. The experiment was replicated three times for each isolate, and repeated twice.

Assessment of aggressiveness

Aggressiveness, as another quantitative component of pathogenicity, was investigated for each fungal isolate on detached leaves of wheat plants (cv. Falat) in laboratory conditions using the methods described by Malihipour *et al.* (2012) and Pariaud *et al.* (2009). Analysis of aggressiveness was determined based on hours post inoculation (hpi) for disease symptom appearance.

Qualitative analyses of cell wall degrading enzymes

For determining cellulase activity, Glucose Yeast Extract Peptone Agar containing 0.5% carboxy-methylcellulose in Petri plates was used for qualitative investigation of cellulase activity. After 3 to 5 d of fungal colony growth, plates were flooded with 0.2% aqueous Congo red solution and de-stained with 1M NaCl for 15 min. Appearance of yellow areas around fungal colonies in red medium indicated cellulase activity (Hankin *et al.*, 1971).

Pectinolytic activity was determined by growing the fungi in Petri plates containing Pectin Agar (containing 5 g L^{-1} pectin, 1 g L^{-1} yeast extract, and 15 g L^{-1} agar in distilled water; pH 5.0). After incubation of 3 to 7 d at 28°C, pectin utilization was detected by flooding the culture plates with freshly prepared iodine-potassium iodide solution (1.0 g iodine + 5.0 g potassium iodide in 330 mL distilled water) (Hankin *et al.*, 1971). Clear zones formed around fungal colonies indicated pectinolytic activity.

In each assay, Petri plates each inoculated with a PDA plug without fungus were used as as negative controls.

Quantitative analysis of cell wall degrading enzymes

For cellulase assays, fungal cultures were prepared in 500 mL capacity Erlenmeyer flasks each containing 250 mL of culture medium, as described by Abdel-Razik (1970). After inoculation, incubation was carried out under shaking (150 rpm) at 27°C and darkness for 10 d. Cellulase activity was assessed using the method of Wood and Bhat (1988). Absorbance was measured at 550 nm, and the amount of reducing sugar released was calculated from the standard curve for glucose. One unit of cellulase activity was defined as the amount of enzyme that catalyzed glucose at 1.0 μ mol min^{-1} during the hydrolysis reaction.

For pectinase assays, the fungal cultures were prepared in 500 mL capacity Erlenmeyer flasks each containing 250 mL culture medium, as described by MacMillan and Voughin (1964). Pectinase activity was determined based on the amount of reducing sugar (D-galacturonic acid) released into the culture supernatant. The amount of D-galacturonic acid was determined using the dinitrosalicylic acid colorimetric method of Colowich (1995), and absorbance was measured at 540 nm. The unit of enzyme activity was defined as the amount of enzyme that released galacturonic acid at 1 μ mol min^{-1}, according to the standard curve. The standard curve was developed based on the absorbance for different concentrations of D-galacturonic acid.

Detection of the genes responsible for production of NIV, DON and zearalenone

For detection of DON, NIV and zearalenone genes, the *Tri5*, *Tri13* and *PKS4* (polyketide synthase) genes were amplified by PCR, using the primers pairs *Tri5F* (5′-AGCGACTACAGGCTTCCCTC-3′) and *Tri5R* (5′-AAACCATCCAGTTCTCCATCTG-3′) for *Tri5*, *Tri13F* (5′-TACGTGAAACATTGTTGGC-3′) and *Tri13R* (5′-GGTGTCCCAGGATCTGCG-3′) for *Tri13*, and *PKS4F* (5′-CGTCTTCGAGAAGATGACAT-3′) and *PKS4R* (5′-TGTTCTGCAAGCACTCCGA-3′) for *PKS4* (Doohan *et al.*, 1999; Waalwijk *et al.*, 2003; Meng *et al.*,

2010). The PCR cycles consisted of an initial denaturation step at 94°C for 2 min, followed by 30 cycles of denaturation (95°C for 35 s), annealing (60°C for 30 s), extension (72°C for 30 s), and final extension at 72°C for 7 min. The PCR products were detected on 1% agarose gels.

Mycotoxin analyses in laboratory cultures

Mycotoxin production by the *Fusarium* isolates in laboratory cultures was investigated using the method of Alvarez *et al.* (2009). Briefly, 25 g of rice grains (*Oryza sativa* L.) were soaked in 100 mL of sterile distilled water for 6 h in 500 mL capacity flasks. Water was then drained and the rice grain was autoclaved twice. Five mL of inoculum suspension (1×10^5 conidia mL^{-1}) of each isolate was added to each flask and incubated at 26 ± 1°C in darkness for 3 weeks. The rice-fungus mixtures were each ground in a mortar and then dispensed in an Erlenmeyer flask with 75 mL of acetonitrile:methanol:water (80:5:15, v/v/v). HPLC analysess was carried out on a Waters Alliance 2695 separations module coupled to a Waters 474 scanning fluorescence detector (Waters Corporation) that was set at 365 nm excitation and 440 nm emission. To perform choromatographic separations, 500 µL of water:methanol (86:14, v/v) was added to each extract and cleaned with a C18 Spherisorb 5 µm (250 × 4.6 mm; Merck). The mobile phase was water:acetonitrile:methanol (78:12:10 v/v/v) with a flow rate of 2.5 mL min^{-1}. NIV, DON and zearalenone production were measured in µg per kg of sample. Standards of the DON, NIV and ZEA were used to construct a five-point calibration curve of peak areas versus concentrations. The injection volume was 50 µL for both the standard solutions and sample extracts.

Statistical analyses

All experiments were set up in completely randomized designs. The data were analyzed by one-way analysis of variance (ANOVA), and comparison of means was carried out using the Duncan's Multiple Range Test ($P \leq 0.05$. Statistical analyses and correlation tests were performed using software Statistical Package for the Social Sciences (SPSS; version 22).

RESULTS

Virulence and aggressiveness assays

Comparison of the data obtained from inoculation of *Fusarium* spp. isolates on wheat leaf segments revealed that different isolates had different virulence capabilities (Table 1). Significant differences in disease index were recorded among the isolates tested. Leaf assays revealed that the greatest lesion length was produced by *F. solani* isolate De1 and *F. flocciferum* isolate As2. The least disease was observed for the *F. equiseti* isolate Ta1 and *F. oxysporum* isoklate Meh1. Other isolates tested fell between these isolates with various levels of virulence on wheat leaf segments (Table 1, Figure 1). The results of the aggressiveness test on detached leaves showed more rapid development of disease symptoms by *F. solani* isolate De1 and *F. flocciferum* isolate As2, compared to the other isolates tested (Table 1).

Analysis of cell wall degrading enzymes

Qualitative and quantitative analysis of CWDEs showed that all the *Fusarium* isolates were capable of producing pectinase and cellulose enzymes (Table 1). In the quantitative assays, the amounts of CWDE activity among isolates varied from 443.5 to 1167 µg mL^{-1} for cellulase and 3546.5 to 5641.5 µg mL^{-1} for pectinase. Quantitative results of CWDE assays agreed with the qualitative results. The *F. equiseti* isolate Mey1 and *F. pseudograminearum* isolate Ta5 had the greatest *in vitro*

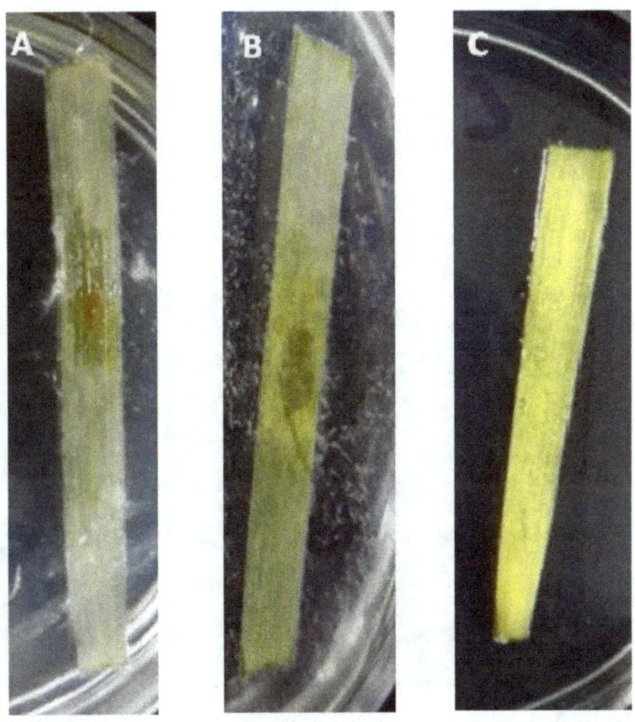

Figure 1. Disease symptoms caused by *Fusarium solani* isolate De1 and *F. flocciferum* isolate As2 on wheat leaves (A and B) and negative control (C).

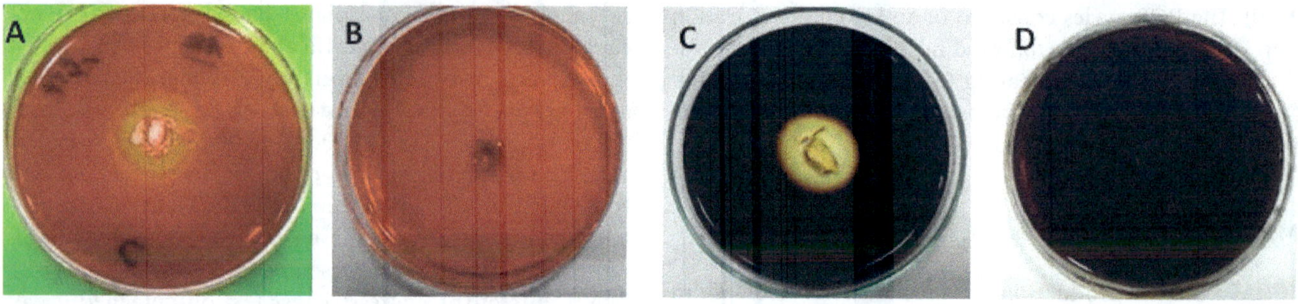

Figure 2. Cellulolytic activity of *Fusarium flocciferum* isolate As2 on GYP medium (A) and its negative control (B). Pectinolytic activity of this isolate on pectin agar medium (C) and its negative control (D).

cellulase activity of all the isolates. The least cellulase activity was measured for *F. equiseti* Ta1 and *F. oxysporum* Meh1. The *Fusarium oxysporum* Ya4 and *F. solani* De1 had the greatest pectinase activities. Least pectinase was measured for *F. flocciferum* Ta6 and *F. equiseti* Ta1. Based on the size of clear culture medium zones for cellulase activity, maximum activity was observed for *F. solani* De1 and *F. flocciferum* As2, while least activity was measured for *F. equiseti* Ta1, *F. oxysporum* Meh1 and *F. equiseti* Ba5.The least pectinase activity was measured for *F. equiseti* Ta1 and *F. pseudograminearum* Mey3, and *F. solani* isolate De1 and *F. flocciferum* isolate As2 gave the greatest pectinase activities (Figure 2).

Detection of trichothecene and zearalenone genotypes by PCR

The *Tri5* gene encodes trichodiene synthase, which catalyses the first step in trichothecene biosynthesis. This gene was detected using the primer set TRI5 (F)/TRI5 (R), which produced a unique PCR product of 544 bp for the isolates which contained the *Tri5* gene (Figure 3A, Table 2) (Doohan *et al.*, 1999; Covarelli *et al.*, 2015). The results obtained from PCR of the *Tri5* gene showed amplification of this gene in 43% of the isolates, which produced either NIV or DON.

Also, the Tri13F/Tri13R primers for amplification of *Tri13* gene amplified a fragment in the range of 200 to 300 bp from DON producers and 400 to 450 bp from NIV producing isolates of *Fusarium* (Figure 3B, Table 2). Among the isolates producing trichothecenes, results obtained from PCR of the *Tri13* gene showed amplification of this gene for 35% of DON-producing isolates, and 65% of NIV- producing isolates.

The *PKS4* (polyketide synthase) gene of *F. graminearum* has been reported to be essential in production of ZEA (Lysøe *et al.*, 2006). The size of PCR products obtained in detecting this gene was approx. 280 bp (Fig-

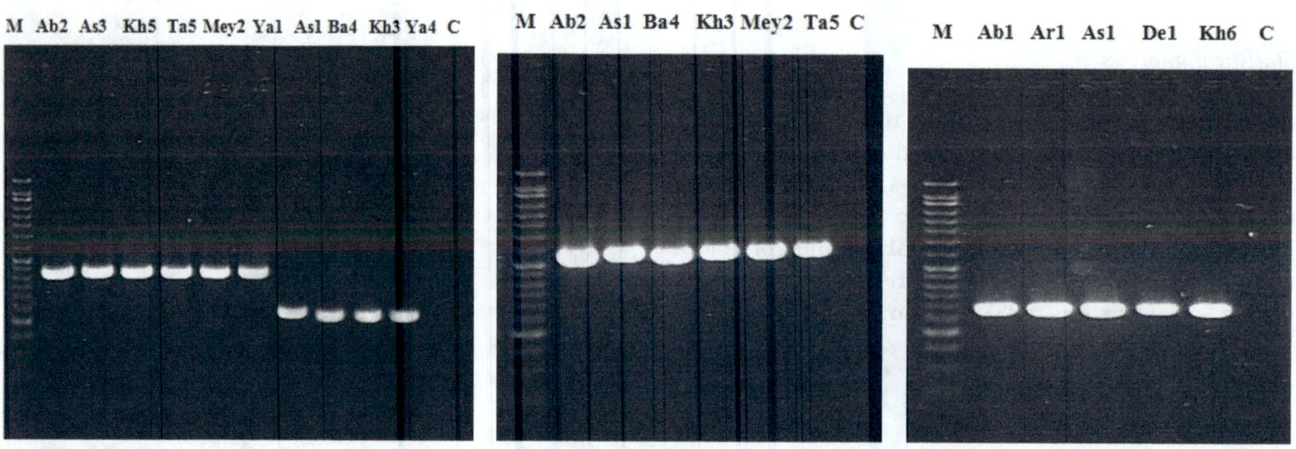

Figure 3. Amplification products using the primer pairs *Tri5F/Tri5R* (A), *Tri13F/Tri13R* (B), *PKS4F/PKS4R* (C), M: marker (1,500 bp). The primer set *Tri5F/Tri5R* produced a unique PCR product with the size of 544 bp for the isolates which contain the *Tri5* gene. The *Tri13F/Tri13R* primers amplified a fragment in the range of 200 to 300 bp from DON-producing *Fusarium* isolates and 400 to 450 bp from NIV-producing isolates. The *PKS4F/PKS4R* primer set produced a fragment of 280 bp for the zearalenone-producing isolates.

Table 2. Origin and species of *Fusarium* isolates obtained from wheat plants in Yazd province of Iran, together with the presence of trichothecene- (NIV and DON) and zearalenone- (ZEA) specific markers detected in the isolates by PCR assays, and their capability for mycotoxin production analyzed by HPLC.

Isolate code	Sample site	Species	Trichotecene genotype	PCR assay results			Toxin production *in vitro* (μg kg−1)		
				Tri5	Tri13	PKS4	NIV	DON	ZEA
Ab1	Abarkuh	*F. acuminatum*	-	-	-	+	-	-	ND
Ab2	Abarkuh	*F. culmorum*	NIV	+	+	+	298	ND	186
Ab3	Abarkuh	*F. culmorum*	DON	+	+	+	ND	1411	252
Ar1	Ardakan	*F. pseudograminearum*	NIV	+	+	+	1007	ND	3075
Ar2	Ardakan	*F. flocciferum*	-	-	-	-	-	-	-
As1	Ashkezar	*F. acuminatum*	DON	+	+	+	ND	379	ND
As2	Ashkezar	*F. flocciferum*	-	-	-	+	-	-	ND
As3	Ashkezar	*F. equiseti*	NIV	+	+	+	850	ND	ND
Ba1	Bafq	*F. solani*	-	-	-	+	-	-	ND
Ba2	Bafq	*F. acuminatum*	-	-	-	+	-	-	ND
Ba3	Bafq	*F. proliferatum*	-	-	-	+	-	-	ND
Ba4	Bafq	*F. culmorum*	DON	+	+	+	ND	1093	114
Ba5	Bafq	*F. equiseti*	-	-	-	-	-	-	-
Ba6	Bafq	*F. proliferatum*	-	-	-	+	-	-	67
Ba7	Bafq	*F. pseudograminearum*	DON	+	+	+	ND	65	3220
Ba8	Bafq	*F. equiseti*	-	-	-	+	-	-	ND
De1	Taft	*F. solani*	-	-	-	+	-	-	ND
De2	Taft	*F. flocciferum*	-	-	-	-	-	-	-
Kh1	Khatam	*F. flocciferum*	-	-	-	-	-	-	-
Kh3	Khatam	*F. equiseti*	DON	+	+	+	ND	103	ND
Kh4	Khatam	*F. culmorum*	NIV	+	+	+	1135	ND	230
Kh5	Khatam	*F. acuminatum*	NIV	+	+	+	350	ND	73
Kh6	Khatam	*F. pseudograminearum*	NIV	+	+	+	1083	ND	1760
Meh1	Mehriz	*F. oxysporum*	-	-	-	+	-	-	53
Meh2	Mehriz	*F. equiseti*	-	-	-	+	-	-	ND
Meh3	Mehriz	*F. solani*	-	-	-	-	-	-	-
Mey1	Meybod	*F. equiseti*	-	-	-	+	-	-	105
Mey2	Meybod	*F. equiseti*	NIV	+	+	+	480	ND	ND
Mey3	Meybod	*F. pseudograminearum*	NIV	+	+	+	258	ND	1385
Mey4	Meybod	*F. oxysporum*	-	-	-	+	-	-	ND
Ta1	Taft	*F. equiseti*	-	-	-	+	-	-	ND
Ta2	Taft	*F. equiseti*	-	-	-	+	-	-	96
Ta4	Taft	*F. pseudograminearum*	NIV	+	+	+	954	ND	1670
Ta5	Taft	*F. pseudograminearum*	NIV	+	-	+	1246	ND	2525
Ta6	Taft	*F. flocciferum*	-	-	-	-	-	-	-
Ya1	Yazd	*F. culmorum*	NIV	+	+	+	743	ND	365
Ya2	Yazd	*F. equiseti*	-	-	-	+	-	-	ND
Ya3	Yazd	*F. equiseti*	-	-	-	+	-	-	78
Ya4	Yazd	*F. oxysporum*	DON	+	+	+	ND	45	ND
Ya5	Yazd	*F. oxysporum*	-	-	-	-	-	-	-

ure 3C, Table 2). Overall, 83% of the *Fusarium* isolates produced this amplicon in the PCR assays.

Mycotoxin analyses using HPLC

Data of detection of trichothecenes produced by *Fusarium* isolates on rice grain showed that among 17 isolates which amplified the *Tri5* gene, 657% produced NIV and 35% produced DON. The levels of NIV ranged from 258 to 1,246 μg kg^{-1}, of DON from 45 to 1,411 μg kg^{-1}, and of ZEA from 53 to 3,220 μg kg^{-1} (Table 2).

Table 3. Correlation analyses between activity of cell wall degrading enzymes (in quantitative and qualitative assays) produced by *Fusarium* spp. isolates and their virulence on wheat leaves.

Correlation	Virulence on leaf segments	*P* value (two-tailed)
Pectinase (quantitative)	0.78231	< .0001
Pectinase (qualitative)	0.94502	< .0001
Cellulase (quantitative)	0.71337	< .0001
Cellulase (qualitative)	0.90916	< .0001

DISCUSSION

This study was a detailed investigation of trichothecene genotypes and quantification of NIV, DON and ZEA by HPLC from *Fusarium* species associated to wheat crown and root rot in Iran. Activities of CWDEs and their relationships with virulence factors, aggressiveness and pathogenicity of *Fusarium* spp. isolates to wheat leaf tissues were also evaluated.

Different types of virulence factors can be produced by different *Fusarium* species pathogenic on cereals (Taheri 2018). Mycotoxins, such as trichthecenes and ZEA, and CWDEs, are among the main factors involved in virulence and aggressiveness of *Fusarium* spp. on host plants. We conclude that the two most virulent *Fusarium* isolates in this study (*F. solani* isolate Del and *F. flocciferum* AS2), which do not produce tricothecenes, may produce other types of virulence factors, such as CWDEs as demonstrated here. Involvement of other virulence factors, such as lipases, xylanases, protein kinases, other proteins and various transcription factors (reviewed by Taheri, 2018), in pathogenicity of these fungi, needs to be investigated in the future studies.

Quantitative and qualitative activities of CWDEs, including cellulase and pectinase which are involved in the infection processes of *Fusarium* spp. on wheat, were investigated here. Aggressiveness of *Fusarium* spp. involves different mechanisms or components, such as production and release of extracellular enzymes which degrade host plant cell walls. The CWDEs are crucial in the processes of pathogen colonization and establishment of disease. Once infection is established, mycotoxins are released and these interfere with the metabolism, physiological processes and structural integrity of host cells (Ortega *et al.*, 2013). These enzymes are particularly important for phytopathogenic fungi without specialized penetration structures (Gibson *et al.*, 2011). The activities of CWDEs produced by the *Fusarium* isolates, which caused maximum or minimum virulence on wheat leaves, were compared for possible associations between the CWDEs and virulence. The isolates *F. solani* Del () and *F. flocciferum* As2, which showed the greatest virulence on wheat leaves, had greater enzyme activities at different time points investigated. In contrast, the isolates Tal and Mehl had the least virulence capability, and the lowest levels of CWDE activity. Correlation analysis revealed high levels of direct association between the capability of *Fusarium* spp. in producing CWDEs and their virulence on the wheat leaves (Table 3). Similarly, Khaledi *et al.* (2016) demonstrated the association of aggressiveness and virulence of *Fusarium* spp. isolates causing head blight of wheat with the levels of CWDEactivity.

In the present study, trichothecene genotype detection revealed 50% amplification of the *Tri5* gene for *F. acuminatum* isolates, 27% amplification for *F. equiseti* isolates, and 25% amplification for those of *F. oxysporum*. *Fusarium pseudograminearum* and *F. culmorum* isolates had 100% amplification of the *Tri5* gene. However, this gene was not detected in isolates of *F. solani*, *F. proliferatum* and *F. flocciferum*. In accordance with our data, Tan and Niessen (2003) showed that *F. solani* was not capable of producing trichothecenes and ZEA, and this species lacked the *Tri5* gene required for biosynthesis of trichothecenes. Khaledi *et al.* (2016) reported that some *F. proliferatum* isolates causing wheat head blight amplified the *Tri5* gene and this species has the ability of trichothecenes biosynthesis, which is in agreement with the findings from the present study.

The *Tri13* gene from the *Fusarium* trichothecene biosynthetic gene cluster is responsible for conversion of DON to NIV (Lee *et al.*, 2001). Our results showed that the NIV was produced by 65% of the isolates, whereas 35% of the isolates produced DON. There are few reports on the geographical distribution of trichothecene chemotypes produced by *Fusarium* spp. in different regions of Iran. Our data showed that the distribution of DON and NIV was not equal in different parts of the studied province, and that NIV was the dominant chemotype. Other investigations in Iran showed dominance of the NIV chemotype in Mazandaran (Haratian *et al.*, 2008) and Golestan provinces (Abedi-Tizaki *et al.*, 2013; Khaledi *et al.*, 2016). Other studies in different regions of the world such as Africa, Asia and Europe have confirmed the presence of NIV and DON chemotypes, but only the DON type has been detected in North America (Miedaner *et al.*, 2000). Both NIV and DON chemotypes have been identified together, in Europe and South America, and the DON chemotype was dominant in these regions. In Asian countries such as Korea and Japan, the NIV chemotype had the greatest distribution (Gale *et al.*, 2011; Lee *et al.*, 2002).

The results of the present study relating to virulence of *Fusarium* isolates on wheat leaf segments showed that all isolates were pathogenic to wheat (cv. Falat), and differences in virulence were observed. Some reports have showned that trichothecenes are virulence factors in plants, and they may contribute to colonization of wheat crowns by the pathogen (Mudge *et al.*, 2006). Maier *et al.*, (2006) showed that NIV and DON act as virulence factors on wheat, while only the NIV chemotype is virulent on maize (. In general in the present study, the isolates with NIV chemotype were more aggressive than the other chemotypes of trichothecenes produced by *Fusarium* spp. This is in agreement with the observations of other researchers (Cumagun *et al.*, 2004; Khaledi *et al.*, 2016).

The polyketide synthase gene *PKS4*, which is involved in ZEA biosynthesis, was used in our study for developing a PCR-based assay to detect ZEA-producing *Fusarium* isolates (Meng *et al.*, 2010). The isolates of *F. flocciferum* and *F. solani* did not produce ZEA. The results of our study showed that the levels of ZEA were not correlated with virulence and aggressiveness of *Fusarium* spp. isolates. This is similar to results of Kuhnem *et al.* (2015), who found that the level of ZEA produced by *F. graminearum* was not related to severity of the disease caused by this species on maize.

Finding novel and effective ways to prevent or decrease production of different types of virulence factors by *Fusarium* spp. may be helpful in management of destructive diseases caused by these important and commonly occurring phytopathogenic fungi.

ACKNOWLEDGMENTS

The Ministry of Science, Research and Technology, and Ferdowsi University of Mashhad, Iran, financially supported this research (project number 3/41181 approved on 2/6/2016). The authors have no conflicts of interest to declare.

LITERATURE CITED

Abdel-Razik A.A., 1970. The parasitism of white *Sclerotium cepivorum* Berk.the incitant of white rot of onion. PhD thesis, Faculty of Agriculture, Assiut University, Assiut, Egypt

Abedi-Tizaki M., Sabbagh S.K., Mazaheri Naeini S.K., Sepehrikia S., 2013. Chemotyping of *Fusarium graminearum* using *Tri13* trichothecene biosynthetic gene. *Journal of Plant Protection* 2: 487-500.

Alvarez C.A., Azcarate M.P., Fernandez P.V., 2009. Toxigenic potential of *Fusarium graminearum sensu stricto* isolates from wheat in Argentina. *International Journal of Food Microbiology* 135: 131–135.

Besharati Fard M., Mohammadi A., Darvishnia M., 2017. *Fusarium* species associated with Wheat crown and root tissues in the Eastern Iran. *Archives of Phytopathology and Plant Protection* 50: 123-133.

Browne R.A., Cooke B.M., 2004. Development and evaluation of an *in vitro* detached leaf assay for pre-screening resistance to *Fusarium* head blight in wheat. *European Journal of Plant Pathology* 110: 91–102.

Colowich S.P., 1995. Methods in Enzymology. London: Academic Prees INC.

Covarelli L., Beccari G., Prodi A., Generotti S., Etruschi F.J.C., ... Mañesc J., 2015. *Fusarium* species, chemotype characterisation and trichothecene contamination of durum and soft wheat in an area of central Italy. *Journal of the Science of Food and Agriculture* 95: 540–551.

Cumagun C.J., Bodwen R.L., Jurgenson J.E., Leslie J.F., Miedaner T., 2004. Genetic mapping of pathogenicity and aggressiveness of *Gibberella zeae* (*Fusarium graminearum*) toward wheat. *Phytopathology* 94: 520–526.

Davari M., Wei S.H., Babay-Ahari A., Arzanlou M., Waalwijk C., ... van Diepeningen A.D., 2013. Geographic differences in trichothecene chemotypes of *Fusarium graminearum* in the North-West and North of Iran. *World Mycotoxin Journal* 6: 137–150.

Desjardins A.E., Hohn T.M., McCormick S.P., 1993. Trichothecene biosynthesis in Fusarium species: chemistry, genetics, and significance. *Microbiology Review* 57: 595-604.

Desjardin A. E, 2006. *Fusarium mycotoxins: Chemistry. Genetics and Biology*. APS Press, St. Paul, MN, USA.

Doohan F.M., Weston G., Rezanoor H.N., Parry D.W., Nicholson P., 1999. Development and use of a reverse transcription-PCR assay to study expression of Tri5 *in vitro* and *in planta*. *Applied and Environmental Microbiology* 65: 3850–3854.

Ehsani M., Chaichi Nosrati A., Shirazi-Beheshtiha S.H., 2014. Measuring toxin Zearalenone in wheat harvested from three regions of North, West and South of Iran. *Bulletin of Environment, Pharmacology and Life Sciences* 3: 69-76.

Ezekiel C.N., Odebod A.C., Fapohunda S.O., 2008. Zearalenone production by naturally occurring *Fusarium* species on maize, wheat and soybeans from Nigeria. *Journal of Biodiversity and Environmental sciences* 2: 77-82.

Gaffoor I., Trail F., 2006. Characterization of two polyketide synthase genes involved in zearalenone biosynthesis in *Gibberella zeae. Applied and Environmental Microbiology* 72: 1793–1799.

Gale L.R., Harrison S.A., Ward T.J., O'Donnel K., Milus E.A., … Kistler H.C., 2011. Nivalenol type populations of *Fusarium graminearum* and *F. asiaticum* are prevalent wheat in southern Louisiana. *Phytopathology* 101: 124–134.

Gibson D.M., King B.C., Hayes M.L., G.C. Bergstrom, 2011. Plant pathogens as a source of diverse enzymes for lignocellulose digestion. *Current Opinion in Microbiology* 14, 264–270.

Gilbert J., Abramson D., McCallum B., Clear R., 2002. Comparison of Canadian *Fusarium graminearum* isolates for aggressiveness, vegetative compatibility, and production of ergosterol and mycotoxins. *Mycopathologia* 153: 209-215.

Hankin L., Zucker M., Sands D.C., 1971. Improved solid medium for the detection and enumeration of pectolytic bacteria. *Journal of Applied Microbiology* 22: 205-209.

Haratian M., Sharifnabi B., Alizadeh A., Safaie N., 2008. PCR analysis of the *Tri13* gene to determine the genetic potential of *Fusarium graminearum* isolates from Iran to produce Nivalenol and Deoxynivalenol. *Mycopathologia* 166: 109–116.

Jiménez M., Máñez M., Hernández E., 1996. Influence of water activity and temperature on the production of zearalenone in corn by three *Fusarium* species. *International Journal of Food Microbiology* 29: 417-421.

Karami-Osboo R., Mirabolfathy M., 2008. Natural zearalenone contamination of wheat from golestan province, northern Iran. *Iranian Journal of Plant Pathology* 44: 60-65.

Khaledi N., Taheri P., Falahati Rastegar M., 2017. Identification, virulence factors characterization, pathogenicity and aggressiveness analysis of *Fusarium* spp., causing wheat head blight in Iran. *European Journal of Plant Pathology* 147: 897-918.

Kikot G.E., Hours R.A., Alconada T.M., 2010. Extracellular enzymes of *Fusarium graminearum* isolates. *Brazilian Archives of Biology and Technology* 53: 779–783.

Kuhnem P.R., Del Ponte E.M., Dong Y., Bergstrom G.C., 2015. *Fusarium graminearum* isolates from wheat and maize in New York show similar range of aggressiveness and toxigenicity in cross- species pathogenicity test. *Phytopathology* 105: 441-448.

Lee T., Oh D.W., Kim H.S., Lee J., Kim Y.H., … Lee Y.W., 2001. Identification of deoxynivalenol and nivalenol producing chemotypes of *Gibberella zeae* by using PCR. *Applied and Environmental Microbiology* l67: 2966–2972.

Lee T., Han Y.K., Kim K.H., Yun S.H., Lee Y.W., 2002. *Tri13* and *Tri7* determine deoxynivalenol and nivalenol producing chemotypes of *Gibberella zeae. Applied and Environmental Microbiology* 68: 2148-2154.

Li Z.F., He C.L., Wang Y., Li M.J., Dai Y.J., … Lin W., 2016. Enhancement of trichothecene mycotoxins of *Fusarium oxysporum* by ferulic acid aggravates oxidative damage in *Rehmannia glutinosa* Libosch. *Scientific Reports* 6: 33962.

Lysøe E., Klemsdal S.S., Bone K.R., Frandsen R.J.N., Johansen T., … Giese H., 2006. The *PKS4* gene of *Fusarium graminearum* is essential for zearalenone production. *Applied and Environmental Microbiology* 72: 3924–3932.

Ma L.J., Geiser D.M., Proctor R.H., Rooney A.P., O'Donnell K., … Kazan K., 2013. *Fusarium* pathogenomics. *Annual Review of Microbiology* 67: 399–416.

MacMillan J.D., Voughin R.H., 1964. Purification and properties of a polyglacturonic acid- transeliminase produced by *Clastridium multiformentans. Biochemistry* 3: 564–572.

Mahmoud M.A., Shehata S.M., 2017. Molecular identification and characterization of *Fusarium* spp. associated with wheat grains. *International Journal of Advanced Research in Biological Sciences* 4: 77-87.

Maier F.J., Miedaner T., Hadeler B., Felk A., Salomon S., … Schaefer W., 2006. Involvement of trichothecenes in fusarioses of wheat, barley and maize evaluated by gene disruption of the trichodiene synthase (*Tri5*) gene in three field isolates of different chemotype and virulence. *Molecular Plant Pathology* 7: 449–461.

Malihipour A., Gilbert J., Piercey-Normore M., Cloutier S., 2012. Molecular phylogenetic analysis, trichothecene chemotype patterns and variation in aggressiveness of *Fusarium* isolates causing head blight in wheat. *Plant Disease* 96: 1016–1025.

Meng K., Wang Y., Yang P., Luo H., Bai Y., … Yao B., 2010. Rapid detection and quantification of zearalenone-producing *Fusarium* species by targeting the zearalenone synthase gene PKS4. *Food Control* 21: 207–211.

Miedaner T., Reinbrecht C., Schilling A.G., 2000. Association among aggressiveness, fungal colonization, and mycotoxin production of 26 isolates of *Fusarium graminearum* in winter rye head blight. *Journal of Plant Diseases and Protection* 107: 124-134.

Mirocha C. J., Xie W., Filho E.R., 2003. Chemistry and detection of fusarium mycotoxins. In fusarium head blight of wheat and barley. In K. J. Leonard & W. R. Bushnell (Eds.), the American Phytopathological society (Vol. 2003, pp. 144–164). Minnesota, USA: St. Paul.

Mudge A.M., Macky R.D., Dong Y.H., Gardiner D.M., White R.G., Manners J.M., 2006. A role for the mycotoxin deoxynivalenol in stem colonisation during crown rot disease of wheat caused by *Fusarium graminearum* and *Fusarium pseudograminearum*. *Physiological and Molecular Plant Pathology* 69: 73-85.

Ortega L.M., Kikot G.E., Astoreca A.L., Alconada T.M., 2013. Screening of *Fusarium graminearum* isolates for enzymes extracellular and deoxynivalenol production. *Journal of Mycology* 358140: 1–7.

Panthi A., Hallen-Adams H., Wegulo S.N., Nopsa J.H., Baenziger P.S., 2014. Chemotype and aggressiveness of isolates of *Fusarium graminearum* causing head blight of wheat in Nebraska. *Canadian Journal of Plant Pathology* 36: 447–455.

Pariaud B., Ravigné V., Halkett F., Goyeau H., Carlier J., Lannou C., 2009. Aggressiveness and its role in the adaptation of plant pathogens. *Plant Pathology* 58: 409–424.

Paulitz T.C., Smiley R.W., Cook R.J., 2002. Insights into the prevalence and management of soil-borne cereal pathogens under direct seeding in the Pacific Northwest, USA. *Canadian Journal of Plant Pathology* 24: 416-428.

Roncero M.I.G., Hera C., Ruiz-Rubio M., Maceira F.I., Madrid M.P., … Dipietro A., 2003. *Fusarium* as a model for studying virulence in soilborne plant pathogens. *Physiological and Molecular Plant Pathology* 62: 87–98.

Sarver B.A., Ward T.J., Gale L.R., Broz K., Kistler H.C., … O'Donnell K., 2011. Novel Fusarium head blight pathogens from Nepal and Louisiana revealed by multilocus genealogical concordance. *Fungal Genetetics and Biology* 48: 1096-1097.

Smiley R.W., Collins H.P., Rasmussen P.E., 1996. Diseases of wheat in long-term agronomic experiments at Pendleton, Oregon. *Plant Disease* 80: 813-820.

Smiley R.W., Gourlie J., Easley S., Patterson L.M., Whittaker R., 2005. Crop damage estimates for crown rot of wheat and barley in the Pacific Northwest. *Plant Disease* 89: 595-604.

Taheri P. 2018. Cereal diseases caused by *Fusarium graminearum*: from biology of the pathogen to oxidative burst-related host defense responses. *European Journal of Plant Pathology* 152: 1–20.

Tan M.K., Niessen L.M., 2003. Analysis of rDNA ITS sequences to determine genetic relationships among, and provide a basis for simplified diagnosis of *Fusarium* species causing crown rot and head blight of cereals. *Mycological. Research* 107: 811-821.

Van der Lee T., Zhang H., van Diepeningen A., Waalwijk C., 2015. Biogeography of *Fusarium graminearum* species complex and chemotypes: a review. *Food Additives & Contaminants: Part A. Chem Anal Control Expo Risk Assess* 32: 453–460.

Waalwijk C., Kastelein P., de Vries I., Kerenyi Z., van der Lee T., Hesselink T., 2003. Major changes in *Fusarium* spp. in wheat in the Netherlands. *European Journal of Plant Pathology* 109: 743–754.

Wanyoike W.M., Kang Z., Buchenauer H., 2002. Importance of cell wall degrading enzymes produced by *Fusarium graminearum* during infection of wheat head. *European Journal of Plant Pathology* 108: 803–810.

Wood T.M., Bhat M., 1988. Methods for measuring cellulase activities. *Methods in Enzymology* 160: 87–112.

Yli-Mattila T., 2010. Ecology and evolution of toxigenic *Fusarium* species in cereals in northern Europe and Asia. *Journal of Plant Pathology* 92: 7-18.

Yu J.B., Bai G.H., Zhou W.C., Dong Y.H., Kolb F.L., 2008. Quantitative trait loci for fusarium head blight resistance in recombinant inbred population of Wangshuibai/Wheaton. *Phytopathology* 98: 87–94.

Zeller K.A., Bowden R.L., Leslie J.F., 2003. Diversity of epidemic populations of *Gibberella zeae* from small quadrats in Kansas and North Dakota. *Phytopathology* 93: 874-880.

Zhang H., Zhang Z., van der Lee T., Chen W.Q., Xu J., … Feng J., 2010. Population genetic analyses of Fusarium asiaticum populations from barley suggest a recent shift favoring 3ADON producers in southern China. *Phytopathology* 100: 328–336.

Zhang J.B., Wang J.H., Gong A.D., Chen F.F., Song B., … Liao Y.C., 2013. Natural occurrence of *Fusarium* Head Blight, mycotoxins and mycotoxin producing isolates of *Fusarium* in commercial fields of wheat in Hubei. *Plant Pathology* 62: 92–102.

4

An integrative approach for the selection of *Pochonia chlamydosporia* isolates for biocontrol of potato cyst and root knot nematodes

MARIA CLARA VIEIRA DOS SANTOS[1,3,*], JOANA HORTA[1], LUÍSA MOURA[2], DAVID V. PIRES[2], ISABEL CONCEIÇÃO[1], ISABEL ABRANTES[1], SOFIA R. COSTA[2,3]

[1] *Centre for Functional Ecology, Department of Life Sciences, University of Coimbra, Calçada Martim de Freitas, P 3000 456 Coimbra, Portugal*
[2] *Mountain Research Centre (CIMO)/ Escola Superior Agrária (ESA), Instituto Politécnico de Viana do Castelo, Refóios, 4990-706 Ponte de Lima, Portugal*
[3] *CBMA- Centre of Molecular and Environmental Biology, Department of Biology, University of Minho, Campus de Gualtar, 4710-057 Braga, Portugal*
*Corresponding author: mcvs@sapo.pt

Competing Interests: The Author(s) declare(s) no conflict of interest.

Editor: Philippe Nicot, Plant Pathology Research Unit INRA, Avignon, France.

Abstract. The nematophagous fungus *Pochonia chlamydosporia*, a natural enemy of plant parasitic nematodes (PPN), has been exploited in the development of sustainable management strategies for PPN control. The intrinsic variation among *P. chlamydosporia* isolates affects biocontrol potential, so sound characterisation is required. Biological, molecular and metabolic analyses can be determinant in the screening and selection of these potential biological control agents. This study aimed to provide integrative characterisation of *P. chlamydosporia* isolates that can support isolate selection for biological control purposes. Eight Portuguese isolates, associated with *Meloidogyne* spp., were used as a case study. Isolates were identified and characterised: i) at the biological level through standard bioassays to evaluate their ability for rhizosphere colonisation, to produce chlamydospores, and to parasitise *Globodera pallida* and *Meloidogyne incognita* eggs; ii) at the molecular level, examining genetic variation using ERIC-PCR; and iii) at the metabolic level, assessing their metabolic profiles using Biolog FF MicroPlates, with concurrent reads of fungal utilisation of 95 different carbon sources. Molecular data and metabolic characterisation were reduced using Principal Component Analysis and compared with the biological characteristics. Molecular profiles could only be related to isolate geographical origin, but the original substrate (eggs or roots), parasitism of *M. incognita* eggs and ability for rhizosphere colonisation were correlated with metabolic profiles, indicative of utilisation of specific carbon sources. *Meloidogyne* egg parasitism and rhizosphere colonisation were related to each other. This integrative characterisation offers novel perspectives on the biology and biocontrol potential of *P. chlamydosporia*, and, once tested on a broader set of isolates, could be used to assist rapid isolate selection.

Keywords. Biolog, isolate selection, metabolic characterisation, nematophagous fungus, rhizosphere interactions.

INTRODUCTION

Plant-parasitic nematodes (PPN) are widespread pests and are severe threats to agricultural crops (Nicol *et al.*, 2011). The most economically important PPN, which have impacts on crops worldwide, are root knot nematodes (RKN), *Meloidogyne* spp., and potato cyst nematodes (PCN), *Globodera* spp. (Jones *et al.*, 2013). The facultative nematophagous fungus *Pochonia chlamydosporia* is an endophyte that parasitises sedentary females and eggs of economically important PPN, including cyst nematodes and RKN, and has been found associated with nematode suppressive soils (Manzanilla-Lopez *et al.*, 2013).

Growth and proliferation of *P. chlamydosporia* in soil can be slow, and only achieved after multiple applications of the fungus (Vieira dos Santos *et al.*, 2014). However, impact of the fungus on nematode reproduction is more affected by the ability to colonise the rhizosphere rather than to proliferate in soil (Manzanilla-Lopez *et al.*, 2013). The fungus is rhizosphere competent and proliferates in this niche using nutrients released in root exudates. Although mostly confined to the rhizosphere, an endophytic behaviour has been reported (Maciá-Vicente *et al.*, 2009). When the colonising *P. chlamydosporia* contacts nematode egg masses in the rhizosphere, the fungus ceases its saprophytic stage and switches to parasitism. Appressoria formed in response to contact with the egg surfaces enable the attachment to, and penetration of, the eggs, which involves mechanical and enzymatic actions. Secreted enzymes destroy the outer vitelline egg membranes exposing the chitin layers, which are then penetrated by infection pegs. These grow into hyphae destroying egg contents and degrading the egg shells from the inside (Mauchline *et al.*, 2004, Monteiro *et al.*, 2017). Since RKN and PCN egg shells are constitutionally different, nematode host preference may be involved in the switch between fungal saprophytic and parasitic phases, but nutrient availability, either released by plants or available in nematode eggs, may play an important role (Mauchline *et al.*, 2004; Esteves *et al.*, 2009). Readily metabolised carbon sources and pH could influence the trophic phase switch as they interfere with the production of an alkaline serine protease (VCP1), known to be involved in the early stages of nematode egg infection (Segers *et al.*, 1994; Ward *et al.*, 2012).

Isolates of *P. chlamydosporia* differ markedly in their rhizosphere competence and ability to parasitise nematode eggs. The outcome of these variations in the regulation of nematode populations is difficult to predict, increasing the need for discriminative screening and selection of isolates. Standard screening encompasses *in vitro* assays for three criteria evaluating the ability of isolates to produce chlamydospores, colonise the rhizosphere, and parasitise nematode eggs (Abrantes *et al.*, 2002).

Chlamydospores are resistant structures that can be produced using different solid substrates. These spores can withstand harsh conditions in soil for long periods without additional nutrients, and initiate fungal growth when in favourable environmental conditions. Chlamydospores are usually regarded as the preferred source of inoculum for biocontrol (Kerry, 2001). Isolates that successfully parasitise nematode eggs are not always competent in establishing in soil, as idiosyncratic results have been obtained for soil establishment (Vieira dos Santos *et al.*, 2014). Inability to colonise the rhizosphere may hinder contact with nematode galls and egg masses, and thus affect biocontrol potential of particular isolates (Manzanilla-López *et al.*, 2013).

Variation in *P. chlamydosporia* isolates has been related to host nematodes and the soil environment from which particular isolates originated (Zavala-González *et al.*, 2015). Field biocontrol efficacy of isolates depends on several factors, including the nematode host, soil biotic and abiotic factors, and geographical origin (Nagesh *et al.*, 2007). Molecular studies, required to identify isolates and provide fingerprinting, have been used to characterise isolates and detect genetic variation (Medina-Canales and Rodríguez-Tovar, 2017).

Enterobacterial Repetitive Intergenus Consensus sequence (ERIC)-PCR as a technique for PCR-based DNA fingerprinting provides rapid evaluation of isolates genetic variation. ERIC primers act randomly and amplify repetitive sequences in fungus genomes, and produce PCR products with multiple band profiles, and this allows isolate discrimination (Morton *et al.*, 2003). However, ERIC-PCR profiling has only been informative for geographical origin and host preference of isolates.

The Biolog FF MicroPlate system (Biolog Inc.) has been used to characterise filamentous fungi according to their metabolic profiles. These profiles are obtained from fungus utilisation of 95 carbon sources from different chemical groups (amines/amides, amino acids, carbohydrates, carboxylic acids, polymers and miscellaneous compounds). This method has been used to identify and aid the definition of new species of fungi with small morphological variation, when molecular techniques are of limited application (Rice and Currah, 2005). In parallel, Biolog FF MicroPlates have been used to assess fungal functional diversity in environmental samples (Lucas *et al.*, 2013), and to support fungal isolate characterisation, particularly if they are closely related (Pawlik *et al.*, 2015; Singh, 2009). The method can also complement bioassays in the selection and characterisation of isolates for biological control (Lopes *et al.*, 2012). To our knowledge, *P. chlamydosporia* has not been characterised using this system.

The main goal of the present study was to provide an integrative methodology that can assist the selection of *P. chlamydosporia* isolates for the biocontrol of PPN. Here, we focus on *G. pallida* and *M. incognita* as, respectively, representative species of PCN and RKN. We anticipate that knowledge gained by the integration of molecular techniques, metabolic profiling and standard *in vitro* bioassays will support isolate selection, and provide new insights into the isolate functional characters that predict their success as biological control agents.

MATERIALS AND METHODS

Pochonia chlamydosporia isolates

Root knot nematode females from infected tomato (*Solanum lycopersicum* L.) roots, obtained in a greenhouse in Alcochete, Setúbal, Portugal, were identified by esterase phenotyping according to Esbenshade and Triantaphyllou (1985) and Pais *et al.* (1986). *Pochonia chlamydosporia* isolation was carried out by plating nematode eggs and host roots on semi-selective medium (De Leij and Kerry, 1991). After 14 d incubation at 25°C, colonies morphologically similar to *P. chlamydosporia* were transferred onto 1.7% corn meal agar (CMA) (Oxoid) for observation of characteristic diagnostic features (Abrantes *et al.*, 2002).

The resulting isolates, from either RKN eggs (PE1, PE2, PE3, PE4, PE5) or tomato roots (PR1, PR2, PR3), were identified by PCR using specific primers. Isolate Pc2, from *G. rostochiensis* eggs (Vieira dos Santos *et al.*, 2013), and a non-native isolate, Vc10 (IMI 331547) originally from *M. incognita* eggs from Brazil supplied by Rothamsted Research, United Kingdom, were also included (Table 1). Isolate Vc10 was used as a standard for *P. chlamydosporia* var. *chlamydosporia*. DNA was extracted from mycelium of *P. chlamydosporia* isolates grown for 10 d on 1.7% potato dextrose agar (PDA) (Difco™) using the E.Z.N.A. Fungal DNA miniKit (OMEGA bio-tek) according to the manufacturer's instructions. The identities of all the isolates were confirmed by PCR. DNA was amplified with the β-tubulin primers specific for *P. chlamydosporia* var. *chlamydosporia* (Hirsch *et al.*, 2000). PCR products were separated in 1% agarose gels stained with GreenSafe Direct Load (Nzytech genes & enzymes), and visualised in a transilluminator (Vilbert Lourmat). The relative mobilities of the PCR bands were calculated with the Bio-print Mega software version 12.15 for Windows (Vilbert Loumart).

Table 1. *Pochonia chlamydosporia* var. *chlamydosporia* isolates, geographic origin, substrate and associated nematode species.

Pc isolate	Geographic origin	Substrate	Nematode species associated
PE1	Setúbal, Portugal	Eggs	*Meloidogyne hispanica*
PE2		Eggs	*M. javanica*
PE3		Eggs	*M. javanica*
PE4		Eggs	*M. incognita*
PE5		Eggs	*M. incognita*
PR1		Roots	*M. hispanica*
PR2		Roots	*M. javanica*
PR3		Roots	*M. javanica*
Pc2	Guarda, Portugal	Eggs	*Globodera rostochiensis*
Vc10 (IMI 331547)	Brasil	Eggs	*M. incognita*

Biological characterisation

The abilities to produce chlamydospores, to colonise the rhizosphere of tomato plants, and to parasitise PPN eggs were tested, based on the methods reviewed by Abrantes *et al.* (2002), with modifications.

Chlamydospore production

Production and viability of chlamydospores produced by the eight isolates and of isolates Pc2 and Vc10 were assessed using the method described by De Leij and Kerry (1991). Five conical flasks per isolate, each containing 100 g of sterilised barley:sand substrate (1:1), were inoculated, each with three agar plugs from 10 d old colonies on CMA. After 30 d incubation at 25°C, the number of chlamydospores per three 1 g sub-samples of the colonised medium per flask was estimated using a haemocytometer. Chlamydospore viability was assessed in sorbose agar with antibiotics (12 g L^{-1} technical agar, 2 g L^{-1} of sorbose and 50 mg L^{-1} each of streptomycin sulphate, chlortetracycline and chloramphenicol). After 2 d at 25°C, the percentage of germinated chlamydospores was estimated using a stereomicroscope (Abrantes *et al.*, 2002). Three biological replicates were carried out.

Rhizosphere colonisation

To study the ability of isolates to colonise the rhizosphere, roots of tomato cv. Tiny Tim roots were used, since tomato is considered to be a good host for *P. chlamydosporia* (Manzanilla-López *et al.*, 2013). Tomato seeds were surface sterilised with a sterilisation solution (1.6% sodium hypochlorite with 0.02% Tween) for 10 min, and rinsed three times with sterile distilled water. The seeds were then germinated in Petri dishes containing Gamborg's B5 medium supplemented with 15 g L^{-1} sucrose 1.5% and 8 g L^{-1} agar, adjusted to pH 6.4. The plates were incubated in the dark, at 25°C. Fungus inoculum was obtained by washing 10-d-old colonies grown on 1.7% CMA with 10 mL Gamborg's B5 medium. The number of spores (chlamydospores and conidia) was counted using a haemocytometer. Five 25 mL capacity flasks per isolate, containing 15 ml of autoclaved sand amended with 5 mL of half-strength Gamborg's B5 medium, were each inoculated with 10^5 spores, and planted with a 3 d-old single germinated tomato seed. The flasks were covered with transparent adhesive film and incubated for 21 d at 25°C and 12 h photoperiod. The plants were then up-rooted, shaken free of sand and placed onto 0.8% water agar for 3 d at 25°C. The percentage of roots colonised by each isolate was estimated using a stereomicroscope. Experimental controls consisted of flasks without fungal inoculum. These assays were conducted twice.

Nematode egg parasitism

The ability of each isolate to parasitise nematode eggs was assessed using standard *in vitro* bioassays (Abrantes *et al.*, 2002). Ten-d-old colonies of each isolate, growing on 1.7% CMA, were washed with 10 mL of sterile distilled water. The numbers of chlamydospores and conidia per mL were quantified using a haemocytometer. Plates containing 0.8% technical agar and antibiotics (streptomycin sulphate, chloramphenicol and chlortetracycline, 50 mg L^{-1} of each) were each inoculated with a 0.2 mL of a 10^5 spores mL^{-1} suspension of each isolate, and were then incubated for 2 d at 25°C. Five plates per isolate were prepared.

A *G. pallida* population, originally from infected potato fields in Cantanhede, Coimbra, was selected from the collection at the NEMATO-lab, Centre for Functional Ecology (CFE), Department of Life Sciences, University of Coimbra (UC), and was propagated on roots of potato cv. 'Desirée' grown in 900 cm^3 plastic pots with a sterilised mixture of sandy loam soil and sand (1:2). Pots were inoculated with a small polyester bag containing cysts, to give approx. 5 eggs g^{-1} of soil. The number of eggs per cyst was assessed by counting five replicates of a suspension of eggs obtained by crushing 50 cysts in water. The pots were placed in a glasshouse at 20°C, with a 16 h photoperiod, and were watered regularly. After three months, new cysts were extracted from the soil using a modified Fenwick can (Shepherd, 1986). Eggs within cysts were released using forceps, and then suspended in water.

A RKN population of *M. incognita*, originally from potato roots and maintained in pot cultures of tomato cv. 'Coração de Boi' at the NEMATO-lab, CFE, UC, was selected. The population was propagated on susceptible tomato cv. 'Tiny Tim', grown in autoclaved sandy loam soil and sand mixture (1:1) in a greenhouse. The culture was transferred periodically to new tomato seedlings inoculated with ten egg masses per plant, placed beneath the tomato roots. Handpicked egg masses were mechanically disrupted, and the released eggs were suspended in water by agitation.

Egg suspensions were sieved through 75 μm and 20 μm sieves to separate eggs from debris, and were concentrated to obtain 2 eggs per μL. Two hundred and fifty eggs of each nematode species were transferred to each Petri plate colonised by the fungus, and were gently spread using a sterilised glass rod. Plates were incubated at 25°C for a further 3 d, after which the number of parasitised eggs was counted using standard methods (De Leij and Kerry, 1991). Experimental controls consisted of plates inoculated with nematode eggs, without the fungus. The bioassays were repeated three times.

Molecular characterisation

All the *P. chlamydosporia* isolates were characterised by ERIC-PCR, as described by Arora *et al.* (1996), and all reactions were performed three times to assess PCR pattern reproducibility. DNA was extracted from mycelium as described above. PCR products were separated in 2% agarose gels stained with GreenSafe Direct Load and visualised in a transilluminator. ERIC-PCR fingerprints produced distinct bands that were visually scored for presence (1) or absence (0) to produce a binary matrix from gel photographs recorded with the Bio-print Mega software version 12.15 for Windows.

Metabolic characterisation

The global phenotypes and utilisation by each of the isolates of 95 low molecular weight carbon sources (plus a negative control) were evaluated using the Biolog FF MicroPlate (Biolog Inc.), following manufacturer instructions. Pure cultures of each isolate were grown on CMA at 25°C in the dark for 3 weeks, until spore formation was visible. The inoculum for the 96 well FF Micro-

Plates was prepared by first soaking a sterile swab in inoculating fluid and then gently rolling over the surface of the agar plates. The obtained spores were suspended in FF inoculating fluid supplied by Biolog in glass tubes (Cat. N° 1006), mixed gently by hand and adjusted to approx. 75 % transmittance at 590 nm using a Biolog Turbidimeter, previously calibrated using a FF Biolog Turbidity standard (Cat. N° 3426). Spore suspension (100 μL) was added to each well, and the FF MicroPlates were then incubated at 25°C in the dark. The optical density at 490 nm (mitochondrial activity) was determined using an ASYS UVM 340 microplate reader (Hitech GmbH) for each plate at 24 h intervals over the next 7 d. Carbon sources were considered not utilised in wells in which colour development was less than, or equal to, that of negative controls.

Statistical analyses

Data of rhizosphere colonisation, numbers and viability of chlamydospores and nematode parasitism variables were analysed using IBM SPSS Statistics v.22. Descriptive statistics were used to explore data and identify extreme outliers that were, in most cases, due to technical error, and five data points were subsequently removed from analyses. Generalised linear mixed models (GLMM) were run for all variables using a normal distribution and an identity link function to establish statistical differences between treatments ($P < 0.05$). 'Isolate' was a fixed factor, and biological repeats and model hierarchy were both included as random factors. Satterthwaite approximations were used due to the small number of data points. Pairwise contrasts between estimated means per isolate adjusted for multiple comparisons were obtained by Least Significant Differences. A similar model structure was used to compare the parasitism of PCN eggs with the parasitism of RKN eggs by each isolate, combining results and adding parasitised PPN egg genus as a fixed factor. Estimated means per isolate for each variable resulting from the GLMM were explored with paired-data Pearson correlation analysis using PAST v.3.18. The highly significant correlation between rhizosphere colonisation and parasitism of *M. incognita* eggs was further analysed using Reduced Major Axis (RMA) analysis.

Multivariate analyses were performed in PAST v.3.18 (Hammer *et al.*, 2001) to reduce the number of variables resulting from the molecular (ERIC band matrix) and metabolic (Biolog- carbon source utilisation) profiles, using Principal Component Analysis (PCA). For the metabolic profiles, the average well colour development (AWCD) of the different isolates were calculated over the 7 d incubation, where AWCD equals the sum of the difference between the OD of the blank well (control) and substrate wells, divided by 95 (the number of substrate wells in the FF MicroPlates). The AWCD values were stabilised only at 7 d, therefore OD readings obtained in each of the 95 carbon sources at this incubation time were then reduced to a smaller, easily interpretable, number of explanatory variables with PCA.

The relationships between synthetic gradients on molecular and metabolic profiles obtained through PCA and the GLMM-estimated means of biological data were assessed through correlation analysis in PAST v.3.18. Pearson correlations were used for the comparison between the PC axes containing over 5% of the variance, and the number and viability of chlamydospores produced, rhizosphere colonisation and parasitism of PCN and RKN eggs. Spearman correlations were used for the comparisons of the same axes with nominal data on isolate geographical origin, substrate (eggs or roots) and nematode species associated.

RESULTS

Pochonia chlamydosporia *isolates*

Eight *P. chlamydosporia* isolates (PE1, PE2, PE3, PE4, PE5, PR1, PR2 and PR3) were obtained, associated with three RKN species: *M. hispanica*, *M. incognita* and *M. javanica* (Table 1).

All isolates presented a specific band (ca. 270 bp) for *P. chlamydosporia* var. *chlamydosporia* (data not shown).

Biological characterisation

Chlamydospore production

The ability to produce chlamydospores *in vitro* was variable among the isolates (Figure 1a). Isolates PR1 and Pc2 produced significantly greater numbers of spores (respectively, 171×10^5 g^{-1} and 187×10^5 g^{-1}) when compared with isolates PE1, PE3, PE5, PR2, PR3 and Vc10 (Figure 1a). Chlamydospore viability was less than 85% for isolates PE3, PR1 and PR3 (Figure 1b).

Rhizosphere colonisation

Root colonisation varied from 60% (PR3) to 92% (Pc2). Root colonisation by isolates PE1, Pc2 and Vc10 was significantly greater compared with root colonisation by isolate PR3 (Figure 2).

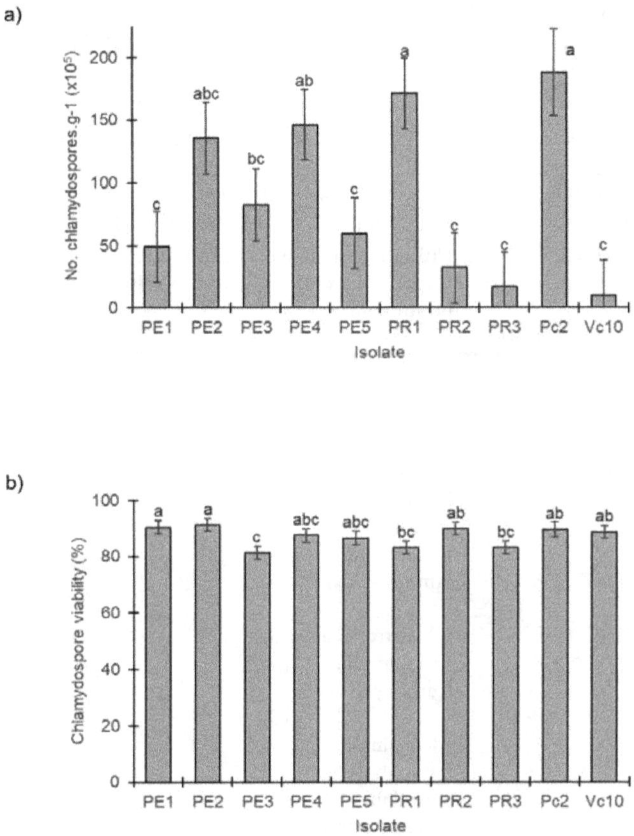

Figure 1. Number of chlamydospores (a) and chlamydospore viability (b) of ten *Pochonia chlamydosporia* isolates (PE1, PE2, PE3, PE4, PE5, PR1, PR2, PR3, Pc2 and Vc10) grown for 30 d on barley:sand substrate (1:1) (refer to Table 1 for isolate information). Bars represent standard errors of means. Columns accompanied by the same letter are not significantly, different (LSD test; $P < 0.05$).

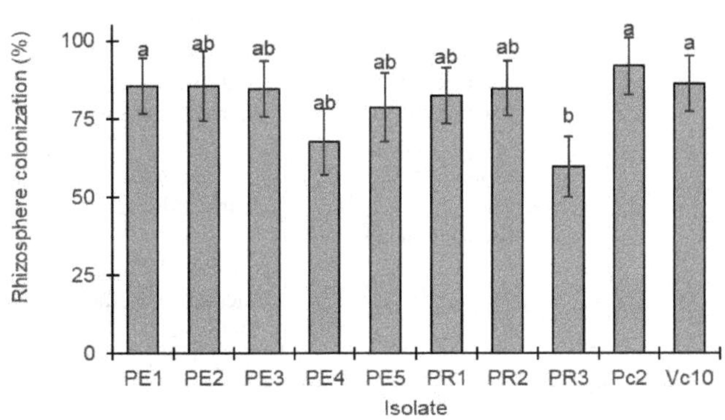

Figure 2. Mean percentages of colonised tomato roots for ten *Pochonia chlamydosporia* isolates (PE1, PE2, PE3, PE4, PE5, PR1, PR2, PR3, Pc2 and Vc10) grown for 21 d on sterilised sand (refer to Table 1 for isolate information). Bars represent standard errors of means. Columns accompanied by same letter are not significantly different (LSD test; $P < 0.05$).

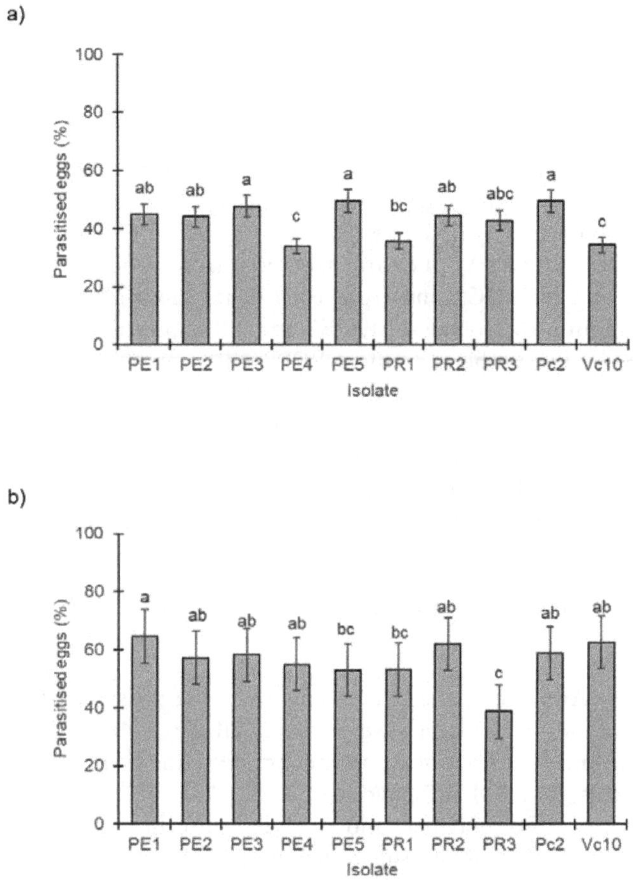

Figure 3. Mean percentages of parasitised eggs of *Globodera pallida* (a) and *Meloidogyne incognita* (b) by ten *Pochonia chlamydosporia* isolates (PE1, PE2, PE3, PE4, PE5, PR1, PR2, PR3, Pc2 and Vc10) after incubation at 25°C for 3 d (refer to Table 1 for isolate information). Bars represent standard errors of means. Columns accompanied by the same letter are not significantly different (LSD test; $P < 0.05$).

Nematode egg parasitism

After 3 d at 25°C, the percentages of parasitised eggs of PCN and RKN were low for all isolates (Figure 3). Pathogenicity of the isolates varied between 34 and 49 % against *G. pallida* eggs (Figure 3a) and 38 to 65 % against *M. incognita* eggs (Figure 3b). Parasitism of PCN eggs was less than parasitism of RKN eggs ($P < 0.01$), with the exception of isolate PR3, which had a greater parasitism rate of PCN eggs than of RKN eggs. Statistically significant differences in egg parasitism between PPN genera were not found for isolates PE1 and PE4. RMA analysis revealed a highly significant positive linear relationship ($r = 0.818$; $P < 0.01$) between parasitism of RKN eggs and rhizosphere colonisation by the fungus.

Molecular characterisation and integration with biological characterisation

Isolates associated with *Meloidogyne* spp. had very similar ERIC patterns. A different band pattern was observed for isolate Pc2 associated with *G. rostochiensis* eggs (Supplementary Material). The PCA analysis separated the fungal isolates into four quadrants, with isolates Pc2 and Vc10 clearly separated from other isolates along PCERIC2 and from each other along PCERIC1 (Figure 4). Isolates PE1, PE2 and PR3 were also separated by synthetic gradient PCERIC1 from isolates PE3, PE4, PE5, PR1 and PR2. Although significant associations were not found between biological characteristics and PC axes PCERIC1 and PCERIC2 (loadings presented in Table 2), the geographical origins of the isolates were correlated with PCERIC1 ($r = 0.707$, $P < 0.05$).

Metabolic characterisation and integration with biological characterisation

Altogether, the ten isolates used the 95 substrates, but only 72 substrates were used by all the isolates: these included four out of six amines/amides; nine of 13 amino acids, 38 of 44 carbohydrates, 12 of 17 carboxylic acids, three of five polymers and six of ten miscellaneous compounds. Six of the substrates, namely n-acetyl-d-mannosamine, sedoheptulosan, d-lactic acid methyl ester, α-cyclodextrin and β-cyclodextrin, were used by fewer than half of the isolates. Isolate PE5 was the only isolate able to utilise all 95 carbon sources, followed by Vc10 (92 carbon sources), isolate PR2 (91) and isolate Pc2 (90). Isolate PR3 utilised the least, with only 77 wells developing colour, as corrected to the water control (Figure 5). The different profiles of carbon source utilisation positioned

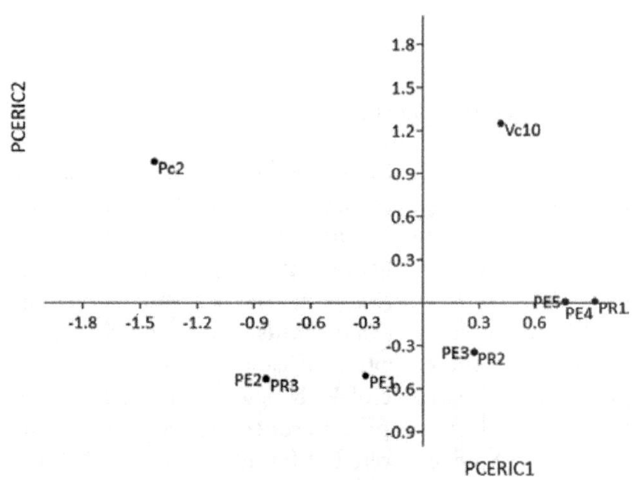

Figure 4. Principal Component Analysis of Enterobacterial Repetitive Intergenus Consensus sequence (ERIC) PCR profiles obtained from *Pochonia chlamydosporia* isolates (PE1, PE2, PE3, PE4, PE5, PR1, PR2, PR3, Pc2 and Vc10; refer to Table 1 for isolate information). PCERIC1 eigen value 0.651; PCERIC2 eigen value 0.400.

Table 2. Synthetic gradients generated by Principal Component Analysis of molecular and metabolic profiling of *Pochonia chlamydosporia* isolates.

Axis	Variance (%)	Positive loading	Negative loading
PCERIC1	49.22	ERIC bands 2, 4 and 6	---
PCERIC2	30.25	ERIC band 8	ERIC band 7
PCBLG1	45.484	Utilisation of most carbon sources; prominent: 2-aminoethanol (amines/amides) glycyl-L-glutamic acid L-glutamic acid L-ornithine L-serine L-threonine (amino acids) β-hydroxy-butyric acid succinic acid (carboxylic acids) adenosine-5'-monophosphate amygdalin succinic acid mono-methyl ester uridine (miscellaneous)	D-ribose (carbohydrates)
PCBLG2	28.58	α-methyl-D-glucoside β-methyl-D-galactoside n-acetyl-D-galactosamine n-acetyl-D-mannosamine (carbohydrates) D-galacturonic acid D-glucuronic acid L-lactic acid (carboxylic acids) D-lactic acid methyl ester (miscellaneous)	2-aminoethanol (amines/amides) L-aspartic acid L-glutamic acid (aminoacids) α-D-glucose L-rhamnose maltose maltotriose (carbohydrates) L-malic acid (carboxylic acids) glycerol succinic acid mono-methyl ester (miscellaneous)

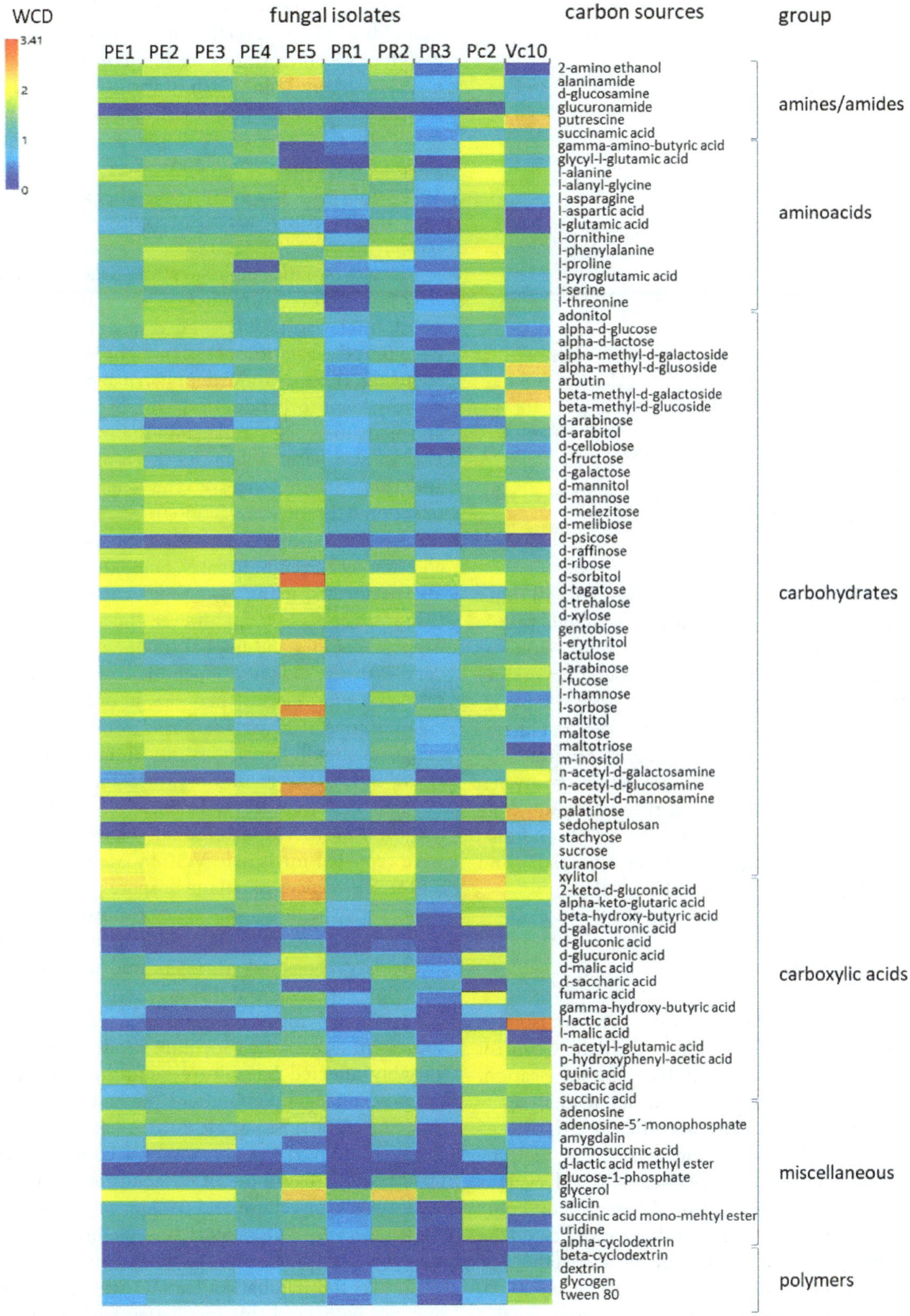

Figure 5. Heat map of carbon source utilisation by *Pochonia chlamydosporia* isolates (PE1, PE2, PE3, PE4, PE5, PR1, PR2, PR3, Pc2 and Vc10; refer to Table 1 for isolate information), assessed through Well Colour Development (WCD) at 7 d, corrected to control, of each of 95 carbon sources/chemical group, using Biolog FF MicroPlates.

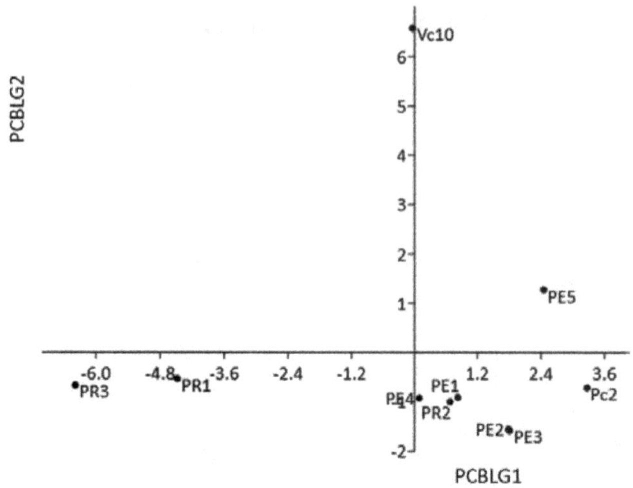

Figure 6. Principal Component Analysis of carbon utilisation profiles obtained through Biolog FF MicroPlates for *Pochonia chlamydosporia* isolates (PE1, PE2, PE3, PE4, PE5, PR1, PR2, PR3, Pc2 and Vc10, refer to Table 1 for isolate information). PCBLG1 eigen value 9.470; PCBLG2 eigen value 5.950.

isolates Vc10 and PE5 in different quadrants to all other isolates, along the two PCA axes of the metabolic profile (PCBLG1 and PCBLG2; Table 2 and Figure 6). This was particularly evident for isolate Vc10, depending mainly on separation obtained by synthetic gradient PCBLG2. Isolates PR1 and PR3 also each had a distinct metabolic profile from all other isolates, being separated by PCBLG1. Significant correlations were detected between the synthetic gradient PCBLG1 and the isolate substrate of origin (eggs or roots; r = -0.646, $P < 0.05$), and also with parasitism of RKN eggs (r = 0.700, $P < 0.05$) and rhizosphere colonisation (r = 0.666, $P < 0.05$).

DISCUSSION

The potential of eight *P. chlamydosporia* isolates associated with *Meloidogyne* spp. as biological control agents was evaluated using integrative characterisation, based on biological, molecular, and metabolic analyses.

Although few statistically significant differences were detected between the evaluated isolates, molecular identification and characterisation allowed evaluation of their variability, affected by different geographic substrate origins (Table 1). This information could be helpful for the potential registration of a specific isolate as a biological control agent (Abrantes *et al.*, 2002). Detection and quantification of particular fungi in soil can be achieved with the use of specific primers, which distinguish the two varieties of *P. chlamydosporia*, var. *chlamydosporia* and var. *catenulata* (Hirsch *et al.*, 2000; Mauchline *et al.*, 2002). The β-tubulin gene allowed design of specific primers for *P. chlamydosporia* var. *chlamydosporia*, and since it has an intron that is not present in other fungi, specific primers do not amplify DNA from other species of *Pochonia* or from other fungal genera (Mauchline *et al.*, 2002; Kerry and Hirsch, 2011). This confirmed the identity of all Portuguese isolates used in this study as *P. chlamydosporia* var. *chlamydosporia*.

Rapid *in vitro* bioassays are necessary for precise selection before introduction of a biological control agent in the soil. Although few statistical differences were found, variation between eight isolates was detected using three standard *in vitro* bioassays, including chlamydospore production, rhizosphere colonisation, and nematode egg parasitism. The inherent variation of *P. chlamydosporia* isolates has been previously described, and this variation stresses the need for precise isolate selection (Abrantes *et al.*, 2002).

Ability to produce chlamydospores *in vitro* was variable among the eight isolates. Standard bioassays are usually performed using agar plugs colonised with fungus (Abrantes *et al.*, 2002). However, the amount of chlamydospores, conidia and hyphal fragments in each plug may vary with the isolates, depending on individual growth rates and sporulation. Therefore, in order to assess differences among *P. chlamydosporia* isolates, equal amounts of spores should be used as inoculum (Vieira dos Santos *et al.*, 2013). The results obtained for reference isolates Pc2 and Vc10 were less than those reported by Vieira dos Santos *et al.* (2013). Additionally, fewer chlamydospores were reported for isolate Vc10 by Medina-Canales *et al.* (2014). Differences in chlamydospore production between isolates may be due to the type of substrate used, as chlamydospore production is greater in substrates with low available carbon and nitrogen, and no sulphur (Mo *et al.*, 2005). Although ability to produce chlamydospores is assessed in the standard characterisation, to ensure success in large-scale inoculum production, this may be variably affected by culture conditions, including carbon and nitrogen contents of the culture substrate (Liu and Chen, 2003; Mo *et al.*, 2005; Vieira dos Santos *et al.* 2012). Moreover, the success of a commercially developed product based on an oil emulsion containing chlamydospores, mycelium and conidia was recently reported (Sellitto *et al.*, 2016). Regarding viability, some isolates have low germination proportions (< 85%), which can be due to the incubation time (3 weeks). Some isolates may not withstand this incubation period, and although they continue to produce chlamydospores, the viability tends to decrease with time (Abrantes *et al.*, 2002). In the present study, all isolates produced adequate numbers of chlamydospores for potential commercial scale production ($\geq 1 \times 10^6$ g^{-1} of substrate) (Medina-Canales *et al.*, 2014). Therefore, we anticipate that, in isolates that produce numbers of chlamydospores greater than accepted threshold of 10^6 g^{-1} of substrate, the ability to produce viable chlamydospores, rather than the exact number of chlamydospores produced, may be an important characteristic for the selection of these biological control agents.

Differences among isolates were detected in their abilities to colonise tomato rhizospheres. The best coloniser of tomato roots was the standard isolate Pc2 with 92% of colonisation. Isolates PE4, PE5 and PR3 colonised less than 80% of tomato roots, failing to achieve the 80% colonisation threshold for selection for biocontrol (Abrantes *et al.*, 2002). Rhizosphere colonisation is a key feature for these biocontrol agents, as it is critical for the control of nematodes by *P. chlamydosporia* (Bourne *et al.*, 1996). The biocontrol efficacy of the fungus depends on fungal rhizosphere colonisation to facilitate colonisation of galls (Manzanilla-López *et al.*, 2013). This assumption has now gained experimental support by our finding of a strong relationship between rhizosphere colonisation and parasitism of RKN eggs. A link between tomato rhizosphere colonisation and parasitism of PCN eggs was not detected. The proportions of colonisation of tomato roots by reference isolates Pc2 and Vc10 were greater than previous results for colonisation of barley roots (Vieira dos Santos *et al.*, 2013). It has been established that plant species differ in their ability to support the fungus (Bourne *et al.*, 1996; Kerry and Bourne, 1996). Whether this can be further related to fungal parasitism of eggs of nematodes that are able to parasitise those plant species is unknown. We speculate that the fungal isolates might emulate nematode host plant preferences, i.e. a virulent parasite of a given nematode population could be a good root coloniser of the host plant preferred by the nematode. This, to our knowledge, has not been investigated.

The low proportion of egg infection obtained through the standard bioassay has also been observed by other authors. This may be related to changes in the mucilage surrounding the eggs, which are considered as important sources of nutrition for the fungi and ensure high rates of egg colonisation. Additionally, isolates parasitise immature eggs more actively than mature eggs containing second-stage juveniles (Irving and Kerry, 1986). Spontaneous hatching detected in both nematode species during the bioassays could also contribute to the low parasitism levels observed. Our results do not support previous findings suggesting that *P. chlamydosporia* isolates have host preferences for host species from which the isolates were initially isolated (Mauchline *et al.*, 2004). Although isolate PR3 was originally associated with RKN, it was more virulent to PCN than RKN; conversely, Pc2, isolated from PCN, was more virulent to RKN eggs than to those of PCN. Seven of the ten isolates tested were more virulent to RKN eggs than to PCN eggs.

Principal Component Analysis of ERIC profiles obtained for *P. chlamydosporia* isolates separated them according to their geographic origin (Figure 4). Isolate Pc2 was the only one from PCN eggs, and was here compared with nine isolates associated with *Meloidogyne* spp. This may have hindered the detection of relationships among ERIC-PCR profiles and nematode origins. Genetic variability among isolates revealed that isolates can be grouped according to their geographical and the host nematode species. The importance of this genetic variation in the regulation of a host population is, however, unknown (Morton *et al.*, 2003).

Isolates PE5, Vc10, PR2 and Pc2 were able to utilise most of the carbon sources (≥ 90), whereas isolate PR3 utilised the least (77) (Figure 5). Isolate PR3 clearly had different carbon requirements as it produced the fewest chlamydospores in barley. Growth and sporulation are influenced by carbon and nitrogen sources, but are also determined by isolate preferences (Mo *et al.*, 2005; Liu and Chen, 2003). Isolate original substrate (eggs or roots), parasitism of *Meloidogyne* eggs, and rhizosphere colonisation ability were significantly correlated with metabolic profiles, indicating that specific carbon sources (amines/amides, amino acids, carboxylic acids and miscellaneous carbon sources) could be used to predict isolate ability for rhizosphere colonisation and for RKN egg parasitism. High proline contents are characteristic of RKN egg shells (Bird and McClure, 1976). For PCN, although proline was the most abundant, 64% of the total amino acids were aspartic acid, glycine and serine (Clarke *et al.*, 1967). Molecular studies have also found that genes encoding proteases known to be involved in nematode egg penetration processes are also highly expressed during endophytic colonisation of barley roots (Lopez-Moya *et al.*, 2017).

The inherent variation of *P. chlamydosporia* isolates has been widely described, and this variation increases the need to carefully select potential biocontrol isolates. Laboratory *in vitro* tests, that evaluate the three biological characteristics considered fundamental for a good fungus performance as biological control agents, enable rapid and economical selection of isolates, before testing their soil activity in pot tests ahead of field trials (Abrantes *et al.*, 2002).

Pochonia chlamydosporia isolate variability was assessed using Biolog FF MicroPlates for the first time.

It is not clear how the metabolic profiles of fungal isolates, or their ability to use particular carbon sources, ultimately affect biocontrol performance in the field. Nevertheless, use of this metabolic approach may facilitate future isolate selection, as the use of specific carbon sources provides complete information on isolate ability for rhizosphere colonisation, and for nematode egg parasitism. These studies may lead to new perspectives for understanding the intricate multitrophic life styles characteristic of *P. chlamydosporia*.

ACKNOWLEDGEMENTS

This work was financed by: FEDER - "Fundo Europeu de Desenvolvimento Regional" funds through the COMPETE 2020 - Programa Operacional Competitividade e Internacionalização (POCI), and by Portuguese funds through FCT - "Fundação para a Ciência e a Tecnologia" in the framework of the project POCI-01-0145-FEDER-016611 (PTDC/AGR-PRO/3438/2014), UID/BIA/04004/2013 in CFE, UID/AGR/00690/2013 in CIMO and UID/BIA/04050/2013 (POCI-01-0145-FEDER-007569) in CBMA; IATV-Instituto do Ambiente, Tecnologia e Vida; and Project ReNATURE - Valorization of the Natural Endogenous Resources of the Centro Region (Centro 2020, Centro-01-0145-FEDER-000007). Grants to M.C. Vieira dos Santos (SFRH/BPD/92308/2013) and S. Costa (SFRH/BPD/102438/2014) were supported by national funds FCT/MCTES and the European Social Fund through the "Programa Operacional do Capital Humano" - POCH of the National Strategic Reference Framework.

LITERATURE CITED

Abrantes I., Santos S., Bourne J., Ciancio A., Lopez-Llorca L. … Verdejo-Lucas S., 2002. *A Manual for Research on Verticillium chlamydosporium, a Potential Biological Control Agent of Root Knot Nematodes* (B.R. Kerry, J.M. Bourne, eds.). International Organization for Biological and Integrated Control for Noxious Animals and Plants/West Paleartic Regional Section (IOBC/WPRS), Gent, Belgium, 84 pp.

Arora, D.K., Hirsch P.R., Kerry B.R., 1996. PCR-based molecular discrimination of *Verticillium chlamydosporium* isolates. *Mycological Research* 7: 801–809.

Bird A., McClure M., 1976. The tylenchid (Nematoda) egg shell: Structure, composition and permeability. *Parasitology* 72: 19–28.

Bourne J.M., Kerry B.R., de Leij F.A.A.M., 1996. The importance of the host plant on the interaction between root knot nematodes (*Meloidogyne* spp.) and the nematophagous fungus, *Verticillium chlamydosporium* Goddard. *Biocontrol Science and Technology* 6: 539–548.

Clarke A.J., Cox P.M., Shepherd A.M., 1967. The chemical composition of the egg shells of the potato cyst-nematode, *Heterodera rostochiensis* Woll. *Biochemical Journal* 104: 1056–1060.

De Leij F.A.A.M., Kerry B.R., 1991. The nematophagous fungus, *Verticillium chlamydosporium* Goddard, as a potential biological control agent for *Meloidogyne arenaria* (Neal) Chitwood. *Revue de Nématologie* 14: 157–164.

Esbenshade P.R., Triantaphyllou A.C., 1985. Use of enzyme phenotypes for identification of *Meloidogyne* species. *Journal of Nematology* 17: 6-20.

Esteves I., Peteira B., Atkins S., Magan N., Kerry B., 2009. Production of extracellular enzymes in different isolates of *Pochonia chlamydosporia*. *Mycological Research* 113: 867–876.

Hammer Ø., Harper D.A.T., Ryan P.D., 2001. PAST: Paleontological statistics software package for education and data analysis. *Palaeontologia Electronica* 4: 9 pp. http://palaeo-electronica.org/2001_1/past/issue1_01.htm

Hirsch P.R., Mauchline T.H., Mendum T.A., Kerry B.R., 2000. Detection of the nematophagous fungus *Verticillium chlamydosporium* in nematode-infested plant roots using PCR. *Mycological Research* 104: 435–439.

Irving F., Kerry B.R., 1986. Variation between strains of the nematophagous fungus, *Verticillium chlamydosporium* Goddard. II. Factors affecting parasitism of cyst nematode eggs. *Nematologica* 32: 474–485.

Jones J.T., Haegeman A., Danchin E.G., Gaur H.S., Helder J., … Perry R.N., 2013. Top 10 plant-parasitic nematodes in molecular plant pathology. *Molecular Plant Pathology* 14: 946–961.

Kerry B.R., 2001. Exploitation of the nematophagous fungal *Verticillium chlamydosporium* Goddard for the biological control of root knot nematodes (*Meloidogyne* spp.). In: *Fungi as Biocontrol Agents - Progress, Problems and Potential* (M.T. Butt, C.W. Jackson, N. Magan, eds.), CAB International, London, UK, 155–167.

Kerry B.R., Bourne J.M., 1996. The importance of rhizosphere interactions in the biological control of plant parasitic nematodes – a case of study using *Verticillium chlamydosporium*. *Pesticide Science* 47: 69–75.

Kerry B.R., Hirsch P.R., 2011. Ecology of *Pochonia chlamydosporia* in the rhizosphere at the population, whole organism and molecular scales. In: *Biological control of plant-parasitic nematodes: building coherence between microbial ecology and molecular mechanisms* (K. Davies, Y. Spiegel, eds.). Springer Netherlands, 171–182.

Liu X.Z., Chen S.Y., 2003. Nutritional requirements of *Pochonia chlamydosporia* and ARF18, fungal parasites of nematode eggs. *Journal of Invertebrate Pathology* 83: 10–15,

Lopes F.A.C., Steindorff A.S., Geraldine A.M., Brandão R.S., Monteiro V.N., … R.N. Silva, 2012. Biochemical and metabolic profiles of *Trichoderma* strains isolated from common bean crops in the Brazilian Cerrado and potential antagonism against *Sclerotinia sclerotiorum. Fungal Biology* 116: 815–824.

Lopez-Moya F., Escudero N., Lopez-Llorca L.V., 2017. *Pochonia chlamydosporia*: multitrophic lifestyles explained by a versatile genome. In: *Perspectives in sustainable nematode management through* Pochonia chlamydosporia *applications for root and rhizosphere health* (R.H. Manzanilla-López, L.V. Lopez-Llorca, eds.). Springer International Publishing AG, 197–207.

Lucas J.A., García-Villaraco A., Ramos B., García-Cristobal J., Algar E. Gutierrez-Mañero J., 2013. Structural and functional study in the rhizosphere of *Oryza sativa* L. plants growing under biotic and abiotic stress. *Journal of Applied Microbiology* 115: 218–235.

Maciá-Vicente J.G., Jansson H., Talbot N.J., Lopez-Llorca L.V., 2009. Real-time PCR quantification and live-cell imaging of endophytic colonization of barley (*Hordeum vulgare*) roots by *Fusarium equiseti* and *Pochonia chlamydosporia. New Phytologist* 182: 213–228.

Manzanilla-López, R.H., Esteves I., Finetti-Sialer M.M., Hirsch P.R., Ward E., … Hidalgo-Díaz L., 2013. *Pochonia chlamydosporia*: Advances and challenges to improve its performance as a biological control agent of sedentary endo-parasitic nematodes. *Journal of Nematology* 45: 1-7.

Mauchline T.H., Kerry B.R., Hirsch P.R., 2002. Quantification in soil and the rhizosphere of the nematophagous fungus *Verticillium chlamydosporium* by competitive PCR and comparison with selective plating. *Applied and Environmental Microbiology* 68: 1846–1853.

Mauchline T.H., Kerry B.R., Hirsch P.R., 2004. The biocontrol fungus *Pochonia chlamydosporia* shows nematode host preference at the infraspecific level. *Mycological Research* 108: 161–169.

Medina-Canales M.G., Rodríguez-Tovar A.V., 2017. Molecular diagnostics of *Pochonia chlamydosporia*. In: *Perspectives in sustainable nematode management through* Pochonia chlamydosporia *applications for root and rhizosphere health* (R.H. Manzanilla-López, L.V. Lopez-Llorca, eds.). Springer International Publishing AG, 197–207.

Medina-Canales M.G., Rodríguez-Tovar A.V., Manzanilla-López R.H., Zúñiga G., Tovar-Soto A., 2014. Identification and molecular characterisation of new Mexican isolates of *Pochonia chlamydosporia* for the management of *Meloidogyne* spp. *Biocontrol Science and Technology* 24: 1-21.

Mo M., Xu C., Zhang K., 2005. Effects of carbon and nitrogen sources, carbon-tonitrogen ratio, and initial pH on the growth of nematophagous fungus *Pochonia chlamydosporia* in liquid culture. *Mycopathologia* 159: 381–387.

Morton C.O., Mauchline T.H., Kerry B.R., Hirsch P.R., 2003. PCR-based DNA fingerprinting indicates host-related genetic variation in the nematophagous fungus *Pochonia chlamydosporia. Mycological Research* 107: 198–205.

Monteiro T.S.A., Lopes E.A., Evans H.C., Freitas L.G. de, 2017. Interactions between *Pochonia chlamydosporia* and nematodes. In: *Perspectives in Sustainable Nematode Management through* Pochonia chlamydosporia *Applications for Root and Rhizosphere Health* (R.H. Manzanilla-López, L.V. Lopez-Llorca, eds.). Springer International Publishing AG, 77–96.

Nagesh M., Hussaini S.S., Ramanujam B., Rangeswaran R., 2007. Molecular identification, characterization, variability and infectivity of Indian isolates of the nematophagous fungus *Pochonia chlamydosporia. Nematologia Mediterranea* 35: 47–56.

Nicol J.M., Turner S.J., Coyne D.L., den Nijs L., Hockland S., Maafi Z.T., 2011. Current nematode threats to world agriculture. In: *Genomics and Molecular Genetics of Plant-Nematode Interactions* (J.T. Jones, G. Gheysen, C. Fenoll, eds.). Springer, Heidelberg, Germany, 21–44.

Pais C.S., Abrantes I.M. de O., Fernandes M.F.M., Santos M.S.N. de A., 1986. Técnica de electroforese aplicada ao estudo das enzimas dos nemátodes-das-galhas-radiculares, *Meloidogyne* spp. *Ciência Biológica Ecology and Systematics* (Portugal) 6: 19–34.

Pawlik A., Janusz G., Debska I., Siwulski M., Frac M., Rogalski J., 2015. Genetic and metabolic intraspecific biodiversity of *Ganoderma lucidum. BioMed Research International*, Article ID 726149, 13 pages. http://dx.doi.org/10.1155/2015/726149

Rice A.V., Currah R.S., 2005. Profiles from Biolog FF plates and morphological characteristics support the recognition of *Oidiodendron fimicola* sp. nov. *Studies in Mycology* 53: 75–82.

Segers R., Butt T.M., Kerry B.R., Peberdy J.F., 1994. The nematophagous fungus *Verticillium chlamydosporium* produces a chymoelastase-like protease\which hydrolyses host nematode proteins *in situ*. *Microbiology* 140, 2715–2723.

Sellitto V.M., Curto G., Dallavalle E., Ciancio A., Colagiero M., Pietrantonio L., Bireescu G., Stoleru V., Storari M., 2016. Effect of *Pochonia chlamydosporia*-based formulates on the regulation of root-knot nematodes and plant growth response. *Frontiers in Life Science* 9: 177–181.

Shepherd A.M., 1986. Extraction and estimation of cyst nematodes. In: *Laboratory Methods for Work with Plant and Soil Nematodes* (J.F. Southey, ed.). Ministry of Agriculture, Fisheries and Food. Her Majesty's Stationary Office London, UK, 31–49.

Singh M.P., 2009. Application of Biolog FF MicroPlate for substrate utilization and metabolite profiling of closely related fungi. *Journal of Microbiological Methods* 77: 102–108.

Vieira dos Santos M.C., Esteves I., Abrantes I., 2012. *In vitro* water stress bioassays with the nematophagous fungus *Pochonia chlamydosporia*: Effects on growth and parasitism. *Biological Control* 63: 310–319.

Vieira dos Santos M.C., Esteves I., Kerry B., Abrantes I., 2013. Biology, growth parameters and enzymatic activity of *Pochonia chlamydosporia* isolated from potato cyst and root-knot nematodes. *Nematology* 15: 493–504.

Vieira dos Santos M.C., Esteves I., Kerry B., Abrantes I., 2014. Interactions between *Pochonia chlamydosporia* and *Meloidogyne chitwoodi* in a crop rotation scheme. *Nematropica* 44, 37–46.

Ward E., Kerry B.R., Manzanilla-López R.H., Mutua G., Devonshire J., ... Hirsch P., 2012. The *Pochonia chlamydosporia* serine protease gene vcp1 is subject to regulation by carbon, nitrogen and pH: Implications for nematode biocontrol. *Plos One* 7: e35657.

Zavala-González E.A., Escudero N., Lopez-Moya F., Aranda-Martinez A., Exposito A., ... Lopez-Llorca L.V., 2015. Some isolates of the nematophagous fungus *Pochonia chlamydosporia* promote root growth and reduce flowering time in tomato. *Annals of Applied Biology* 166, 472–483.

5

Identification of SSR markers linked to Botrytis grey mould resistance in chickpea (*Cicer arietinum*)

Supriya SACHDEVA[1], Shruti DAWAR[2], Upasana RANI[2], Basavanagouda Siddanagouda PATIL[1], Khela Ram SOREN[3], Sarvjeet SINGH[2], Satish Kumar SANWAL[4], Sanjeev K. CHAUHAN[1], Chellapilla BHARADWAJ[1,*]

[1] *Division of Genetics, Indian Agricultural Research Institute (ICAR), Pusa, New Delhi 110012, India*
[2] *Principal Pulse Breeder, Division of Plant Breeding and Genetics, Punjab Agricultural University, Ludhiana, Punjab, India*
[3] *ICAR-Indian Institute of Pulse Research, Kanpur, UP, India*
[4] *ICAR-CSSRI, Karnal, Haryana, India*
*Corresponding author: drchbharadwaj@gmail.com

Summary. Botrytis grey mould (BGM), caused by *Botrytis cinerea*, is emerging as an important disease of chickpea in the northern and eastern parts of the Indian Subcontinent, including Nepal, Bangladesh, Pakistan, and in Australia. This fungus has a very broad host range, and sources of complete resistance to the disease have not been found in *Cicer arietinum* L. germplasm. Resistance to this pathogen has been identified in some wild *Cicer* species. A set of 371 lines, including 164 landraces and 207 interspecific derivative lines (derived from crosses of cultivated chickpea with *C. pinnatifidum, C. judaicum* or *C. reticulatum*) have been screened against Botrytis grey mould under field conditions, and using the cut twig method at the Punjab Agricultural University (PAU), Ludhiana, in 2015-16 and 2016–17. Strong correlations between the two screening methods were indicated by paired-t tests. The Bulked Sample Analysis (BSA) approach was used to screen DNA of the five most resistant and five most susceptible host lines using 300 simple sequence repeat (SSR) markers. Eighty-eight markers were polymorphic. Chi-square statistic values showed strong correlations of TA144, GA102, TA194, TA140 and TR2 with the resistant bulks, signifying their usability as putative markers linked to BGM resistance, and for development of BGM tolerant genotypes in chickpea. Future studies should rapidly ascertain marker trait associations, and identify and develop diagnostic markers that provide an accurate method of molecular tagging BGM resistant genes in chickpea.

Keywords. Chickpea, Grey mould resistance, SSR markers, Bulked sample analysis.

Competing Interests: The Author(s) declare(s) no conflict of interest.

Editor: Diego Rubiales, Institute for Sustainable Agriculture, (CSIC), Cordoba, Spain.

INTRODUCTION

Chickpea (*Cicer arietinum* L.) is a highly nutritious food legume, ranking as third in world importance after peas and drybeans (Bharadwaj *et al.*, 2010). Chickpea is grown on 14.56 million ha in 50 different countries, and average yields are 1.12 t ha^{-1}(FAOSTAT, 2017). Various biotic stresses affect chickpea production globally. These include fungal, viral, and bacterial diseases, nematodes, insect pests and parasitic weeds. Transferring resistance genes to elite cultivars is one of the most efficient approaches for overcoming biotic stresses (Li *et al*., 2015). Most traits of agronomic importance are complex in nature, governed by various genes and environmental interactions (EL-Soda *et al*., 2014). The identification of linked QTLs is crucial, therefore, for understanding the molecular basis of these traits (Xu *et al*., 2010).

Botrytis grey mould (BGM), caused by *Botrytis cinerea* Pers. ex. Fr., is an economically important disease of chickpea. The occurrence of BGM has been reported in many countries, including Argentina, Australia, Canada, Columbia, Bangladesh, Nepal, Pakistan, India, Spain and the United States of America (Haware *et al*., 1998). This disease may cause yield losses upto 100% under favourable conditions (Pande *et al*., 2006a; Pande *et al*., 2002). The disease may occur at any stage of chickpea development (Hawthorne *et al*., 2006) but usually occurs at the time of flowering. Temperatures of 20-30°C and relative humidity from 70 to 100% favour BGM infections. The pathogen infects flowers most easily (Pande *et al*., 2006b, c), and infected pods may carry the infection through infected seeds to the next growing season (Nene *et al*., 2012; Matthews *et al*., 2014). Genotypes with profuse seedling growth and early flowering and canopy closure tend to develop BGM infections more often compared with other cultivars.

Botrytis cinerea has a very broad host range. Source of complete resistance to the pathogen have not been found in *C. arietinum* germplasm (Singh and Reddy, 1991), but wild *Cicer* sp. have been identified as good sources of BGM resistance. Wild *Cicer* species, including *C. judaicum, C. bijugum, C. echinospermum,* and *C. pinnnatifidum* (Singh *et al*., 1991; Haware 1998; Pande *et al*., 2002), have shown higher levels of resistance to BGM in comparison to the cultivated species. Among these wild species, *C. echinospermum* is being used for transfer of BGM resistance to the cultivated species (ICRISAT, 2007). Kaur *et al*. (2013) developed interspecific derivative lines from crosses of *C. arietinum* and *C. pinnatifidum* that were found to be highly tolerant to BGM, and this resistance can be transferred to elite lines for development of high yielding and BGM tolerant chickpea cultivars. Very few studies have been undertaken to exploit wild species and the level of diversity amongst different host plant accessions.

The present study used wild *Cicer* spp. accessions and inter-specific hybrids to identify BGM resistant genotypes and putative markers linked to BGM resistance in chickpea.

Different methods have been used for screening for BGM resistance under laboratory, greenhouse and field conditions (Gurha *et al*., 2003; Pande *et al*., 2006a, b). The cut-twig method is a non-destructive procedure for sampling, and this has been very useful in inter-specific hybridization (Kaur *et al*., 2013). In comparison to the traditional methods of breeding, time saving and cost effective approaches are now widely used in genomics and crop improvement (Sun *et al*., 2010). Marker-assisted selections and selective phenotyping can be simplified using such approaches (Xu and Crouch, 2008). Bulked BSA uses individuals with extreme phenotypes and these variants are then pooled as bulks. Bulked BSA or the Sampling-bulking method for marker development and trait mapping has been named differently as bulked segregant analysis (Michelmore *et al*., 1991) and DNA pooling (Giovannoni *et al*., 1991), and is usually achieved with molecular marker systems (Gillman *et al*., 2011; Asnaghi *et al*., 2004). Genomic regions with large differential allele frequencies between the bulks reveal association of the regions with the QTLs associated with particular trait (Deokar *et al*., 2019) (Figure 1). These QTLs are then subjected to statistical tests to verify the confidence intervals for their location (Takagi *et al*., 2013). Next Generation Sequencing-based BSA has been used for QTL mapping for different traits due to improved efficiency and affordability of NGS platforms (Deokar *et al*., 2019; Chen *et al*., 2017; Pandey *et al*., 2017; Singh *et al*., 2016; Das *et al*., 2016; Illa-Berenguer *et al*., 2015; Kaminski *et al*., 2015; Lu *et al*., 2014). BSA coupled with molecular breeding techniquesal-

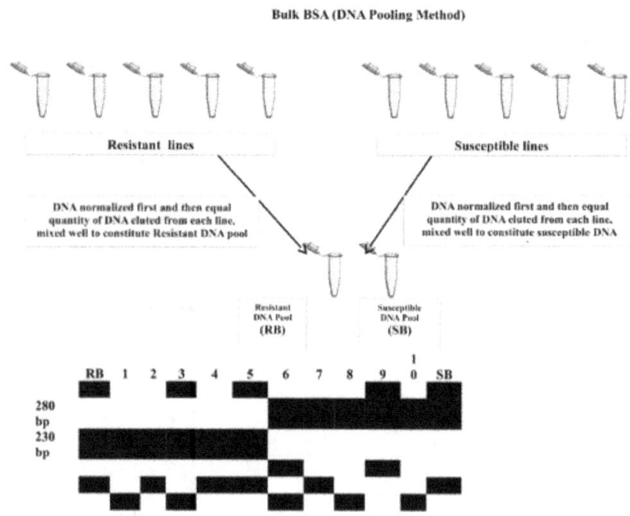

Figure 1. Figure depicting bulking sampling method where equal amount of DNA is eluted from each line after normalization and corresponding bulks generated are used for mapping genes for BGM resistance.

low rapid identification of markers, and is a promising approach for trait mapping and candidate gene discovery in plant breeding.

MATERIALS AND METHODS

Plant material

A total of 371 *Cicer* accessions, including 164 landraces and 207 F_6 interspecific derivatives lines (derived from crosses of cultivated chickpea with *C. pinnatifidum, C. judaicum* or *C. reticulatum*) were screened under field conditions and controlled conditions against *Botrytis* grey mould. The cut-twig method is non-destructive and simple, and is widely used for laboratory studies. Theaccessions were screened at the Punjab Agricultural University (PAU), Ludhiana, during 2015-16 and 2016-17. Of the 371 lines, three landraces and 23 extreme Recombinant Inbred Lines (RILs) were selected, for identification of markers linked to BGM resistance given in Table 1.

Field evaluations

The test lines were sown at plant spacings of 40 cm in 2 m length rows in fields at three replications, in a randomized complete block design (Figure 2a). The chickpea line L550 was used as a susceptible check after every eight rows in the trials. The resulting plants, at the flowering stage, were sprayed with a local isolate of *Botrytis cinerea*, in the first week of February in the evening hours. The inoculum suspension contained 10^5 conidia mL^{-1}. Isolate 24, race 510, of *B. cinerea* (Singh and Bhan, 1986) was used for screening the plants against BGM. The isolate was preserved on slants of potato dextrose agar (20g dextrose, 20g agarose, 200g potato and 1L distilled water), and were multiplied in potato dextrose broth and stored at 25°C. Following inoculation, water was applied using a sprinkler, to maintain high relative

Table 1 Plant material used in this study with, their parentage.

Name of genotype	Parentage/ Cross
GLW 42	(ICCV96030 × *C. pinnatifidum* 188) × ICCV96030
GLW 67	(ICCV96030 × *C. pinnatifidum* 188) × ICCV96030
GLW 69	(ICCV96030 × *C. pinnatifidum* 188) × ICCV96030
GLW107	(ICCV96030 × *C. pinnatifidum* 188) × ICCV96030
GLW108	(ICCV96030 × *C. pinnatifidum* 188) × ICCV96030
GLW115	(ICCV96030 × *C. pinnatifidum* 188) × ICCV96030
GLW174	(ICCV96030 × *C. pinnatifidum* 188) × ICCV96030
GLW185	(ICCV96030 × *C. pinnatifidum* 188) × ICCV96030
GLW501	GPF2 × [PBG 1 × (ICCV96030 × *C. pinnatifidum* 188) × ICCV96030]
GLW502	GL769 × *C. reticulatum*ILWC 129
GLW503	[PBG 1 × (ICCV96030 × *C. pinnatifidum* 188) × ICCV96030]
GLW504	[(ICCV96030 × *C. pinnatifidum* 188) × ICCV96030] PBG 1
GLW505	GL769 × *C. reticulatum*ILWC 129
PAU7007	(GL769 × *C. reticulatum*129)× GL769
PAU7014	(GL769 × *C. reticulatum* 129)× GL769
C. judaicum ILWC-0*	-
C. judaicum ILWC-223*	-
GLW22	(ICCV96030 × *C. pinnatifidum* 188) × ICCV96030
GLW25	(ICCV96030 × *C. pinnatifidum* 188) × ICCV96030
GLW183	(ICCV96030 × *C. pinnatifidum* 188) × ICCV96030
GLW186	(ICCV96030 × *C. pinnatifidum* 188) × ICCV96030
C. reticulatum ILWC-292*	-
GL1001	JG62 × ICCV05530
GL1002	JG62 × ICCV05530
GLW91	(ICCV96030 × *C. pinnatifidum* 188) × ICCV96030
PBG-7	GPF2 × BG1084

*Chickpea landraces.

Figure 2. Figure showing screening of wild *Cicer* accessions under field conditions and laboratory conditions.

humidity. Symptoms of the disease became visible about 10 dafter inoculations, and disease severity was assessed for five plants per replication 7 d following inoculation. Disease severity was scored using a 1–9 scale, where 1 = no infection on any plant part; 3 = 1–2 lesions visible on leaves; 5 = burnt leaves with stem rotting; 7 = stem rotting with 50% dead leaves; and 9 = extensive stem rotting with fungal infection and 100% leaf death (Gurha *et al.*, 2003). An average disease score was calculated.

Cut twig screening technique

Three twigs from each wild accession were cut and placed in a tray containing water in a completely randomized design in three replications. The twigs were then wrapped in moist cotton plugs and placed in test tubes (15 × 10 cm) freshly filled with tap water. Inoculation of the twigs was achieved by spraying conidium suspensions of *B. cinerea* (10^5 conidia mL^{-1}). Following inoculation, the test tubes were covered with wet polythene (Figure 2b). Incubation followed in a growth chamber, and BGM severity was recorded using the 1–9 severity scale (see above). Paired t-tests were performed to examine whether the BGM severity scores obtained under field and laboratory conditions were correlated (Table 2).

Genomic DNA extraction and pooled DNA analysis using SSRs

The CTAB method (Kumar *et al.*, 2013) was used to extract DNA from young leaves of the twenty six extreme RILs. The concentration and purity of the DNA from the genotypes was further checked on 0.8% agarose gels and a nanodrop 1000 spectrophotometer (Thermo Scientific). Equal amounts of DNA from each line were taken after normalization and mixed well to constitute the resistant and susceptible bulks for bulked BSA. These bulks were screened for polymorphism using 300 simple sequence repeat (SSR) markers (Varshney *et al.* 2014; Gupta *et al.*, 2012; Gaur *et al.*, 2011); Bharadwaj *et al.*, 2010). These primers were custom synthesized from G-Biosciences, and the PCRs were carried out in the Chickpea Molecular Breeding Laboratory, Division of Genetics, ICAR-IARI. The 10 μL PCR mix consisting of 1 μL of 20 ng genomic DNA, 1.6 μL of 10× TBE buffer, 1μL of 10 mM dNTP mix, 1 μL forward and 1μL reverse primer and 0.3 μL of 3U μL^{-1}Taq polymerase (Genei), was amplified using a G-STORM thermal cycler (Labtech). The PCR reaction was set as per Yadav *et al.*(2011), with an initial denaturation at 90°C for 1 min and 30 seconds followed by 38 cycles including three different steps, including denaturation at 94°C for 20 sec, annealing at 50–58°C for 50 sec, and extension at 72°C for 50 seconds. This was then followed by a final extension step at 72°C for 7 min, before cooling to 4°C.Amplicons were resolved on 3% Agarose (Lonza) using 1.0× TBE buffer. The amplified products were separated on a horizontal gel electrophoresis system (Biorad) at 120 V for 3 h. Gel staining was by ethidium bromide (10 mg mL^{-1}), and visualized using the UVITECH Gel Documentation system (UVITECH Imaging System). Amplicons were scored as alleles for each locus. Allele sizes were determined by comparing with a standard 100 bp DNA ladder (Genei). Band patterns of the extreme bulks were compared with those of resistant and susceptible lines to confirm linkage of SSR markers with BGM resistance. Chi-square tests were performedto determine goodness of fit of the test lines for the phenotypic and SSR data, by comparing the observed frequency (O) with the expected frequency (E).

RESULTS

Identification of BGM resistance in chickpea accessions

The screening results for BGM resistance of the 26 extreme lines are presented in Table 2. The scoring results of field screening were in close agreement with those from laboratory assessments, similar scores were obtained for each accession under both conditions. Lines ILWC-0, ILWC-223, PBG-7 and nine RILs showed moderately susceptible to susceptible reactions to BGM under field and laboratory conditions. Line ILWC292 and thirteen RILs, including: GLW91, PAU7014, GLW42, GLW67, GLW 115, GLW174, GLW183, GLW186, GLW501, GLW504, GLW 506, GL1001 and GL1002, were resistant to moderately resistant to BGM,

Table 2 Mean Bortytis grey mould (BGM) severity scores for test host lines under field conditions and controlled conditions, and paired-t test analysis for correlating the respective BGM severity scores for the two screening methods (field and greenhouse).

Name of genotype	Disease reaction	Mean BGM score (1-9 scale)		Difference	D^2
		Field screening[a]	Lab screening[b]		
GLW42	Moderately resistant	5.0	5.5	-0.5	0.25
GLW67	Moderately resistant	5.0	4.5	0.5	0.25
GLW69	Moderately susceptible	8.0	8.0	0.0	0
GLW107	Susceptible	9.0	9.0	0.0	0
GLW108	Susceptible	9.0	9.0	0.0	0
GLW115	Moderately resistant	6.0	6.0	0.0	0
GLW174	Moderately resistant	5.5	4.5	1.0	1
GLW185	Susceptible	9.0	9.0	0.0	0
GLW501	Moderately resistant	5.5	5.0	0.5	0.25
GLW502	Moderately susceptible	7.0	7.5	-0.5	0.25
GLW503	Moderately susceptible	7.0	8.0	-1.0	1
GLW504	Moderately resistant	5.5	5.5	0.0	0
GLW505	Moderately susceptible	8.0	8.0	0.0	0
PAU7007	Moderately resistant	5.5	5.0	0.5	0.25
PAU7014	Resistant	4.0	4.0	0.0	0
C. judaicum ILWC-0	Susceptible	9.0	9.0	0.0	0
C. judaicum ILWC-223	Susceptible	9.0	9.0	0.0	0
GLW-22	Susceptible	9.0	9.0	0.0	0
GLW-25	Susceptible	9.0	9.0	0.0	0
GLW-183	Moderately resistant	5.0	5.5	-0.5	0.25
GLW-186	Moderately resistant	5.0	5.5	-0.5	0.25
C. reticulatum ILWC-292	Resistant	3.0	3.5	-0.5	0.25
GL1001	Moderately resistant	5.5	4.0	1.5	2.25
GL1002	Moderately resistant	5.5	5.0	0.5	0.25
GLW91	Resistant	3.0	3.0	0.0	0
PBG-7	Susceptible	9.0	9.0	0.0	0
$\bar{D}$					0.03846
$(\Sigma D-\bar{D})^2$					8.71153
SE of $\bar{D}$					0.34846
t					0.11037
df					25
Table value					2.06

[a]Average score from five plants per accession per replication, screened under field conditions.
[b]Average score from 12 plants (three plants per accession per replication), screened under laboratory conditions.
D = Difference between the field screening scores and laboratory screening scores.
$\bar{D}$ = Mean difference.
SE = standard error.
df = degrees of freedom.

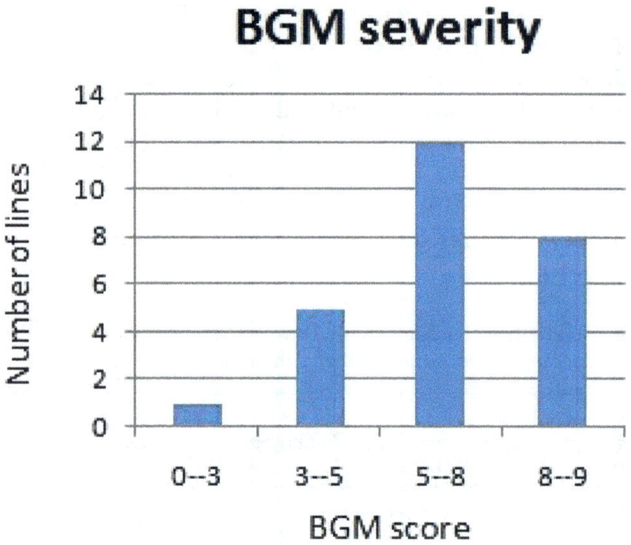

Figure 3. Frequency distribution of BGM scores of all wild *Cicer* spp. screened under field and laboratory conditions.

and could be used as resistant donors. The observed t-value of 0.1103 was much less than 2.06, ($P = 0.05$; d.f. = 25), showing that there was no statistically significant difference between the two screening methods (Table 2). The frequency distributions of the extreme lines for disease severity (recorded on the 1–9 scales) demonstrated a normal distribution, signifying that resistance to BGM was quantitative in nature (Figure 3).

Pooled DNA analysis using SSRs

Twenty six accession variants were selected from 371 wild chickpeas for genotyping. Of these, DNA from the five most resistant lines (ILWC 292, GLW91, GL1001, PAU7007 and PAU7014) and the five most susceptible accessions (*C. judaicum* ILWC-223, GLW22, GLW25, GLW69 and PBG7) were pooled and bulks were generated. Eighty eight markers were found to be polymorphic between the contrasting bulks, and these are list-

Table 3 List of polymorphic SSR markers used for bulked BSA in the five most Botrytis grey mould resistant and five most susceptible lines, and their corresponding bulks with their linkage groups (LGs)

Serial No.	Primer Name	LG	Serial No.	Primer Name	LG	Serial No.	Primer Name	LG	Serial No.	Primer Name	LG
1	TR43	1[a]	26	TR19	2[b]	51	NCGR209	8[d]	76	NCPGR248	*
2	TA196	3[b]	27	TS82	2[b]	52	NCPGR210	8[a]	77	NCPGR249	7[b]
3	TA144	3[c]	28	TR58	2[c]	53	NCPGR215	*	78	NCPGR250	*
4	TS5	3[b]	29	TA37	2[b]	54	NCPGR216	*	79	NCPGR252	6[d]
5	TA140	2[c]	30	GA16	2[c]	55	NCPGR218	*	80	NCPGR253	*
6	TAA104	5[b]	31	TA96	2[b]	56	NCPGR219	*	81	NCPGR254	6[a]
7	TR2	3[a]	32	TA27	2[b]	57	NCPGR220	6[a]	82	NCPGR255	6[d]
8	GA6	8[b]	33	TA21	5[b]	58	NCPGR221	3[a]	83	NCPGR267	6[b]
9	TA18	7[b]	34	CaSTMS24	2[b]	59	NCPGR224	4[a]	84	NCPGR268	7[a]
10	H3A10	5[b]	35	GA102	5[c]	60	NCPGR225	3[a]	85	NCPGR269	*
11	H2120	5[d]	36	CaSTMS22	5[b]	61	NCPGR226	6[a]	86	NCPGR272	3[b]
12	TS58	3[b]	37	NCPGR76	6[a]	62	NCPGR227	*	87	NCPGR274	6[a]
13	TAA170	4[b]	38	NCPGR78	*	63	NCPGR228	5[b]	88	NCPGR275	*
14	TA206	2[b]	39	NCPGR79	6[b]	64	NCPGR229	6[a]			
15	TA194	2[c]	40	NCPGR82	6[b]	65	NCPGR232	5[a]			
16	GAA47	4[b]	41	NCPGR84	*	66	NCPGR235	*			
17	TA3a	2[b]	42	NCPGR91	4[a]	67	NCPGR236	4[b]			
18	TR31	3[c]	43	NCPGR93	6[b]	68	NCPGR237	*			
19	TA42	7[b]	44	NCPGR95	7[b]	69	NCPGR238	6[a]			
20	TA89	4[b]	45	NCPGR96	*	70	NCPGR240	3[a]			
21	TA110	2[c]	46	NCPGR97	*	71	NCPGR241	3[a]			
22	STMS28	3[b]	47	NCPGR98	2[b]	72	NCPGR242	3[a]			
23	TR20	4[b]	48	NCPGR99	7[b]	73	NCPGR244	*			
24	CaSTMS2	4[b]	49	NCPGR100	3[a]	74	NCPGR246	*			
25	TA42	5[b]	50	NCPGR101	1[b]	75	NCPGR247	4[d]			

[a] Choudharyet *al.*, 2012; [b] Varshney *et al.*, 2014; [c] Bharadwaj *et al.*, 2010; [d] Gauret *al.*, 2011; *Choudhary, unpublished.

Table 4 Chi-square (χ^2) test results of wild chickpeas screened using Bulked BSA for Botrytis grey mould screening.

Sno.	Marker	Observed (O)	Expected (E)	χ^2 =(Obs-Exp)^2/ Exp	χ^2 (P = 0.05)	
1	TR2	9	11	0.363636364	3.841	Strong correlation
2	TA194	11	12	0.083333333		Strong correlation
3	TA144	12	12	0.00		Strong correlation
4	GA102	11	12	0.083333333		Strong correlation
5	TA140	10	11	0.090909091		Strong correlation

Low χ^2 values indicate strong correlation with resistant allele.

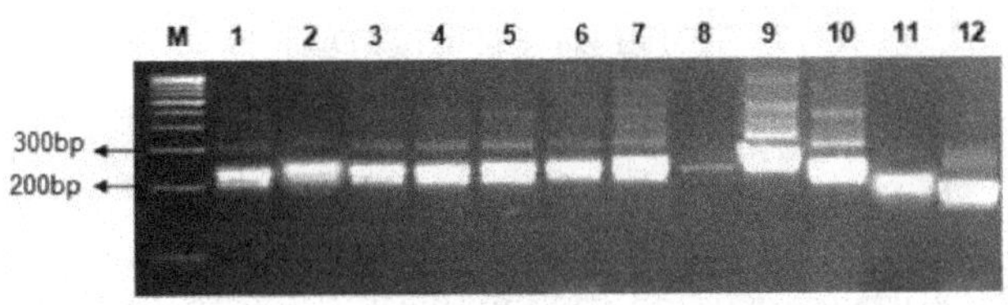

Figure 4. Gel image showing Bulked sample analysis with DNA pools generated with TA144, M: 100bp Banglore Genei ladder, 1:ILWC292,2:GL1001, 3:PAU7007, 4:PAU7014, 5:GLW91, 6:*C. judaicum* ILWC223, 7: GLW22, 8: GLW25, 9: PBG7, 10: GLW69, 11: Resistant bulk, 12: Susceptible bulk.

ed in Table 3. Most of these markers were found to be located on linkage groups (LGs) 2, 3 and 6. Anuradha *et al.* (2011) also identified three QTLs linked with BGM resistance in chickpea on LG 3 and LG 6. The BSA and chi-square statistic indicated strong correlations of TA144, GA102, TA194, TA140 and TR2 with the resistant bulks (Table 4 and Figure 4). These markers may be associated with resistance against BGM, and they could assist plant breeders in speedy development of BGM resistant cultivars.BSA combined with advanced technologies can be used to identify and develop diagnostic and constitutive markers improving the efficiency of breeding programmes and lead to animproved understanding of the molecular basis of BGM resistance.

DISCUSSION

Wild accessions serve as excellent sources of resistance to biotic and abiotic stresses, according to previous reports, and many chickpea accessions have now been identified with resistance to diseases (Madrid *et al.*, 2008). Accessions including *C. judaicum, C. pinnatifidum* and *C. echinospermum* have shown good resistance to BGM at PAU, Ludhiana, so these may be used as donors for introgressing BGM resistance in chickpea. These sources of BGM resistance were therefore used in the present study through interspecific hybridization.

Botrytis grey mould is an economically important disease of chickpea, which may cause complete yield losses under heavy rains and high humidity. It is the major production constraint limiting sustainable chickpea yields. There is an urgent need to develop varieties with resistance to BGM with greater yield stability. Screening under field and controlled conditions are time-consuming and selection of BGM-resistant genotypes may take more than 1 year or season if conventional methods are used. Screening for disease resistance may also be influenced by environmental interactions, density of pathogen inoculum, presence of some pathogens, and pathogen virulence, resulting in variable disease outcomes. Furthermore, discrepancies in scoring of disease reactions may affect the introgression of BGM resistance into elite chickpea genotypes. On the other hand, marker-assisted selection saves time in comparison to phenotypic field or greenhouse evaluations.

Bulked sample analysis and molecular markers together help discern markers associated with genes governing disease resistance in a number of plant species (Ballini *et al.*, 2008). Owing to scarcity of polymorphic markers in chickpea, BSA provides a rapid method for identifying markers linked to BGM resistance. Resistant genotypes may be identified within short periods, inferior genotypes can be excluded from the next cycle of selection, increasing the efficiency of breeding programmes.

Microsatellites combined with BSA have been used to identify molecular markers linked to genes of interest (Shoba *et al.*, 2012), and comparison of pooled DNA samples is much easier than evaluating all the individuals of different populations (Hallden *et al.*, 1997; Sweeney and Dannebeger, 1994). BSA has been successfully used in fine mapping of QTLs for *Ascochyta*blight resistance in *C. arietinum*, genes controlling powdery mildew resistance in pea (Fondevilla *et al.*, 2008), cotyledon seed colour in *Glycine max*, QTLs for rust resistance including VuUGM02, VuUGM08 and VuUGM19 in cowpea (Uma *et al.*, 2016), *Phaeoisariopsis griseola* resistance genes in *Phaseolus vulgaris* (Alzate-Marin *et al.*, 2001), 73 blast resistance genes and 350 QTLs in rice (Ghaley *et al.*, 2012),and discovery of SNPs for agronomically important traits in *Arachis hypogea* (Pandey *et al.*, 2017). NGS-assisted BSA identified six candidate genes in the QTL regions on chromosomes Ca2 and Ca4 and validated for their association with *Ascochyta* blight resistance in the CPR02 population in chickpea (Deokar *et al.*, 2019). A QTL-seq approach coupled with BSA identified candidate genes (Ca_04364 and Ca_04607) for 100seed weight, and one gene (Ca_04586) for total root dry weight to total plant dry weight ratio using CAPS markers in chickpea.

Anuradha *et al.* (2011) developed 126 F_{10} derived RILs) derived from a cross between a moderately BGM resistant kabuli cultivar (ICCV2) and a highly BGM susceptible desi chickpea cultivar (JG62) atthe International Crops Research Institute for the Semi-Arid Tropics (ICRISAT), Patancheru, India. This used the single seed descent (SSD) method, and identified two QTLs for BGM resistance on LG3 and one on LG6, which after validation, can be used for marker-assisted breeding. Kaur *et al.* (2013) also developed an inter-specific population for BGM resistance using the cut twig technique. Sixty two F9 BGM resistant lines derived were evaluated for agronomically important and yield traits, and identified four lines, (GL 29029, GL29206, GL29212 and GL29081) with high degrees of resistance to BGM. These lines were crossed with the BGM-susceptible high yielding cultivar BG256 for molecular analysis, and genotyping of F_2 populations identified SSR markerspotentially linked with *Ascochyta* blight and BGM resistance genes. Of the 120 markers used, six SSRs (TA2, TA110, TA139, CaSTMS7, CaSTMS24 and TR29) were found to be polymorphic. These markers can be used for identification of markers linked to BGM resistance, and assist in marker-assisted backcrossing for resistance breeding.

In the present study, comprehensive evaluation of the test host lines and paired-t test analysis both revealed that the field and laboratory screening methods gave similar results for BGM screening of the test lines. This demonstrates that laboratory screening methods coupled with molecular marker techniques can serve as powerful tools in genetics and crop improvement. The polymorphic SSR markers identified in the present study can be used to develop chickpea cultivars with high levels of BGM resistance, that has been a challenging task due to lack of sources of high levels of resistance in cultivated chickpea. Low chi-square statistic values of five SSRs (TA144, GA102, TA194, TA140 and TR2) in comparison to the critical value at $P = 0.05$ indicated their strong correlation with the resistant bulk and BGM resistance. We therefore conclude that bulked BSA is simple and accurate method for rapidly ascertaining marker-trait associations rapidly, and may be value for molecular tagging of BGM resistant genes in chickpea.

ACKNOWLEDGEMENTS

The authors acknowledge the ICAR-IARI, Incentivizing Research on Agriculture (Component IV: Chickpea) Project funded by ICAR for the financial, field and laboratory support.

LITERATURE CITED

Alzate-Marin A.L., Costa, M.R., Sartorato, A., Rava, C.A., de Barros, E.G., Moreira, M.A., 2001. Use of markers as a tool to investigate the presence of disease resistance genes in common bean cultivars. *Embrapa Arroz e Feijão-Artigoemperiódicoindexado (ALICE).*

Anuradha C., Gaur P.M., Pande S., Gali K.K., Ganesh M., Kumar, J., Varshney, R.K., 2011. Mapping QTL for resistance to botrytis grey mould in chickpea. *Euphytica* 182(1): 1–9.

Asnaghi C., Roques D., Ruffel S., Kaye C., Hoarau J.-Y., Télismart H., Girard J.C., 2004. Targeted mapping of a sugarcane rust resistance gene (Bru1) using bulked segregant analysis and AFLP markers. *Theoretical and Applied Genetics* 108: 759–764.

Ballini E., Morel J.B., Droc G., Price A., Courtois B., Notteghem J.L., Tharreau D., 2008. A genome-wide meta-analysis of rice blast resistance genes and quantitative trait loci provides new insights into partial and complete resistance. *Molecular Plant-Microbe Interactions* 21(7): 859–868.

Bharadwaj C., Chauhan S.K., Rajguru G., Srivastava R., Satyavathi, T.C., ... Solanki R.K., 2010. Diversity analysis of chickpea (*Cicer arietinum*) cultivars using STMS markers. *Indian Journal of Agricultural Sciences* 80(11): 947–951.

Chen Q., Song J., Du W.P., Xu L.Y., Jiang Y., Zhang J., Xiang X.L., 2017. Identification, mapping, and molecular marker development for Rgsr8.1: a new quantitative trait locus conferring resistance to *Gibberella* stalk rot in maize (*Zea mays* L.). *Frontiers in Plant Science* 8: 1355.

Choudhary S., Gaur R., Gupta S., Bhatia S., 2012. EST-derived genic molecular markers: development and utilization for generating an advanced transcript map of chickpea. *Theoretical and Applied Genetics* 124(8): 1449–1462.

Das S., Singh M., Srivastava R., Bajaj D., Saxena M.S., Rana J.C., Bansal K.C., 2016. mQTL-seq delineates functionally relevant candidate gene harbouring a major QTL regulating pod number in chickpea. *DNA Research* 23: 53–65.

Deokar A., Sagi M., Daba K., Tar'an B., 2019. QTL sequencing strategy to map genomic regions associated with resistance to *Ascochyta* blight in chickpea. *Plant Biotechnology Journal* 17(1): 275–288.

El-Soda M., Malosetti M., Zwaan B.J., Koornneef M., Aarts M.G., 2014. Genotype× environment interaction QTL mapping in plants: lessons from *Arabidopsis. Trends in Plant Science* 19(6): 390–398.

Fondevilla S., Rubiales D., Moreno M.T., Torres A.M., 2008. Identification and validation of RAPC and SCAR markers lined to the gene *Er3* conferring resistance to *Erysiphe pisi* DC in pea. *Molecular Breeding* 22(2): 193–200.

Food and Agriculture Organization of the United Nations, 2017. FAOSTAT statistics database. http://www.fao.org/faostat/en/#home

Gaur R., Sethy N.K., Choudhary S., Shokeen B., Gupta V., Bhatia S., 2011.Advancing the STMS genomic resources for defining new locations on the intraspecific genetic linkage map of chickpea (*Cicer arietinum* L.). *BMC genomics* 12(1): 117.

Ghaley B.B., Christiansen, J.L., Andersen, S.B., 2012. Genetic diversity in blast resistance of Bhutan rice landraces. *Euphytica* 184(1): 119–130.

Gillman J.D., Tetlow A., Lee J.-D., Shannon J.G., Bilyeu K., 2011. Loss-of-function mutations affecting a specific *Glycine max* R2R3 MYB transcription factor result in brown hilum and brown seedcoats. *BMC Plant Biology* 11: 155.

Giovannoni J.J., Wing R.A., Ganal M.W., Tanksley S.D., 1991. Isolation of molecular markers from specific chromosomal intervals using DNA pools from existing mapping populations. *Nucleic Acids Research* 19(23): 6553–6568.

Gupta S., Choudhary S., Bhatia S., Gaur R., 2012. EST-derived genic molecular markers: development and utilization for generating an advanced transcript map of chickpea. *Theoretical and Applied Genetics* 124(8): 1449–1462.

Gurha S.N., Singh G., Sharma Y.R., 2003. Diseases of chickpea and their management. In: *Chickpea research in India*' (Massod Ali, Shiv Kumar, NB Singh, eds.), pp. 195–227.

Halldén C., Säll T., Olsson K., Nilsson N.O., Hjerdin A., 1997. The use of bulked segregant analysis to accumulate RAPD markers near a locus for beet cyst nematode resistance in *Beta vulgaris. Plant Breeding* 116: 18–22.

Haware M.P., 1998. Diseases of chickpea. In: *The pathology of food and pasture legumes*' (Allen D.J., Lenne J.M., eds.), pp. 473–516.

Hawthorne W., Davidson J.A., McMurray L., Lindbeck K.D., Brand J., 2006. Chickpea disease management strategy-southern region. *Disease Management Guide Series.(Pulse Australia: Sydney).*

ICRISAT 2007. ICRISAT Archival Report 2007. International Crops Research Institute for the Semi-Arid Tropics, Patancheru 502 324, Andhra Pradesh, India.

Illa-Berenguer E., Van Houten J., Huang Z., Knaap E., 2015. Rapid and reliable identification of tomato fruit weight and locule number loci by QTL-seq. *Theoretical and Applied Genetics* 128: 1329–1342.

Kaminski K.P., Kørup K., Andersen M.N., Sonderkaer M., Andersen M.S., … Nielsen K.L., 2015. Cytosolic glutamine synthetase is important for photosynthetic efficiency and water use efficiency in potato as revealed by high-throughput sequencing QTL analysis. *Theoretical and Applied Genetics* 128: 2143–2153.

Kaur L., Sirari A., Kumar D., Singh Sandhu J., Singh S., … Sharma M., 2013. Combining Ascochyta blight and Botrytis grey mould resistance in chickpea through interspecific hybridization. *Phytopathologia Mediterranea* 52(1): 157–165.

Kumar T., Bharadwaj C., Satyavathi C.T., Jain P.K., 2013.A high throughput, improved rapid and reliable genomic DNA extraction protocol from chickpea (*Cicer arietinum* L.). *Vegetos* 26(2): 185–190.

Li H., Rodda M., Gnanasambandam A., Aftab M., Redden R., … Slater A.T., 2015. Breeding for biotic stress resistance in chickpea: progress and prospects. *Euphytica* 204(2): 257–288.

Lu H., Lin T., Klein J., Wang S., Qi J., Zhou Q., Sun J., 2014. QTL-seq identities an early flowering QTL located near Flowering Locus T in cucumber. *Theoretical and Applied Genetics* 127: 1491–1499.

Madrid E., Rubiales D., Moral A., Moreno M.T., Millán T., … Rubio J., 2008. Mechanism and molecular markers associated with rust resistance in a chickpea interspecific cross (*Cicer arietinum* x *Cicer reticulatum*). *European Journal of Plant Pathology* 121: 43–53.

Mathews P., McCaffery D., Jenkins L., 2014. Winter crop variety sowing guide 2014. *NSW Department of Primary Industries: Sydney.*

Michelmore R.W., Paran I., Kesseli R.V., 1991. Identification of markers linked to disease-resistance genes by bulked segregant analysis: a rapid method to detect markers in specific genomic regions by using segregating populations. *Proceedings of the National Academy of Sciences* 88(21): 9828–9832.

Nene Y.L., Reddy M.V., Haware M.P., Ghanekar A.M., Amin K.S., Pande S., Sharma M., 2012. *Field Diag-*

nosis of Chickpea Diseases and their Control. Information Bulletin No. 28 (revised). International Crops Research Institute for the Semi-Arid Tropics.

Pande S., Galloway J., Gaur P.M., Siddique K.H.M., Tripathi H.S., … Kishore G.K., 2006a. *Botrytis* grey mould of chickpea: a review of biology, epidemiology, and disease management. *Australian Journal of Agricultural Research* 57(11): 1137–1150.

Pande S., Kishore G.K., Upadhyaya H.D., Rao J.N., 2006b. Identification of sources of multiple disease resistance in mini-core collection of chickpea. *Plant Disease* 90(9): 1214–1218.

Pande S., Ramgopal D., Kishore G.K., Mallikarjuna N., Sharma M., … Rao J.N., 2006c. Evaluation of wild *Cicer* species for resistance to *Ascochyta* blight and *Botrytis* gray mold in controlled environment at ICRISAT, Patancheru, India. *Journal of SAT Agricultural Research* 2(1): 1–3.

Pande S., Singh G., Rao J.N., Bakr M.A., Chaurasia P.C.P., … Gowda C.L., 2002. Integrated management of *Botrytis* gray mold of chickpea. *India: International Crops Research Institute for the Semi-Arid Tropics.*

Pandey M.K., Khan A.W., Singh V.K., Vishwakarma M.K., Shasidhar Y., … Garg V., 2017. QTL-seq approach identified genomic regions and diagnostic markers for rust and late leaf spot resistance in groundnut (*Arachis hypogaea* L.). *Plant Biotechnology Journal* 15: 927–941.

Shoba D., Manivannan N., Vindhiyavarman P., Nigam S.N., 2012. SSR markers associated for late leaf spot disease resistance by bulked segregant analysis in groundnut (*Arachis hypogaea* L.). *Euphytica* 188(2): 265–272.

Singh G., Bhan L.K., 1986. Physiological Races of *Botrytis cinerea* causing gray mold of chickpea. *Plant Disease Research* 1(1-2): 69–74.

Singh K.B., Holly L., Bejiga G., 1991. *A catalog of kabuli chickpea germplasm-an evaluation report of winter-sown kabuli chickpea land races, breeding lines and wild cicer species* (No. SB351. C45 S61).

Singh K.B., Reddy M.V., 1991. Advances in disease-resistance breeding in chickpea. *Advances in Agronomy* 45: 191–222.

Singh V.K., Khan A.W., Jaganathan D., Thudi M., Roorkiwal M., … Garg V., 2016. QTLseq for rapid identification of candidate genes for 100-seed weight and root/total plant dry weight ratio under rainfed conditions in chickpea. *Plant Biotechnology Journal* 14: 2110–2119.

Sun Y., Wang J., Crouch J.H., Xu Y., 2010. Efficiency of selective genotyping for genetic analysis of complex traits and potential applications in crop improvement. *Molecular breeding* 26(3): 493–511.

Sweeney P.M., Danneberger T.K., 1994. Random amplified polymorphic DNA in perennial ryegrass: a comparison of bulk samples vs. individuals. *Horticultural Science* 29(6): 624–626.

Takagi H., Abe A., Yoshida K., Kosugi S., Natsume S., … Innan, H., 2013. QTL-seq: rapid mapping of quantitative trait loci in rice by whole genome resequencing of DNA from two bulked populations. *The Plant Journal* 74(1): 174–183.

Uma M.S., Hegde N., Hittalmani S., 2016. Identification of SSR marker associated with rust resistance in cowpea (*Vigna unguiculata* L.) using bulk segregant analysis. *Legume Research: An International Journal* 39(1): 39–42.

Varshney R.K., Mir R.R., Bhatia S., Thudi M., Hu Y., … Riera-Lizarazu, O., 2014. Integrated physical, genetic and genome map of chickpea (*Cicer arietinum* L.). *Functional & Integrative Genomics* 14(1): 59–73.

Xu Y., Crouch J.H., 2008. Marker-assisted selection in plant breeding: from publications to practice. *Crop Science* 48(2): 391–407.

Xu Y. 2010. *Molecular Plant Breeding.* CABI, Wallingford, UK, 734 pp.

Yadav S., Bharadwaj C., Chauhan S.K., Rizvi A.H., Kumar J., Satyavathi C.T., 2011. Analysis of genetic diversity in *Cicer* species using molecular markers. *The Indian Journal of Genetics and Plant Breeding* 71(3): 272–275.

6

Population structure of the faba bean blight pathogen *Ascochyta fabae* (teleomorph, *Didymella fabae*) in Tunisia

Noura OMRI BEN YOUSSEF[1,*], Lydie KERDRAON[2], Lucie MIEUZET[3], Imen HALILA[1], Noura JAMMEZI[1], Ahlem MBAZIA[1], Mohamed KHARRAT[1], Christophe LE MAY[3,4,*]

[1] *Laboratoire des Grandes Cultures, Institut National de la Recherche Agronomique de Tunisie, Université de Carthage, Rue Hédi Karray, EL Manzah, 1004, Tunis, Tunisie*
[2] *INRA, UMR, UMR BIOGER, INRA-AgroParisTech, Avenue Lucien Brétignières BP 01 78850 Thiverval-Grignon, France*
[3] *INRA, UMR 1349 IGEPP, Domaine de la Motte, 35653 Le Rheu, France*
[4] *Agrocampus Ouest, UP ESP, 65 rue de Saint Brieuc, 35043 Rennes, France*
*Corresponding authors: lemay@agrocampus-ouest.fr; noura.mori@gmail.com

Competing Interests: The Author(s) declare(s) no conflict of interest.

Editor: Jenny Davidson, South Australian Research and Development Institute (SARDI), Adelaide, Australia.

Abstract. Ascochyta blight, caused by *Ascochyta fabae* (teleomorph: *Didymella fabae*) has decreased faba bean production in Tunisia and worldwide. The teleomorph has recently been observed in Tunisia, raising new questions of how to control this major disease. Isolates (317) of the pathogen were obtained between 2011 and 2013, from four geographical regions in Tunisia (Beja, Bizerte, Jendouba and Tunis). The 240 isolates obtained in 2012 were compared for mating type frequency and genetic variation by using ten polymorphic SSR markers. Of these isolates, MAT1-2 was more common (2:1) in Tunisia than MAT1-1, but this ratio can change according to population. Low to high genetic variation was detected between locations and among pathogen populations. Beja showed the greatest genotypic richness (R = 0.42), followed by Tunis (R = 0.25), Bizerte (R = 0.13), and Jendouba (R = 0.11). Indices of association (I_A) and *Rd* were significantly different from 0 in all the populations, suggesting high multilocus linkage disequilibrium and confirming clonal populations. Population structure of the isolates was inferred using Bayesian analyses, Principal Component Analysis (PCA), and Minimum Spanning Networks, which all revealed that the populations from each location were not distinct. Evaluating changes in seasonal genetic diversity showed low to high variances of F_{ST} values between the two cropping seasons in all regions. However, the PCA analysis failed to separate the *A. fabae* isolates sampled during the two successive seasons into two groups, indicating that these populations did not constitute distinct genetic groups. These results suggest that gene flow was limited among populations, even those separated by short geographic distances. Future studies should enlarge the number of samples of representative populations, to overcome the limitations of a small sample size and to provide a more accurate assessment of *A. fabae* population structure.

Keywords. Ascochyta blight, *Ascochyta fabae*, faba bean, genetic differentiation, population genetics.

INTRODUCTION

The faba bean disease, Ascochyta blight, caused by *Ascochyta fabae* Speg. (anamorph) and its teleomorph *Didymella fabae* Jellis & Punith, is a widespread disease, particularly in temperate regions, including Canada, Argentina, Europe, the Mediterranean region, the Middle East, Australia, New Zealand, Japan, Korea and China (Maurin *et al.*, 1990). This is also one of the most important foliar diseases affecting faba beans in Tunisia (Kharrat *et al.*, 2006). The fungus attacks all aerial parts of host plant (leaves, stems, pods and seeds) causing necrotic lesions, and leads to faba bean yield and quality damage. In Western Europe, depending on weather conditions during the cropping seasons, Ascochyta blight causes averages of 10 to 30% yield losses (Hanounik and Robertson, 1989). In Tunisia, yield losses are estimated to 15%, and may reach 95% in disease favourable conditions (Kharrat *et al.*, 2006).

Ascochyta fabae is an haploid heterothallic pathogen that can survive during intercropping seasons on infested seeds and plant debris. Seeds constitute the major source of primary inoculum, and seed exchanges between and within countries are considered the main sources of introduction and mechanism of disease expansion or dispersal (Kaiser *et al.* 1997). *Ascochyta fabae* is maintained on crop stubble as asexual and sexual forms for at least one season (Rubiales and Trapero-Casas, 2002; Omri Ben Youssef *et al.*, 2012). Due to the stage of maturation of pseudothecia, the release of ascospores may overlap with the vegetative development of faba bean (Rubiales and Trapero-Casas, 2002).

Although applying fungicides to control *A. fabae* on seeds and faba bean plants gives satisfactory results, plant resistance remains the best way to control this disease. Considerable efforts have been made to identify sources of resistance in faba bean germplasm, particularly in Syria, Egypt, Canada, Morocco, France and Tunisia (Hanounik and Robertson, 1989; Rashid *et al.* 1991; Maurin and Tivoli, 1992; Kharrat *et al.*, 2006). Partial host resistance has contributed to reducing chemical intervention to control the disease (Avila *et al.*, 2004; Kharrat *et al.*, 2006).

Pathogen variability represents a major challenge for the development and use of resistant cultivars. In general, new virulent isolates of pathogens emerge through selection imposed by their hosts. These shifts in virulence necessitate the adoption of disease control strategies that target pathogen populations rather than individuals (McDonald and Linde, 2002). Studying plant pathogen diversity principally aims to identify which forces will heavily influence the evolution of pathogen populations, and to assess their ability to evolve (Milgroom and Peever, 2003; de Meeûs *et al.*, 2007). Several studies have evaluated population diversity of some *Ascochyta* spp. using different molecular tools, in particular for *A. rabiei* (Bayraktar *et al.*, 2007; Ali *et al.*, 2012; Atik *et al.*, 2013) and *A. pinodes* (Le May *et al.*, 2012; Padder *et al.*, 2012; Laloi *et al.*, 2016; Le May *et al.*, 2017). Studies conducted on populations of *A. rabiei* collected in several countries showed high diversity levels that were attributed to genetic flow provided by the exchange of seeds between farmers (Nourollahi *et al.*, 2011). For *A. fabae* populations, very few studies evaluated the genetic diversity of this fungus. A study using Random Amplified Polymorphic DNA (RAPD) markers, which included 36 isolates taken from different countries (Tunisia, Morocco, Italy, France, Spain, Australia and Algeria), showed high genotypic diversity, with 26 different genotypes observed within these isolates (Rouaissi *et al.*, 2001). More recently, Ozkilink *et al.* (2011, 2015), using a set of 18 Simple Sequence Repeat (SSR) markers, showed that *A. fabae* populations from Syria displayed a high level of polymorphism. They showed high diversity but no differentiation, and rejected a null hypothesis of random mating in the studied population, which is indicative of predominating asexual reproduction.

Ascochyta blight has decreased faba bean production in Tunisia, a country characterized by variable weather conditions, diverse crop areas, various seed origins (produced by farmers or otherwise), and cultural practices. Since *D. fabae* has been observed recently in Tunisia (Omri Ben Youssef *et al.*, 2012), it has become especially important to study the genetic diversity of the pathogen populations in the main faba bean growing areas affected by Ascochyta blight. The present study was undertaken to: (i) collect information about the genetic diversity of *A. fabae* populations in the various areas of faba bean production in Tunisia; (ii) assess the mating type distribution in this country using mating-type (MAT) specific primers; and (iii) determine whether the diversity of *A. fabae* populations evolves from one cropping season to another.

MATERIALS AND METHODS

Fungal isolates

A total of 317 *A. fabae* isolates were used in this study to analyze genetic variability. All isolates were collected between 2011 and 2013, from several faba bean fields in four different geographical locations of Tunisia, including Beja (three to nine fields, 122 isolates), Bizerte (two to seven fields, 87 isolates), Jendouba (three to eight

Table 1. Origin of *Ascochyta fabae* isolates (location, sampling date) used in the study.

Location	Sampling date	Number of populations	Number of isolates collected
Beja	2011	3	13
	2012	9	109
Bizerte	2012	2	71
	2013	7	16
Jendouba	2011	8	28
	2012	3	39
Tunis	2012	2	21
	2013	2	20

fields, 67 isolates), and Tunis (two fields, 41 isolates) (Table 1). All isolates were collected from infected host leaves, and as single-conidium isolates. For each isolation, approx. 5 mm^2 of diseased leaf tissue was surface sterilized for 1 min in 70% ethanol, rinsed three times in sterile water, placed on sterile filter paper to remove excess water. The tissue piece was then cultured on a V8 medium (99 mL V8 vegetable juice (Campbell), 35 g agar, and 901 mL distilled water, autoclaved at 105°C for 30 min) in Petri dishes for 14 d. Pycnidiospores from resulting cultures were spread on 2% malt agar and incubated for 12h, as described by Onfroy *et al.* (1999). Single germinating pycnidiospores were transferred (with dissecting microscope magnification) to fresh potato dextrose agar (PDA) plates, and cultures were incubated at 20°C with a 12h photoperiod under cool white fluorescent lamps. These single-spore cultures were then maintained on malt agar slants and stored in the dark at 4°C.

DNA extraction

Each isolate was grown in 75 mL of Tripton (LT) liquid medium supplemented with streptomycin (1.5 g) and penicillin (0.75 g). Each culture was raised from four pieces (each approx. 1 cm^2) cut from the margin of an actively growing culture on malt agar. Inoculated vials were incubated, under agitation, for 14 d at 20°C, with a 12h photoperiod under cool white fluorescent lamps. Mycelia were harvested by vacuum filtration through two layers of sterilized Miracloth (Calbiochem CN Biosciences, Inc.), rinsed twice in sterile water, and then stored at -80°C until lyophilized. DNA was extracted from mycelium and isolated using Nucleospin Plant II Kit (Macherey-Nagel) according to the manufacturer's instructions.

Mating type assays

The mating type of all the 240 *A. fabae* isolates collected in 2012 was determined using the multiplex MAT-specific PCR assay (Cherif *et al.,* 2006). Primer combinations AL2p2SeqF4 (5'GCAACATCCTAGCATGATG3') specific to MAT1-1, AL1p1SeqF5 (5'CTGTCTCACCCAAGGCAAAC3') specific to MAT1-2, and ACom1A1AvAfAp (5'CACATCACCCCACAAGTCAG3') were used, specific to an aligning flanking 3′ region of *A. lentis, A. viciae-villosae, A. fabae* and *A. pisi.* Single PCR was carried out in 25 μL containing 10 ng of genomic DNA, 1× PCR buffer (containing 1.5 mM $MgCl_2$), 0.2 mM dNTPs, 1 unit of *Taq* DNA polymerase (Promega) and 0.2 μM each of the primers. Amplification was performed in a BioRad Cycler thermal cycler (Bio-Rad Laboratories), and cycling conditions consisted of an initial denaturation at 95°C for 3 min followed by 35 cycles, each of 94°C for 20 sec, 58°C for 20 sec, and 72°C for 40 sec, with a final extension at 72°C for 10 min. DNA amplicons were separated in 1.5% ethidium bromide-stained agarose gels, and were visualized under UV light on a gel documentation system ChemiDOCTM XRS (Bio-Rad). Amplicon size was estimated using a DNA ladder (Hyperladder II).

SSR amplification and analyses

Ten SSR loci and their primer sequences (Table 2) were obtained from a published report on *A. fabae* (Ozkilink *et al.,* 2011). The M13-tailed primer was used to detect genetic diversity of all *A. fabae* isolates of the collection. The M13 sequence (CACGACGTTGTAAAACGAC) was added to the 5′ end of the forward primers, which were synthesized by Sigma-Genosys Ltd. PCR was performed in a 12 μL reaction mixture containing 0.2 mM dNTP, 1× PCR buffer, 1.5 mM $MgCl_2$, and 1 unit of *Taq* DNA polymerase (Promega), with the addition of 1ng genomic DNA, 0.24 μL (10 nM) of the forward primer labeled at the 5′ end with M13 sequence, 0.3 μL (10 nM) of the reverse primer and 0.06 μL (10 nM) of the fluorescent dye FAM (Applied Biosystems). Following an initial denaturation step of 2 min at 94°C, PCR was performed for a total of 50 cycles, each of 20 sec at 94°C, 25 sec at 58°C, and 23 sec at 67°C, with a final extension for 10 min at 72°C. For analysis on the genetic analyzer, 3 μL of the 25 μL diluted and pooled PCR products were mixed with 9.9 μL of formamide and 0.1 μL of LIZ-445 size standard (Applied Biosystem) in a 96 well PCR plate (GeneMate). The mixture was heated for 5 min at 95°C then chilled on ice and analyzed with the DNA analyzer.

Table 2. Primer sequences of the ten SSRs markers used for genetic study of *Ascochyta fabae* isolates.

Locus	Repeating motif	Primer sequences (5' to 3')
CAA46	F: CTA CAT TTC CCG TGC CTG AC R: GGC AGC CAG AGT TTG AGA AC	(CAA)17
CAA28	F: GAG TCA GTG GCG AGT GTG G R: GTC CGT TGC CCG TCT TTC	(GTGTAGT)3N(94) (TGT)8
AFCAA12	F: TCT TGG ACG CGT CTC TCT TG R: GCC AGT CTG GTT CAT CTA CC	(GTT)8N(32) (CACTG)6
AFCAA13	F: TTC GGC AGC ACA TCC TTC AG R: TGA GCA ATC TGA GCG GTT GG	(GCA)6N(13) (CAGCAACAA)3N(27) (CAA)4N(15)(CAA)(16)
AFCAA5	F: ATC ATC GCG TAC GTC GAC AC R: AAG ATG CTG GAG GGT GTC AG	(CTGCCACTGACACAGCTA)6
CA31	F: GAGCGTACCCAAACGCTATC R: GCTTCTTCGGCCTCAGTATG	(TGTGAGCG)11
AFCAA1	F: TCT ACT GAC GAT GCA TAG CG R: TAC CCA AAC GCT ATC GAA GC	(TTG)8
CA10	F: GCT TGT GCT TGT GCT TGT TC R: ACA TTC GTC CAT TGC ACC TT	(GTGTTGTGC)3N(103) (TGTGT)3
CA3	F: AGC AAC AAC AAG ACG CAG TG R: AGC TTG GGA TCT GCT TCC TT	(GTGTGCAGTGTGTA)7
CAA57	F: GAG CGT ACC CAA ACG CTA TC R: GCT TCT TCG GCC TCA GTA TG	(TGTGAGCG)9

Data analyses

As indicated in Table 1, isolates from the various locations were sampled over several seasons. Two different analyses were performed on the collection of *A. fabae* isolates. These were: i) gene diversity, genetic differentiation and genetic diversity, carried out for 240 isolates obtained in 2012; and ii) seasonal population differentiation, carried out for the isolates (317) sampled during the following seasons in the four locations (Beja, Bizerte, Jendouba and Tunis).

Gene diversity and genetic differentiation

SSR data were used to define MultiLocus Genotypes (MLGs), and were checked for repeated MLGs. The number of repeated MLGs (G) was identified using GENCLONE 2.0 (Arnaud-Haond and Belkir, 2007). Genotypic evenness was evaluated using the index R= (G-1)/ (N-1), with G as the number of distinct multilocus genotypes and N the number of isolates (Grünwald *et al.*, 2003). Genetic diversity was estimated by allelic richness (*Ar*) using POPULATIONS 1.2.32 software (Langella, 1999). *Ar* was corrected for unequal sample size in each datasetby standardizing allelic richness to the smallest sample size, set at six individuals. Expected heterozygosity (H′) was computed using Arlequin software 3.1 (Excoffier *et al.*, 2005). Clonality was assessed with the index of association (I_A) and the R_d statistic, a measure of the multilocus linkage disequilibrium, and was calculated using Multilocus software version 3.1b (Agapow and Burst, 2001). The association between the scored alleles was estimated by comparing the variance of the genetic distances in the data set to the mean variance of 1,000 artificial re-sampled datasets. The R_d statistic is much less dependent on the number of loci than the index of association (Montarry *et al.*, 2010). I_A and R_d values are low for recombining population, and high for clonal or selfed populations (Burt *et al.*, 1996).

Population structure analyses

Partition of molecular diversity among and within regions, as well as among and within populations, was studied for the 240 isolates collected in 2012, using an analysis of molecular variance (AMOVA) (Excoffier *et al.*, 1992; Lynch and Milligan, 1994). AMOVA was performed using Arlequin 3.1 (Excoffier *et al.*, 2005). Pairwise F_{ST} values were calculated with Arlequin software, for sub-populations of *A. fabae* sampled in the different locations. A Bonferroni correction (adjusted alpha = 0.00316; 15 comparisons) was applied to take into account multiple testing.

A principal component analysis (PCA) was performed using the procedure available in the package ADEGENET (Jombart *et al.*, 2008) for the statistical freeware R version 3.1.1. PCA has an important advantage over other methods, such as the Bayesian clustering algorithm implemented in STRUCTURE (Pritchard *et al.*, 2000), because it does not require strong assumptions about an underlying genetic model, such as the Hardy-Weinberg equilibrium or the absence of linkage disequilibrium between loci (Jombart *et al.*, 2008).

STRUCTURE software version 2.2 was also used without any assumptions about population structure or assigning individuals to populations. The analysis was performed using 5 ′ 10^5 burn-in replicates and a run length of 1 ′ 10^6 Markov chain Monte Carlo (MCMC) replicates, adopting the admixed model and the correlated allele frequencies option. The number of genetic groups (K value) was estimated using the model developed by Evanno *et al.* (2005), which provides an estimate of the posterior probability of the data for a given K, Pr(X/K) (Perrier and Jacquemoud-Collet, 2006). The height of the modal value of the distribution was used as an indicator of the strength of the signal detected by STRUCTURE software. Five independent runs were performed for each analysis in order to verify the convergence of parameter estimates, and isolates were classified by location in order to determine if the same genotypes were present in the different populations and locations.

A Minimum Spanning Network (MSN) is an excellent way to visualize relationships among individuals. By using the package Poppr from R statistical software, version 3.1.2 (©2014, The R Foundation for Statistical Computing), MSN was calculated using Nei's distance for the different *A. fabae* populations. Multilocus genotypes (MLGs) were collapsed to multilocus haplotypes, represented by circles each containing the number of associated isolates and sized in proportion to haplotype frequency. Haplotype information could lead to more powerful tests of genetic association than single-locus analyses, but it is not easy to estimate haplotype frequencies from genotype data due to phase ambiguity. The challenge is compounded when individuals are pooled to save costs or to increase sample size. By collapsing the total allele frequencies of each pool suitably, the maximum likelihood estimates of haplotype frequencies based on the collapsed data can be calculated very quickly regardless of pool size and haplotype length (Kuk *et al.*, 2013).

Seasonal population differentiation

Pairwise comparisons were carried out between populations collected over two successive seasons, independently for each location (Beja, Bizerte, Jendouba, and Tunis). Multilocus genotype variability (N_T) and allelic richness (R) were estimated (described above), and pairwise F_{ST} values were calculated using Arlequin software. APCA was also performed to detect the distributions of isolates collected in the two successive growing seasons.

Mating type frequency analyses

Chi-square was used to test mating type ratios for all the locations sampled in the 2012 growing season. The ratios were expected to be 1:1 for randomly mating populations (Milgroom, 1996). Mating type ratios were considered significantly different from 1:1 at $P < 0.01$.

RESULTS

Population diversity of Ascochyta fabae *in Tunisia*

The genetic diversity and structure of Tunisian *A. fabae* populations were investigated using ten SSRs and mating type markers. The multiplex PCR easily differentiated two mating types, and the isolates were unambiguously assigned to one of those types. Of the 240 Tunisian isolates sampled in 2012, MAT1-2 was more common in Tunisia than MAT1-1 with a ratio of 2:1 (Table 3). At Beja and Tunis, the two mating types occurred in a 1:1 ratio ($P = 0.925$). For the other two locations, all the populations showed skewed mating type distributions, indicating that MAT1-2 was more common in all these populations (Table 3). The overall distribution showed statistically significant deviation from the 1:1 ratio in only two locations: Bizerte and Jendouba ($P < 0.001$).

All ten microsatellite loci were found to be polymorphic for the 240 isolates. The greatest heterozygosity were obtained for AFCAA13 and CAA28. Low levels of genetic diversity were detected among and within the locations, as populations from different locations shared many common alleles at all the SSR loci (percentage of common alleles greater than 35%) (Table 4). Narrow genetic diversity was found within each of the sixteen

Table 3. Isolates of *Ascochyta fabae* collected from different locations of Tunisia in the 2012 cropping season, and their mating types (MAT 1-1 and MAT 1-2) as determined by PCR.

Location	Number of isolates	*MAT 1-1*	*MAT 1-2*	c^{2e}	P^f
Beja	109	55	54	0.008	0.925
Bizerte	71	15	56	28.928	< 0.001
Jendouba	39	1	38	38.348	< 0.001
Tunis	21	10	11	0.154	0.694

Table 4. Distribution of private (P) and common (C) alleles among populations of *Ascochyta fabae* sampled in 2012 from four main locations of faba bean production in Tunisia (Beja, Bizerte, Jendouba, Tunis).

Location		Loci										Total
		CAA46	CAA28	AFCAA13	AFCAA12	AFCAA5	CA31	AFCAA1	CA10	CA33	CAA57	
Beja	C	21	7	7	10	8	11	12	15	12	11	114
	P	3	1	0	1	0	0	2	0	3	1	11
	Total	24	8	7	11	8	11	14	15	15	12	125
	%P	*12.5*	*12.5*	*0.0*	*9.1*	*0.0*	*0.0*	*14.3*	*0.0*	*20.0*	*8.3*	8.8
Bizerte	C	4	4	4	3	3	3	3	4	4	3	35
	P	4	1	1	3	0	4	7	2	2	4	28
	Total	8	5	5	6	3	7	10	6	6	7	63
	%P	*50.0*	*20.0*	*20.0*	*50.0*	*0.0*	*57.1*	*70.0*	*33.3*	*33.3*	*57.1*	*44.4*
Jendouba	C	2	3	4	3	3	2	2	3	2	2	26
	P	5	2	0	1	0	0	1	1	4	1	15
	Total	7	5	4	4	3	2	3	4	6	3	41
	%P	*71.4*	*40.0*	*0.0*	*25.0*	*0.0*	*0.0*	*33.3*	*25.0*	*66.7*	*33.3*	*36.6*
Tunis	C	4	4	4	3	4	3	3	3	5	3	36
	P	0	0	0	0	0	0	1	0	0	1	2
	Total	4	4	4	3	4	3	4	3	5	4	38
	%P	*0.0*	*0.0*	*0.0*	*0.0*	*0.0*	*0.0*	*25.0*	*0.0*	*0.0*	*25.0*	5.3

populations, reflecting abundant intraregional gene flow. At a location level, the percentage of common alleles shared among populations ranged from 55.6% to 94.7% (Table 4). Isolates from Beja had four loci (AFCAA13, AFCAA5, CA31 and CA10) that were common among the isolates of the different populations sampled at this location. At Bizerte, only one locus (AFCAA5) was common between the isolates. At Jendouba, three loci were only found among the isolates sampled in the different fields at this location (AFCAA13, AFCAA5, and CA31). At Tunis, of the ten loci, only two alleles were shared among isolates of the two populations sampled (AFCAA1, CAA57) (Table 4).

Low to high genetic and genotypic diversity was detected, depending on the location and the population (Table 5). GENCLONE analysis detected 67 different multilocus genotypes (MLG) of *A. fabae* among all the isolates sampled in 2012, with only six genotypes present in more than one location. Twenty genotypes were detected only once in the overall population. No MLG was detected in all four locations (Table 5). MLGs 28, 4, 60, and 62 represented 12.4%, 7.8%, 6.6% and 5.4% of the studied isolates, respectively. MLGs 27 and 42 represented, 3.9% and 3.1% of the isolates, respectively. The other MLGs included less than 2.7% of isolates. By location, genotypic richness (R) ranged from 0.11 (Jendouba) to 0.42 (Beja). Genotypic diversity was significantly greater than 0 ($P <$ 0.05) in all four locations. Isolates from Beja showed the greatest genotypic diversity (D = 0.982), followed by those from Bizerte (D = 0.809), Tunis (D = 0.803) and Jendouba (D = 0.726) (Table 5). At Beja, the frequency of the most common genotype (MLG 27) represented 8.8% of all the isolates sampled at that location. At Bizerte, the main MLG (MLG 28) represented 39.7% of all the isolates. At Jendouba and Tunis, the main MLGs (MLG 4 and MLG 42) represented 43.5 and 30.8% of all the isolates sampled at those two locations (Table 5). Different MLGs were recorded at different locations, respectively (Table 5). MLGs 13, 28, and 62 were detected at Beja and Bizerte, MLG 56 at Beja and Tunis, MLG 42 at Jendouba and Tunis, and MLG 60 was detected at Bizerte and Jendouba.

Genetic diversity indices also showed low to high variability among isolates within populations at the different locations studied (Table 6). Genotypic richness (R) ranged from 0 to 1 within the different populations studied. Isolates from Beja showed the greatest genotypic richness (R = 0.42), followed by Tunis (R = 0.25), Bizerte (R = 0.13) and Jendouba (R = 0.11). Allelic richness (*Ar*) was greatest at Beja (*Ar* = 5.84), followed by Bizerte (*Ar* = 3.95), Tunis (*Ar*= 3.35) and Jendouba (*Ar* = 1.63). As the populations studied were of unequal sample size, *Ar* was corrected for each dataset. This showed that Beja still had the greatest allelic richness (*Ar* = 5.04), followed by Tunis (*Ar* = 3.2), Bizerte (*Ar*= 3.15) and Jendouba (*Ar* = 1.49). Statistics on the Index of association (I_A) and *Rd* were significantly different from 0 at all the locations (I_A ranged from 1.827 ($P < 0.001$) to 7.223 ($P < 0.001$); *Rd* ranged from 0.182 ($P < 0.001$) to 0.642 ($P < 0.001$)). This

Table 5. Occurrence of multilocus genotypes (MLGs) among sixteen populations of *Ascochyta fabae* sampled at four locations (Beja, Bizerte, Jendouba, Tunis) in Tunisia, during the 2012 cropping season.

MLG	Beja	Bizerte	Jendouba	Tunis	Total occurence
MLG1-MLG2	X				3; 2
MLG3		X			1
MLG4-MLG5			X		20; 1
MLG6-MLG12	X				1; 2; 3; 1; 3; 2; 2
MLG13	X	X			2; 1
MLG14-MLG18	X				4; 4; 3; 2; 1
MLG19-MLG20			X		2; 1
MLG21-MLG22		X			3; 7
MLG23				X	3
MLG24		X			7
MLG25				X	2
MLG26-MLG27	X				1; 10
MLG28	X	X			3; 29
MLG29	X				1
MLG30		X			7
MLG31-MLG37	X				1; 1; 5; 3; 2; 1; 2
MLG38		X			4
MLG39-MLG40	X				4; 1
MLG41			X		1
MLG42				X	8
MLG43-MLG47	X				1; 2; 1; 4; 1
MLG48			X		4
MLG49-MLG50	X				1; 3
MLG51			X		1
MLG52-MLG55	X				2; 3; 1; 1
MLG56	X			X	1; 3
MLG57	X				3
MLG58				X	3
MLG59	X				5
MLG60		X	X		1; 16
MLG61				X	7
MLG62	X	X			1; 1"
MLG63-MLG67	X				2; 2; 4; 2; 3; 13
Ni	109	71	39	21	
Nu	48	10	8	6	
Fm	8.8%	39.7%	43.5%	30.8%	
D	0.982	0.809	0.726	0.803	

Nu = Number of unique multilocus genotypes in each population.
Fm = Frequency of the most common genotype
D = Genotypic diversity (Pielou, 1969).
Null hypothesis of no genotypic diversity was tested by comparing *D*-values obtained from 1000 randomized data sets to those estimated from the observed data set. *0.01 < *P* < 0.05, **0.001 < *P* < 0.01, ****P* < 0.001

Table 6. Genetic diversity, multilocus gametic disequilibrium, and test for random mating within *Ascochyta fabae* populations from Tunisia sampled during the 2012 cropping season.

Location	Populations	N	G	R	*Ar*	*Ar* (n=6)	H'	I_A	R_d
Beja	Bej1	8	8	1	5.5	5.3	0.886	0.241	0.029
	Bej2	7	6	0.83	4.1	4.1	0.742	1.531	0.174
	Bej3	12	11	0.91	6.0	5.3	0.850	1.464	0.171
	Bej4	12	10	0.82	6.7	5.7	0.841	2.971	0.349
	Bej5	16	11	0.67	6.5	5.1	0.800	2.677	0.304
	Bej6	6	6	1	4.4	4.4	0.867	1.360	0.175
	Bej7	14	12	0.85	6.8	5.4	0.809	2.192	0.252
	Bej8	25	19	0.75	7.7	5.4	0.806	1.735	0.198
	Bej9	9	8	0.86	4.9	4.7	0.792	2.274	0.271
	Total	*109*	*48*	*0.42*	*5.84*	*5.04*	*0.821*	*1.827*	*0.182*
Bizerte	Biz1	16	4	0.20	3.5	2.9	0.495	7.376	0.820
	Biz2	55	6	0.10	4.4	3.4	0.589	7.071	0.786
	Total	*71*	*10*	*0.13*	*3.95*	*3.15*	*0.542*	*7.223*	*0.641*
Jendouba	Jen1	6	2	0.20	1.9	1.9	0.300	8.000	1.000
	Jen2	22	3	0.10	1.1	1.1	0.089	-	0.511
	Jen3	11	1	0	2.0	1.5	0.018	2.996	-
	Total	*39*	*5*	*0.11*	*1.63*	*1.49*	*0.135*	*5.498*	*0.603*
Tunis	Tun1	15	6	0.36	3.7	3.4	0.480	4.789	0.534
	Tun2	6	4	0.60	3.0	3.0	0.613	4.251	0.476
	Total	*21*	*6*	*0.25*	*3.35*	*3.2*	*0.630*	*4.520*	*0.483*

N = number of isolates.
G = number of distinct multilocus genotypes.
R = genotypic richness.
Ar=, allelic richness
Ar (n=6), allelic richness corrected for samples size
H': expected heterozygosity without bias (Nei, 1978).
I_A = association index.
Rd statistic is a measure of the multilocus linkage disequilibrium (Agapow and Burt, 2001)

indicates high multilocus linkage disequilibrium and clonal populations (Table 6).

Structure of Ascochyta fabae *populations in Tunisia*

Partition of molecular diversity among and within locations, as well as among and within populations, was studied using an analysis of molecular variance (AMOVA) (Excoffier *et al.*, 1992; Lynch and Milligan, 1994). AMOVA revealed that 74.34% (*P* < 0.001) of the total genetic variance was partitioned within populations. A relatively low proportion of genetic variability was attributable to differences among populations within locations (14.52 %, *P* < 0.001)) and among locations (11.14 %, *P* = 0.0146).

High variances of F_{ST} values were detected, ranging from -0.055 to 0.982. Pairwise F_{ST} s were greater between

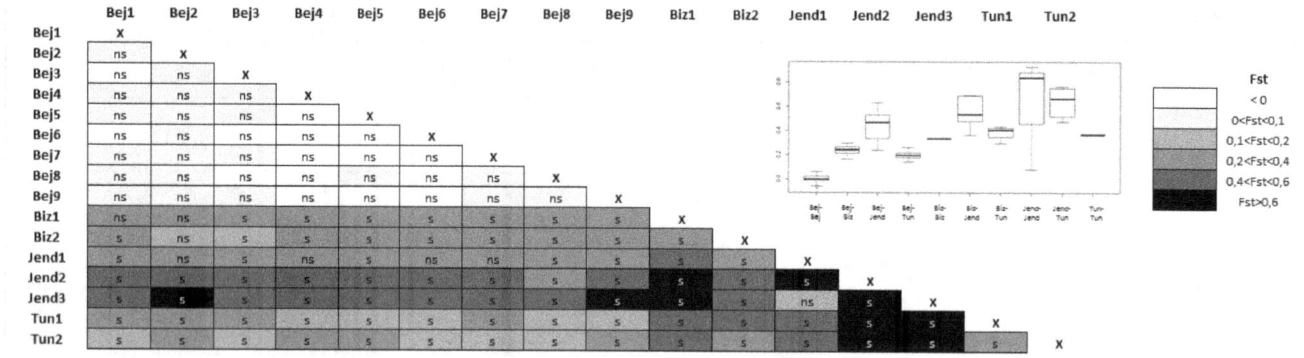

Figure 1. Matrix of pairwise F_{ST} among *Ascochyta fabae* sub-populations sampled in 2012. "ns" indicates non-significant differentiation between two populations (p: sub-population) from different locations (Bej = Beja, Biz = Bizerte, Jend = Jendouba, Tun =: Tunis). The distribution of pairwise F_{ST}s among each sub-population is indicated by the box plots.

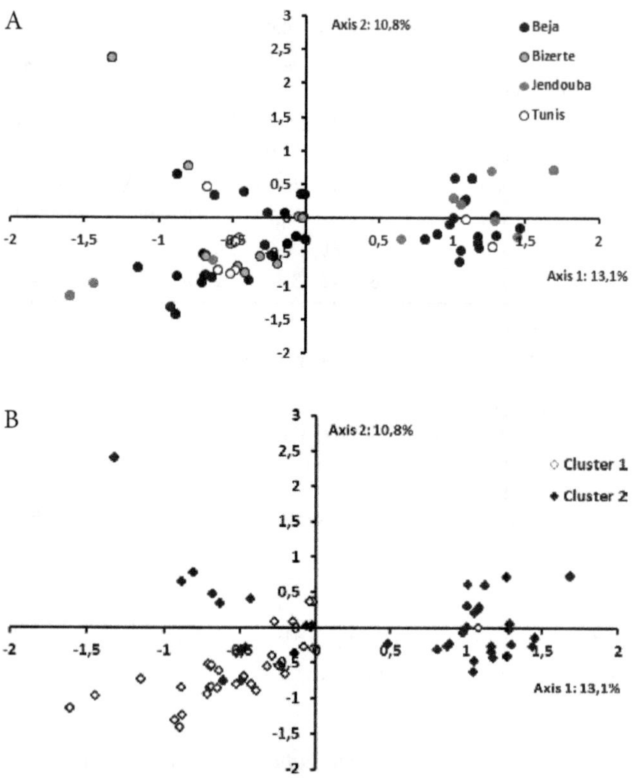

Figure 2. Principal component analysis (PCA) realized on the *Ascochyta fabae* sub-populations sampled from the five locations in Tunisia in 2012. **A**: isolates were assigned to their respective location (Beja, Bizerte, Jendouba, Kef, or Tunis), **B**: isolates were assigned to their respective clusters (C1, C2, or C1/C2) estimated with STRUCTURE (k = 2, deltak = 10).

locations than between populations within specific locations (Figure 1). The isolation for distance test did not show significant correlation (data not shown). The F_{ST} value was not correlated to the geographic distance separating two populations.

Lack of genetic differentiation of *A. fabae* populations was confirmed by the STRUCTURE and PCA analyses (Figure 2). PCA failed to separate the *A. fabae* isolates of the different populations sampled at Beja, Bizerte, Jendouba and Tunis into different groups (Figure 2A). The low percentage of genetic diversity explained by the two principal axes of the PCA (respectively, 13.1% and 10.8%) suggests that all the isolates belonged to the same population. The STRUCTURE analysis also failed to separate the isolates from the different locations into different groups (k = 2, *deltaK* = 10) (Figure 2B). The Bizerte populations were homogenic compared to the other locations.

A minimum spanning network (MSN) using Nei's distance for the different populations of *A. fabae* was calculated. As with the two other analyses, the MSN failed to clearly classify and structure the different populations (Figure 3). Except for the isolates from the population Jen 2 that constituted a single node, all others were distributed throughout the network, and the isolates were not grouped according to origin.

Seasonal genetic structure change in different Tunisian locations

Seasonal change of genetic diversity was monitored at the four locations (Beja, Bizerte, Jendouba and Tunis), and with the 317 isolates. Except for Tunis, all the locations showed different genotypic richness (R) in the two seasons. R values were greater in the second season than in the first (0.40 to 0.83 at Beja, 0.14 to 0.47 at Bizerte, and 0.16 to 0.30 at Jendouba) (Table 7). Multilocus genotypes occurred in from four to 46 isolates for the different populations. Only Beja and Jendouba showed common genotypes between the two successive seasons (Table 7).

Low to high variances of F_{ST} values were detected between the two successive seasons at the different loca-

Table 7. Genetic differentiation, test for random mating (I_A, R_d), and pairwise genetic comparison F_{ST} between two successive seasons at four Tunisian locations.

	Beja		Bizerte		Jendouba		Tunis	
	Season 1	Season 2	Season 1	Season 2	Season 1	Season 2	Season 1	Season 2
N^a	13	109	71	16	28	39	21	20
N_T	11	46	11	8	9	8	6	4
R	0.83	0.40	0.14	0.47	0.30	0.16	0.20	0.16
N_C	2		-		3		-	
I_A	1.473***		4.797***		4.929***		4.541***	
R_d	0.167***		0.535***		0.548***		0.506***	
F_{ST}	0.0177 ns		0.1882***		0.0645*		0.3269***	

A = Number of isolates analyzed.
N_T = Number of MLG.
N_C = Number of common MLG.

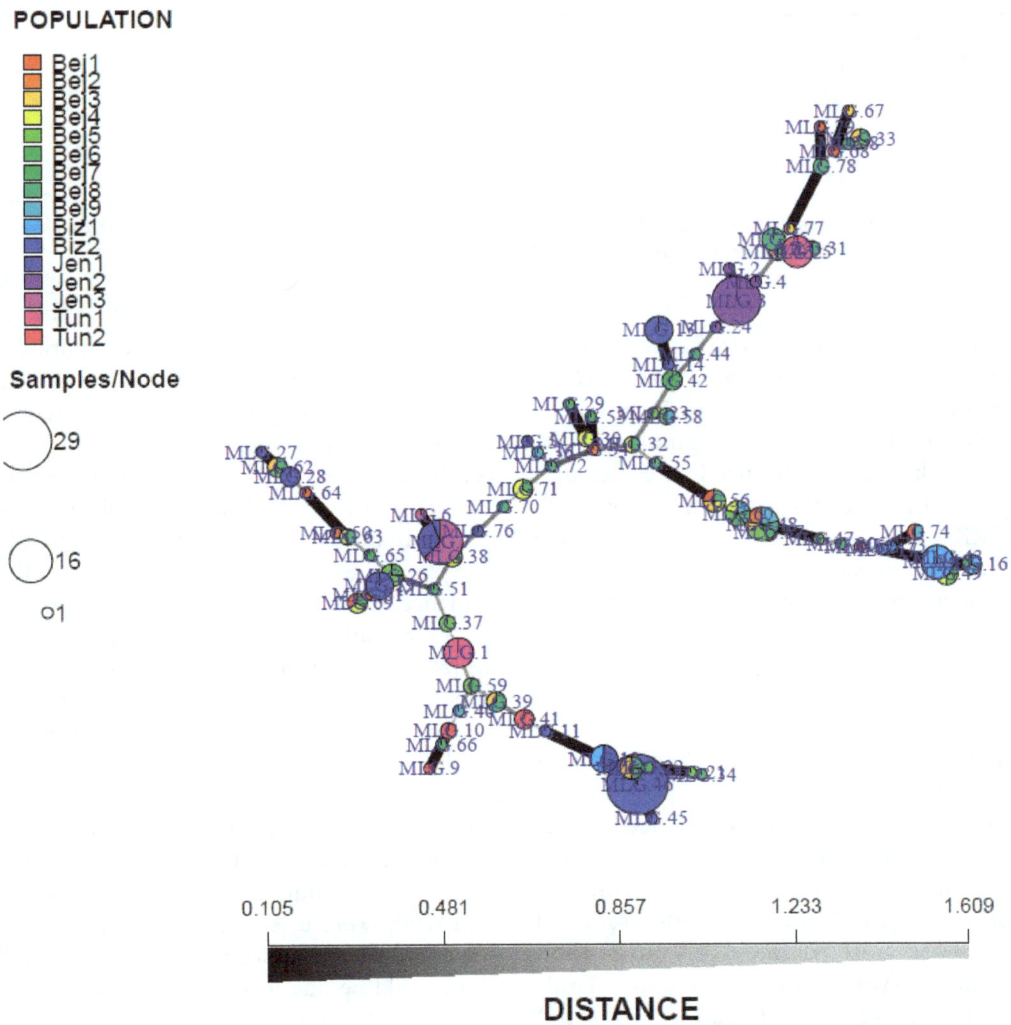

Figure 3. Minimum spanning network (MSN) of 67 haplotypes detected in the *Ascochyta fabae* collection of 2012. Each circle represents a unique haplotype and the different colours represent the sampling fields. The circle sizes represent the haplotype frequencies and the numbers of isolates is indicated in circles. Line widths and the shading represent relatedness of the haplotypes, based on Nei's genetic distance.

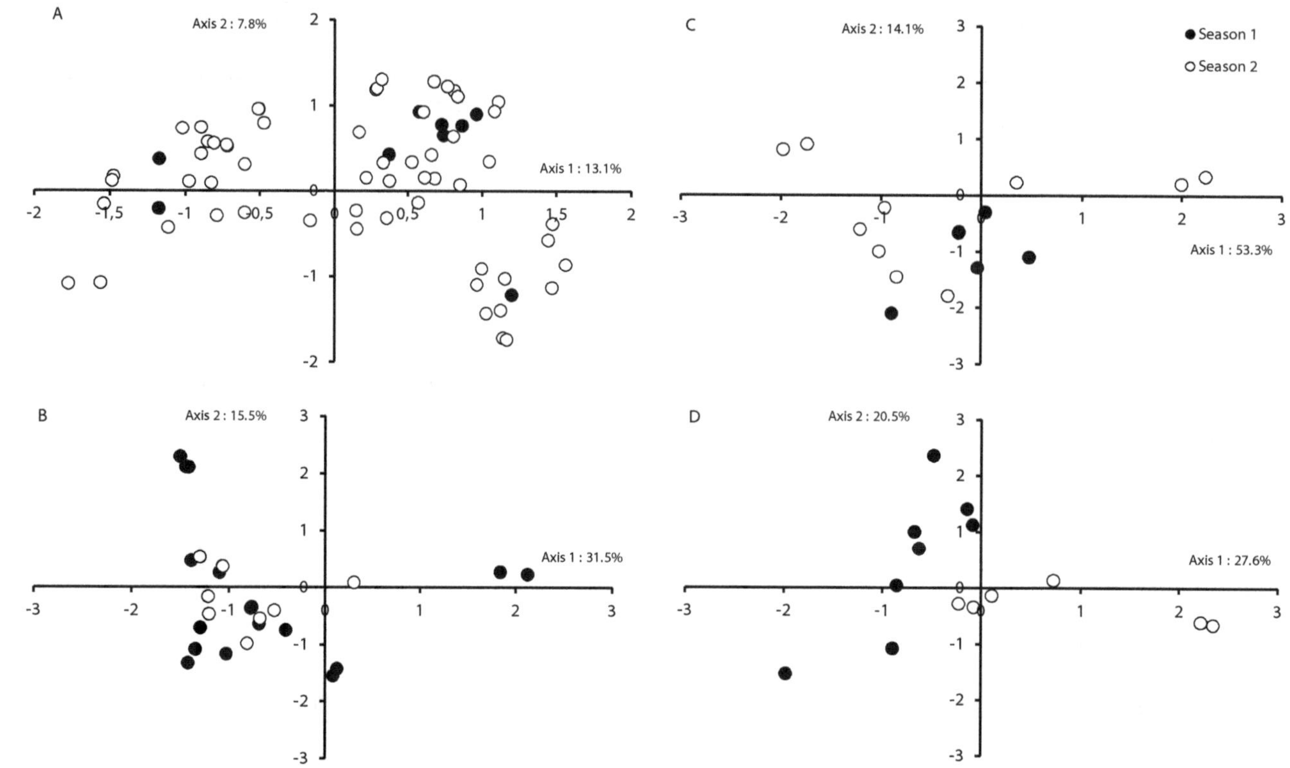

Figure 4. Principal component analysis (PCA) realized on the *Ascochyta fabae* sub-populations sampled during two successive cropping seasons at four locations in Tunisia. (A = Beja; B = Bizerte; C = Jendouba; D = Tunis).

tions. Pairwise F_{ST} were significantly different between the two seasons in Tunis (F_{ST}= 0.3269), Bizerte (F_{ST}= 0.1882) and Jendouba (F_{ST}= 0.0645) (Table 7). The PCA failed to separate the *A. fabae* isolates sampled in the two seasons from each location into two different groups, with the exception of the isolates from Tunis (Figure 4). The STRUCTURE analysis was also performed on these data, but failed to separate isolates from the different seasons into different groups (data not shown).

DISCUSSION

This study is only the second published report of examination of the diversity of *A. fabae* populations. Each of the ten pairs of SSR primers produced a single amplicon for each locus of each isolate with a high level of polymorphism, as established by Ozkilink *et al.* (2011). Among the 240 Tunisian isolates sampled in 2012, MAT1-2 was more common in Tunisia than MAT1-1, and mating type results showed that Tunisian populations have skewed distribution (1:2 ratio) for MAT1-1:MAT1-2. Of the four locations studied, two (Beja and Tunis) gave a 1:1 distribution while for the other two locations, MAT1-2 was more common than MAT1-1. This skewed distribution was confirmed by a chi-square test, suggesting that sexual reproduction may not occure at these two locations. These results were also confirmed by statistics on the Index of association (I_A) and *Rd*, both of which were significantly different from 0. This indicates that multilocus gametic disequilibrium and clonal (genotypic) diversity analyses did not provide evidence for randomly mating populations.

Our results were consistent with those of Ozkilinc *et al.* (2015) for the number of MLGs observed in *A. fabae* populations, but not for the distribution of mating types. The occurrence of sexual reproduction in the life cycle of Tunisian *A. fabae* populations is not likely to be regular, and ascospores may not be important primary inoculum, unlike for other Ascochyta blight diseases (Tivoli and Banniza, 2007). This conclusion should be confirmed, as the sample sizes of the populations in the present study were unequal , and therefore insufficient for the appropriate statistical tests. The observed disequilibrium could be due to sampling phenomena, particularly small sample sizes.

In their study, Ozkilinc *et al.* (2011) sampled *D. fabae* from chickpeas in southeastern Turkey and Israel between 2004 and 2007. They showed high gene diversity, but with gametic disequilibrium associated with small sample sizes. This result suggests that the gametic disequilibrium observed could be caused by admixtures of genetically distinct populations and/or sampling error, as all populations were assigned to a single genetic population in the STRUCTURE analysis, PCA and MNS analyses (Milgroom, 1996).

Despite the lack of differentiation *into* populations, many rare genotypes and a significant number of private alleles were detected. Only five genotypes of the pathogen were present in more than one location. This contradiction may be explained by the characteristics of the used marker, and suggests that new introductions of the pathogen had been made, either through mutation as microsatellite loci mutate rapidly or through migration or recombination. Between 2003 and 2012, Tunisia annually imported an average of 92.6 tons of faba bean seeds, for either multiplication by seed companies or for farmer use (ONAGRI, 2015). These imports, from countries such as Syria, may be a potential source of new pathogen genotypes. Faba beans imported for consumption average approx. 392 tons annually (ONAGRI, 2015). They are less rigorously controlled for diseases and could be a potential introduction source of new genotypes.

The limited dispersal of *A. fabae* isolates within fields may also lead to gametic disequilibrium. Even if random mating had occurred within restricted genotype profiles, this would have resulted in population admixture, which is one of the causes of gametic disequilibrium (Milgroom, 1996). The absence of common MLGs to all four locations suggests that populations from different locations may originate from multiple population sources. This hypothesis is supported by the F_{ST} values significantly different from zero. Migration could take place through seed movement. Farmers' use of their self-produced seeds may lead to genotypic differentiation at the field scale within the same region according to the values of F_{ST} calculated for pairs of fields; this appeared particularly at Bizerte and Jendouba. The hypothesis is plausible given that agricultural systems in Tunisia are characterized by the predominance of small-holder farmers relying on subsistence farming. In addition, lack of appropriate disease management strategies can result in pathogen multiplication and spatial dispersion through rain episodes and favourable temperatures (Tivoli and Banniza, 2007).

Furthermore, the greater genetic diversity detected at Beja could also be due to climatic conditions being more conducive to the pathogen in this region. *Ascochyta fabae* is favoured by low temperatures, and severe rainfall episodes that result in splash dispersal of pycnidiospores (Maurin *et al.*, 1990). The different locations had similar temperatures during the cropping season (Beja; Tmean = 14.3°C, Tmin = 8.9°C, Tmax = 19.6°C: Bizerte; Tmean = 14.9°C, Tmin = 10.5°C, Tmax = 19.3°C: Jendouba; Tmean = 14.6°C, Tmin = 8.6°C, Tmax = 20.7°C: Tunis; Tmean = 15.0°C, Tmin = 10.1°C, Tmax = 19.9°C). However, differences in rainfall were observed. Beja recorded annual rainfall of 479 mm, compared to 293 mm at Tunis, 345 mm at Jendouba, and 368 mm at Bizerte. This greater higher rainfall at Beja provided a conducive environment for *A. fabae* to spread and rapidly reproduce (Pritchard *et al.*, 1989; Tivoli and Banniza, 2007).

Selection of *A. fabae* by the host may be ruled out as having any significant impact. Research was carried out in Tunisia attempting to identify sources of resistance (Kharrat *et al.*, 2006). Some faba bean cultivars were released, but none are currently sold in their pure original states at seed markets. Due to the mixing of seed types, cultivars lose their distinctive characteristics after a few years, and the pathogen would face a host population characterized by high levels of genetic diversity. This would not necessarily expose the pathogen to directional selection, but would likely maintain pathogen diversity (McDonald and Linde, 2002).

Monitoring seasonal changes of the genetic structure of *A. fabae* at Beja, Bizerte, Jendouba and Tunis showed that new genotypes were detected in the second season in each region. This was unlikely to be derived from an event of sexual reproduction, since the allelic richness did not correlate with the number of MLGs identified in each region. A high level of inbreeding, revealed by significant values of F_{ST} in the two seasons, as well as the absence or scarcity of common genotypes, support the hypothesis that genotypic variation may be the result of genotype migration through seed exchange between different regions, as reported by Kaiser *et al.* (1997), and/or differential genotype adaptive potential to environmental conditions. As shown in the case study of Frenk el *et al.* (2010) from wild and domesticated *Cicer* species, a summer cropping system can act as a climatic factor for diversification of *A. rabiei* populations. These findings weaken any hypothesis of sexual reproduction in the different regions, but strengthen that of involvement of other evolutionary forces of diversity such as genotypic migration. This emphasizes the important role of seeds in disease dispersion at a regional scale (Tivoli and Banniza, 2007; Ali *et al.*, 2012).

In conclusion, our study has shown that neither the genetic diversity observed within the Tunisian populations of *A. fabae*, nor multilocus gametic disequilibrium, provide evidence for randomly mating populations of the pathogen in Tunisia. Lack of differentiation could be attributable to migration or mutation. The role of sexual reproduction should be confirmed by extending the sampling of isolates in these different locations, by calculating the P_{sex} index (GENETIX) and by evaluating the locus compatibility parameters (Arnaud-Haond *et al.*, 2005). The extent of asexual compared to sexual reproduction can affect disease management.

Sexual reproduction plays a role in the overwintering/adaptability of *A. fabae* by generating pathogen variability and facilitating long-distance dispersal via airborne ascospores (Kaiser, 1992; Trapero-Casas and Kaiser, 1992; Kaiser *et al.*, 1997). This can shape the structure and change genetic variability in populations. Despite recent advances, the role of gene flow (and of its two mains triggers, reproduction and dispersal) in the local evolution of pathogen populations is still poorly assessed. Many fungal plant pathogens alternate cycles of asexual multiplication with single annual episodes of sexual reproduction (Barrès *et al.*, 2012). The number of cycles of asexual multiplication has demographic impact as this corresponds mostly to the epidemic phase of a disease, and will also affect the genetic characteristics of pathogen populations.

Some studies of population genetics of airborne plant pathogens have used nested hierarchical sampling strategies. Gobbin *et al.* (2005) explained the arrival of new genotypes and the erosion of clonal structure in *Plasmopara viticola* populations by the continual input of sexual spores. For *A. fabae*, no data of changes in clonal structure during epidemic seasons is available. Studying the temporal changes in genes and genotypic diversity between the beginning and the end of an epidemic would be informative for the different epidemiological processes involved (Prugnolle and De Meeûs, 2010). This could provide insights into the balance between auto- and allo-infection processes.

To fully determine the degree of variability contributed by sexual recombination in pathogen populations, sampling on small spatial, but large temporal scales would help to address this question. Furthermore, analyzing progeny from naturally occurring pseudothecia for segregation would further indicate variability due to recombination (Milgroom, 1996). Studying the biology (temperature optima) of the two mating types and their fitness may further clarify this point. Migration of clones driven by seed movement at national and international levels may play a key role in obtaining high levels of genotypic diversity (Kaiser *et al.*, 1997). On the other hand, multiple cycles of asexual reproduction result in rapid increases of secondary inoculum and aggressive pathogen clones (Kaiser *et al.*, 1997; Tivoli and Banniza, 2007).

Strategies to better control Ascochyta blight will prioritize a focus on seeds. Several rules will need to be developed, including: i) imposition of standards for certified faba bean seed production, and inciting farmers to use certified seeds (to reduce seed transmitted diseases), and to avoid the use of seeds sold in local markets; ii) establish a strict system to control imported faba beans for local consumption; and iii) advocating systematic fungicide seed treatments to avoid risks of disease transmission through seeds during growing seasons.

ACKNOWLEDGEMENTS

This study was co-funded by the Ministry of Agriculture, Hydraulic Resources and Fisheries, the Ministry of Higher Education and Scientific Research of Tunisia, INRA (UMR IGEPP 1349 Rennes, France) and the MEDILEG-ARIMNET project (2012-2015; Proposal ID 396: Breeding, agronomic and biotechnological approaches for reintegration and revalorization of legumes in Mediterranean agriculture). We thank Dr Tobin Peever (Pullman University, USA) and Hilal Ozkilinc (ICARDA) for their support and advice in using SSR markers.

LITERATURE CITED

Ali H., Alam S.S., Attanayake R.N., Rahman M., Chen W., 2012. Population structure and mating type distribution of the chickpea blight pathogen *Ascochyta rabiei* from Pakistan and the United States. *Journal of Plant Pathology* 94: 99–108.

Agapow P.M., Burst A., 2001. Indices of multilocus linage disequilibrium. *Molecular Ecology Notes* 1: 101–102.

Arnaud-Haond S., Belkhir K., 2007. Genclone: A Computer Program to Analyse Genotypic Data,Test for Clonality and Describe Spatial Clonal Organization. *Molecular Ecology Notes* 7 (1): 15–17.

Arnaud-Haond S., F. Alberto, Procaccini G., Serrao E.A., Duarte C.M., 2005. Assessing genetic diversity in clonal organisms: low diversity or low resolution? Combining power and cost-efficiency in selecting markers. *Journal of Heredity* 96: 1-8.

Arnaud-Haond S., Duarte C.M., Alberto F., Serrao E.A., 2007. Standardizing methods to address clonality in population studies. *Molecular Ecology* 16: 5115–5139.

Atik O., Ahmed M.M., Abang M.M., Imtiaz M., Hamwieh A., Baum M., El-Ahmed A., Murad S., Yabrak M.M., 2013. Pathogenic and genetic diversity of *Didymella rabiei* affecting chickpea in Syria. *Crop Protection* 46: 70–79.

Avila C.M., Satovic Z., Siller J.C., Rubiales D., Moreno M.T., Torres A.M., 2004. Isolate and organ-specific QTLs for ascochyta blight resistance in faba bean. *Theoretical Applied Genetics* 108: 1071–1078.

Barrès B., C. Dutech, A. Andrieux, F. Halkett and P. Frey, 2012. Exploring the role of asexual multiplication in poplar rust epidemics: impact on diversity and genetic structure. *Molecular Ecology* 21, 4996–5008.

Bayraktar H., Dolar F.S., Maden S., 2007. Mating type groups of *Ascochyta rabiei* (teleomorph: *Didymella rabiei*), the causal agent of chickpea blight in Central Anatolia. *Turkish Journal of Agriculture & Forestry* 31: 41-6.

Burt A., Carter D.A., Koenig G.L., White T.J., Taylor J.W., 1996. Molecular markers reveal cryptic sex in the human pathogen *Coccidiodes immitis. Proceedings of the National Academy of Sciences, USA 93*:770–773.

Chérif M., Chilvers M.I., kamatsu H.A, Peever T.L., Kaiser W.J., 2006. Cloning of the mating type locus from *Ascochyta lentis* (teleomorph: *Didymella lentis*) and development of a multiplex PCR mating assay for Ascochyta species. *Current Genetics* 50: 203–215.

De MêeusT., McCoy K.D., Prugnolle F., Chevillon C., Durand P., Hurtrez-Boussès S., Renaud F., 2007. Population genetics and molecular epidemiology or how to "débusquer la bête".*Infection, Genetics and Evolution* 7: 308–332.

Evanno G., Regnaut S., Goudet J., 2005. Detecting the number of clusters of individuals using the software STRUCTURE: a simulation study. *Molecular Ecology* 14: 2611–2620.

Excoffier L., Smouse P.E., Quattro J.M., 1992. Analysis of molecular variance inferred from metric distances among DNA haplotypes application to human mitochondrial DNA restriction data. *Genetics* 131: 479–491.

Excoffier L., Laval G., Schneider S., 2005. Arlequin (version 3.0): An integrated software package for population genetics data analysis. *Evolutionary Bioinformatics* 1: 47–50.

Frenkel O., Peever T.L., Chilvers M.I., Ozkilinc H., Can C., Abbo S., Shtienberg D., Sherman A., 2010. Ecological genetic divergence of the fungal pathogen *Didymella rabiei* on sympatric wild and domesticated *Cicer* spp. (Chickpea). *Applied and Environmental Microbiology* 76 (1): 30–39.

Gobbin D., Jermini M., Loskill B., Pertot I., Raynal M., Gessler C., 2005. Importance of secondary inoculum of *Plasmopara viticola* to epidemics of grapevine downy mildew. *Plant Pathology* 54: 522–534.

Grünwald N.J., Goodwin S.B., Milgroom M.G., Fry W.E., 2003. Analysis of genotypic diversity data from populations of microorganisms. *Phytopathology* 93: 738–746.

Hanounik S.B., Robertson I.D., 1989. Resistance in *Vicia faba* germplasm to blight caused by *Ascochyta fabae. Plant Disease* 73: 202–205.

Jombart T., Devillard S., Dufour A.B., Pontier D., 2008. Revealing cryptic spatial patterns in genetic variability by a new multivariate method. *Heredity* 101 (1): 92–103.

Kaiser, W.J., 1992. Epidemiology of *Ascochyta rabiei.* In: Disease resistance breeding in kabuli chickpeas. (Singh, K.B., Saxena, M.C. eds). ICARDA, Aleppo, Syria. pp: 117–143

Kaiser W.J., Wang B.C., Rogers J.D., 1997. *Ascochyta fabae* and *A. lentis*: host specificity teleomorphs (*Didymella*), hybrid analysis, and taxonomic status. *Plant Disease* 81: 809- 816.

Kharrat M., Le Guen J., Tivoli B., 2006. Genetics of resistance to 3 isolates of *Ascochyta fabae* on faba bean (*Vicia faba* L.) in controlled conditions. *Euphytica* 151: 49–61.

Kuk A.Y., Li X., Xu J., 2013. A fast collapsed data method for estimating haplotype frequencies from pooled genotype data with application to the study of rare variants. *Statistics in Medicine* 32 (8): 1343–1360.

Laloi G., Montarry J., Guibert M., Andrivon D., Michot D., Le May C., 2016. Aggressiveness changes in populations of *Didymella pinodes* over winter and spring pea cropping seasons. *Applied and Environmental Microbiology* 8: 4330–4339.

Langella O., 1999. Population Genetic Software. Available at: http://bioinformatics.org/~tryphon/populations

Le May C., Guibert M., Leclerc A., Andrivon D., Tivoli B., 2012. A Single, Plastic Population of *Mycosphaerella pinodes* Causes Ascochyta Blight on Winter and Spring Peas (*Pisum sativum*) in France. *Applied and Environmental Microbiology* 78 (23): 8431–8440.

Le May C., Chilvers M.I., Saucereau A.L., Guibert M., Peever T.L., 2017. Spatiotemporal distribution of *Ascochyta pinodes* and *Ascochyta pinodella* during the winter growing season in France. *Plant Pathology*, DOI: 10.1111/ppa.12823

Lynch M., Milligan B.G., 1994. Analysis of population genetic structure with RAPD markers. *Molecular Ecology* 3: 91–99.

Maurin N., Tivoli B., Onfroy C., 1990. Mieux connaître les maladies et leurs agents pathogènes pour mieux les combattre : exemple de l'anthracnose de la féverole. *Perspectives Agricoles* 146: 36–45.

Maurin N., Tivoli B., 1992.Variation in the resistance of *Vicia faba* to *Ascochyta fabae* in relation to disease development in field trials. *Plant Pathology* 41: 737–744.

McDonald B.A., Linde C., 2002. Pathogen population genetics, evolutionary potential, and durable resistance. *Annual Review of Phytopathology* 40: 349–379.

Milgroom M.G., 1996. Recombination and the multilocus structure of fungal populations. *Annual Review of Phytopathology* 34: 475–477.

Milgroom M.G., Peever T.L., 2003. Population biology of plant pathogens: the synthesis of plant disease epidemiology and population genetics. *Plant Disease* 89: 608–617.

Montarry J., Andrivon D., Glais I., R. Corbière, Mialdea G., Delmotte F., 2010. Microsatellite markers reveal two admixed genetic groups and an ongoing displacement within the French population of the invasive plant pathogen *Phytophthora infestans. Molecular Ecology* 19: 1965–1977.

Nei M., 1978. Estimation of Average Heterozygosity and Genetic Distance from a Small Number of Individuals. *Genetics* 89(3): 583–590.

NourollahiK., Javannikkhah M., Naghavi M.R., Lichtenzveig J., Okhovat S.M., Oliver R.P., Ellwood S.R., 2011. Genetic diversity and population structure of *Ascochyta rabiei* from the western Iranian Ilam and Kermanshah provinces using MAT and SSR markers. *Mycological Progress* 10: 1-7.

Omri Ben Youssef N., Le May C., Mlayeh O., Kharrat M., 2012. First report of *Didymella fabae*, teleomorph of *Ascochyta fabae*, on faba bean crop debris in Tunisia. *Phytopathologia Mediterranea* 51: 369–373.

ONAGRI.2015.ONAGRI DATABASE 2015 http://www.onagri.nat.tn/

Onfroy C., Tivoli B., Corbière R., Bouznad Z., 1999. Cultural, molecular and pathogenic variability of *Mycosphaerella pinodes* and *Phoma medicaginis* var. *pinodella* isolates from dried pea (*Pisum sativum*) in France. *Plant Pathology* 48 (2): 218–229.

Ozkilinc H.K., Akamatsu H., Abang M., Thomas K., Chilvers M.I., Peever T.L., 2011. Development, characterization and linkage analysis of microsatellite loci for the Ascochyta blight pathogen of faba bean, *Didymella fabae. Journal of Microbiological Methods* 87: 128–130.

Ozkilinc H.K., Thomas K., Abang M., Peever T.L., 2015. Population structure and reproductive mode of *Didymella fabae* in Syria. *Plant* Pathology 64: 1110–1119. DOI: 10.1111/ppa.12359

Padder B.A., Kapoor V., Kaushal R.P., Sharma P.N., 2012. Identification and Genetic Diversity Analysis of Ascochyta Species Associated with Blight Complex of Pea in a Northwestern Hill State of India. *Agricultural Research* 1(4): 325–337.

Perrier X, Jacquemoud-Collet J.P., 2006. DARwin software, http://darwin.cirad.fr/darwin

Pielou E.C., 1969. An introduction to mathematical ecology. Wiley- Interscience, New York, USA, 286 pp.

Pritchard, J.K., Stephens M., Donnelly P., 2000. Inference of population structure using multilocus genotype data. *Genetics* 155 (2): 945–959.

Pritchard P R., Rowe P S., ossall R S., 1989. A comparison of infection of resistant and susceptible lines of field bean (*Vicia faba*) by *Ascochyta fabae. Plant Pathology* 38: 266–270.

Prugnolle F., De Meeûs T., 2010. Apparent high recombination rates in clonal parasitic organisms due to inappropriate sampling design *Heredity.* 104: 135–140

Rashid K.Y., Bernier C.C., Conner R.L., 1991. Genetics of resistance in faba bean inbred lines to five isolates of *Ascochyta fabae. Canadian Journal of Plant Pathology* 13: 218–225.

Rouaissi M., Hajlaoui M.R., Kharrat M., 2001. Caractérisation moléculaire par RAPD et étude de la diversité génétique d'*Ascochyta fabae*, agent causal de l'anthracnose des fèves. *Actes des 8èmes Journées Nationales sur les Résultats de la recherche Agronomique - Nabeul, 13 & 14 Nov*. 2001-GC.

Rubiales D., Trapero-Casas A., 2002. Occurrence of *Didymella fabae*, the Teleomorph of *Ascochyta fabae*, on Faba Bean Straw in Spain. *Journal of Phytopathology* 150: 146–148.

Tivoli B. and Banniza S., 2007. Comparison of the epidemiology of ascochyta blights on grain legumes. *European Journal of Plant Pathology* 119: 59–76.

Trapero-Casas, A., Kaiser W.J., 1992. Development of *Didymella rabiei*, the teleomorph of *Ascochyta rabiei*, on chickpea straw. *Phytopathology* 82: 1261–1266.

7

Control of bottom rot in hydroponic lettuce, caused by strains of *Botrytis cinerea* with multiple fungicide resistance

Michael CHATZIDIMOPOULOS[1], Athanassios C. PAPPAS[2]

[1] *Department of Agriculture, Crop Production and Rural Environment, Laboratory of Plant Pathology, University of Thessaly, Fitokou Street, 384 46, N. Ionia, Volos, Greece*
[2] *Present address: 17 Bizaniou, 1156 69 Papagos, Greece*
*Corresponding author: mxatzid@agr.uth.gr

Summary. For two consecutive growing periods, fungicide-resistant *Botrytis cinerea* strains were detected in high proportions in glasshouse-grown lettuce, but at variable frequencies. Pre-transplanting fungicide sprays applied on two successive occasions reduced disease severity and increased the number of healthy plants without leaving detectable residues above accepted MRLs at harvest. In some instances, the disease was further decreased when pre-transplanting applications were combined with one or two further sprays applied soon after transplanting. The fungicide mixture of fludioxonil + cyprodinil was the most effective against the disease and provided better control of *B. cinerea* isolates *in situ*. These treatments gave satisfactory disease control despite the predominance of multi-fungicide resistant *B. cinerea* populations.

Keywords. Fungicides, grey mould, fludioxonil, cyprodinil, chlorothalonil.

Competing Interests: The Author(s) declare(s) no conflict of interest.

Editor: Jean-Michel Savoie, INRA Villenave d'Ornon, France.

INTRODUCTION

Bottom rot of butterhead lettuce (*Lactuca sativa* L.) caused by *Botrytis cinerea* Pers. :Fr. is the most common disease problem in hydroponic lettuce production in Greece, during the late autumn to early spring period. Infections can start in nurseries and spread systemically (endophytically) in plants, without early visible symptoms (Sowley *et al.*, 2010). *Botrytis* head rots are less common and, in most cases, follow the appearance of 'tip burn' symptoms, due to inadequate transport of calcium into emerging leaves. These rots can be avoided by keeping calcium in balanced nutrient solutions, using cultivars which are less susceptible to 'tip burn', and manipulating the environment (Morgan, 1999; 2012). Good ventilation practices reducing excess of moisture combined with application of fungicides give adequate control of infections caused by *Botrytis cinerea* (Dik and Wubben, 2007).

In addition to multi-site fungicides such as thiram, compounds with site-specific modes of action against grey mould in lettuce crops are currently registered in Greece. These include anilinopyrimidines (cyprodinil and pyrimethanil), the phenylpyrrole fludioxonil, the succinate dehydrogenase inhibitor (SDHI) boscalid and the quinone outside inhibitor (QoI) pyraclostrobin. Two commercial fungicide formulations with widespread use against *B. cinerea* in lettuce are Signum® (26.7% boscalid + 6.7% pyraclostrobin; BASF) and Switch® (25% fludioxonil + 37.5% cyprodinil; Syngenta). However, the use of fungicides for *B. cinerea* control in various crops has been associated with the development of fungicide resistance (Hahn, 2014). High levels of resistance against site-specific fungicides are the result of gene mutations at positions encoding their target sites. For example, the point mutations G143A, H272R, and F412S, which lead to changes in the target proteins CytB, SdhB, and Erg27, confer high resistance of the pathogen to, respectively, the QoI, SDHI, and hydroxyanilide fungicide classes (Leroux, 2007).

Multi-drug resistance (MDR) is another mechanism associated with fungicide resistance in *B. cinerea*. This involves mutations leading to over-expression of efflux transporters such as the ATP-binding cassette (ABC) and the major facilitator superfamily (MFS), allowing weak resistance towards fungicides with unrelated modes of action (Kretschmer *et al.*, 2009). Very often, MDR and specific fungicide resistance types are coupled (Leroch *et al.*, 2013; Fernández-Ortuño *et al.*, 2014; Rupp *et al.*, 2016). The presence of *B. cinerea* strains with multiple fungicide resistance to all site-specific classes of fungicides have been reported in different parts of the world, especially for small fruits (Weber, 2011; Amiri *et al.*, 2013; Fernández-Ortuño *et al.*, 2014). Recent surveys made on lettuce crops in Greece and Germany have also demonstrated increasing threats from emergence of multiple fungicide resistance in *B. cinerea* populations (Chatzidimopoulos *et al.*, 2013; Weber and Wichura, 2013).

Current trends in agriculture demand fewer chemical applications, while maintaining profitable high-quality production with low pesticide residues. The limited number of registered fungicide formulations against bottom rot of lettuce forces growers to make repeated seasonal sprays with the one fungicide. Some studies have also shown that pesticide residues are detected in greater amounts in leafy vegetables compared to other crops (Skovgaard *et al.*, 2017). Multiple applications may compromise reduced pesticide strategies, which aim to delay the development of resistance and reduce pesticide residues.

The present study was undertaken: (i) to evaluate the efficacy and timing of applications with current botryticides against multi-resistant *B. cinerea* strains; (ii) to detect and measure possible fungicide residues at harvest; and (iii) to determine effects of different compounds against selected resistant isolates of the pathogen *in situ*.

MATERIALS AND METHODS

Host plant material

The 2-year experiments were carried out in a commercial lettuce glasshouse located at Krokion, Magnesia, Greece. The glasshouse was surrounded by cereal crops and olive trees, which were unlikely to be sources of *B. cinerea* inoculum. Pelletized lettuce seeds (*Lactuca sativa* 'Penelope'; 'butterhead' type, Rijk Zwaan), pre-treated with thiram were used in all tests. Seeding, germination and emergence of nursery plants took place in 4 × 4 cm horticultural cells filled with a peat-based substrate. The young seedlings were transplanted at the 4- to 5-leaf stage, about 5 weeks after sowing, into Hortiplan hydroponic gutters (nutrient film technique, NFT). For plant nutrition, a dense aqueous (bore water) solution was prepared, composed of (mg L^{-1}): Ca 200, Mg 40, K 210, P (PO_4^{3-}) 50, N (NH_4^+) 25, N (NO_3^-) 165, Fe 5, Mn 0.5, Cu 0.1, Zn 0.1, B 0.5, Mo 0.05. The pH of the nutrient solution was maintained between 5.5 and 5.8, and conductivity between 1.5 and 2.0 mS (Resh, 2012). The aqueous solution was supplied to plant every 15 min during daytime.

Experimental design and treatments

Experiments were organized in randomized blocks with three replicates for each treatment. Each plot consisted of 50 plants spaced 20 cm apart in one row. Fungicide applications was carried out using a hand-operated sprayer at 1,120 L ha $^{-1}$ at 10 to 20-day intervals. To minimize the effects on neighbouring treatments, plots were separated with a plastic frame (100 × 50 cm), during the spray applications. The last application was made at least 4 weeks before harvest. Plants sprayed with water were used as experimental controls.

The fungicides used, at the standard recommended labelled rates for vegetable crops, were as follows: chlorothalonil (Daconil® 50 SC, Syngenta Ltd) at 3 mL L^{-1}; fenhexamid (Teldor ® 50 WG, Bayer CropScience) at 1.5 g L^{-1}; boscalid + pyraclostrobin (Signum® 26.7 + 6.7 WG, BASF SE) at 1.5 g L^{-1}; and fludioxonil + cyprodinil (Switch® 25 + 37.5 WG, Syngenta Crop Protection AG) at 0.5 g L^{-1}.

In the first trial (2012-2013 season), lettuce was sown on November 8, 2012, was transplanted into the NFT system on December 7, and was harvested on January 24, 2013. In this trial, the effectiveness of chlorothalonil and the mixture of fludioxonil + cyprodinil was evaluated in two, three or four spray programmes (Table 1). Fungicides were applied on two occasions during the

Table 1. Comparison of fungicide spray programs against bottom rot in hydroponic lettuce (Trial 1).

Applications			Disease		AUDPC	Healthy plants at harvest (%)	Fungicide residues[a] (mg Kg^{-1})	Phenotypes[b] recovered (%)		
Pre-Transplanting (2)	Post-Transplanting (1 or 2)		Severity (%)	Incidence (%)				I	II	W
Fludioxonil+ cyprodinil	-	-	0.6 c[c]	5.3 d	4.19 b	92.67 b	<LoQ[d]	0	0	0
	Fludioxonil+ cyprodinil	-	0.4 c	2.7 d	2.80 b	85 b	Fludioxonil:<LoQ Cyprodinil:0.053	0	0	0
	Fludioxonil+ cyprodinil	Fludioxonil+ cyprodinil	0.1 c	0.7 d	0.66 c	93.67 b	Fludioxonil:0.06 Cyprodinil:0.2	0	0	0
Chlorothalonil	-	-	20.7 a	40.7 b	267.14 a	64.67 ab	<LoQ	33	0	67
	Chlorothalonil	-	14.2 ab	26.0 c	201.86 a	64.67 ab	0.061	50	0	50
	Chlorothalonil	Chlorothalonil	9.2 bc	25.3 c	120.26 a	73.33 b	0.073	75	0	25
Control			21.8 a	55.3 a	261.91 a	37.33 a	-	62	15	23

[a] Maximum residue levels as determined by the European Community. Fludioxonil: 15; Cyprodinil:15; Chlorothalonil: 0.01; Fenhexamid: 40; Boscalid: 30; Pyraclostrobin: 2.
[b] *B. cinerea* phenotypes: I=$QoI^{R}Bos^{R}Ani^{R}Ben^{HR}Dic^{MR}$, II=$Hyd^{R}QoI^{R}Bos^{R}Ani^{R}Phen^{MR}Ben^{HR}Dic^{MR}$, W=Wild type.
[c] Means followed by the same letter in each column do not significantly differ (P = 0.05, Student-Newman-Keuls).
[d] LoQ: limit of quantification (0.005 mg Kg^{-1}).

nursery stage (November 19, December 1), followed by one or two more applications after transplanting (December 13 and 24).

In the second trial (2013–2014 season), lettuce was sown on October 17, 2013, transplanted on November 23, and harvested on January 27, 2014. Following the results from the first trial, the addition of fungicides from unrelated chemical groups was also evaluated in two-, three- or four-spray programmes. Two basal applications were carried out during the nursery stage (October 27 and November 17) with either chlorothalonil or the fludioxonil + cyprodinil mixture. After transplanting, one (December 6) or two (December 30) more applications were made with either fenhexamid, fludioxonil + cyprodinil, boscalid + pyraclostrobin or chlorothalonil, in alternating applications (Table 2).

Efficacy of the fungicide programmes was evaluated at harvest by recording the number of healthy plants from each plot. The disease incidence and severity were also recorded every week by counting the number and estimating the proportion (%) lesion area on plants infected by *B. cinerea* in each treatment. The sample size for all assessments was 50 plants per plot. At the end of each trial, the area under the disease progress curve (AUDPC) was calculated, based on the formula:

$$AUDPC = \sum_{i=1}^{N_i-1} \left(\frac{y_i + y_{i+1}}{2}\right)(t_{i+1} - t_i)$$

where t = the time of each assessment; y = the percent disease severity at each assessment; and n = the number of assessments.

Fungicide residues on lettuce heads were determined at harvest, in randomly collected samples of three plants from each treatment. Following EU directions, highly sensitive and selective multi residue methods were used to detect multiple fungicides (Waziha *et al.*, 2018). Chlorothalonil and fenhexamid residues were determined using gas chromatography with an electron capture detector. Boscalid, pyraclostrobin, cyprodinil and fludioxonil were analyzed using liquid chromatography with tandem mass spectrometry, following acetonitrile extraction/partitioning. Analyses were carried out at the Cadmion accredited Analytical Laboratory, 202 00 Kiato Korinthia, Greece.

Airborne inoculum monitoring

Portable air samplers (Burkard Manufacturing Co Ltd) were used, containing 9 cm Petri plates with selective medium (Edwards and Seddon, 2001), as slightly modified by Chatzidimopoulos *et al.* (2014b). The medium was enriched with a discriminatory concentration of each fungicide and was used to entrap *B. cinerea* propagules in the air. Only fungicide-resistant isolates were able to grow on these media. The different fungicides were dissolved in dimethyl sulphoxide and were added to different plates with the selective medium. The final con-

Table 2. Comparison of fungicide spray programs against bottom rot caused by *Botrytis cinerea* in hydroponic lettuce (Trial 2).

Applications			Disease		AUDPC	Healthy plants at harvest (%)	Fungicide residues[a] (mg Kg^{-1})	Phenotypes[b] recovered (%)		
Pre-Transplanting (2)	Post-Transplanting (1 or 2)		Severity (%)	Incidence (%)				I	II	W
Fludioxonil+ cyprodinil	-	-	0.8 b[c]	2.67 b	7.35 b	90.00 a	<LoQ[d]	0	0	100
	Fludioxonil+ cyprodinil		1.2 b	7.33 b	9.72 b	97.33 a	<LoQ	0	100	0
	Fenhexamid	-	0.7 b	0.67 b	11.64 b	95.33 a	<LoQ	0	100	0
	Boscalid+ pyraclostrobin	-	0.9 b	2.67 b	9.58 b	96.67 a	<LoQ	50	50	0
	Chlorothalonil	-	1.1 b	2.67 b	15.71 b	98.00 a	<LoQ	100	0	0
	Fenhexamid	Fludioxonil+ cyprodinil	2.4 b	10.67 b	21.92 b	94.00 a	Fludioxonil:0.43 Cyprodinil:0.74 Fenhexamid<LoQ	25	75	0
	Fludioxonil+ cyprodinil	Fenhexamid	0.9 b	1.33 b	11.93 b	92.00 a	Fludioxonil <LoQ Cyprodinil<LoQ Fenhexamid:5.95	0	100	0
Chlorothalonil	-	-	1.7 b	3.33 b	23.94 b	83.33 a	<LoQ	100	0	0
	Chlorothalonil	-	0.6 b	1.33 b	6.26 b	96.67 a	<LoQ	0	0	0
	Fenhexamid	-	2.1 b	4.67 b	33.22 b	89.33 a	<LoQ	25	75	0
	Boscalid+ pyraclostrobin	-	0.8 b	1.33 b	13.38 b	95.33 a	<LoQ	75	25	0
	Fludioxonil+ cyprodinil	-	1.7 b	2.00 b	26.84 b	93.33 a	<LoQ	0	75	25
	Fenhexamid	Fludioxonil+ cyprodinil	3.0 b	4.67 b	45.34 b	89.33 a	Fludioxonil:0.43 Cyprodinil:0.74 Fenhexamid<LoQ	0	0	0
	Fludioxonil+ cyprodinil	Fenhexamid	1.6 b	3.33 b	20.62 b	92.00 a	Fludioxonil <LoQ Cyprodinil<LoQ Fenhexamid:5.95	0	100	0
Control			14.2 a	28 a	182.58 a	88.00 a		60	20	20

[a] Maximum residue levels as determined by the European Community. Fludioxonil: 15; Cyprodinil:15; Chlorothalonil: 0.01; Fenhexamid: 40; Boscalid: 30; Pyraclostrobin: 2.
[b]*B. cinerea* phenotypes: I=$QoI^R Bos^R Ani^R Ben^{HR} Dic^{MR}$, II=$Hyd^R QoI^R Bos^R Ani^R Phen^{MR} Ben^{HR} Dic^{MR}$, W=Wild type.
[c] Means followed by the same letter in each column do not significantly differ (P = 0.05, Student-Newman-Keuls).
[d] LoQ: limit of quantification (0.005 mg Kg^{-1}).

centration of the solvent in the growth medium did not exceed 1%. The discriminatory doses used were: 1 mg L^{-1} fenhexamid from the hydroxyanilide (Hyd) class; 10 mg L^{-1} pyraclostrobin from QoI class (strobilurins) plus 100 mg L^{-1} salicylhydroxamic acid; 10 mg L^{-1} boscalid (Bos) from the SDHI class (carboxamides); 10 mg L^{-1} cyprodinil from the anilinopyrimidine (Ani) class; 1 mg L^{-1} fludioxonil from the phenylpyrrole (Phen) class, and 3 mg L^{-1} iprodione from dicarboximide (Dic) class. Plates containing no fungicide were used as experimental controls.

The air samplers were operated simultaneously for 60 min near midday once every 10 d during the experimental periods. To avoid the formation of holes in the medium, the agar layer inside each Petri dish was thick (each dish contained approximately 20 mL of medium). Typical *B. cinerea* sporulating colonies developed on the sampling media, following 6 d of incubation at 20°C in the dark. The number of colonies on each fungicide-containing medium was expressed as the proportion of the total number of colonies on the control plates.

Isolation of the pathogen and definition of the resistance phenotype

Botrytis cinerea was isolated from plants bearing lesions at the stem bases, during the harvest. Infected plant tissues were transferred in separate moist polyethylene bags to the laboratory and stored at room temperature for 24 h. From each sample a single isolate was made onto sterilized potato dextrose agar (PDA) media by slight touching a flamed wire loop onto a freshly sporulating *Botrytis* lesion. In order to identify the phenotype, each isolate was tested for sensitivity response to the fungicides fenhexamid, chlorothalonil, pyraclostrobin, boscalid, cyprodinil, fludioxonil, carbendazim (50% WP, Cequisa SA) or iprodione, with the point inoculation method (Chatzidimopoulos *et al.*, 2013). The same procedure was followed to determine the phenotype of airborne trapped inocula after 7 d of growth in plates containing *B. cinerea* selective medium, in a sample of 100 collected colony forming units (CFUs). The samples were collected at random from control plates and plates amended with fungicides, among different sampling dates.

In situ *pathogenicity assays*

Lettuce plants were grown in 9 cm diam. plastic pots containing peat substrate, in a growth chamber (Sanyo MLR-350HT) with 10 h light period at 18°C, until the 14th true leaf unfolded. Leaf blades (approx. 4 × 4 cm) were excised from the upper half of each plant and immersed in aqueous fungicide suspensions at the same rates used in the field trials. Control leaf blades were immersed in sterilized water. The leaves were allowed to dry for 30–40 min and were then placed in Petri dishes containing water agar (1.5%) with adaxial surfaces uppermost. Three *B. cinerea* isolates, each from the three dominant phenotypes in the glasshouse, were used as inoculum. Mycelium discs (5 mm diam.) were removed from the periphery of 3-d-old colonies grown on PDA, and were aseptically placed upside down over the leaf blades. The Petri dishes with the inoculated leaves were then placed in the growth chamber conditions described above. After 72 h incubation, the mean diameters of the lesion on each leaf blade (minus the 5 mm of mycelial plug) was determined using a measuring rod. Three replicates per isolate/fungicide treatment were made.

Data analyses

Data from the glasshouse trials were analysed by one-way ANOVA and Student-Newman-Keuls test. Tukey's Honestly Significant Difference test was used to assess differences of mean values from the pathogenicity trials. To meet the assumptions of ANOVA, percentage and count values were logarithmically transformed to base 10 where necessary. P-values ≤ 0.05 were considered statistically significant. Statistical analyses were performed using ARM software (Revision 2017.4, Gylling Data Management, Inc.).

RESULTS

Control of disease

In the untreated control lettuce plants, small, hardly visible, brown lesions appeared at the base of the petioles of the bottom leaves, 14 d after transplanting. *Botrytis cinerea* invaded the basal stems *via* the senescent cotyledons or leaf petioles within 3 weeks after transplanting, at the 13 true unfolded leaf growth stage. Early assessments at 1 to 3 weeks after transplanting indicated that when the infection was initiated at an earlier stage of development, the disease progressed more rapidly and eventually the plant stem bases rotted within 10 to 14 d. All the plants which showed early symptoms of infections within this period eventually rotted. The overall disease severity ranged from 0.11 to 0.3 %, and incidence from 2 to 8 %, in both trials at 14 d after transplanting. When the infection was initiated at a later stage, the stem rot progressed very slowly. The disease was more severe in the first trial than the second. By the time of the last assessment at harvest, disease severity and incidence were up to 21.8 % and 55.3 % for trial 1 and up to 14.2 % and 28 % for trial 2 (Tables 1 and 2).

In first trial, two pre-transplanting applications of fludioxonil + cyprodinil reduced disease incidence and severity compared with unsprayed plants (Table 1). Disease control was further improved when one or two more applications were made after transplanting. The AUDPC values for these treatments were very low ranging from 0.66 to 4.19 and a significantly increased number of healthy plants was observed at harvest compared to the untreated control. Chlorothalonil, although providing some control of the disease, was the least effective fungicide. Although disease incidence and severity incidence were reduced, the mean AUDPC values were high, ranging from 120.26 to 267.14 for all treatments, and these were not significantly different from the untreated control. However, the programme with two post-transplanting sprays of chlorothalonil significantly increased the number of healthy plants at harvest compared to the untreated control. With the exception of post-transplanting applications with chlorothalonil, no fungicide residues exceeding the maximum residue level (MRL) defined by the European Community (EC) were detected at harvest. These are 0.01 mg kg^{-1} for chlorothalonil, 15 mg kg^{-1} for fludioxonil and 15 mg kg^{-1} for cyprodinil. For the pre-transplanting applications alone, the residues at harvest were below the adopted analytical reporting limits of quantification (LoQs) of 0.005 mg kg^{-1} (Table 1).

In the second trial, with lower disease pressure, two applications of fungicide at the pre-transplanting stage, with either fludioxonil + cyprodinil or chlorothalonil, decreased disease incidence and severity. One or two more post-transplanting fungicide applications with alternating treatments did not improve disease control (Table 2). All the treatments were of high efficacy, and the disease progress (AUDPC) was significantly reduced compared to the untreated control (Table 2). Although increased numbers of healthy plants were observed from most treatments compared to the untreated control, the differences between treatments were not statistically significant. With the exception of two cases (four-spray programmes), in which the fungicides fenhexamid and fludioxonil + cyprodinil were applied at 27 d before harvest, no fungicide residues were detected. Residues of fenhexamid were 5.95 mg kg^{-1} and of fludioxonil + cyprodinil were 0.43 + 0.74 mg kg^{-1}. However, the residue amounts were much less than the European MRLs, at 40 mg kg^{-1} of fenhexamid, 15 mg kg^{-1} of fludioxonil and 15 mg kg^{-1} of cyprodinil (Table 2).

Fungicide resistant airborne inoculum

Fungicide resistant *B. cinerea* inocula in the air of the lettuce glasshouse were detected at all the sampling dates, for both of the trials. The numbers of trapped CFUs in each of the six fungicide-amended substrates are shown in Figure 1. From the beginning until the end of the experimental periods, iprodione-, pyraclostrobin- and cyprodinil-resistant CFUs were detected at frequencies comparable to the CFUs trapped in control plates. CFUs resistant to fenhexamid (Fen^{R}) were trapped on all sampling dates but at variable frequencies. An increase of the Fen^{R} population was observed during the second trial. However, the total numbers of CFUs trapped in the media from the mid-December to mid-January were generally low. Boscalid resistant (Bos^{R}) inocula were detected at variable frequency during both trials. The Bos^{R} populations reached peaks in the middle of both growing periods. By contrast, with the exception of the late sampling dates during the second trial, fludioxonil resistant (Phen^{MR}) CFUs were rarely trapped (Figure 1).

Approximately 90% of the isolates recovered from the control plates exhibited multiple resistance to fungicides (Figure 2). The three prevalent resistant phenotypes in decreased frequency were: QoI^{R}Bos^{R}Ani^{R}Ben^{HR}Dic^{MR} (phenotype I; 57% frequency), Hyd^{R}QoI^{R}Bos^{R}Ani^{R}Phen^{MR}Ben^{HR}Dic^{MR} (phenotype II; 33%) and wild type (phenotype W; 10%). Only phenotype-II CFUs were detected in media amended with fenhexamid or fludioxonil (Figure 2). In media amended with pyraclostrobin, cyprodinil, iprodione and boscalid, the most

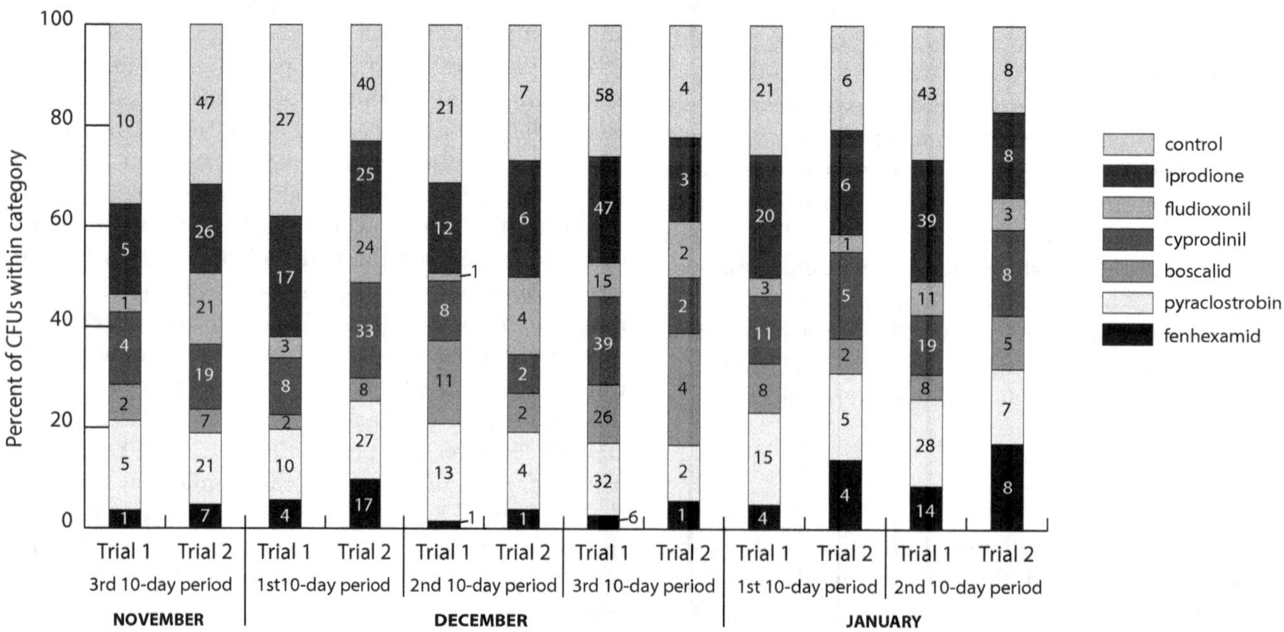

Figure 1. Number of airborne *B. cinerea* conidia trapped in different fungicide amended media during 2012-13 (Trial 1) and 2013-14 (Trial 2) growing periods. The number of CFUs on fungicide enriched selective media are indicated in each bar and expressed as % of the total number of CFUs on all the plates including controls (top bars). The discriminatory concentrations used were: 1 mg L^{-1} fenhexamid, 10 mg L^{-1} pyraclostrobin + 100 mg L^{-1} SHAM, 10 mg L^{-1} boscalid, 10 mg L^{-1} cyprodinil, 1 mg L^{-1} fludioxonil and 3 mg L^{-1} iprodione.

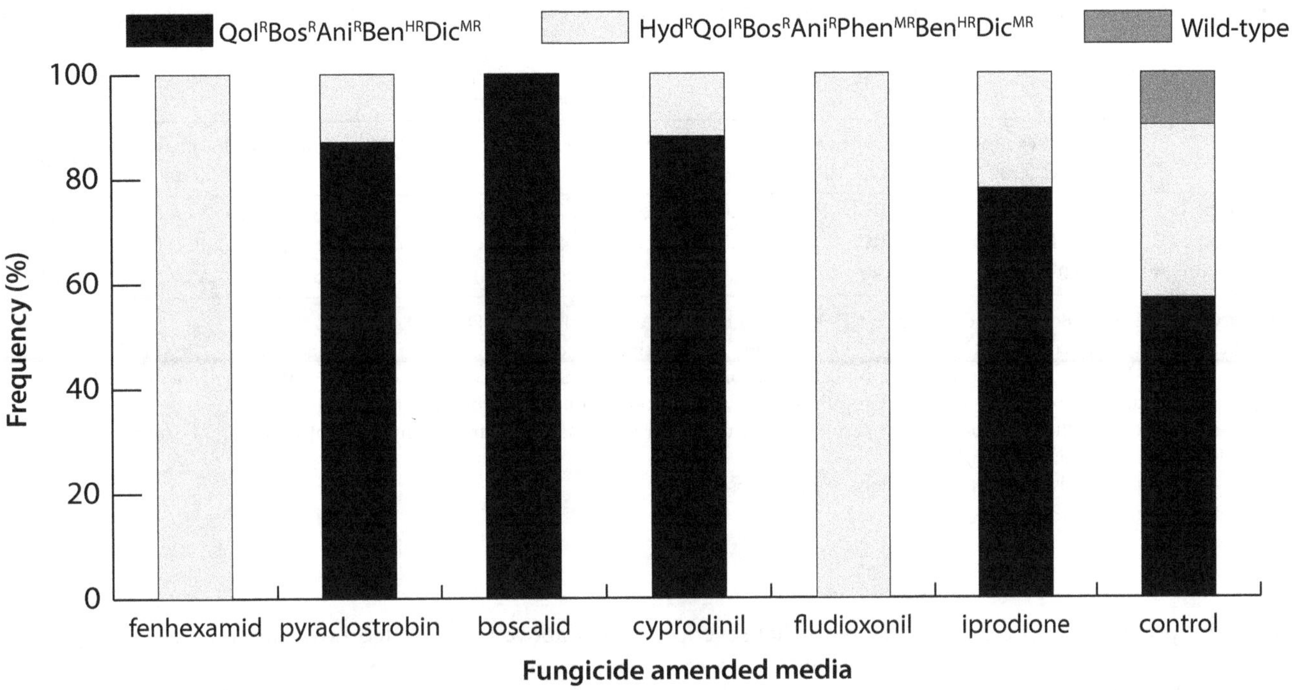

Figure 2. *Botrytis cinerea* phenotypes recovered from fungicide amended selective media. The sensitivity response of each CFU to the discriminatory doses of the fungicides, was examined by the point inoculation method.

prevalent phenotype was type I at frequencies ranging from 78 to 100%. CFUs of the phenotype II were detected at lower frequencies (12 to 22%) on media containing pyraclostrobin, cyprodinil and iprodione (Figure 2). Those three were the only phenotypes detected on all media tested.

Detection of fungicide resistant strains in diseased plants

All the isolates recovered from infected plants were classified in the following three phenotypes: $QoI^R Bos^R Ani^R Ben^{HR} Dic^{MR}$ (phenotype I; 58%), $Hyd^R QoI^R Bos^R Ani^R Phen^{MR} Ben^{HR} Dic^{MR}$ (phenotype II; 24%) and wild-type (phenotype W; 18%). The multiple resistant phenotype II was detected in plants treated at least once after transplanting with either fenhexamid, boscalid + pyraclostrobin or fludioxonil + cyprodinil (Table 2). This phenotype was not detected in plants treated only with chlorothalonil. On the other hand, phenotype I prevailed by 60% in the isolations made from the infected plants in the controls (Table 2). Phenotype-I strains were also most frequently detected in treatments where the QoI fungicide was included. No diseased plants were observed from the spray programmes with the mixture of fludioxonil + cyprodinil in the first trial (Tables 1 and 2).

Pathogenicity of fungicide resistant isolates in situ

Fenhexamid, pyraclostrobin, boscalid and cyprodinil failed to inhibit the development of *B. cinerea* lesions *in vitro* when the isolate used in inoculations was characterized as resistant to the respective fungicide. No differences were observed in lesion size compared to controls (Table 3). In contrast, fludioxonil was more effective, even against the isolate CR-32 which was characterized as moderately resistant to this fungicide. Chlorothalonil gave variable effects against the isolates with multiple resistances to other fungicides (Table 3). The mean lesion sizes ranged from 2.6 to 12.2 mm, although these strains had previously been characterized as sensitive *in vitro* to chlorothalonil.

DISCUSSION

Selection of resistant individuals in fungal populations subjected to selective pressure due to fungicides is an evolutionary mechanism that promotes advantageous genotypes (Walker *et al.*, 2013). In the present study, during a 2-year monitoring schedule in a glasshouse, multiple resistant isolates were detected after two to four fungicide applications per year. In addition, the phenotypic characterization of the isolates obtained in this

Table 3. *In situ* lesion development on lettuce leaves treated with fungicides and inoculated with multiple resistant isolates of *Botrytis cinerea*.

Isolate[a]	Size of lesion (mm)						
	Fenhexamid (1.5 g)[b]	Pyraclostrobin (0.4 g)	Boscalid (0.8 g)	Cyprodinil (0.375 g)	Fludioxonil (0.5 g)	Chlorothalonil (2.5 mL)	Control
C-01 (Type I)	0.0 **Aa** (S)[d]	18.0 **BCb** (R)	27.0 **Cb** (R)	15.4 **Bb** (R)	4.6 **Aa** (S)	3.4 **Aa** (S)	20.0 **BCb**
CR-32 (Type II)	21.0 **BCb**[c] (R)	19.8 **BCb** (R)	24.2 Cb (R)	18.4 **BCb** (R)	10.9 **Ab** (MR)	12.2 **Bb** (S)	23.8 **Cb**
A-56 (Wild type)	3.0 **Aa** (S)	4.6 **Aa** (S)	0.0 **Aa** (S)	5.4 **Aa** (S)	2.6 **Aa** (S)	2.6 **Aa** (S)	15.4 **Ba**

[a] Phenotype: **C-01**=$QoI^R Bos^R Ani^R Ben^{HR} Dic^{MR}$; **CR-32**=$Hyd^R QoI^R Bos^R Ani^R Phen^{MR} Ben^{HR} Dic^{MR}$; **A-56**=wild-type.
[b] Fungicide treatments (concentration per L). The fungicide formulations used were: fenhexamid as Teldor 50 WG (Bayer CropScience); pyraclostrobin as F500 25 EC (Syngenta Ltd); boscalid as 510F 50 WG (BASF SE); cyprodinil as Chorus 50 WG (Syngenta Crop Protection AG); fludioxonil as Geoxe 50 WG (Syngenta Crop Protection AG); chlorothalonil as Daconil 500 SC (Syngenta Ltd).
[c] Numbers followed by the same bold upper-case letters in rows and low-case letters in columns do not differ significantly according to Tukey's HSD post hoc test; $P = 0.05$.
[d] Sensitivity group: R=resistant, MR=moderately resistant, S=sensitive.

study, and previous genetic analyses, showed that several resistance alleles to different fungicide molecules were selected, due to fungicide pressure (Chatzidimopoulos *et al.*, 2013; Chatzidimopoulos *et al.*, 2014a; Chatzidimopoulos *et al.*, 2014b). Furthermore, next generation sequencing data revealed that an underlying MDR mechanism was also present in these strains (Chatzidimopoulos *et al.*, 2016). The presence of such strains in an isolated area after repeated use of fungicides suggests that a stepwise accumulation of resistances occurred over time, and that no pathogen migration took place from neighbouring crops, as has occurred in other cases (Rupp *et al.*, 2016).

Different strategies have been adopted to avoid or reduce the risks of production losses due to fungicide resistance. These have included applications only in pre-formulated or tank fungicide mixes, or in rotations with effective non-cross-resistant fungicides, preferably multi-site inhibitors with low risk and limited numbers of treatments (Brent, 2012). In the present study, despite the existence of multi-resistant airborne inoculum, all the application programs improved disease control and left fungicide residues below the defined European Community MRLs. Few multi-site inhibitors with activity against *B. cinerea* are now available. Use of chlorothalonil, a multi-site inhibitor that is still available, is restricted to certain crops due to deposition of undesirable residues. However, when applied in this study on lettuce in the nursery, satisfactory disease control was achieved without detectable fungicide residues at harvest. Applications with chlorothalonil-based fungicides after transplanting should be avoided, however, since there is then risk of the remaining residues on lettuce being above the accepted LoQ limits. Due to recent decision of the European Standing Committee (SCoPAFF) against the renewal of chlorothalonil registration in EU countries, growers are likely to have (from spring 2020) one fewer vital tool to combat fungicide resistance.

Recent surveys made on lettuce revealed infections by *B. cinerea* strains that are multi-resistant to most available botryticides (Chatzidimopoulos *et al.*, 2013; Weber and Wichura, 2013). By using an air sampler with a selective medium, as proposed by Edwards and Seddon (2001), modified and enriched with appropriate doses of fungicides (Chatzidimopoulos *et al.*, 2014b)2014b, we revealed the prevalence of *B. cinerea* resistant strains in the air of the glasshouse throughout two experimental periods. Most trapped isolates exhibited multiple resistances to specific fungicides. High degrees of resistance to carbendazim (benzimidazole class) and moderate resistance to iprodione (dicarboximide class) were always present, even when neither of these fungicides was included in the spray programmes. Two dominant resistant phenotypes ($QoI^R Bos^R Ani^R Ben^{HR} Dic^{mR}$ and $Hyd^R QoI^R Bos^R Ani^R Phen^{MR} Ben^{HR} Dic^{MR}$) were detected in the air of the experimental glasshouse and isolated from infected plants. These phenotypes constituted 84% of the total *Botrytis* population in the glasshouse, and had also been isolated from diseased plants originating from the same site in previous years (Chatzidimopoulos *et al.*, 2013).

Recent studies have shown that in the absence of fungicide selection pressure, resistance to fenhexamid (Billard *et al.*, 2012), cyprodinil (Bardas *et al.*, 2008) or boscalid (Veloukas *et al.*, 2014) may entail a fitness costs in *B. cinerea*. However, Rupp *et al.* (2016) concluded that multiple resistant strains are likely to possess high fitness in the field, and that they are essentially immune to sprays with any of the current botryticides. In the present study, significant decreases of the airborne resistant populations were observed at the beginning of each growing period. Moreover, no strains highly resistant to fludioxonil have been detected in the field, although this phenylpyrrole fungicide has been used for over two decades (Baroffio *et al.*, 2003; Chatzidimopoulos *et al.*, 2013). Similarly, Fernández-Ortuño *et al.* (2012) reported the high efficacy of the mixture boscalid + pyraclostrobin against *B. cinerea* in strawberry, even though a resistant population to SDHIs and QoIs was present at high frequency. This information may explain the good efficacy of the fungicides in our field trials. The high frequency of resistant strains within the pathogen population in the atmosphere of the glasshouse may be the consequence of good efficacy of the fungicides against the wild-type *B. cinerea* strains. The present assays have shown that under strong disease pressure, (as for the mycelial plugs used in the *in situ* assays) fungicides lose their efficacy against resistant strains. Under conditions of low disease pressure, however, as for airborne spores in the second trial, the fungicides may retain their efficacy. When conidia of the selected strains were used to check the *in situ* efficacy of the fungicides, lesions were not formed in most cases (unpublished data). The phytoalexin lettucenin A in young plants may act preventively on infections triggered by spores (Bennett *et al.*, 1994).

All fungicide applications only at the host nursery stage reduced disease incidence and severity, and increased the number of healthy plants at harvest in a trial with high disease pressure, and to a lesser extent (not significantly) under lower disease pressure. According to Sowley *et al.* (2010), the infections could be initiated at nursery stages and spread systemically throughout the plants without the development of visible symptoms. This disease progress can be arrested by fungicide protection of the young seedlings at the nursery or at early transplanting stages. The pre-transplanting applications provided improved disease control when these were combined with one or two post-transplanting applications. However, the additional applications did not significantly improve the number of healthy plants at harvest. In the second trial, under low disease pressure, the number of healthy plants in control plots was similar with the number of the healthy plants in the treated plots. The initial symptoms of infection were observed at 3 weeks after transplanting, but the disease progressed very slowly because of unfavourable climatic conditions in the glass house. Furthermore, recent studies have shown that lettuce crops become less susceptible to infections as they age (develop thick bases and resistant leaves), and some cultivars, such as iceberg and romaine types, are more prone to infections than others (Shim *et al.*, 2014).

The mixture of fludioxonil + cyprodinil was the most effective of the different fungicides tested here against lettuce bottom rot at both rates of application. The high efficacy of this mixture in lettuce crops was also reported by Matheron and Porchas (2008). Kilani and Fillinger (2016) also reported that high resistance to fludioxonil does not exist among *B. cinerea* populations worldwide, and this status is not expected to change in the future. In addition, our results have shown that fludioxonil provided better control of *B. cinerea* resistant isolates *in situ*, in comparison with other fungicides. Similar results were observed by Rupp *et al.* (2016) in *in planta* tomato assays. The inclusion of this compound in the fungicide programmes could explain the effective disease control in field trials.

In conclusion, it was shown that appropriate selection and timing of fungicide sprays is fundamental for control of bottom rot of lettuce grown in hydroponic systems. In a trial with high disease pressure, two fungicide applications during plant growth in the nursery combined with one further application at transplanting, provided the best disease control without leaving detectable fungicide residues at harvest. However, under less disease pressure, the two applications at the nursery stage were enough to protect the plants from bottom rot caused by *B. cinerea*. Fungicide selection was made according to the risk for resistance development. Fludioxonil was the most effective compound against the multiple resistant strains of *B. cinerea*. Primary applications were evaluated, with a multi-site inhibitor (chlorothalonil) or mixtures of non-cross resistant compounds (e.g. fludioxonil + cyprodinil) in alternation with other effective compounds with different modes of action (such as fenhexamid or boscalid). These treatments gave satisfactory disease control, despite the predominance of multi-fungicide resistant populations of *B. cinerea*.

ACKNOWLEDGEMENTS

The authors thank Crocus Flora S. A. for providing the necessary facilities and help in carrying out the glasshouse experiments, and the Cadmion Analytical Laboratory for fungicide residue analyses. Dr K. J. Brent read the manuscript of this paper and provided considerable review suggestions.

LITERATURE CITED

Amiri A., Heath S.M., Peres N.A., 2013. Phenotypic characterization of multifungicide resistance in *Botrytis cinerea* isolates from strawberry fields in Florida. *Plant Disease* 97: 393–401.

Bardas G.A., Myresiotis C.K., Karaoglanidis G.S., 2008. Stability and fitness of anilinopyrimidine-resistant strains of *Botrytis cinerea*. *Phytopathology* 98: 443–450.

Baroffio C.A., Siegfried W., Hilber U.W., 2003. Long-term monitoring for resistance of *Botryotinia fuckeliana* to anilinopyrimidine, phenylpyrrole, and hydroxyanilide fungicides in Switzerland. *Plant Disease* 87: 662–666.

Bennett M.H., Gallagher M.D.S., Bestwick C.S., Rossiter J.T., Mansfield J.W., 1994. The phytoalexin response of lettuce to challenge by *Botrytis cinerea*, *Bremia lactucae* and *Pseudomonas syringae* pv. *phaseolicola*. *Physiological and Molecular Plant Pathology* 44: 321–333.

Billard A., Fillinger S., Leroux P., Lachaise H., Beffa R., Debieu D., 2012. Strong resistance to the fungicide fenhexamid entails a fitness cost in *Botrytis cinerea*, as shown by comparisons of isogenic strains. *Pest Management Science* 68: 684–691.

Brent K.J., 2012. Historical perspectives of fungicide resistance. In: *Fungicide resistance in crop protection: Risk and management* (T.S. Thind, ed.) CAB International, Wallingford, UK, 3–18.

Chatzidimopoulos M., Ganopoulos I., Madesis P., Vellios E., Tsaftaris A., Pappas A.C., 2014a. High-resolution melting (HRM) analysis for rapid detection and characterization of *Botrytis cinerea* phenotypes resistant to fenhexamid and boscalid. *Plant Pathology* 63: 1336–1343.

Chatzidimopoulos M., Ganopoulos I., Vellios E., Madesis P., Tsaftaris A., Pappas A.C., 2014b. Development of a two-step high-resolution melting (HRM) analysis for screening sequence variants associated with resistance to the QoIs, benzimidazoles and dicarboximides in airborne inoculum of *Botrytis cinerea*. *FEMS Microbiology Letters* 360: 126–131.

Chatzidimopoulos M., Papaevaggelou D., Pappas A.C., 2013. Detection and characterization of fungicide resistant phenotypes of *Botrytis cinerea* in lettuce crops in Greece. *European Journal of Plant Pathology* 137: 363–376.

Chatzidimopoulos M., Psomopoulos F., Malandrakis E.E., Ganopoulos I., Madesis P., Vellios E.K., Drogoudi P., 2016. Comparative genomics of *Botrytis cinerea* strains with differential multi-drug resistance. *Frontiers in Plant Science* 7: 554.

Dik A.J., Wubben J.P., 2007. Epidemiology of *Botrytis cinerea* diseases in greenhouses. In: *Botrytis: Biology, pathology and control* (Y. Elad, B. Williamson, P. Tudzynski, N. Delen, eds.), Springer, Dordrecht, The Netherlands, 319–333.

Edwards S.G., Seddon B., 2001. Selective media for the specific isolation and enumeration of *Botrytis cinerea* conidia. *Letters in Applied Microbiology* 32: 63–66.

Fernández-Ortuño D., Chen F., Schnabel G., 2012. Resistance to pyraclostrobin and boscalid in *Botrytis cinerea* isolates from strawberry fields in the Carolinas. *Plant Disease* 96: 1198–1203.

Fernández-Ortuño D., Grabke A., Li X., Schnabel G., 2014. Independent emergence of resistance to seven chemical classes of fungicides in *Botrytis cinerea*. *Phytopathology* 105: 424–432.

Hahn M., 2014. The rising threat of fungicide resistance in plant pathogenic fungi: *Botrytis* as a case study. *Journal of Chemical Biology* 7: 133–141.

Kilani J., Fillinger S., 2016. Phenylpyrroles: 30 years, two molecules and (nearly) no resistance. *Frontiers in Microbiology* 7: 2014.

Kretschmer M., Leroch M., Mosbach A., Walker A.S., Fillinger S., … Hahn M., 2009. Fungicide-driven evolution and molecular basis of multidrug resistance in field populations of the grey mould fungus *Botrytis cinerea*. *Plos Pathogens* 5: e1000696.

Leroch M., Plesken C., Weber R.W.S., Kauff F., Scalliet G., Hahn M., 2013. Gray mold populations in German strawberry fields are resistant to multiple fungicides and dominated by a novel clade closely related to *Botrytis cinerea*. *Applied and Environmental Microbiology* 79: 159–167.

Leroux P., 2007. Chemical control of *Botrytis* and its resistance to chemical fungicides. In: *Botrytis: Biology, pathology and control* (Y. Elad, B. Williamson, P. Tudzynski, N. Delen, eds.), Springer, Dordrecht, The Netherlands, 195–222.

Matheron M.E., Porchas M., 2008. Effect of fungicides and lettuce cultivar on severity of *Botrytis* gray mold: 2007 Study. *Vegetable Report* P-152: 20–22.

Morgan L., 1999. *Hydroponic lettuce production: a comprehensive, practical and scientific guide to commercial hydroponic lettuce production*. Casper Publications, Narrabeen, Australia, 111 pp.

Morgan L., 2012. *Hydroponic salad crop production*. Suntec publications, Tokomaru, New Zealand, 246 pp.

Resh H.M., 2012. *Hydroponic food production: A definitive guidebook for the advanced home gardener and the commercial hydroponic grower*. 7th ed. CRC Press, Taylor & Francis Group, Boca Raton, USA, 560 pp.

Rupp S., Weber R.W.S., Rieger D., Detzel P., Hahn M., 2016. Spread of *Botrytis cinerea* strains with multiple fungicide resistance in German horticulture. *Frontiers in Microbiology* 7: 2075.

Shim C.K., Kim M.J., Kim Y.K., Jee H.J., 2014. Evaluation of lettuce germplasm resistance to gray mold disease for organic cultivations. *The Plant Pathology Journal* 30: 90–95.

Skovgaard M., Encinas S.R., Jensen O.C., Andersen J.H., Condarco G., Jørs E., 2017. Pesticide residues in commercial lettuce, onion, and potato samples from Bolivia - A threat to public fealth? *Environmental health insights* 11: 1178630217704194.

Sowley E.N.K., Dewey F.M., Shaw M.W., 2010. Persistent, symptomless, systemic, and seed-borne infection of lettuce by *Botrytis cinerea. European Journal of Plant Pathology* 126: 61–71.

Veloukas T., Kalogeropoulou P., Markoglou A.N., Karaoglanidis G.S., 2014. Fitness and competitive ability of *Botrytis cinerea* field isolates with dual resistance to SDHI and QoI fungicides, associated with several *sdh*B and the *cyt*b G143A mutations. *Phytopathology* 104: 347–356.

Walker A.S., Micoud A., Remuson F., Grosman J., Gredt M., Leroux P., 2013. French vineyards provide information that opens ways for effective resistance management of *Botrytis cinerea* (grey mould). *Pest Management Science* 69: 667–678.

Farha W., Abd El-Aty A.M., Rahman M.M., Jeong J.H., Shin H.C., ... Shim J.H., 2018. Analytical approach, dissipation pattern and risk assessment of pesticide residue in green leafy vegetables: A comprehensive review. *Biomedical Chromatography* 32: e4134.

Weber R.W.S., 2011. Resistance of *Botrytis cinerea* to multiple fungicides in northern German small-fruit production. *Plant Disease* 95: 1263–1269.

Weber R.W.S., Wichura A., 2013. Fungicide resistance of *Botrytis cinerea* on lettuce in northern Germany. *Journal of Plant Diseases and Protection* 120: 115–121.

8

Genetic and agronomic characterization of chickpea landraces for resistance to *Fusarium oxysporum* f. sp. *ciceris*

MARCO FIERRO[1], DAVIDE PALMIERI[1], FILIPPO DE CURTIS[1], DOMENICO VITULLO[1], JOSEFA RUBIO[3], JUAN GIL[2], GIUSEPPE LIMA[1,*], TERESA MILLAN[2]

[1] *Dipartimento Agricoltura, Ambiente e Alimenti, Università del Molise, via De Sanctis, 86100 Campobasso, Italy*
[2] *Universidad de Córdoba, Departamento de Genética-ETSIAM, Campus de Rabanales, 14071 Córdoba, Spain*
[3] *Instituto de Investigación y Formación Agraria y Pesquera de Andalucía (IFAPA), Area de Genómica y Biotecnología, Alameda del Obispo, 3092 Córdoba, Spain*
*Corresponding author: lima@unimol.it

Competing Interests: The Author(s) declare(s) no conflict of interest.

Editor: Sabine Banniza, University of Saskatchewan, Canada.

Summary. Chickpea, with annual grain production of approx. 15 million tons, is the third largest world pulse crop, which is an important source of protein for human and animal diets. In Italy, chickpea production is mainly based on landraces cultivated on small farms. However, the attention that consumers give to local products stimulates farmers to extend the use of chickpea landraces by reintroducing them in crop rotations. Production of chickpea using landraces can be adversely affected by agronomic factors and, particularly, plant diseases such as Fusarium wilt, caused by the widespread fungal pathogen *Fusarium oxysporum* f. sp. *ciceris* (*Foc*). Therefore, studies of agronomic adaptation of landraces and on disease resistance are needed. The most important agronomic traits and levels of resistance to *Foc* were evaluated for 18 chickpea landraces collected from Central Italy. These landraces were also characterized for their genetic traits in comparison to some of the main Spanish cultivars, and to two reference cultivars with worldwide distribution. Molecular characterization showed variability in genetic and phenotypic traits among the Italian landraces. In particular, landrace 203 locally known as "Longano" was resistant to *Foc* and could be considered in chickpea breeding programmes. Comparative analyses based on molecular markers showed, with some exceptions, that the Italian landraces are genetically different compared to the main Spanish cultivars analyzed in this study.

Keywords. Fusarium wilt, chickpea biodiversity, germplasm conservation, molecular markers, disease resistance improvement.

INTRODUCTION

Chickpea (*Cicer arietinum* L.) is the only cultivated species of the genus *Cicer*, and originated in East Turkey (van der Maesen, 1987). It is a self-pollinated diploid annual plant species (2n = 2x = 16) with a small genome (740

Mbp) (Arumuganathan and Earle, 1991). Plant breeders have distinguished two chickpea types: desi and kabuli. The seeds of the desi type are normally small, angular and dark, and are produced by plants with purple flowers, while seeds of the kabuli type are large, round, white to cream, and are produced by plants with white flowers (Taylor and Ford, 2007).

Chickpea, is the third largest pulse crop in the world after dry bean and pea, with annual production of approx. 15 million tons. It is mainly cultivated on the Indian subcontinent. In Europe, the major chickpea producing country is Spain, followed by Italy, Portugal and the Balkan Countries. In Italy, chickpea is the third largest cultivated dry legume, grown on 20,025 ha, with grain production of 33,541 t (FAO, 2017). The crop is mainly grown on small farms, and landraces are particularly appreciated for local markets. The landrace varieties usually take their names from the location where they have been traditionally cultivated (Negri, 2003).

Although Italy is the second largest European producer of chickpea, cultivation is still restricted because of adverse environmental and/or agronomic factors responsible for variability in yields (Rossini, 2008). Nationally produced and imported chickpeas are mainly for human food use. The introduction of chickpea as animal feed could represent a viable alternative to soybean, enlarging its use in crop rotations, especially in arid and non-irrigated areas (Crinò and Saccardo, 2008). The use of landraces could have an important economic role, as well as social and cultural significance. Chickpea cultivation is sometimes connected to ethnic preferences of particular linguistic minorities, and also enhances plant biodiversity preservation programmes implemented by germplasm banks aimed at avoiding loss of crop plant ecotypes (Laghetti *et al.*, 2011). Cultivated landraces can be affected by a wide range of pathogens, in particular *Fusarium oxysporum* f. sp. *ciceris* (*Foc*), *Ascochyta rabiei* (Pande *et al.*, 2005; Jiménez-Díaz *et al.*, 2015), and other emerging pathogens (De Curtis *et al.*, 2014), which can cause significant yield losses.

Selection of landraces is mainly based on commercial characteristics of the grain while disease resistance is not often taken into account (Zaccardelli *et al.*, 2012).

Fusarium oxysporum f. sp. *ciceris* is the causal agent of chickpea Fusarium wilt with worldwide distribution. This Ascomycete fungus has been characterized into eight races (0, 1A, 1B/C, 2, 3, 4, 5 and 6), which are differentiated by genetic compatibility with the host and by geographical distribution. In addition to the pathogenic variability of the fungus, two distinct types, referred to as yellowing and wilting syndromes, have been distinguished on the basis of symptoms on infected plants (Trapero-Casas and Jiménez-Díaz, 1985; del Mar Jiménez-Gasco *et al.*, 2004). The pathogen is soil-borne, and can survive in the soil 6 or more years, even in the absence of the host plants, because of its durable survival structures. The infection of host plants can occur at different phenological stages with greatest incidence during the pod-forming phase, mainly when crops are subjected to water stress and sudden temperature increases. Infections occurring during vegetative and reproductive stages can result in complete yield losses. For these reasons, disease management is critical for chickpea production (Arunodhayam *et al.*, 2014).

Chickpea resistance towards the different races of *Foc* is of the "gene-for-gene" type, and different molecular markers associated with resistance are available. A gene cluster, located on Linkage Group (LG) 2 of the chickpea genetic map, confers resistance against races 0, 1, 2, 3, 4, 5 (Sharma and Muehlbauer, 2007). A second gene conferring resistance against race 0 is located on LG5 (Halila *et al.*, 2008). Molecular markers tightly linked to *Foc* resistance genes are useful tools for characterisation and genotype selection. The microsatellite TA59, located on LG2, is the marker most closely associated with resistance to *Foc* race 5 (*Foc*5) (Jendoubi *et al.*, 2017).

Flowering time is another agronomic trait defining adaptation and included in this study. Classical genetic analyses and conventional mapping studies have resulted in the identification of Quantitative Trait Loci (QTLs) on different LGs showing that genes governing this trait are distributed throughout the chickpea genome (Mallikarjuna *et al.*, 2017). Two conserved major QTLs have been identified, QTL_{DF3} on LG3 (Cobos *et al.*, 2009; Aryamanesh *et al.*, 2010) and QTL_{DF1} on LG4 (Cobos *et al.*, 2007; Varshney *et al.*, 2014).

In the present study, we focused on race *Foc*5, which is important in the Mediterranean basin (Jiménez-Díaz *et al.*, 2015). We characterised 18 landraces of chickpea from Central Italy for resistance to *Foc*5 and for some agronomic traits. Molecular markers, distributed along the chickpea genome, were used to study genetic variability among the landraces. Flowering time, an agronomic trait defining ecotype adaptation, was also investigated.

MATERIALS AND METHODS

Plant material

Eighteen Italian chickpea landraces previously collected by *Agenzia Regionale per lo Sviluppo Agricolo*

Rurale e della Pesca, Campobasso, Italy, during its institutional activity aimed at collecting and preserving traditional crops from different regional locations for a germplasm bank, were examined in the present study (Table 1). All landraces were the kabuli type, except accession number 184 which was the desi type. Forty-eight Spanish cultivars and two undomesticated *Cicer* species used in a previous study by Castro *et al.* (2010a), were also included as reference cultivars for comparative analyses, using the same molecular markers adopted for the Italian landraces.

Morphological and agronomic assessments

For each landrace, ten seeds were sown in each of three replicate plots in an experimental field at *Instituto de Investigación y Formación Agraria y Pesquera de Andalucía*, Córdoba, Spain, on January 2015. The following morphological traits were recorded: percentage of emerged plants 15 d after sowing, flowering time (assessed as 50% full opened flowers), growth habit and seed size (determined as 100-seed weights). Phenotypic evaluation of growth habit was assessed on three replicates, each of ten adult plants, for each landrace seeded in the field, based on the scale: 1 = semi erect; 2 = erect (Ali *et al.*, 2015).

Evaluation of resistance to Fusarium oxysporum *f. sp.* ciceris *Race 5*

The landraces were evaluated for wilt reaction to race 5 of *Foc* under controlled condition. The pathogenicity test was conducted in a growth chamber (daily cycle of 12 h light at 25 ± 2°C and 12 h dark at 22 ± 2°C). Lines ILC3279 and WR315 from the International Centre for Agricultural Research in the Dry Areas, which are, respectively, susceptible and resistant to *Foc*5, were included as experimental controls. Ten seeds of each landrace and control line were sown into plastic trays (60×40×10 cm - five lines per tray) that were filled with perlite. Three replicates were sown of each seedline. The trays were irrigated with nutrient solution. Conidia of *Foc*5 were obtained by growing the fungus in potato dextrose broth at 25°C and 100 rpm for at least 7 d. After incubation, fungal mycelium was removed, conidia were collected by centrifugation and their concentration adjusted to 1×10^6 conidia mL^{-1}. When plants were approx. 9 cm height they were removed from perlite and their roots were cut to approx. 5 cm lengths, and the plants were then dipped in conidial suspension for 5 min. The plants were then replanted in the same trays from which they were previously removed. Disease incidence, as percentage of dead plants, was recorded, commencing with appearance of the first symptoms on the susceptible

Table 1. Geographical origins and agronomic traits of Italian chickpea landraces assessed in this study.

Landraces	Town/Province[a]	Geographical coordinates	Type[b]	Germination (%)	Growth habit[c]	100 seeds weight (g)	Flowering time (days)
62	Cercemaggiore/CB	41° 28′ N/14° 43′ E	K	90	1	27.1	87
64	Cercemaggiore/CB	41° 28′ N/14° 43 ′E	K	80	1	36.7	87
73	Salcito/CB	41° 45′ N/14° 31′ E	K	90	1	37.3	86
76	S. Elia a Pianisi/CB	41° 37′ N/14° 53′ E	K	100	1	38.4	87
83	Casacalenda/CB	41° 44′ N/14° 51′ E	K	90	1	36.2	88
97	S. Angelo del Pesco/IS	41° 53′ N/14° 15′ E	K	100	1	35.3	86
99	Venafro/IS	41° 29′ N/14° 02′ E	K	90	1	31.7	84
111	Ripabottoni/CB	41° 41′ N/14° 49′ E	K	100	1	52.4	84
125	Morrone del Sannio/CB	41° 43′ N/14° 47′ E	K	60	1	48.6	84
147	Riccia/CB	41° 29′ N/14° 50′ E	K	80	1	42.5	84
148	Filignano/IS	41° 32′ N/14° 03′ E	K	100	1	30.5	84
160	Miranda/IS	41° 39′ N/14° 15′ E	K	90	1	35.8	87
184	Cercemaggiore/CB	41° 28′ N/14° 43′ E	D	100	1	21.9	87
203	Longano/IS	41° 31′ N/14° 15′ E	K	90	1	47.9	67
228	Riccia/CB	41° 29′ N/14° 50′ E	K	100	1	35.4	84
237	Montagano/CB	41° 39′ N/14° 40′ E	K	100	1	34.6	79
241	Riccia/CB	41° 29′ N/14° 50′ E	K	100	1	31.2	83
245	Capracotta/IS	41° 50′ N/14° 16′ E	K	90	2	27.2	87

[a] CB= Campobasso (IT), IS= Isernia (IT); [b] K=Kabuli, D=Desi; [c] 1= semi-erect, 2 = erect.

control ILC3279. Disease severity on each plant of all landraces was assessed each week for 3 weeks.

Foc symptoms and pathogen resistance were evaluated by using the following empirical disease scale: 0–10% of plants wilting = high resistance (R), 11–89% of plants wilting = intermediate resistance (I), >90% of plants wilting = high susceptibility (S) (Sharma *et al.*, 2005).

Molecular marker analyses

DNA extraction was carried out on young leaflets from five different plants of each landrace. Approximately 0.1 g of the mixed tissues was frozen in liquid nitrogen and stored at -80°C. DNA was isolated using the Plant DNAzol® Reagent (Invitrogen™). DNA was quantified by Nanodrop and used in PCR reactions to amplify the 12 microsatellite markers listed in Table 2. These molecular markers have been selected for their distribution across different linkage groups of the chickpea genome (Winter *et al.*, 2000; Millan *et al.*, 2010). Among the used markers, nine (those marked with TA prefix) were reported by Winter et al. (2000), two (CaGM14822 and CaGM07922) were reported by CAGM (http://cegresources.icrisat.org/CicArMiSatDB/, and one (H2I20) was reported by Lichtenzveig et al. (2005).

In the present study a high number of markers for LG2 (TA27, TA59, CaGM07922) was included, to target the resistance gene associated to *Foc*5 (Castro *et al.*, 2010b). The microsatellite H2I20 located on LG5 has been associated with a gene conferring resistance to *Foc*0 (Jendoubi *et al.*, 2016). The markers CaGM14822 associated with QTL_{DF1} on LG4 and TA142 linked to QTL_{DF3} on LG3 were used for their associations with flowering time (Ali, 2015).

Microsatellite alleles were visualized by electrophoresis in 2.5% (w/v) agarose, and polyacrylamide (10%, C2, 67%) gels, or with capillary electrophoresis using an automatic capillary sequencer (ABI 3130 Genetic Analyzer Applied Biosystems /HITACHI, Madrid, Spain) at the Central Research Support Service, University of Córdoba, Spain. Data from Fragment Analysis were analyzed using the GeneMapper and the Peak Studio V2.2 software packages (McCafferty *et al.*, 2012).

Statistical analyses

Assessment of Fusarium wilt resistance

In the pathogenicity tests, the number of plants showing; i) no symptoms (healthy), ii) light symptoms (yellowing and/or loss of leaf turgidity), iii) heavy symptoms (withering), or iv) dead plant, were periodically assessed. Disease symptom data were used to calculate the Area Under the Disease Progress Curve (AUDPC) with the following equation:

$$AUDPC = \sum_{i=1}^{n-1} \left(\frac{Y_i + Y_{i+1}}{2}\right)(t_{i+1} - t_i)$$

where Y_i is Fusarium wilt severity at the ith observation, t_i is time (d) at the ith observation, and n is the total number of observations (Campbell and Madden, 1990).

The AUDPC data from the different assessments were subjected to ANOVA using the SPSS statistics software v.25. Means were separated by Tukey's tests.

Analyses with molecular markers

Allele frequencies of data obtained in the diversity analyses were calculated and used to determine; i) size range, ii) number of alleles, and iii) Polymorphism Information Content (PIC) of each marker (Shete *et al.*, 2000). The alleles were scored as present (1) or absent (0) to create a binary data matrix. This matrix was used to calculate the degree of genetic similarity between all pairwise combinations, using the Dice coefficient of similarity. Clustering of the genotypes was determined using the Unweighted Pair Group Method with Arithmetic Mean (UPGMA). Statistical analyses was performed using the NTSYS-pc 2.02j software.

RESULTS

Agronomic traits

In the field assessments, seed germination was greater than 80% for all the chickpea lines except accession 125, for which 60% of seeds germinated. Only accession 245 had an erect growth habit (value 2), whereas the other landraces had semi-erect growth habits (value 1). Landraces 111, 125, 147 and 203 had large seed sizes (100-seed weights from 42.5 to 52.4 g) while landraces 62, 245 and 184 had smaller seed sizes (100-seed weights, respectively, 27.1, 27.2, and 21.9 g). Flowering began 83 d after sowing in all landraces except for landraces 237 and 203, which flowered at, respectively, 79 and 67 d (Table 1).

Evaluation of Fusarium wilt resistance

Data collected from the pathogenicity tests were used to calculate the AUDPC index, which allows disease progression to be compared, and reveals the presence of susceptible and resistant phenotypes. Resistance in control genotype WR315 was confirmed by the absence of *Foc* wilt symptoms (AUDPC = 0). In contrast, the *Foc* susceptible control ILC3279 showed wilt symptoms at the first assessment date, and had a final average AUDPC = 14. Among all the tested landraces, accession 203 showed complete resistance to the pathogen, and disease symptoms were absent at all the assessment dates. All the other landraces developed symptoms and were susceptible to the pathogen, with AUDPC values from 10 to 15 (Figure 1).

Molecular marker analyses

The analyses were performed using twelve microsatellite markers (Table 2). All markers revealed high levels of polymorphisms, displaying a total of 100 different alleles with fragment sizes ranging from 150 to 350 bp. The number of alleles per locus varied from two to 15, with an average value of 8.33. STMS TA142 and TA78 amplified, respectively, the minimum (two) and maximum (15) number of alleles (Table 2). Eighteen out of the 100 alleles detected in the chickpea landraces were classified as 'rare' because of their low frequency (<0.03), 69 as 'common' (0.03-0.20) and 13 were considered the 'most frequent' alleles (>0.20). Only common alleles were detected at all the 12 STMS loci studied. Rare alleles per locus ranged from one to five in TA27, TA59, TA11, TA14, TA78 and TA144. The number of common alleles per locus ranged from one (CaGM07922 and TA142) to 12 (TA11 and TA78). For the most frequent alleles, zero, one or two such alleles were detected in the majority of the STMS loci, except for TA135 in which three alleles were detected (Table 2). Based on PIC values obtained, most STMS, except for CaGM07922, TA135, TA142, CaGM14822 and H2I20, were considered informative markers (PIC >0.63). The most polymorphic marker was TA78 with a PIC value of 0.88 and 15 alleles (Table 2).

All landraces were genotyped with two markers previously associated with flowering time (CaGM14822 and TA142). CaGM14822 had three alleles of 300, 320 and 350 bp, and TA142 two alleles of 150 and 160 bp. All these alleles showed clear association with early/late flowering time. The control line WR315 (early flowering) had alleles 380 and 133 for both markers, whereas the second control line ILC3279 (late flowering) had alleles 350 and 144 for both markers (Table 3).

For analyses of the *Foc*5 resistance genes located on LG2, the markers TA27, TA59 and CaGM07922 were considered. The resistant landrace 203 showed a 230/233 bp allele for TA27, a 225 bp allele for TA59 and a 350 bp

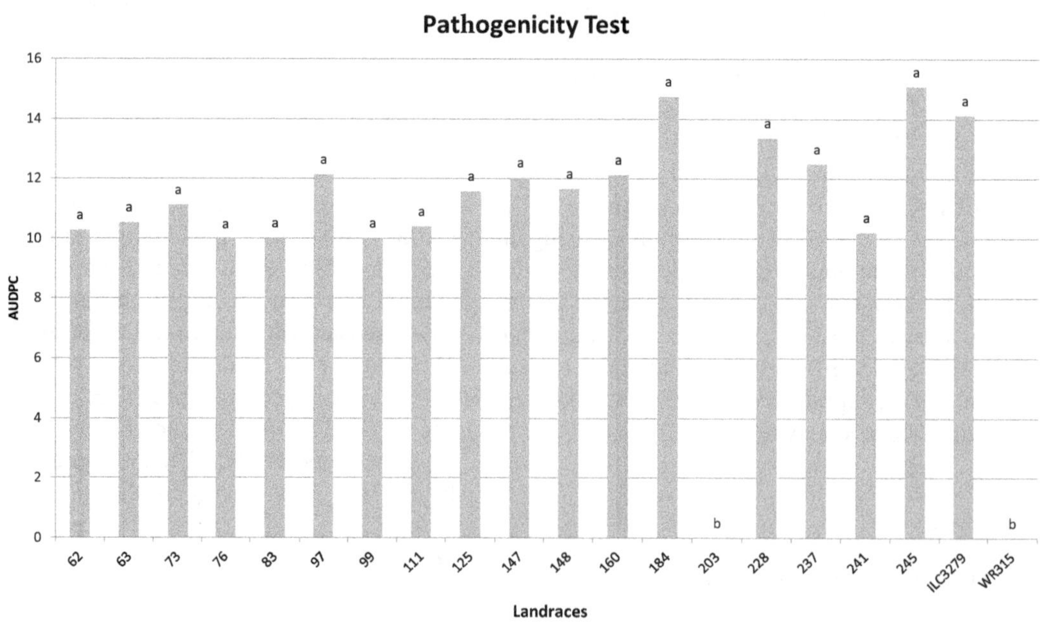

Figure 1. Average Area Under the Disease Progress Curve (AUDPC) obtained from pathogenicity tests carried out in controlled conditions on Italian chickpea landraces inoculated with *Fusarium oxysporum* f. sp. *ciceris* race 5 (*Foc*5). *Foc*5-susceptible ILC3279 and *Foc*5-resistant WR315 were included as controls. Accessions with letters in common above the columns are not significantly different ($P = 0.05$), according to Tukey's Test.

Table 2. Size ranges, numbers and frequencies of alleles and Polymorphism Information Content (PIC) observed in 18 Italian chickpea landraces, studied with 12 microsatellite markers.

Marker	Linkage Group	Size range (bp)	N° of alleles	Rare alleles (<0.03)	Common alleles (0.03-0.2)	Most frequent alleles (> 0.2)	PIC
TA113	1	169-217	11	0	10	1	0.81
TA27	2	218-248	11	4	7	0	0.84
TA59	2	222-273	12	5	6	1	0.80
CaGM07922	2	300-350	3	0	1	2	0.34
TA142	3	150-160	2	0	1	1	0.21
TA135	3	187-199	5	0	2	3	0.59
CaGM14822	4	300-350	3	0	2	1	0.35
H2I20	5	180-230	5	0	3	2	0.39
TA11	5	220-262	14	2	12	0	0.83
TA14	6	242-278	11	3	8	0	0.82
TA78	7	191-236	15	3	12	0	0.88
TA144	8	230-254	8	1	5	2	0.73
Total			100	18	69	13	
Mean			8.33	1.5	5.75	1.08	0.63

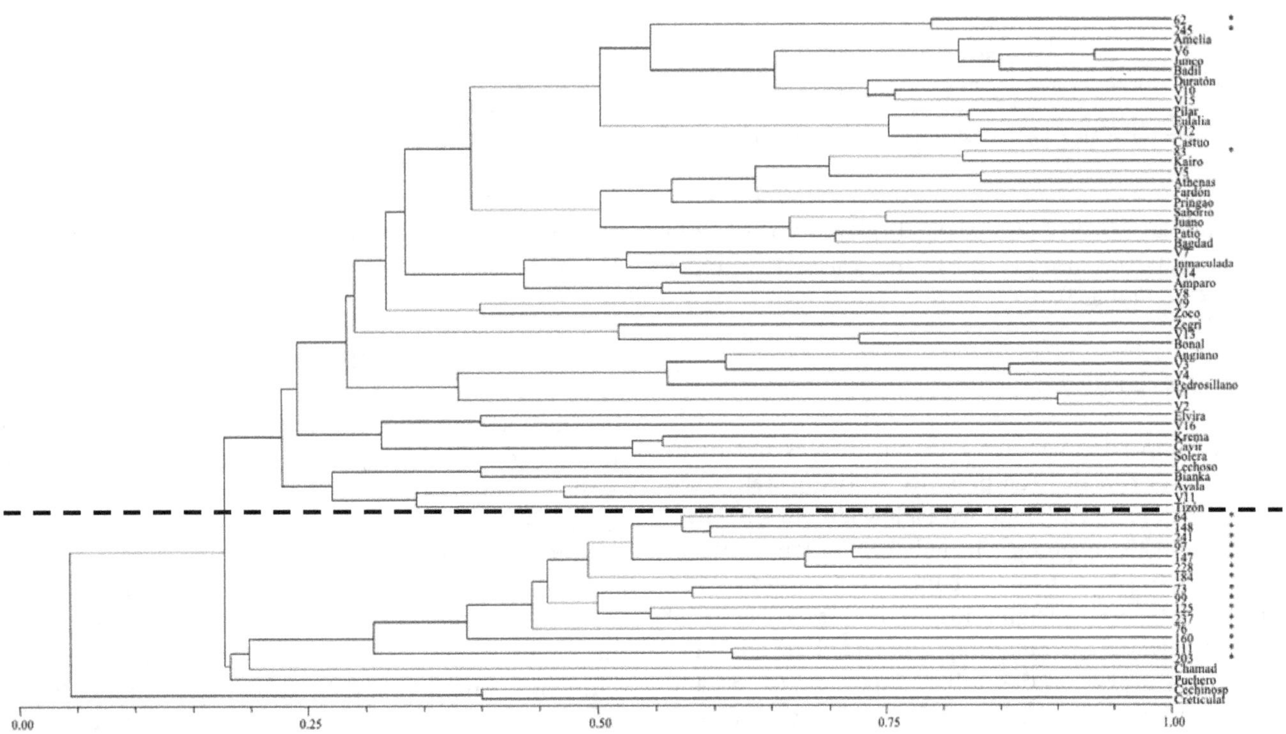

Figure 2. UPGMA dendrogram obtained from cluster analyses of 18 Italian chickpea landraces (marked with asterisks) and 48 Spanish cultivars, based on Dice coefficients of similarity, using 12 microsatellite markers selected for their Polymorphism Information Content (PIC) values. The dotted line separates two Subgroups, which mainly include Italian (Sub-group 1) and Spanish (Sub-group 2) chickpea lines. *Cicer reticulatum* and *Cicer echicnospermum* were included as outgroup controls.

Table 3. Associations between phenotypic data for flowering and resistance to *Foc*5 with microsatellite marker alleles found in the 18 Italian chickpea landraces. Microsatellite markers associated with flowering time were CaGM14822 and TA142, and with resistance to *Foc* were TA27, TA59 and CaGM07922.

Landraces	Flowering time[a]	Flowering time (days)		Resistance to *Foc*5[b]	Foc		
		CaGM14822	TA142		TA27	TA59	CaGM07922
62	T	300	150	S	236-239	246-255-273	300
64	T	300	150-160	S	221-233-242	225-234	300
73	T	300	150	S	221-227-239	231-234-237	350
76	T	300	150	S	227-230	228-234	300
83	T	350	160	S	221-236-239	243-246	350
97	T	300	150	S	221-239-242	231-234	350
99	T	300-350	150-160	S	221-227	231-234-237	300
111	T	320	150	S	221-227-230	225	300
125	T	300	150	S	221-233-236	234	350
147	T	300	150	S	239-242	222-225-234	350
148	T	300	150	S	221-227-233-236	228	300
160	T	300	150	S	221-224-230	237-240	300
184	T	300-350	150	S	221-236-239-242	246	300
203	P	320	150	R	230-233	225	350
228	T	300	150	S	221-239-242-245-248	234	320
237	T	300	150-160	S	221-230-233-236	234-237	300-350
241	T	300	150	S	218-221-227-239	225-234-237	300
245	T	300	150	S	236	252-255	300
WR315	P	380	133	R	221	225	270
ILC3279	T	350	144	S	216	234	300

[a] T = late, P = early; [b] R = high resistance, S = high susceptibility.

allele for CaGM07922. The allele displayed for TA59 in landrace 203 was also present in the resistant reference line WR315 (Table 3).

Genetic diversity analyses

The comparison between Italian landraces and Spanish cultivars revealed the presence of two genetic subgroups (Figure 2). Although with a low level of similarity of close to 0.20, subgroup 1 included 15 Italian landraces and the Spanish cultivars 'Chamad' (of unknown origin) and 'Puchero' (originating from mass selection of Spanish germplasm). Subgroup 2 included all the other Spanish cultivars and the Italians landraces 62, 83 and 245. Landraces 62 and 245, with a similarity coefficient of 0.79, were similar to the Spanish cultivars 'Amelia', 'Badil', 'Junco' and 'Duraton', with which they are grouped with a similarity coefficient of 0.55. Landrace 83 and the Spanish variety "Kairo", with a similarity coefficient of 0.83, were similar to each other, and both were grouped with the Spanish varieties "Athenas" and V5 (similarity coefficient = 0.70), that were subsequently grouped with 'Fardón', 'Pringao', 'Juano', 'Saborio', 'Bagdad' and 'Patio' (Figure 2).

DISCUSSION

Chickpea landraces could be valuable resources for improving sustainability of production on small farms of southern Italy (Negri, 2003; Crinò and Saccardo, 2008). The most important environmental and agronomic factors affecting yields are the length of cultural cycles and growth habits. Currently, the short cycle with spring sowing of chickpea crops is adopted in Italy. Although the long cultural cycle with winter sowing is potentially more productive, crop yields are often affected by adverse factors such as low seed germination rates, high disease incidence and early appearance of weeds (Rossini, 2008).

In cultivated chickpea varieties, bushy growth habit similar to that of the wild relative *Cicer reticulatum* is typical of varieties or landraces with a low amounts of selection (Cobos *et al.*, 2009). As expected, most of the tested landraces showed semi-erect growth habits (Table

1). Most of these landraces were selected in mountainous or hilly areas, where chickpeas are usually produced in marginal and minimally mechanized cropping systems. As a consequence, growers primarily selected chickpea seeds based on seed size and yield. In modern agriculture, growth habit is important for mechanical harvesting (Cobos *et al.*, 2009) and to escape weeds which are common in winter production (Rossini, 2008).

Landrace 203 ('Longano') flowered 67 d after sowing and was the earliest flowering line (Table 3). Early flowering, not affected by photoperiod and typical of plants selected and developed at low latitudes, results in a biological cycle that is 1 month shorter than for other genotypes. Early flowering genotypes escape summer drought stress at high latitudes, resulting in increased yields (Cobos *et al.*, 2009). Based on the knowledge acquired so far, there are no other Italian ecotypes with a flowering trait as early as 67 d. Flowering time is an important trait to increase profitability from chickpea crops, and early flowering allows the plants to escape biotic and abiotic stresses (Semere Mallu *et al.*, 2014). Furthermore, in the Mediterranean basin which is characterized by frequent water stress in the summer, the early flowering phenotypes can allow farmers to obtain increased yields (Siddique and Loss, 2003; Rubio *et al.*, 2004).

Morphological analyses carried out in the present study showed the potential improving chickpea landraces for traits needed to increase yields, particularly growth habit and flowering time. These improvements could be pursued by crossing different landraces to introduce required characteristics without losing the unique genetic characteristics of these ecotypes.

Most of the evaluated landraces were very susceptible to *Foc*5. The exception was landrace 203 ('Longano'), which was highly resistant to the pathogen (AUDPC = 0), and similar to the resistant ecotype WR315 used as control (Table 3 and Figure 1). Genetic resistance to *Foc*5 is conferred by a single gene located on LG2 of the chickpea map (Castro *et al.*, 2010a). High resistance to *Foc*5 in landrace 203 was observed previously in preliminary assays carried out in naturally infested experimental fields at two different locations in southern Spain (Córdoba and Escacena del Campo) (data not shown).

Differences in disease severity among the other landraces studied here are unlikely to be associated with different degrees of genetic resistance to *Foc*5, but are more likely to be from the effect of different responses to specific functions involved in the wilt stress. These may include uptake of iron and other nutrients, and response to water deficiency (Blum, 2017).

Among the tested landraces, accession 203 ('Longano') showed valuable agronomic traits, including large seed size, early flowering and, particularly, a high level of genetic resistance to *Foc*5, which is considered the most aggressive chickpea pathogen worldwide. Based on our findings, landrace 203 could satisfy the current high demand for local products by consumers, and could be a good candidate for large-scale and wide-spread use in agriculture.

Our results with STMS markers did not allow establishment of relationships between different alleles and phenotypic traits, such as flowering time with resistance to *Foc*5. The prediction of the resistance alleles with STMS markers is complex because they show extensive polymorphisms within species. Therefore, their use is not recommended for screening germplasm collections (Madrid *et al.*, 2014). Haplotypes with SNP markers characteristic of the main sources of disease resistance should also be used (Caballo *et al.*, 2019). However, in the case of TA59, tightly linked to *Foc* resistance and used previously in breeding programmes (Castro *et al.*, 2010b), the alleles present in resistant and susceptible controls, respectively, were also present in the resistant landrace 203 and the susceptible line 228 (Table 3).

The study of similarity among Italian landraces and Spanish cultivars, based on molecular markers, showed that the two subgroups, in most of cases, correlated with geographic origins. Some exceptions were found in subgroup 1, where two Spanish cultivars joined with Italian landraces, and in subgroup 2, where three Italian landraces joined with the Spanish cultivars (Figure 2). This genetic similarity could be explained by historic germplasm exchanges during the Spanish domination of southern Italy during the 15th to 18th Centuries.

Future chickpea improvement programmes aimed at conservation and promotion of genetic resources should consider these landraces as good genetic resources for selection and/or breeding. In addition, our results highlight the potential of local varieties to be improved and exploited as productive and profitable crops even in marginal areas, stimulating the farmers of these areas to expand the cultivation of chickpea, and to provide alternative crops in the cropping systems.

ACKNOWLEDGEMENTS

We thank the *Agenzia Regionale per lo Sviluppo Agricolo Rurale e della Pesca*, Campobasso, Italy, and particularly Dr Michelina Colonna, for providing the Italian chickpea landraces used in this study. The research was supported by INIA project RTA2017-00041-00-00 (co-financed by the European Union through the ERDF2014–2020, *Programa Operativo de Crecimiento Inteligente*).

LITERATURE CITED

Ali L., 2015. Búsqueda de genes candidatos asociados a caracteres de interés agronómico en garbanzo (*Cicer arietinum* L.). PhD thesis, *Universidad de Cordoba*, Cordoba - ES, https://helvia.uco.es/handle/10396/12752.

Ali L., Azam S., Rubio J., Kudapa H., Madrid E., … Millán T., 2015. Detection of a new QTL/gene for growth habit in chickpea CaLG1 using wide and narrow crosses. *Euphytica* 2: 473–485.

Arumuganathan K., Earle E.D., 1991. Nuclear DNA content of some important plant species. *Plant Molecular Biology Reporter* 9: 208–218.

Arunodhayam K., Eswara Reddy N.P., Madhuri V., 2014. Pathogenicity and management of *Fusarium* wilt of chickpea, *Cicer arietinum* L. - A review. *Current Biotica* 7: 343–358.

Aryamanesh N., Nelson M.N., Yan G., Clarke H.J., Siddique K.H.M., 2010. Mapping a major gene for growth habit and QTLs for ascochyta blight resistance and flowering time in a population between chickpea and *Cicer reticulatum*. *Euphytica* 173: 307–319.

Blum A., 2017. Osmotic adjustment is a prime drought stress adaptive engine in support of plant production. *Plant, Cell & Environment* 40: 4–10.

Caballo C., Madrid E., Gil J., Chen W., Rubio J., Millan T., 2019. Saturation of genomic region implicated in resistance to Fusarium (*F. oxysporum* f. sp. *ciceris*) race 5 in chickpea. *Molecular Breeding* 39: 16.

Campbell C.L., Madden L.V., 1990. Temporal analysis of epidemics. I. Description and comparison of disease progress curves. In: *Introduction to Plant Disease Epidemiology*, John Wiley & Sons, New York, NY, USA, 161–202.

Castro P., Millan T., Gil J., Merida J., Garcia M.L., Rubio J., Fernandez-Romero M.D., 2010a. Identification of chickpea cultivars by microsatellite markers. *The Journal of Agricultural Science* 149: 451–460.

Castro P., Piston F., Madrid E., Millan T., Gil J., Rubio J., 2010b. Development of chickpea near-isogenic lines for Fusarium wilt. *Theoretical and Applied Genetics* 121: 1519–1526.

Cobos M.J., Rubio J., Fernández-Romero M.D., Garza R., Moreno M.T., Millán T., Gil J., 2007. Genetic analysis of seed size, yield and days to flowering in a chickpea recombinant inbred line population derived from a Kabuli x Desi cross. *Annals of Applied Biology* 151: 33–42.

Cobos M.J., Winter P., Kharrat M., Cubero J.I., Gil J., Millan T., Rubio J., 2009. Genetic analysis of agronomic traits in a wide cross of chickpea. *Field Crops Research* 111: 130–136.

Crinò P., Saccardo F., 2008. Cece e lupino italiani ancora una risorsa. *L'informatore Agrario* 17: 47–52.

De Curtis F., Palmieri D., Vitullo D., Lima G., 2014. First report of *Fusarium oxysporum* f. sp. *pisi* as causal agent of root and crown rot on chickpea (*Cicer arietinum*) in southern Italy. *Plant Disease* 98: 995 (https://doi.org/10.1094/PDIS-09-13-0941-PDN).

del Mar Jiménez-Gasco M., Milgroom M.G., Jiménez-Díaz R.M., 2004. Stepwise evolution of races in *Fusarium oxysporum* f. sp. *ciceris* inferred from fingerprinting with repetitive DNA sequences. *Phytopathology* 94: 228–235.

FAO, 2017. FAOSTAT - http://www.fao.org/faostat/en/#search/Chick%20peas (accessed, January 23, 2019).

Halila I., Cobos M.J., Rubio J., Millán T., Kharrat M., Marrakchi M., Gil J., 2008. Tagging and mapping a second resistance gene for Fusarium wilt race 0 in chickpea. *European Journal of Plant Pathology* 124: 87–92.

Jendoubi W., Bouhadida M., Millan T., Kharrat M., Gil J., Rubio J., Madrid E., 2016. Identification of the target region including the Foc0 1 /foc0 1 gene and development of near isogenic lines for resistance to Fusarium Wilt race 0 in chickpea. *Euphytica* 210: 119–133.

Jendoubi W., Bouhadida M., Boukteb A., Béji M., Kharrat M., 2017. Fusarium wilt affecting chickpea crop. *Agriculture* 7: 23 (Open access: doi:10.3390/agriculture7030023).

Jiménez-Díaz R.M., Castillo P., Jiménez-Gasco M. d. M., Landa B.B., Navas-Cortés J.A., 2015. Fusarium wilt of chickpeas: Biology, ecology and management. *Crop Protection* 73: 16–27.

Laghetti G., Pignone D., De Lisi A., Cifarelli S., Faslia N., Hammer K., 2011. Collecting crop genetic resources in Italian towns of Albanian origin across the Molise, Calabria and Sicily regions. *Genetic Resources and Crop Evolution* 58: 139–152.

Lichtenzveig J., Scheuring C., Dodge J., Abbo S., Zhang H.B., 2005. Construction of BAC and BIBAC libraries and their applications for generation of SSR markers for genome analysis of chickpea, *Cicer arietinum* L. *Theoretical and Applied Genetics* 110: 492–510.

Madrid E., Seoane P., Claros MG., Barro F., Rubio J., Gil J., Millán T., 2014. Genetic and physical mapping of the QTLAR3 controlling blight resistance in chickpea (*Cicer arietinum* L). *Euphytica* 198: 69–78.

Mallikarjuna B.P., Samineni S., Thudi M., Sajja S.B., Khan A.W., … Gaur P.M., 2017. Molecular mapping of flowering time major genes and QTLs in chickpea (*Cicer arietinum* L.). *Frontiers in Plant Science* 8: 1140 (Open access: https://www.frontiersin.org/articles/10.3389/fpls.2017.01140/full).

McCafferty J., Reid R., Spencer M., Hamp T., Fodor A., 2012. Peak studio: a tool for the visualization and analysis of fragment analysis files. *Environmental Microbiology Reports* 4: 556–561.

Millan T., Winter P., Jüngling R., Gil J., Rubio J., ... Muehlbauer F.J., 2010. A consensus genetic map of chickpea (*Cicer arietinum* L.) based on 10 mapping populations. *Euphytica* 175: 175–189.

Negri V., 2003. Landraces in central Italy: where and why they are conserved and perspectives for their on-farm conservation. *Genetic Resources and Crop Evolution* 50: 871–885.

Pande S., Siddique K.H.M., Kishore G.K., Bayaa B., Gaur P.M., ... Crouch J.H., 2005. Ascochyta blight of Chickpea (*Cicer arietinum* L.): a review of biology, pathogenicity and disease management. *Australian Journal of Agricultural Research* 56: 317–332.

Rossini F., 2008. Soluzione ai problemi della coltura del cece. *L'Informatore Agrario* 17: 45–47.

Rubio, J., Flores F., Moreno M.T., Cubero J.I., Gil J., 2004. Effects of the erect/bushy habit, single/double pod and late/early flowering genes on yield and seed size and their stability in chickpea. *Field Crops Research* 90: 255–262.

Semere Mallu T., Githiri Mwangi S., Bernard Nyende A., Ganga Rao N.V.P.R., Achieng Odeny D., ... Kumar A., 2014. Assessment of genetic variation and heritability of agronomic traits in chickpea (*Cicer arietinum* L). *International Journal of Agronomy and Agricultural Research* 5: 76–88.

Sharma K.D., Chen W., Muehlbauer F.J., 2005. Genetics of chickpea resistance to five races of Fusarium wilt and a concise set of race differentials for *Fusarium oxysporum* f. sp. *ciceris*. *Plant Disease* 89: 385–390.

Sharma K.D., Muehlbauer F.J., 2007. Fusarium wilt of chickpea: physiological specialization, genetics of resistance and resistance gene tagging. *Euphytica* 157: 1–14.

Shete S., Tiwari H., Elston R.C., 2000. Estimating the heterozygosity and polymorphism information content value. *Theoretical Population Biology* 57: 265–271.

Siddique K.H.M., Loss S., 2003. Cool season grain legumes in dryland Mediterranean environments of Western Australia: significance of early flowering. In: *Management of Agricultural Drought - Agronomic and Genetic Options* (N.P. Saxena, ed.), New Delhi, India: Oxford University Press 151–162.

Taylor P.W.J., Ford R., 2007. Chickpea. In: *Genome Mapping and Molecular Breeding in Plants* (C. Kole, ed.), Springer-Verlag, Berlin Heidelberg 111–121.

Trapero-Casas A., Jiménez-Díaz R.M., 1985. Fungal wilt and root rot diseases of chickpea in southern Spain. *Phytopathology* 75: 1146–1151.

van der Maesen L.J.G., 1987. Origin, history and taxonomy of chickpea. In: *The Chickpea* (M.C. Saxena, K.B. Singh, ed.), C.A.B. International Cambrian News Ltd, Aberystwyth, UK, 11–34.

Varshney R.K., Thudi M., Nayak S.N., Gaur P.M., Kashiwagi J., ... Viswanatha K.P., 2014. Genetic dissection of drought tolerance in chickpea (*Cicer arietinum* L.). *Theoretical and Applied Genetics* 127: 445–462.

Winter P., Benko-Iseppon A.M., Hüttel B., Ratnaparkhe M., Tullu A., ... Muehlbauer F.J., 2000. A linkage map of the chickpea (*Cicer arietinum* L.) genome based on recombinant inbred lines from a *C. arietinum*×*C. reticulatum* cross: localization of resistance genes for Fusarium wilt races 4 and 5. *Theoretical and Applied Genetics* 101: 1155–1163.

Zaccardelli M., Sonnante G., Lupo F., Piergiovanni A.R., Laghetti G., ... Lioi L., 2012. Characterization of Italian chickpea (*Cicer arietinum* L.) germplasm by multidisciplinary approach. *Genetic Resources and Crop Evolution* 60: 865–877.

9

Isolation of atypical wheat-associated xanthomonads in Algeria

Bahria KHENFOUS-DJEBARI[1,*], Mohamed KERKOUD[2], Marion FISCHER-LE SAUX[3], Traki BENHASSINE[4], Zouaoui BOUZNAD[1], Ralf KOEBNIK[5], Claude BRAGARD[6]

[1] *Laboratoire de Phytopathologie et de Biologie Moléculaire, Département de Botanique, ENSA El Harrach, Alger, Algérie*

[2] *DiagGene, 8, Rue le Nôtre 49066 Angers, France*

[3] *IRHS, INRA, Agrocampus-Ouest, Université d'Angers, SFR 4207 QuaSaV, 49071, Beaucouzé, France*

[4] *Département de Biologie Cellulaire et de Biologie Moléculaire, USTHB, Bab Ezzouar, Alger, Algerie*

[5] *IRD, Cirad, Université de Montpellier, IPME, Montpellier, France*

[6] *Applied Microbiology-Phytopathology, Earth & Life Institute, UC Louvain, Louvain-la-Neuve, Belgium*

*Corresponding author: b.djebari@ensa.dz

Competing Interests: The Author(s) declare(s) no conflict of interest.

Editor: Roberto Buonaurio, University of Perugia, Italy.

Summary. Bacterial leaf streak and black chaff are important bacterial diseases of wheat, which have been reported to be caused by *Xanthomonas translucens*. In 2016, symptoms of bacterial leaf streak and black chaff were observed in Algeria, at experimental wheat breeding stations and in farmers' fields under sprinkler irrigation on two wheat cultivars, 'Hiddab' and 'Simeto'. Yellow *Xanthomonas*-like bacterial colonies were isolated from plant material, including leaves, spikes and post-harvest crop residues. Initial characterisation using biochemical, physiological and pathogenicity tests identified the bacteria as *Xanthomonas*. Diagnostic PCR targeting the 16S-23S rRNA intergenic region indicated that the strains were *X. translucens*, a clade-1 xanthomonad. However, partial DNA sequences of the housekeeping genes *gyrB* and *rpoD* revealed that the strains belong to clade 1, but likely represent a new *Xanthomonas* species that has not been previously described on wheat or other *Gramineae*. The most closely related strain, NCPPB 2654, was isolated from a bean plant in the United Kingdom in 1974. Further characterization is required to clarify the taxonomic status of the Algerian *Xanthomonas* isolates from wheat, and to determine their host ranges and impacts on plant cultivation.

Keywords. *Xanthomonas*, bacterial leaf streak, black chaff, wheat.

INTRODUCTION

Wheat is a major crop worldwide, affected by some bacterial diseases, among which bacterial leaf streak (BLS) is the most important. BLS is caused by *Xanthomonas translucens* (ex Jones *et al.* 1917) (Vauterin *et al.*, 1995). When symptoms occur on the wheat plant glumes, the disease caused by

X. translucens is called black chaff (BC) (Duveiller *et al.*, 1997). BLS is widely distributed in the world and the disease is prevalent in most regions where small-grain cereals are cultivated (Paul and Smith, 1989; Duveiller and Maraite, 1994). However, these diseases have not been extensively studied in North Africa, despite sporadic reports of occurrence in countries close to Algeria, including Libya (Bragard *et al.*, 1995), Morocco and Tunisia (Sands and Fourest, 1989).

Yield losses caused by BLS are generally not considered problematic. However, losses of up to 40% have been recorded under conditions that are conducive for the pathogen (Duveiller *et al.*, 1997). Algeria cultivates cereals on an area of 3.3 million ha, 1.3 million of which are irrigated and thus likely to be vulnerable to BLS. Furthermore, more than 20% of yield may be lost if 50% of flag leaf area is affected by the disease (Duveiller and Maraite, 1993). BLS generally appears late in the growing season in regions with temperate climate or in warmer environments characterized by cool nights and frequent temperature variations (Duveiller *et al.*, 2002). The symptoms are usually more obvious after heading (Wiese, 1987). All aerial parts of host plants can be affected, but the leaves and glumes are more often affected than the other parts. In cases of severe damage, the seeds can be blackened and wrinkled (Zillinsky, 1983). BC is characterized by many black longitudinal stripes on the upper portions of the glumes (Smith, 1917), that can be identified by greasy appearance. Yellow bacterial droplets exude along the lesions, particularly in wet weather.

Xanthomonas species are known for their ability to adhere to and colonize host leaf surfaces as epiphytes before invading the intercellular spaces (Boulanger *et al.*, 2014; Dutta *et al.*, 2014; Zarei *et al.*, 2018). Xanthomonads have evolved several strategies for successful infection, including mechanisms to suppress host plant resistance and access nutrients from host cells (Büttner *et al.*, 2010; Fatima and Senthil-Kumar, 2015; Jacques *et al.*, 2016). However, much less is known for *X. translucens*, which belongs to xanthomonad clade 1, while most functional research has been carried out with clade-2 xanthomonads (Parkinson *et al.*, 2007). The importance of the bacterial type III secretion system and TAL effectors for pathogenicity has been demonstrated (Wichmann *et al.*, 2013; Peng *et al.*, 2016; Falahi Charkhabi *et al.*, 2017; Pesce *et al.*, 2017).

Since no data are available on bacterial diseases of small grain cereals in Algeria, the objective of this study was to assess the presence of wheat-associated xanthomonads in this country, to verify their ability to cause BLS on cereals and to determine their identity. This research was based on morphological, biochemical and physiological characterization of bacterial isolates, complemented by pathogenicity tests on host plants and DNA-based molecular diagnostics.

MATERIALS AND METHODS

Bacterial strains

Bacteria were isolated from symptomatic wheat leaf and spike samples and from post-harvest plant residues originating from fields that had shown symptoms of BLS and BC in the previous growing season (Table 1) (Kar-

Table 1. Details of Algerian bacterial isolates obtained from wheat.

Isolate	Gene	GenBank Accession number	Sample	Symptoms[a]	Wheat cultivar	Location	Year
X1	*gyrB* *rpoD*	MF142045 MF142046	Leaf	BLS	'Simeto'	El Goléa	2016
X2	*gyrB*	MF142047	Leaf	BLS	'Simeto'	El Goléa	2016
X3	*gyrB*	MF142048	Leaf	BLS	'Simeto'	El Goléa	2016
X4	*gyrB*	MF142049	Leaf	BLS	'Simeto'	El Goléa	2016
X5	*gyrB*	MF142050	Crop residue	No visible symptoms	'Hiddab' (HD1220)	Algiers experimental station	2016
X8	*gyrB* *rpoD*	MF142051 MF142052	Spike	BC	'Simeto'	El Goléa	2016
X12	*gyrB* *rpoD*	MF142053 MF142054	Crop residue	No visible symptoms	'Hiddab' (HD1220)	Algiers experimental station	2010
X13	*gyrB* *rpoD*	MF142055 MF142056	Crop residue	No visible symptoms	'Hiddab' (HD1220)	Algiers experimental station	2010
X16	*gyrB*	MF142057	Spike	BC	Breeding line	Algiers experimental station	2016
X17	*gyrB*	MF142058	Spike	BC	Breeding line	Algiers experimental station	2016

[a]BLS = bacterial leaf streak, BC = black chaff.

avina *et al.*, 2008). Samples (10 g each) of plant organs or post-harvest stubble residues were each ground to a fine powder and mixed with 50 mL phosphate-buffered saline (PBS). The resulting cell suspension was shaken for 20 min and allowed to settle for 3 h. Series of tenfold dilutions were prepared from the suspension supernatants and appropriate amounts of each dilution were plated on standard nutrient agar medium and on semi-selective Wilbrink's medium supplemented with 0.75 g L^{-1} of boric acid, 10 mg L^{-1} of cephalexin and 75 mg L^{-1} of cycloheximide (WBC medium) to reduce fungal growth (Duveiller, 1990). For comparison, *X. translucens* pv. *undulosa* strain UPB753, which had been isolated from wheat in Brazil (Bragard *et al.*, 1995), was used as a reference strain. Pure cultures were obtained upon cultivation at 30°C. Based on morphological characteristics, *Xanthomonas*-like bacterial strains were assigned a designation number and kept at 4°C on glucose, yeast extract-calcium carbonate agar (GYCA) tubes for short-term conservation and at -80°C in 20% glycerol for long-term storage.

Phenotypic characterization

Established tests were used to identify the bacteria (Bradbury, 1986; Van den Mooter and Swings, 1990; Schaad *et al.*, 2001). These included formation of mucoid colonies on GYCA medium, Gram reaction using the KOH test, oxidase test using tetramethyl-*p*-phenylenediamine dihydrochloride reagent, oxidative and fermentative metabolism of glucose, hydrolysis of Tween 80, aesculin and starch, production of levane sucrase, production of catalase, liquefaction of gelatine, H_2S production from cysteine, nitrate reduction, growth at 35°C, and growth on 2% and 5% NaCl media.

DNA manipulations

For rapid tests, bacterial cells were lysed by brief boiling and then directly used for PCR, as described previously (Maes *et al.*, 1996). For PCR amplification of housekeeping genes, DNA was isolated using the Wizard® Genomic DNA Purification Kit (Promega Corp.), following the manufacturer's instructions.

PCR amplification of the alanine-specific tRNA gene in the 16S-23S rRNA intergenic region was performed as described previously (Maes *et al.*, 1996), with modifications. The reaction mixture was prepared in a total volume of 25 μL comprising 5 μL of 5 × PCR buffer, 2.5 μL $MgCl_2$ (25 mM), 0.75 μL dNTP mixture (25 mM), 0.5 μL of each PCR primer at 10 μM (T1, 5'-CCGCCATAGGGCGGAGCACCCCGAT; T2, 5'-GCAGGTGCGACGTTTGCAGAGGGATCTGCAAATC), 2.5 μL DNA sample (50 ng μL^{-1}), 0.2 μL Taq polymerase (Promega), and 13.05 μL distilled water. PCR was performed with the following conditions: 90°C for 2 min, 29 cycles of 93°C for 30 s, 50°C for 30 s, 72°C for 30 s, and a final extension for 10 min at 72°C. The PCR products were separated by electrophoresis on 2% agarose gel in TAE buffer, stained with Midori green (Nippon Genetics Europe) and visualized under UV light.

Previously published MLSA primers were used for PCR amplification and partial DNA sequencing of two housekeeping genes, *gyrB* and *rpoD* (Fargier and Manceau, 2007). PCR amplifications were performed as recommended (Mhedbi-Hajri *et al.*, 2013) in a 50 μL reaction mixture containing 1 × GoTaq® buffer, 200 μM dNTP, 0.5 μM of each primer, 0.4 U of GoTaq® DNA polymerase (Promega), and 3 ng of genomic DNA, with an initial denaturation at 94°C for 2 min, 30 cycles of denaturation for 1 min at 94°C, annealing for 1.5 min at 60°C, extension for 1.5 min at 72°C, and a final extension for 10 min at 72°C. 8 μL of PCR products reaction mixtures were analysed by electrophoresis on 1.5% agarose gel in TAE buffer, stained with Midori green direct and visualized under UV light. The remaining amplified PCR products were purified with the Wizard® PCR clean-up kit (Promega) and sequenced with reverse and forward primers using BigDye Terminator v3.1 cycle sequencing kit (Applied Biosystems). Both forward and reverse sequences were aligned and manually edited before deposition in the GenBank database (Table 1).

Pathogenicity tests

Bacterial cells were suspended in sterile solution of 0.90% (w/v) of NaCl and the concentration of cells was adjusted to 1 × 10^9 CFU mL^{-1} for hypersensitive reaction tests on tobacco plants of the variety Xanthi and to 1 × 10^7 CFU mL^{-1} for pathogenicity tests. Sterile saline solution served as negative controls in the pathogenicity assays.

All strains were tested for pathogenicity, by inoculation on the sensitive wheat cultivar 'Acsad 885', using three different assays. First, after injection of sterile water into plant leaf sheath at 2.5 cm above soil level, three-leaf stage seedlings were puncture inoculated with a sterile needle that had been passed through a bacterial colony ("pricking inoculation") (Bragard and Maraite, 1992). Second, using a needle-less plastic syringe, bacterial suspensions were infiltrated through the upper leaf surfaces until appearance of liquid-soaked areas of about 2 cm length ("leaf infiltration") (Bragard and Maraite, 1992). Third, whole leaves were immersed into bacterial solutions for 20 sec ("dip inoculation") (Darsonval *et al.*, 2009). Plants were incubated at 28°C and 95% relative humidity with a photoperiod of 16h/8h (day/night), and symptoms were scored over time.

To re-isolate bacteria from infected plant material, symptomatic leaf segments were cut into small pieces in sterile physiological saline and plated on standard and semi-selective media.

Bioinformatic analyses

To link the different strains with their respective taxa among the species of *Xanthomonas*, corresponding *gyrB* and *rpoD* gene portions were retrieved from GenBank and PAMDB databases (https://www.ncbi.nlm.nih.gov, http://www.pamdb.org) (Almeida *et al.*, 2010; Sayers *et al.*, 2019). For recently described species and pathovars that are not represented in PAMDB, such as "*Xanthomonas pseudalbilineans*", *X. maliensis*, *X. floridensis*, *X. nasturtii*, and *X. prunicola*, corresponding sequences were extracted from the genome sequences (Supplementary Table 1) (Pieretti *et al.*, 2015; Triplett *et al.*, 2015; Hersemann *et al.*, 2016b; Vicente *et al.*, 2017; López *et al.*, 2018).

Multiple sequence alignments were performed using the MUSCLE algorithm (https://www.ebi.ac.uk/Tools/msa/muscle/) (Edgar, 2004). Phylogenetic trees were generated using the phylogeny.fr pipeline, with default parameters (http://www.phylogeny.fr) (Dereeper *et al.*, 2008). Newick files were generated and the tree was manipulated using the iTOL website (https://itol.embl.de) (Letunic and Bork, 2011) to improve visualization.

RESULTS

Isolation of bacteria from wheat

BLS symptoms were observed in various plots of the Algiers experimental station, at the tillering and at the heading crop growth stages (Figure 1A). Additionally, severe disease symptoms of BC were observed at the heading stage (Figure 1B). Affected varieties in the Algiers area included wheat 'Hiddab' (HD1220), and also several breeding lines and cultivars grown at the Algiers experimental station for studies under the pedoclimatic conditions of the region, to verify their performance, efficiency, specific features and disease resistance. Symptoms were also observed on the cultivar 'Simeto', grown under the sprinkler irrigation at El Ménéa (El Goléa), South Algeria. This is an area characterized by cool nights and high day temperatures. Furthermore, during the 2016 agricultural campaign, the disease was sporadically present in plots of the Algiers experimental station, and more widespread in plots under pivot irrigation, where 'Simeto' was sown. Yellow colonies on media, resembling *Xanthomonas*, were easily isolated from the plant symptoms. In order to evaluate if post-harvest material could serve as reservoirs for infections, crop residues from plots at the Algiers multiplication station that had shown symptoms, were analysed as well, as described previously (Karavina *et al.*, 2008). *Xanthomonas*-like bacteria were also isolated from this material.

Figure 1. Disease symptoms on wheat. (A) Symptoms of bacterial leaf streak at the heading stage, (B) symptoms of black chaff at the heading stage, (C) chlorotic to necrotic lesions from leaf infiltration 7 d after inoculation, (D) disease symptoms 8 d after pricking inoculation of wheat seedlings at the three-leaf stage, and (E) water-soaked spots associated with bacterial exudates 8 d after dip inoculation of leaves.

Phenotypical characterization of bacteria isolated from wheat

Among all the strains obtained from the different origins, 30 were retained (Table 2) with consistent biochemical and physiological test responses corresponding to those obtained with the *X. translucens* reference strain (Bradbury, 1986; Van den Mooter and Swings,

Table 2. Biochemical and physiological tests used to identify bacterial isolates from wheat, and a reference strain (*Xanthomonas* translucens pv. *undulosa* strain UPB753 (Bragard *et al.*, 1995)).

Test	Algerian isolates	Reference strain
Mucoid and yellow colonies on GYCA medium	+	+
Gram staining	-	-
Oxidase	-	-
Metabolism of glucose	oxidative	oxidative
Tween 80 hydrolysis	+	+
Aesculin hydrolysis	+	+
Starch hydrolysis	+	+
Levane sucrase	+	+
Catalase	+	+
Liquefaction of gelatine	+	+
H_2S production from cysteine	+	+
Nitrate reduction to nitrite	-	-
Growth at 35 °C	+	+
Growth in 2% NaCl	+	+
Growth in 5% NaCl	-	-
Hypersensitive reaction on tobacco	+	+
Pathogenicity on wheat cv. Acsad 885	+	+
Diagnostic PCR (Maes *et al.*, 1996)	+	+

1990; Schaad *et al.*, 2001). All these strains grew as mucoid and yellow pigmented colonies, and they were negative for Gram staining, oxidase activity and nitrate reduction. The strains could hydrolyse aesculin, gelatine and starch and produced catalase, levane sucrase, lipase and hydrogen sulphide from cysteine. Bacteria grew at 35°C and in nutrient broth supplemented with 2% sodium chloride, but not when supplemented with 5% sodium chloride. All strains triggered hypersensitive reactions on tobacco. These analyses indicated that the wheat-associated bacteria belonged to the genus *Xanthomonas* and may be related to *X. translucens.*

Pathogenicity assays with xanthomonads isolated from wheat

Three inoculations methods were applied to evaluate the pathogenicity of the bacterial strains. Upon leaf infiltration of a susceptible cultivar, chlorotic to necrotic lesions developed within 7 d after infection (Figure 1C). When pricking three-leaf stage wheat seedlings, similar symptoms were observed 8 d after inoculation (Figure 1D). Dip inoculation of wheat leaves resulted in symptoms that were clearly visible after 8 d, including typical water-soaked spots associated with bacterial exudates from the plant organs (Figure 1E).

Bacteria were re-isolated from infected plant material. Morphological, biochemical and physiological characterization confirmed the identity of the re-isolated bacteria with the inoculum, thus fulfilling the Koch's postulates.

DNA-based diagnostics of xanthomonads isolated from wheat

In order to evaluate whether the strains belong to *X. translucens*, they were subjected to a protocol that had been developed for specific detection of this species, based on a discriminatory region in the 16S-23S intergenic region, which encodes two tRNAs (Maes *et al.*, 1996). A DNA fragment with a size of 139 bp was amplified with PCR primers T1 and T2 for all 30 strains.

To further characterize ten representative strains from wheat, a portion of the *gyrB* gene that was previously used in multiple locus sequence analysis (MLSA) was amplified by PCR and sequenced, using previously developed primers (Fargier and Manceau, 2007; Young *et al.*, 2008). All sequences were identical. Homologous sequences of type, pathotype or other *Xanthomonas* strains were retrieved from the PAMDB database (Almeida *et al.*, 2010). In addition, sequences for those of *Xanthomonas* species that were not available at PAMDB, but had been included in a previous *gyrB*-based phylogenetic study, including sequences for undescribed species ("slc" species-level clades) (Parkinson *et al.*, 2009), were used for comparison. Sequences were aligned using MUSCLE and manually trimmed to 528 bp. A phylogenetic tree was calculated using the Phylogeny.fr pipeline (Figure 2). This analysis revealed that the Algerian sequences clustered with other sequences from clade-1 xanthomonads, including *Xanthomonas hyacinthi*, *X. theicola* and *X. translucens*. The closest sequence corresponded to strain NCPPB 2654 from species-level clade 5. This strain was isolated in 1974 by F. W. Catton, from navy bean (*Phaseolus vulgaris*) in the United Kingdom. No other sequence information is available for this species-level clade.

This result prompted us to partially sequence another housekeeping gene *rpoD*, that was previously used in MLSA studies (Young *et al.*, 2008). Multiple sequence alignment of four 870-bp sequences from representative Algerian strains revealed their identity with each other. A phylogenetic tree was generated including representative haplotypes for clade-1 xanthomonads from PAMDB. In addition, sequences from three additional *X. translucens* pathotype strains and two sequences belonging to the recently suggested clade-1 species "*X. pseudalbilineans*" were included (Pieretti *et al.*, 2015; Hersemann

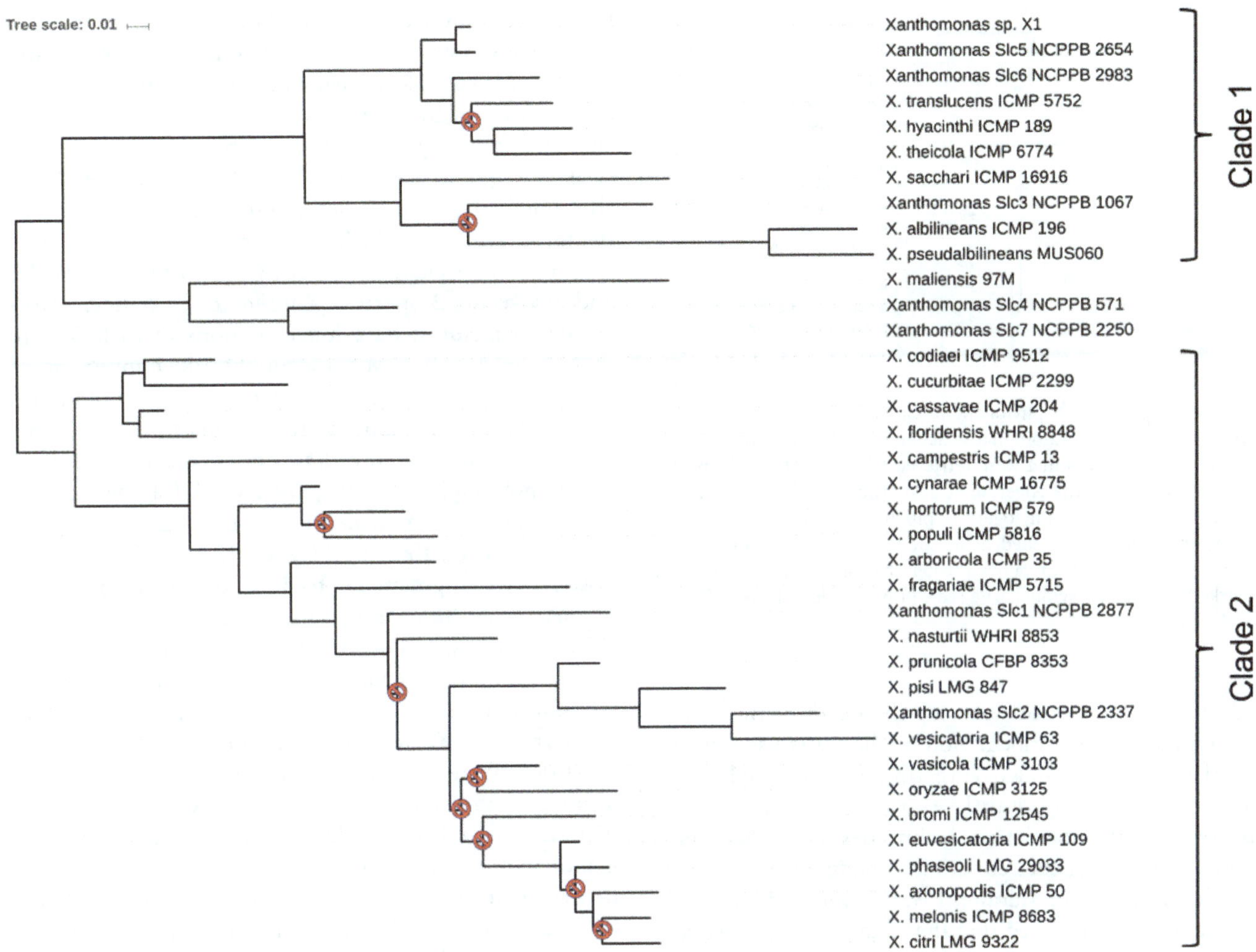

Figure 2. *gyrB*-based phylogenetic classification of the Algerian wheat-associated xanthomonads. Representative strains of all species following taxonomic revisions of *Xanthomonas gardneri* and *X. axonopodis* species complex are included (Constantin *et al.*, 2016; Timilsina *et al.*, 2019). Taxonomically unassigned species-level clades (Slc 1–7) are indicated (Parkinson *et al.*, 2009). All sequences were trimmed to the portion that is available in GenBank for the *Xanthomonas phaseoli* type strain (528 bp, Accession number KT585789). The tree was constructed with the Phylogeny.fr pipeline, using default parameters, and graphically edited using the iTOL suite. All nodes were supported by bootstrap values greater than 0.75, except for those marked with a stop symbol. The scale of branch lengths is indicated at the top left.

et al., 2016b). This analysis confirmed that the Algerian strains belong to clade-1, but did not cluster with any of the six described species. This indicates that the Algeria strains from wheat belong to another species, with strain NCPPB 2654 as the likely founder (Figure 3).

DISCUSSION

Diseased wheat plants were reported from several plots in Algeria, with symptoms on the leaves and spikes as well as melanotic areas on the glumes, that were similar to those described for leaf streak and black chaff (Duveiller *et al.*, 2002). These diseases result from bacterial infections, often in high temperature and high humidity conditions, and the observed symptoms have long been described for these diseases (Smith, 1917; Johnson and Hagborg, 1944). The strains characterized in the present study were mostly obtained from sprinkler-irrigated fields or from breeding stations where genotype behaviour was being assessed.

All strains obtained from affected wheat samples, including leaves, spikes and crop residues, had phenotypic characteristics that corresponded to those described in the literature for *X. translucens*. Moreover, inoculation of wheat seedlings at the three-leaf stage caused water-soaked lesions within 8 d after inoculation, accompanied by signs of necrosis around the inoculat-

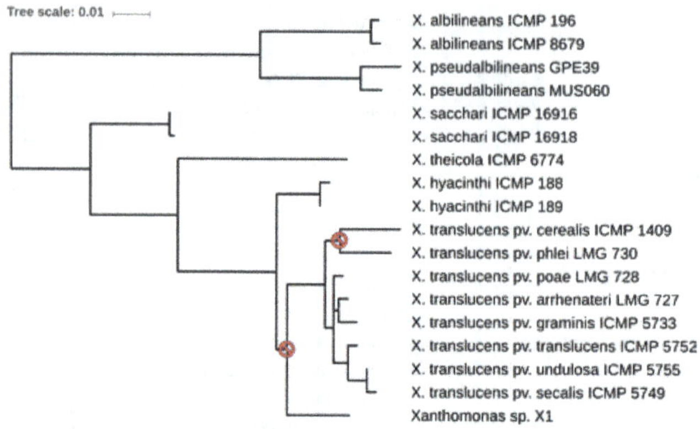

Figure 3. *rpoD*-based phylogenetic classification of the Algerian wheat-associated xanthomonads. Sequences were trimmed to the size of the sequences that were retrieved from PAMDB (855 bp for *X. translucens* and the Algerian strain). The phylogenetic tree was constructed with the Phylogeny.fr pipeline, using default parameters, and graphically edited using the iTOL suite. All nodes were supported by bootstrap values greater than 0.75, except for those marked with a stop symbol. The scale of branch lengths is indicated at the top left.

ed areas and ultimately causing the death of the leaves. When the pricking inoculation method was used, initial local necrosis expanded with time and affected whole leaf blades, concomitantly showing typical exudate droplets. The inoculation techniques mimicked temperature and moisture conditions conducive for the disease (Duveiller and Maraite, 1993; Duveiller *et al*, 1997). These results confirmed that the Algerian *Xanthomonas*-like strains were pathogenic on wheat cultivar 'Acsad 885'. Bacteria could also be isolated from crop residues 3 to 4 weeks after harvest, indicating that the pathogen can survive on plant material. This may serve as reservoirs for new infections in the next cropping cycle. Similar observations have been made for *X. translucens*, which was found to survive on crop debris for more than 30 months under laboratory conditions, and for less than 8 months under field conditions (Malavolta Jr. *et al.*, 2000). Occurrence on, and isolation from, wheat plants and the symptoms observed in the fields and upon artificial inoculation strongly indicated that the strains were *X. translucens*.

Initial molecular characterization of the strains using a PCR assay that was developed for detection of *X. translucens*, including pathovars infecting small-grain cereals (pvs. *cerealis*, *hordei*, *secalis*, *translucens* and *undulosa*) and those infecting forage grasses (pvs. *arrhenatheri*, *graminis*, *phlei*, *phleipratensis* and *poae*), further supported that the Algerian strains were *X. translucens* (Maes *et al.*, 1996). This assay targets the 16S-23S intergenic region, which encodes two tRNAs in all xanthomonads, one for alanine (UGC anticodon) and one for isoleucine (GAU anticodon). The diagnostic primers T1 and T2 anneal immediately upstream and downstream of the tRNA (Ala) gene. Most xanthomonads have short regions of 14 to 19 bp between the transcribed sequences for the tRNA (Ala) and the tRNA (Ile) (Gonçalves and Rosato, 2002), and therefore lack the target region for the T2 primer. However, *X. translucens*, two other clade-1 species (*X. hyacinthi*, *X. theicola*) and two clade-2 species (*Xanthomonas codiaei*, *Xanthomonas melonis*) have longer regions of 75 to 79 bp, that are fairly conserved and might allow annealing of the T2 primer under less stringent conditions (data not shown). Since the region corresponding to the T1 primer is less similar for *X. codiaei* and *X. melonis*, the PCR should not amplify the diagnostic DNA fragment of 139 bp, as had been confirmed for *X. melonis* (Maes *et al.*, 1996). However, that the PCR could amplify the diagnostic DNA fragment for bacteria of the *X. hyacinthi*-*X. theicola*-*X. translucens* subclade cannot be excluded. When the assay was developed by Maes and co-workers, all xanthomonads that were known as pathogens on *Gramineae* (i.e. *X. albilineans*, *X. axonopodis*, *X. bromi*, *X. oryzae* and *X. vasicola*), and representative strains of most described species of *Xanthomonas*, were included, although some species of clade-1 were not tested (e.g. *X. hyacinthi* and *X.theicola*). Yellow disease of hyacinth, a monocot of the *Asparagaceae*, was the first disease described to be caused by *Xanthomonas* (Van Doorn and Roebroeck, 1993). Infection of asparagus by *X. translucens* has been reported (Rademaker *et al.*, 2006). Therefore, the taxonomic status of the Algerian strains remained uncertain based on the diagnostic PCR.

Since we felt that the diagnostic PCR is not able to unambiguously identify strains of *X.translucens*, two housekeeping genes that are included in MLSA schemes were analysed (Young *et al.*, 2008). Partial sequences of both genes, *gyrB* and *rpoD*, clustered with sequences from other clade-1 xanthomonads, such as *X. hyacinthi*, *X. theicola* and *X. translucens*, but were distant enough to question whether they belonged to any of the described species. The *gyrB* gene has been used for exhaustive phylogenetic analyses of *Xanthomonas*, including strains from species-level clades that still await precise taxonomic assignment (Parkinson *et al.*, 2007, 2009). The partial *gyrB* gene sequences from the Algerian isolates were very similar to that of strain NCPPB 2654, which was isolated from a navy bean plant in the United Kingdom (https://www.fera.co.uk/ncppb). According to the NCPPB website, this strain has a fatty acid profile typical of *Xanthomonas*, but is not pathogenic on bean pods. It would be interesting to know whether this strain was isolated near a wheat field, and to test this strain for pathogenicity on wheat plants. Likewise, the next similar sequence corresponded to strain NCPPB 2983, which was deposited as *Xanthomonas campestris* pv. *phormiicola* and belongs to species-level clade 6. This strain was isolated in Japan from New Zealand flax (*Phormium tenax*), a member of the *Asparagales* (*Asphodelaceae*). Three species related to the Algerian strains, *X. hyacinthi*, *X. translucens* and *X. campestris* pv. *phormiicola*, are able to colonize plants in the *Asparagales*, and future work will evaluate whether the Algerian strains from wheat can infect these plants as well.

In conclusion, this is the first description of wheat-pathogenic xanthomonads from Algeria, which were atypical in that they most likely do not belong to *X. translucens*. Further characterization, ideally including whole-genome sequencing, will clarify their taxonomic status and their host range (Peng *et al.*, 2016; Langlois *et al.*, 2017). It will also be important to compare these strains with other bacterial pathogens of cereals, and to elucidate whether candidate type III effectors, phytohormones and/or toxins are involved in pathogenicity and host adaptation (Royer *et al.*, 2013; Gardiner *et al.*, 2014; Hersemann *et al.*, 2016a, 2016b, 2017; Triplett *et al.*, 2016; Nagel and Peters, 2017).

ACKNOWLEDGEMENTS

Part of this research was carried out in the laboratory of the Earth and Life Institute at UC-Louvain. The authors thank the team in this laboratory, with special thanks to Pr H. Maraite and to Mrs L.N.Salhi. Mr Y.A.Merdoukh's assistance with field sampling is gratefully acknowledged.

LITERATURE CITED

Almeida N.F., Yan S., Cai R., Clarke C.R., Morris C.E., … Vinatzer B.A., 2010. PAMDB, a multilocus sequence typing and analysis database and website for plant-associated microbes. *Phytopathology* 100: 208–215.

Boulanger A., Zischek C., Lautier M., Jamet S., Rival P., … Lauber E., 2014. The plant pathogen *Xanthomonas campestris* pv. *campestris* exploits N-acetylglucosamine during infection. *mBio* 5: e01527-14.

Bradbury J.F., 1986. *Guide to Plant Pathogenic Bacteria*. CAB International Mycological Institute, Farnham House, Slough, UK, 332 pp.

Bragard C., Maraite H., 1992. Pathogenic variation in *Xanthomonas campestris* pv. *undulosa*. In: *Proceedings of the 8th International Conference on Plant Pathogenic Bacteria*, June 9-12, 1992, Versailles, France, 807–812.

Bragard C., Verdier V., Maraite H., 1995. Genetic diversity among *Xanthomonas campestris* strains pathogenic for small grains. *Applied and Environmental Microbiology* 61: 1020–1026.

Büttner D., Bonas U., 2010. Regulation and secretion of *Xanthomonas* virulence factors. *FEMS Microbiology Reviews* 34: 107–133.

Constantin E.C., Cleenwerck I., Maes M., Baeyen S., Van Malderghem C, … Cottyn B., 2016. Genetic characterization of strains named as *Xanthomonas axonopodis* pv. *dieffenbachiae* leads to a taxonomic revision of the *X. axonopodis* species complex. *Plant Pathology* 65: 792–806.

Darsonval A., Darrasse A., Durand K., Bureau C., Cesbron S., Jacques M.A., 2009. Adhesion and fitness in the bean phyllosphere and transmission to seed of *Xanthomonas fuscans* subsp. *fuscans*. *Molecular Plant-Microbe Interactions* 22: 747–757.

Dereeper A., Guignon V., Blanc G., Audic S., Buffet S, … Gascuel O., 2008. Phylogeny.fr: robust phylogenetic analysis for the non-specialist. *Nucleic Acids Research* 36: W465–W469.

Dutta B., Gitaitis R., Smith S., Langston Jr D., 2014. Interactions of seedborne bacterial pathogens with host and non-host plants in relation to seed infestation and seedling transmission. *PLoS One* 9: e99215.

Duveiller E., 1990. A seed detection method of *Xanthomonas campestris* pv. *undulosa*, using a modification of Wilbrink's agar medium. *Parasitica* 46: 3–17.

Duveiller E., Maraite H., 1993. Study on yield loss due to *Xanthomonas campestris* pv. *undulosa* in wheat under high rainfall temperature condition. *Journal of Plant Diseases and Protection* 100: 453–459.

Duveiller E., Maraite H., 1994. Effect of temperature and air humidity on multiplication of *Xanthomonas campestris* pv. *undulosa* and symptom expression in susceptible and field tolerant wheat genotypes. *Journal of Phytopathology* 143: 227–232.

Duveiller E., Bragard C., Maraite H., 1997. Bacterial leaf streak and black chaff caused by *Xanthomonas translucens*. In: *The Bacterial Diseases of Wheat. Concepts and Methods of Disease Management* (E. Duveiller, L. Fucikovsky, K. Rudolph, eds.), International Maize and Wheat Improvement Center, Estado de México, Mexico, 25–47.

Duveiller E., Bragard C., Maraite H., 2002. Bacterial leaf streak and black chaff. In: *Bread Wheat. Improvement and Production* (B. C. Curtis, S. Rajaram and H. Gomez Macpherson, ed.), *FAO Plant Production and Protection Series No. 30,* Food and Agriculture Organization of the United Nations, Rome.

Edgar R.C., 2004. MUSCLE: multiple sequence alignment with high accuracy and high throughput. *Nucleic Acids Research* 32: 1792–1797.

Falahi Charkhabi N., Booher N.J., Peng Z., Wang L., Rahimian H., … Bogdanove A.J., 2017. Complete genome sequencing and targeted mutagenesis reveal virulence contributions of Tal2 and Tal4b of *Xanthomonas translucens* pv. *undulosa* ICMP11055 in bacterial leaf streak of wheat. *Frontiers in Microbiology* 8: 1488.

Fargier E., Manceau C., 2007. Pathogenicity assays restrict the species *Xanthomonas campestris* into three pathovars and reveal nine races within *X. campestris* pv. *campestris*. *Plant Pathology* 56: 805–818.

Fatima U., Senthil-Kumar M., 2015. Plant and pathogen nutrient acquisition strategies. *Frontiers in Plant Science* 6: 750.

Gardiner D.M., Upadhyaya N.M., Stiller J., Ellis J.G, Dodds P.N., … Manners J.M, 2014. Genomic analysis of *Xanthomonas translucens* pathogenic on wheat and barley reveals cross-kingdom gene transfer events and diverse protein delivery systems. *PLoS One* 9: e84995.

Gonçalves E.R., Rosato Y.B., 2002. Phylogenetic analysis of *Xanthomonas* species based upon 16S-23S rDNA intergenic spacer sequences. *International Journal of Systematic and Evolutionary Microbiology* 52: 355–361.

Hersemann L., Wibberg D., Blom J., Widmer F., Kölliker R., 2016a. Draft genome sequence of the *Xanthomonas bromi* type strain LMG 947. *Genome Announcements* 8: e00961-16.

Hersemann L., Wibberg D., Widmer F., Vorhölter F.J., Kölliker R., 2016b. Draft genome sequences of three *Xanthomonas translucens* pathovar reference strains (pv. *arrhenatheri*, pv. *poae* and pv. *phlei*) with different specificities for forage grasses. *Standards in Genomic Sciences* 11: 50.

Hersemann L., Wibberg D., Blom J., Goesmann A., Widmer F., … Kölliker R., 2017. Comparative genomics of host adaptive traits in *Xanthomonas translucens* pv. *graminis*. *BMC Genomics* 18: 35.

Jacques M.A., Arlat M., Boulanger A., Boureau T., Carrère S, … Vernière C., 2016. Using ecology, physiology, and genomics to understand host specificity in *Xanthomonas*. *Annual Review of Phytopathology* 54: 163–187.

Johnson T., Hagborg W.A.F., 1944. Melanism in wheat induced by high temperature and humidity. *Canadian Journal of Research* 22: 7–10.

Karavina C., Tigere T.A., Chihiya J., 2008. The contribution of soil and crop debris inocula to the outbreak of bacterial common blight in field beans (*Phaseolus vulgaris* L) under Zimbabwean conditions. *Journal of Sustainable Development in Africa* 10: 221–233.

Langlois P.A., Snelling J., Hamilton J.P., Bragard C., Koebnik R., …Leach J.E, 2017. Characterization of the *Xanthomonas translucens* complex using draft genomes, comparative genomics, phylogenetic analysis, and diagnostic LAMP assays. *Phytopathology* 107: 519–527.

Letunic I., Bork P., 2011. Interactive Tree Of Life v2: online annotation and display of phylogenetic trees made easy. *Nucleic Acids Research* 39: W475–W478.

López M.M., Lopez-Soriano P., Garita-Cambronero J., Beltrán C., Taghouti G., … Marco-Noales E., 2018. *Xanthomonas prunicola* sp. nov., a novel pathogen that affects nectarine (*Prunus persica* var. *nectarina*) trees. *International Journal of Systematic and Evolutionary Microbiology* 68: 1857–1866.

Maes M., Garbeva P., Kamoen O., 1996. Recognition and detection in seed of the *Xanthomonas* pathogens that cause cereal leaf streak using rDNA spacer sequences and polymerase chain reaction. *Phytopathology* 86: 63–69.

Malavolta Jr. V.A., de Oliveira M.A.R, de Oliveira A.R., 2000. Identification and survival of *Xanthomonas translucens* pv. *cerealis* in seeds and crop debris of wheat. *Summa Phytopathologica* 26: 20–23.

Mhedbi-Hajri N., Hajri A., Boureau T., Darrasse A., Durand K., … Jacques M.A., 2013. Evolutionary history of the plant pathogenic bacterium *Xanthomonas axonopodis*. *PLoS One* 8: e58474.

Nagel, R., Peters R.J., 2017. Investigating the phylogenetic range of gibberellin biosynthesis in bacteria. *Molecular Plant-Microbe Interactions* 30: 343–349.

Parkinson N., Aritua V., Heeney J., Cowie C., Bew J., Stead D., 2007. Phylogenetic analysis of *Xanthomonas* species by comparison of partial gyrase B gene sequences. *International Journal of Systematic and Evolutionary Microbiology* 57: 2881–2887.

Parkinson N., Cowie C., Heeney J., Stead D., 2009. Phylogenetic structure of *Xanthomonas* determined by comparison of *gyrB* sequences. *International Journal of Systematic and Evolutionary Microbiology* 59: 264–274

Paul V.H., Smith I.M., 1989. Bacterial pathogens of *Gramineae*: systematic review and assessment of quarantine status for the EPPO region. *EPPO Bulletin* 19: 33–42

Peng Z., Hu Y., Xie J., Potnis N., Akhunova A., … Liu S., 2016. Long read and single molecule DNA sequencing simplifies genome assembly and TAL effector gene analysis of *Xanthomonas translucens*. *BMC Genomics* 17: 21.

Pesce C., Jacobs J.M., Berthelot E., Perret M., Vancheva T., … Koebnik R., 2017. Comparative genomics identifies a novel conserved protein, HpaT, in proteobacterial type III secretion systems that do not possess the putative translocon protein HrpF. *Frontiers in Microbiology* 8: 1177.

Pieretti I., Cociancich S., Bolot S., Carrère S., Morisset A., … Royer M., 2015. Full genome sequence analysis of two isolates reveals a novel *Xanthomonas* species close to the sugarcane pathogen *Xanthomonas albilineans*. *Genes (Basel)* 6: 714–733.

Rademaker J.L., Norman D.J., Forster R.L., Louws F.J., Schultz M.H., de Bruijn F.J., 2006. Classification and identification of *Xanthomonas translucens* isolates, including those pathogenic to ornamental asparagus. *Phytopathology* 96: 876-884.

Royer M., Koebnik R., Marguerettaz M., Barbe V., Robin G.P., … Cociancich S., 2013. Genome mining reveals the genus *Xanthomonas* to be a promising reservoir for new bioactive non-ribosomally synthesized peptides. *BMC Genomics* 14: 658.

Sands D.C., Fourrest E., 1989. *Xanthomonas campestris* pv. *translucens* in North and South America and in the Middle East. *EPPO Bulletin* 19: 127–130.

Sayers E.W., Agarwala R., Bolton E.E., Brister J.R., Canese K., … Ostell J., 2019. Database resources of the National Center for Biotechnology Information. *Nucleic Acids Research* 47: D23–D28.

Schaad W., Jones J.B., Chun W., 2001. *Laboratory Guide for Identification of Plant Pathogenic Bacteria, Third Edition*. American Phytopathological Society Press, St Paul, MN, USA, 373 pp.

Smith E.F., 1917. A new disease of wheat. *Journal of Agricultural Research* 10: 51–53.

Timilsina S., Kara S., Jacques M.A., Potnis N., Minsavage G.V., … Fischer-Le Saux M., 2019. Reclassification of *Xanthomonas gardneri* (ex Šutič 1957) Jones *et al.* 2006 as a later heterotypic synonym of *Xanthomonas cynarae* Trébaol *et al.* 2000 and description of *X. cynarae* pv. *cynarae* and *X. cynarae* pv. *gardneri* based on whole genome analyses. *International Journal of Systematic and Evolutionary Microbiology* 69: 343–349.

Triplett L.R., Shidore T., Long J., Miao J., Wu S., … Leach J.E., 2016. AvrRxo1 is a bifunctional type III secreted effector and toxin-antitoxin system component with homologs in diverse environmental contexts. *PLoS One* 11: e0158856.

Triplett L.R., Verdier V. Campillo T., Van Malderghem C., Cleenwerck I., … Leach J.E, 2015. Characterization of a novel clade of *Xanthomonas* isolated from rice leaves in Mali and proposal of *Xanthomonas maliensis* sp. nov. *Antonie Van Leeuwenhoek* 107: 869–881.

Van Doorn J., Roebroeck E.J.A., 1993. *Xanthomonas campestris* pv. *hyacinthi*: cause of yellow disease in *Hyacinthus*. In: *Xanthomonas* (J. G. Swings, E. L. Civerolo, ed.), Springer Science+Business Media, Dordrecht, The Netherlands, 83–91.

Van den Mooter M., Swings J., 1990. Numerical analysis of 295 phenotypic features of 266 *Xanthomonas* strains and related strains and an improved taxonomy of the genus. *International Journal of Systematic Bacteriology* 40: 348–369.

Vauterin L., Hoste B., Kerter K., Swings J., 1995. The relations ships within the genus *Xanthomonas* and a proposal for a new classification. *International Journal of Systematic Bacteriology* 45: 472–489.

Vicente J.G., Rothwell S., Holub E.B., Studholme D.J, 2017. Pathogenic, phenotypic and molecular characterisation of *Xanthomonas nasturtii* sp. nov. and *Xanthomonas floridensis* sp. nov., new species of *Xanthomonas* associated with watercress production in Florida. *International Journal of Systematic and Evolutionary Microbiology* 67: 3645–3654.

Wichmann F., Vorhölter F.J., Hersemann L., Widmer F., Blom J., … Kölliker R., 2013. The noncanonical type III secretion system of *Xanthomonas translucens* pv. *graminis* is essential for forage grass infection. *Molecular Plant Pathology* 14: 576–588.

Wiese M.V., 1987. *Compendium of Wheat Diseases*. 2nd ed. American Phytopathological Society, St. Paul, MN, USA, 112 pp.

Young J.M., Park D.C., Shearman H.M., Fargier E., 2008. A multilocus sequence analysis of the genus *Xanthomonas*. *Systematic and Applied Microbiology* 31: 366–377.

Zarei S., Taghavi S.M., Hamzehzarghani H., Osdaghi E., Lamichhane J.R., 2018. Epiphytic growth of *Xanthomonas arboricola* and *Xanthomonas citri* on nonhost plants. *Plant Pathology* 67: 660–670.

Zillinsky F.J., 1983. *Common Diseases of Small Grain Cereals. A Guide to Identification.* International Maize and Wheat Improvement Center (CIMMYT), México, D.F., Mexico, 141 pp.

10

Distribution of large-spored *Alternaria* species associated with early blight of potato and tomato in Algeria

Djida AYAD[1,2,*], Djihad ARIBI[1], Bruno HAMON[2], Abdelaziz KEDAD[1], Philippe SIMONEAU[2], Zouaoui BOUZNAD[1]

[1] *ENSA-El Harrach, Laboratoire de Phytopathologie et Biologie Moléculaire. Algiers, Algeria*

[2] *UMR 1345 IRHS, Université d'Angers-INRA-Agrocampus Ouest, SFR QUASAV, 42 Rue G. Morel, F-49071 Beaucouzé, France*

*Corresponding author: medaydji@gmail.com

Competing Interests: The Author(s) declare(s) no conflict of interest.

Editor: Vladimiro Guarnaccia, University of Torino, Italy.

Abstract. Potato and tomato are important crops in Algerian agriculture, and both are threatened by abiotic and biotic stresses, and early blight is a major disease affecting both crops. Surveys carried out from 2012 to 2015 in 12 major growing regions for these crops yielded a total of 247 *Alternaria* isolates having morphological and cultural characteristics of sections *Alternaria* and *Porri*. Since early blight symptoms and morphological characteristics of the isolates did not allow sharp distinction between the different large-spored species of *Alternaria*, the isolates in section *Porri*, often considered primary causes of the diseases, were selected for molecular characterization by diagnostic PCR using specific primers. This allowed species identification of 147 *Alternaria* isolates as *A. solani*, *A. protenta*, *A. grandis* or *A. linariae*. These species were present on potato and tomato crops at varying frequencies, depending on the hosts and on bioclimatic locations. Pathogenicity tests for the four species, on detached leaflets and whole seedlings, showed that all were pathogenic to potato and tomato, with varying virulence. These results suggest that parasitic specialization of these *Alternaria* species on solanaceous plants should be reconsidered.

Keywords. Solanaceous crops, parasitic specialization, prevalence.

INTRODUCTION

Potato (*Solanum tuberosum* L.) and tomato (*S. lycopersicum* L.) are important crops grown in Algeria. They are distributed differently in the various bioclimatic zones of the country, and are grown with different cultural practices. Potato crops occupy significant area in several regions of Algeria, from north (Ain Defla) to south (El Oued), and from east (Skikda, Guelma) to west (Mostaganem, Mascara). Tomato, however, is mainly cultivated in the northern part of the country, less extensively and mostly in plastic houses, as in the Biskra region. Both crops are often cultivated side by side or follow one another in rotations for several years, in many coastal regions of the country. Intensification of these two crops without compliance to good agricultural practices led to the development of important diseases caused by fungi and oomycetes, including early blight and late blight.

The first reference mentioning *A. solani* Sorauer (= *Macrosporium solani* Ellis & G. Martin) as a pathogen causing potato leaf blight was that of Galloway (1891) in Australia. Chester (1892) then reported this pathogen in the United States of America, other solanaceous cultivated plants (Van der Waals *et al.*, 2001; Odilbekov, 2015). A few years later, *A. solani* and *A. alternata* (Fr.) Keissl. were identified on potato leaves but *A. alternata* was considered first as saprophytic (Jones and Grout, 1897). It was not until 1984 that *A. alternata* was recognized as pathogenic on several solanaceous plants (Droby *et al.* 1984). The name "early blight" for the disease caused by *A. solani* was coined by Jones (1893) to distinguish this from late blight. The symptoms of the early blight appear on the leaves as dark, elongated or circular lesions each with concentric rings surrounded by a yellow halo.

On the basis of cultural techniques and morphological characteristics of the conidia (size and beak length), Simmons (2000, 2007) described several new *Alternaria* species on *Solanaceae*, that may be responsible for early blight on potato and tomato. Thus, new species were described, which are morphologically very similar to *A. solani*. These include *A. grandis* E.G. Simmons on *S. tuberosum*, *A. tomatophila* E.G. Simmons on *S. lycopersicum*, *A. cretica* E.G. Simmons & Vakal. On *S. lycopersicum* var. *esculentum*, and *A. subcylindrica* E.G. Simmons & R.G. Roberts on *S. lycopersicum* var. *cerasiforme*.

More recently, molecular and genomics techniques applied to *Alternaria* allowed redefinition of various species with large conidia commonly isolated from *Solanaceae*, and considered as the primary causes of early blight. Woudenberg *et al.* (2014), by synonymizing *A. linariae* (Neerg.) E.G. Simmons with *A. cretica* E.G. Simmons & Vakal., *A. cucumericola* E.G. Simmons & C.F. Hill, *A. subcylindrica*, *A. tabasco* E.G. Simmons & R.G. Roberts, and *A. tomatophila*, have expanded the host range of *A. linariae* to *Solanaceae*, *Cucurbitaceae*, and *Scrophulariaceae*. *Alternaria linariae*, a common pathogen of tomato, was also reported on potato during surveys carried out in Algeria (Ayad *et al.*, 2018). *Alternaria Grandis*, which was confused for a long time with *A. solani*, has been regularly reported in many countries on potato (Lourenço *et al.*, 2009; Cardoso, 2014; Bessadat *et al.*, 2016; Landschoot *et al.*, 2017). *Alternaria grandis* was also recently isolated from tomato (Bessadat *et al.*, 2017) in the northwest of Algeria. *Alternaria protenta*, which is closely related to *A. solani* and whose host range was extended to *Asteraceae*, *Euphorbiaceae*, *Gramineae* and *Solanaceae* by Woudenberg *et al.* (2014), was recently detected on potato in Belgium (Landschoot *et al.*, 2017) and in Algeria (Ayad *et al.*, 2017).

The aims of the present study were, firstly, to identify at the species level the large-spored *Alternaria* isolates obtained from surveys carried out during three cropping years on different potato and tomato fields across the main bioclimatic areas of Algeria. Since early blight symptoms and the morphological characteristics of the isolates do not allow distinction between the different large-spored *Alternaria* spp., specific identification of the isolates was based on PCR techniques. The second objective was to assess the aggressiveness of isolates on their respective potato and tomato hosts, and to clarify the parasitic specialization of replace by large-spored *Alternaria* species by artificial cross inoculations *in vitro* and under conditions similar to those occurring in the field.

MATERIALS AND METHODS

Surveys and sampling

Surveys were carried in 60 locations across 12 regions in Algeria, with different cropping systems (fields and greenhouses). Three to four plots were sampled in each region during three successive cropping years: 2012–2013, 2013–2014 and 2014–2015. The majority of the localities were selected from the main potato- and tomato-producing regions represented by different bioclimatic areas of Algeria: Mediterranean climate (Algiers, Tipaza, Guelma and Skikda), semi-arid climate (Mostaganem, Chlef, Mascara, Ain Defla and Bouira), and arid climate (Laghouat, Biskra and El Oued) (Figure 1). Sampling was carried out from potato and tomato plant organs showing the typical symptoms of early blight: 164 isolates were collected from potato leaves essentially and 83 from leaves, stems and fruits of tomato.

Isolation, purification and conservation of isolates

Isolations were performed according to the method of Van der Waals *et al.* (2004). Small tissue pieces (3 to 4 mm^2) were cut from individual lesion edges, disinfected in 1% active NaOCl for 3 min, washed twice in sterile distilled water, and then dried with absorbent sterile paper. Four pieces of tissues were plated on individual Petri plates containing Potato Dextrose Agar (PDA), and these were incubated at 22°C under continuous light for 1 to 2 weeks. The purification of the *Alternaria* spp. isolates as single conidium cultures was car-

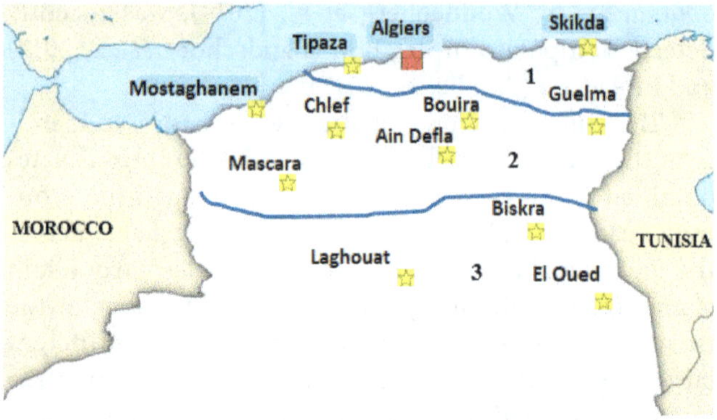

Figure 1. Geographic localities in the different bioclimatic areas: 1 = Mediterranean climate, 2 = semi-arid climate, 3 = arid climate

ried out between the 3rd and the 5th day of incubation. Plates containing the single conidium cultures were then placed under near-UV (12h dark and 12h light) to induce sporulation and allow further morphological characterization.

Single conidium cultures for short-term use were grown in Petri dishes containing a PDA for 12 d then stored at 4°C. For medium term use, 12-d-old PDA cultures in test tubes were stored at 4°C. For long term use, mycelium/agar plugs (8 mm diam.) were stored at 80°C in cryotubes containing sterile glycerol (30% w/w).

Identification of the large-spored Alternaria species using molecular markers

On the basis of the cultural and the morphological characteristics, 155 isolates with large conidia (out of a total 247 isolates) were selected for molecular characterization by PCR using specific primers, for analysis of genomic DNA by PCR/RFLP, using double enzymatic digestion of a portion of the calmodulin gene and sequencing the calmodulin and the RPB2 genes. Isolates were grown on PDA medium at room temperature for 10 d. Total genomic DNA from fungal samples was extracted according to the method described by Goodwin and Lee (1993). The DNA extracts of the 155 isolates were amplified using two sets of specific primer pairs to differentiate between *A. linariae* and *A. solani* and related species (*A. grandis* and *A. protenta*) (Gannibal *et al.*, 2014). The primer pair OasF7 and OasR6 amplified a 164 bp fragment from the Alt al gene of *A. solani*, *A. grandis* and *A. protenta*, and the primer pair OatF4 and OatR2 amplified a 438 pb fragment from the calmodulin encoding gene of *A. linareae*. The PCR conditions were as described by Gannibal *et al.* (2014). To differentiate between *A. grandis* and *A. solani sensu lato* (i.e. including *A. protenta*), a portion of the calmodulin gene was amplified from the isolates that each gave a positive signal with the primer pair OasF7 – OasR6 using the primer pair CALDF1 / CALDR1 (Lawrence *et al.*, 2013). The resulting PCR products were double digested with the restriction enzymes *Rsa*I and *Hae*II. The digestion products were then separated by electrophoresis in 2% agarose gels and visualized under UV light after staining with ethidium bromide. Predicted restriction patterns mainly differed by the size of the larger fragment, i.e. 420 bp for *A. solani sensu lato* and 292 bp for *A. grandis* (Figure 2). Identification of the *Alternaria* species was validated by sequencing the fragment of the calmodulin locus amplified as described above from 42 isolates. To confirm the presence of *A. protenta*, a portion of the RPB2 locus was amplified from the isolates identified as *A. solani* based on their calmodulin sequence, using the primer pair RPB2-5F2 (Sung *et al.*, 2007) and RPB2-7cR (Liu *et al.*, 1999), and the resulting PCR product was sequenced. Sanger sequencing was performed by GATC Biotech. Sequence analyses were carried out using the Phylogeny.fr web service (Dereeper *et al.*, 2008). Multiple sequence alignments were generated with MUSCLE and curated using the Gblocks algorithm. Maximum likelihood (ML) analyses were performed with PhyML. The robustness of the ML topologies was evaluated using the Shimodaira-Hasegawa (SH)-like test for branches.

Pathogenicity tests of large-spored Alternaria species on their respective hosts

To confirm pathogenicity of isolates, inoculations onto detached leaflets and whole plants were performed on susceptible varieties of tomato (cv. Marmande) and

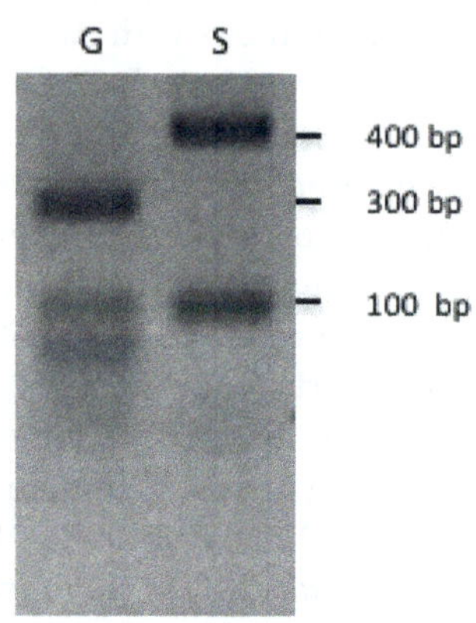

Figure 2. Typical restriction patterns obtained by double digestion of PCR product corresponding to a portion of the calmodulin gene from *A. solani* and *A. grandis*. Size of the larger fragment, 420 bp for *A. solani sensu lato* (S) and 292 bp for *A. grandis* (G)

potato (cv. Spunta). These isolates were selected for their ability to produce conidia. Conidium suspensions were prepared from 15-d-old cultures. The final concentration of the suspensions was adjusted to 10^4 conidia mL^{-1}. The inoculations were performed by depositing drops (20 µL each) on detached leaflets or by spraying suspensions onto whole plants. Disease severity was evaluated at 15 d -post-inoculation, using a visual rating scale from 1 to 9 expressing the extension of necrosis CIP (International Potato Center) (2008).

RESULTS

Characterization of the large-spored Alternaria species according to early blight symptoms on potato and tomato

Isolation from plant samples showing typical early blight symptoms yielded 247 isolates (164 from potato and 83 from tomato). On the basis of the cultural and morphological characteristics of the colonies and conidia, the isolates were divided in three groups. The first (Group A: 92 isolates) included all the isolates with abundant conidium production and small catenulate conidia and identified as species belonging to section *Alternaria*. The second (Group B: 92 isolates) contained all isolates producing large solitary beaked conidia, related to species belonging to the section *Porri*. The third (Group C: 63 isolates) consisted of large-spored isolates that only formed sterile mycelium after culture purification. In relation to the host plants, 89% of the isolates belonging to the section *Porri* originated from potato, and 73% of the non-sporulating isolates originated from tomato. No correlation was found between the presence of yellow halos on symptomatic leaves and the production of a diffusible pigment on growth medium by the isolates from the diseased samples. Similarly, the various forms of symptoms generated colonies producing either large or small conidia. Thus, there was no relationships between necrosis size (small, medium or large necrotic lesions), the types of the colonies in culture medium (sporulating or non-sporulating) and the conidium types (large solitary or small catenulate conidia).

Molecular characterization and identification of large-spored Alternaria species

Given the very similar morphological characteristics of the conidia, and the difficulty to distinguish the different *Alternaria* species within the section *Porri*, molecular markers and DNA-sequencing were used for their identification. The small-spored isolates (Group A) were not included in this examination. The three molecular approaches allowed clear species identification for 147 of 155 isolates from Groups B and C, while eight isolates could not be identified due to problems in DNA extraction and amplification at the selected loci (Table 1). Preliminary tests were carried out for the identification of the 147 isolates using two pairs of specific primers that differentiate *A. solani*, *A. protenta* and *A. grandis* (OasF7/OasR6) from *A. linariae* (OATF4/OATR2). All the tested isolates generated positive signals with one of the two primer pairs, confirming the presence of these species on potato and tomato crops in Algeria. In this analysis, 37 isolates including seven from potato and 30 from tomato, were identified as *A. linariae* by amplification of the 438 pb from the gene encoding for calmodulin (Table 1). A PCR product corresponding to a fragment of the Alt a1 gene (164 bp) was amplified from the remaining 110 isolates. These isolates, of which 86 were isolated from potato and 24 from tomato, could be assigned to either *A. solani* or to phylogenetically related-species, i.e. *A. grandis* and *A. protenta*. Further identification was conducted in order to differentiate between *A. solani* and *A. grandis* isolates using PCR/RFLP, by amplifying the calmodulin locus followed by restriction enzyme digestion (*Hae*II and *Rsa*I). Based on the characteristic electrophoretic profiles of the restricted PCR products, *A. solani* and *A. protenta* were separated from *A. grandis*. Thus, of the 110 isolates that gave positive signals with the OasF7/OasR6 primer pair, 93 were *A. solani*, 12 were *A. grandis* and five showed abnormal enzymatic restriction digestion patterns that may be due to partial digestion with at least one of the restriction enzymes. These five isolates, as well as 37 other representative isolates obtained from the two host plants collected in nine of the twelve surveyed Algerian regions and representing the three identified species (*A. linariae*, *A. grandis* and *A. solani*), were selected for species confirmation by DNA sequencing which was firstly performed at the calmodulin locus. Phylogenetic analyses of sequences derived from the calmodulin locus of the 42 isolates (Figure 3) separated the isolates into two clades: the first (A) contained the isolates previously identified by PCR with specific primers as *A. linariae*, and the second (B) included two sub clades (B1 and B2) regrouping isolates previously identified as *A. grandis* (sub clade B1) and *A. solani* (sub clade B2). DNA sequencing at the RPB2 locus then carried out on the isolates of the sub clade B2 gave the phylogenetic tree shown in Figure 4, in which among the 17 isolates of *A. solani*, three were identified as *A. protenta*, two of which were isolated from the potato and one from tomato.

Table 1. Identification of large-spored of *Alternaria* spp isolates.

Isolate	Species[a]	Host	Geographic origin	PCR OATF4/ OATR2	PCR OAsF7/OAsR6	PCR RFLP Profile[b]	CAL	Rpb2
DA001	***A. linariae***	Potato	Tipaza	+	-	ND	MH243795	–
DA002	A. linariae	Potato	Tipaza	+	-	ND	MH243769	–
DA003	***A. linariae***	Potato	Tipaza	+	-	ND	MH243793	–
DA005	*A. solani s.l*	Potato	Alger	-	+	S	–	–
DA006	***A. linariae***	Potato	Alger	+	-	ND	MH243789	–
DA007	***A. linariae***	Potato	Alger	+	-	ND	MH243794	–
DA008	***A. solani***	Potato	Alger	-	+	S	MH243805	MH243818
DA009	***A. grandis***	Potato	Alger	-	+	?	MH243790	–
DA010	*A. solani s.l*	Potato	Alger	-	+	S	–	–
DA011	***A. solani***	Potato	Alger	-	+	S	MH243806	MH243820
DA012	*A. solani s.l*	Potato	Alger	-	+	S	–	–
DA013	*A. solani s.l*	Potato	Alger	-	+	S	–	–
DA014	*A. solani s.l*	Potato	Alger	-	+	S	MH243796	–
DA015	*A. solani s.l*	Potato	Bouira	-	+	S	–	–
DA016	*A. solani s.l*	Potato	Bouira	-	+	S	–	–
DA017	*A. solani s.l*	Potato	Bouira	-	+	S	–	–
DA018	*A. solani s.l*	Potato	Bouira	-	+	S	–	–
DA019	***A. solani***	Potato	Bouira	-	+	S	MH243808	MH243822
DA020	*A. solani s.l*	Potato	Bouira	-	+	S	–	–
DA021	*A. solani s.l*	Potato	Bouira	-	+	S	–	–
DA022	*A. solani s.l*	Potato	Bouira	-	+	S	–	–
DA025	***A. solani***	Potato	Bouira	-	+	S	MH243786	MH243823
DA026	*A. solani s.l*	Potato	Bouira	-	+	S	–	–
DA028	*A. solani s.l*	Potato	Bouira	-	+	S	–	–
DA029	*A. solani s.l*	Potato	Bouira	-	+	S	–	–
DA030	*A. solani s.l*	Potato	Bouira	-	+	S	–	–
DA031	*A. solani s.l*	Potato	Bouira	-	+	S	–	–
DA032	*A. solani s.l*	Potato	Bouira	-	+	S	–	–
DA033	*A. solani s.l*	Potato	Bouira	-	+	S	–	–
DA034	*A. solani s.l*	Potato	Bouira	-	+	S	–	–
DA035	*A. solani s.l*	Potato	Mascara	-	+	S	–	–
DA036	*A. solani s.l*	Potato	Mascara	-	+	S	–	–
DA037	*A. linariae*	Tomato	Mostaganem	+	-	ND	–	–
DA038	***A. grandis***	Potato	Alger	-	+	G	MH243770	–
DA039	*A. linariae*	Potato	Alger	+	-	ND	–	–
DA040	***A. linariae***	Tomato	Mascara	+	-	ND	MH243792	–
DA041	*A. linariae*	Tomato	Mostaganem	+	-	ND	–	–
DA042	*A. linariae*	Tomato	Mostaganem	+	-	ND	–	–
DA043	*A. solani s.l*	Potato	Mostaganem	-	+	S	–	–
DA045	*A. solani s.l*	Potato	Mascara	-	+	S	–	–
DA046	*A. solani s.l*	Potato	Mostaganem	-	+	S	–	–
DA047	***A. grandis***	Potato	Mostaganem	-	+	G	MH243771	–
DA048	*A. solani s.l*	Potato	Mostaganem	-	+	S	–	–
DA049	*A. solani s.l*	Potato	Chlef	-	+	S	–	–
DA050	***A. solani***	Potato	Chlef	-	+	S	MH243772	MH243809
DA051	***A. grandis***	Potato	Chlef	-	+	G	–	–
DA052	***A. grandis***	Potato	Chlef	-	+	G	MH243773	–

(Continued)

Table 1. (Continued).

Isolate	Species[a]	Host	Geographic origin	PCR OATF4/ OATR2	PCR OAsF7/OAsR6	PCR RFLP Profile[b]	CAL	Rpb2
DA053	*A. solani s.l*	Potato	Chlef	-	+	S	–	–
DA054	***A. grandis***	Potato	Mostaganem	-	+	G	MH243774	–
DA055	***A. grandis***	Potato	Mostaganem	-	+	G	–	–
DA056	*A. solani s.l*	Potato	Mostaganem	-	+	S	–	–
DA057	***A. solani***	Potato	Mascara	-	+	?	MH243791	MH243819
DA058	*A. solani s.l*	Potato	Mascara	-	+	S	–	–
DA059	*A. solani s.l*	Potato	Mostaganem	-	+	S	–	–
DA060	***A. grandis***	Potato	Mostaganem	-	+	G	MH243775	–
DA061	***A. grandis***	Potato	Mostaganem	-	+	G	–	–
DA062	***A. protenta***	Tomato	Mostaganem	-	+	S	MH243797	MH243810
DA063	*A. solani s.l*	Potato	El Oued	-	+	S	–	–
DA064	*A. solani s.l*	Potato	El Oued	-	+	S	–	–
DA065	*A. solani s.l*	Potato	El Oued	-	+	S	–	–
DA066	*A. solani s.l*	Potato	El Oued	-	+	S	–	–
DA067	*A. solani s.l*	Potato	El Oued	-	+	S	–	–
DA068	*A. solani s.l*	Potato	El Oued	-	+	S	–	–
DA069	*A. solani s.l*	Potato	El Oued	-	+	S	–	–
DA070	*A. solani s.l*	Potato	El Oued	-	+	S	–	–
DA071	*A. solani s.l*	Potato	El Oued	-	+	S	–	–
DA072	*A. solani s.l*	Potato	El Oued	-	+	S	–	–
DA073	*A. solani s.l*	Potato	El Oued	-	+	S	–	–
DA074	*A. solani s.l*	Potato	El Oued	-	+	S	–	–
DA075	*A. solani s.l*	Potato	El Oued	-	+	S	–	–
DA076	*A. solani s.l*	Potato	El Oued	-	+	S	–	–
DA077	*A. solani s.l*	Potato	El Oued	-	+	S	–	–
DA078	*A. solani s.l*	Potato	El Oued	-	+	S	–	–
DA079	*A. solani s.l*	Potato	El Oued	-	+	S	–	–
DA080	*A. solani s.l*	Potato	El Oued	-	+	S	–	–
DA081	*A. solani s.l*	Potato	El Oued	-	+	S	–	–
DA082	***A. protenta***	Potato	El Oued	-	+	?	KX870505	KX870507
DA083	*A. solani s.l*	Potato	El Oued	-	+	S	–	–
DA084	*A. solani s.l*	Potato	El Oued	-	+	S	–	–
DA085	*A. solani s.l*	Potato	El Oued	-	+	S	–	–
DA086	***A. protenta***	Potato	El Oued	-	+	S	KX870506	KX870508
DA087	*A. solani s.l*	Potato	El Oued	-	+	S	–	–
DA088	*A. solani s.l*	Potato	El Oued	-	+	S	–	–
DA089	*A. solani s.l*	Potato	El Oued	-	+	S	–	–
DA090	***A. solani***	Potato	El Oued	-	+	S	MH243776	MH243811
DA091	*A. solani s.l*	Potato	El Oued	-	+	S	–	–
DA092	*A. solani s.l*	Potato	El Oued	-	+	S	–	–
DA093	*A. solani s.l*	Potato	Alger	-	+	S	–	–
DA094	***A. solani***	Potato	Alger	-	+	S	MH243798	MH243812
DA096	*A. solani s.l*	Potato	Alger	-	+	S	–	–
DA097	*A. solani s.l*	Potato	Alger	-	+	S	–	–
DA098	*A. solani s.l*	Potato	Alger	-	+	S	–	–
DA099	***A. grandis***	Tomato	Alger	-	+	G	MH243777	–
DA100	***A. linariae***	Tomato	Alger	+	-	ND	MH243807	–

(Continued)

Table 1. (Continued).

Isolate	Species[a]	Host	Geographic origin	PCR OATF4/ OATR2	PCR OAsF7/OAsR6	PCR RFLP Profile[b]	CAL	Rpb2
DA101	***A. linariae***	Tomato	Alger	+	-	ND	MH243778	_
DA102	*A. linariae*	Tomato	Alger	+	-	ND	_	_
DA103	*A. linariae*	Tomato	Alger	+	-	ND	_	_
DA104	*A. linariae*	Tomato	Alger	+	-	ND	_	_
DA107	*A. linariae*	Tomato	Alger	+	-	ND	_	_
DA108	*A. solani s.l*	Potato	Guelma	-	+	S	_	_
DA109	***A. linariae***	Tomato	Tipaza	+	-	ND	MH243779	_
DA110	*A. linariae*	Tomato	Tipaza	+	-	ND	_	_
DA111	***A.linariae***	Tomato	Tipaza	+	-	ND	MH243780	_
DA112	*A. linariae*	Potato	Skikda	+	-	ND	_	_
DA113	***A. solani***	Tomato	Guelma	-	+	S	MH243799	MH243821
DA114	***A. solani***	Potato	Guelma	-	+	?	MH243788	MH243813
DA115	*A. solani s.l*	Potato	Bouira	-	+	S	_	_
DA116	*A. solani s.l*	Potato	Bouira	-	+	S	_	_
DA117	*A. linariae*	Tomato	Biskra	+	-	ND	_	_
DA118	*A. linariae*	Tomato	Biskra	+	-	ND	_	_
DA119	***A. linariae***	Tomato	Biskra	+	-	ND	MH243800	_
DA120	*A. linariae*	Tomato	Biskra	+	-	ND	_	_
DA121	*A. linariae*	Tomato	Biskra	+	-	ND	_	_
DA122	*A. linariae*	Tomato	Biskra	+	-	ND	_	_
DA123	*A. solani s.l*	Tomato	Biskra	-	+	S	_	_
DA124	*A. linariae*	Tomato	Biskra	+	-	ND	_	_
DA125	*A. linariae*	Tomato	Biskra	+	-	ND	_	_
DA126	*A. linariae*	Tomato	Biskra	+	-	ND	_	_
DA127	*A. linariae*	Tomato	Biskra	+	-	ND	_	_
DA128	*A. solani s.l*	Tomato	Biskra	-	+	S	_	_
DA129	*A. solani s.l*	Tomato	Biskra	-	+	S	_	_
DA130	*A. solani s.l*	Tomato	Biskra	-	+	S	_	_
DA131	*A. solani s.l*	Tomato	Biskra	-	+	S	_	_
DA132	*A. solani s.l*	Tomato	Biskra	-	+	S	_	_
DA133	*A. solani s.l*	Tomato	Biskra	-	+	S	_	_
DA134	*A. linariae*	Tomato	Biskra	+	-	ND	_	_
DA135	*A. solani s.l*	Tomato	Biskra	-	+	S	_	_
DA136	***A. solani***	Tomato	Biskra	-	+	S	MH243801	MH243814
DA137	*A. solani s.l*	Tomato	Biskra	-	+	S	_	_
DA138	*A. solani s.l*	Tomato	Biskra	-	+	S	_	_
DA139	*A. solani s.l*	Tomato	Biskra	-	+	S	_	_
DA140	***A. solani***	Tomato	Biskra	-	+	S	MH243802	MH243815
DA141	***A. solani***	Tomato	Biskra	-	+	S	MH243803	MH243816
DA142	***A. solani***	Tomato	Biskra	-	+	S	MH243804	MH243817
DA143	*A. solani s.l*	Tomato	Biskra	-	+	S	_	_
DA144	*A. linariae*	Tomato	Biskra	+	-	ND	_	_
DA145	*A. solani s.l*	Tomato	Biskra	-	+	S	_	_
DA146	*A. solani s.l*	Potato	Tipaza	-	+	S	_	_
DA147	***A. linariae***	Tomato	Tipaza	+	-	ND	MH243781	_
DA148	*A. linariae*	Tomato	Mostaganem	+	-	ND	_	_
DA149	*A .linariae*	Tomato	Mostaganem	+	-	ND	_	_

(Continued)

Table 1. (Continued).

Isolate	Species[a]	Host	Geographic origin	PCR OATF4/ OATR2	PCR OAsF7/OAsR6	PCR RFLP Profile[b]	CAL	Rpb2
DA150	*A. linariae*	Tomato	Mostaganem	+	-	ND	_	_
DA152	***A. grandis***	Tomato	Alger	-	+	G	MH243782	_
DA153	***A. linariae***	Tomato	Alger	+	-	ND	MH243783	_
DA154	***A .grandis***	Tomato	Alger	-	+	?	MH243787	_
DA155	***A. grandis***	Tomato	Alger	-	+	G	MH243784	_
DA156	***A. grandis***	Tomato	Alger	-	+	G	MH243785	_

[a] Species names in bold characters were supported by sequence data; *A. solani s.l* stands for *A. solani sensu lato* (i.e. including *A. protenta*). [b] PCR-RFLP profiles: S = typical of *S. solani*; G = typical of *A. grandis*; ND = not determined; ? = atypical restriction profile

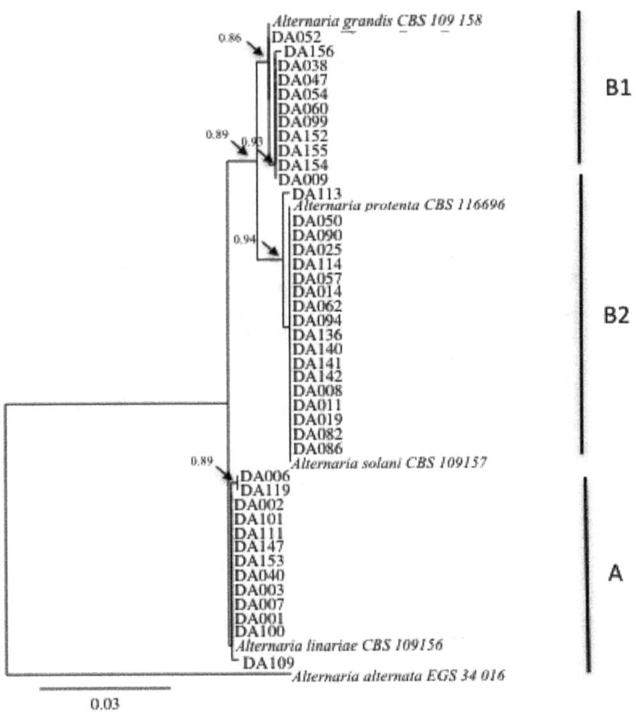

Figure 3. Phylogenetic tree reconstructed by the maximum likelihood method from the alignment of calmodulin sequences of 42 isolates belonging to the *Alternaria* section *Porri*. Bootstrap support values greater than 80 % are indicated by arrows. The calmodulin sequences of the following strains were included as references: *A. linariae* CBS 109156 (GenBank number JQ646257), *A. protenta* CBS 116696 (GenBank number JQ646236), *A. solani* CBS 109157 (GenBank number KJ397981), *A.grandis* CBS 109158 (GenBank number JQ646249). The calmodulin sequence from *A. alternata* EGS 34016 (Genbank number JQ646208) was used to root the tree.

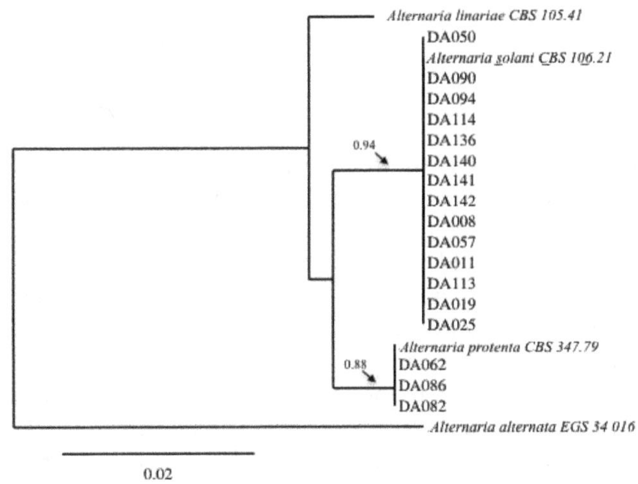

Figure 4. Phylogenetic tree reconstructed by the maximum likelihood method from the alignment of RPB2 sequences of 17 isolates belonging to cluster B2 in the calmodulin phylogeny. Bootstrap support values greater than 80 % are indicated by arrows. The RPB2 sequences of the following strains were included as references: *A. linariae* CBS 105.41 (GenBank number KJ718353), *A. solani* CBS 106.21 (GenBank number KJ718410) *A. protenta* CBS 347.79 (GenBank number KJ718392). The RPB2 sequence from *A. alternata* EGS 34016 (Genbank number JQ646490) was used to root the tree.

Distribution and prevalence of the large-spored Alternaria species in the bioclimatic regions of Algeria

The surveys carried out through the different Algerian potato and tomato growing regions revealed the presence of *A. grandis*, *A. linariae* and the species complex *A. solani-A. protenta* at variable levels. The frequencies of these species varied according to the host plants (potato or tomato) and to the geographical locations (Table 2). *Alternaria solani sensu lato* (i.e. including *A. protenta*) was the most common species in Algeria, representing 65% of the large-spored isolates, followed by *A. linariae* (25%) and *A. grandis* with low isolation frequency (10%). Isolation frequencies of these species varied with the sampled crop, although they were all found on both potato and tomato. *Alternaria solani sensu lato* (80%) and *A. grandis* (64%) were more prevalent on potato than on tomato. Conversely, *A. linariae*

Table 2. Frequencies of the four large-spored *Alternaria* species according to the host plants (potato and tomato) and to the geographical locations

Origin	*A. solani*-*A protenta*	*A. grandis*	*A. linariae*
Host plant origin			
Potato	80.2[a]–82.8[b]	64.3–9.6	18.9–7.5
Tomato	19.8–35.2	35.7–9.2	81.1–55.6
Geographic origin (bioclimatic area)			
Mediterranean	16.7–39.0	50.0–17.0	48.6–44.0
Semi-arid	34.3–70.2	50.0–14.9	18.9–14.9
Arid	49.0–79.7	0–0	32.4–20.3

[a] Percentage of the total number of isolates of the same species according to the host plant or the geographic origin

[b] Percentage of the total number of isolates from the same host or from the same geographic origin

Figure 5. Symptoms on detached leaflets (a, d) and whole tomato (c, e) or potato (b, f) plants inoculated by *A. linariae* isolate DA002 (a, b and c) or by *A. grandis* isolate DA009 (d, e and f)

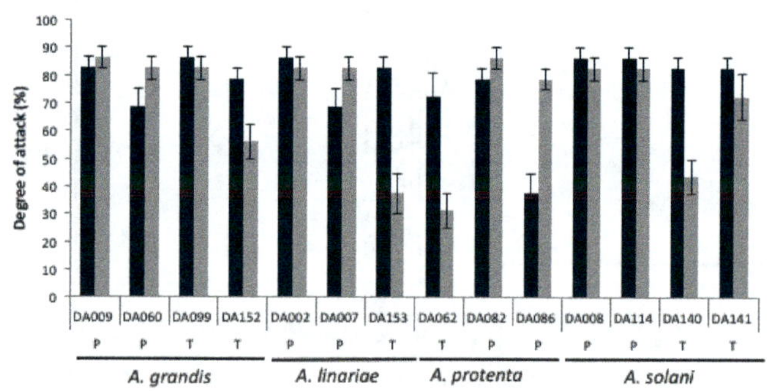

Figure 6. Disease severity evaluated using a visual rating scale (0-100%) at 15 days post-inoculation of tomato (black bars) and potato (grey bars) leaves with isolates representing the four large-spored *Alternaria* species. Letters (P or T) under the isolates identities refer to the host of origin (Potato or Tomato).

was more frequently isolated from tomato than from potato. Locations of sample collection could be divided into three temperature and precipitation zones (Mediterranean, semi-arid or arid). In Mediterranean and arid zones, isolates were almost equally distributed between potato and tomato. *Alternaria linariae* was over-represented in the Mediterranean zone (44% of the collected isolates) compared to the arid zone where it represented only 20% of the collected isolates. The reverse situation was observed for *A. solani sensu lato*. *Alternaria grandis* was not isolated from samples collected from the arid zone, although it represented 15% of isolates from the semi-arid zone and 17% of those from the Mediterranean zone. Only small-spored *Alternaria* isolates were obtained from samples in two locations, i.e. Ain Defla and Laghouat.

Pathogenicity and parasitic specialization of Alternaria species on potato and tomato

Fourteen isolates representing the four large-spored *Alternaria* species were used in the pathogenicity tests. These were: *A. solani* (isolates DA008, DA114, DA140 and DA141), *A. grandis* (isolates DA009, DA060, DA099 and DA152), *A. protenta* (DA062, DA082 and DA086) and *A. linariae* (DA002, DA007 and DA153). The symptoms (necroses surrounded by yellow halos) obtained on detached leaflets in Petri dish assays were confirmed by high degrees of aggressiveness obtained on whole seedling plants of potato and tomato (Figure 5). All the tested isolates produced symptoms on both plant species irrespective of their host plant of origin (Figure 6). The degrees of aggressiveness for each host/pathogen combination varied for the isolates, however. On average, *A. solani* isolates were more aggressive on both tomato and potato, while *A. protenta* isolates were less aggressive with a marked preference for their original host plants.

DISCUSSION

Early blight has long been attributed to large-spored *Alternaria* species, i.e. *A. solani* and *A. grandis* on potato and *A. solani* and *A. linariae* on tomato. Small-spored species related to *A. alternata* are considered to be responsible for the brown spot disease (Nolte, 2008, Tymon *et al.*, 2016). The surveys carried out in Algeria between 2012 and 2015 have shown the presence of several *Alternaria* species producing large conidia in samples from both hosts with typical early blight symptoms, as well as in samples with symptoms that resemble brown spot. On the other hand, many Algerian isolates of the section *Alternaria* collected by Bessadat *et al.* (2017) have caused typical early blight symptoms after artificial inoculations on tomato, and were therefore considered pathogenic but with low aggressiveness. This complicated the attribution of a given symptom to a given *Alternaria* species. Moreover, morphological similarities between species within the *Alternaria* section *Porri* precludes reliable identification at the species level based only on these criteria. In the present work, molecular tools were used to confirm the identity of large-spored isolates originating from tomato and potato leaves. The diagnostic species-specific primers developed by Gannibal *et al.* (2014) were successfully used to quickly characterized 147 isolates from a collection the 155 large-spored isolates, and to distinguish *A. linariae* from isolates that could correspond to either *A. solani*, *A. grandis* or *A. protenta*. No amplification signal was obtained for eight isolates either due to poor quality of the extracted DNA or because these isolates corresponded to another species within section *Porri*. Although never previously observed in Algeria, it has recently been reported that *A. macrospora* may be responsible for leaf spot of tomato in China (Zhang *et al.*, 2017). To discriminate between *A. grandis*, on one hand, and *A. solani* or *A. protenta*, on the other, we took advantage of sequence polymorphism within the calmodulin gene to develop a PCR-RFLP assay. Reliability of these diagnostic methods was checked by sequencing the calmodulin locus, and taken together they allowed us to accurately type more than 90% of the large-spored isolates without the need for nucleotide sequencing. Applying these tools to our isolates, we observed that *A. solani* and *A. linariae* predominate on their respective potato or tomato hosts in most of the surveyed areas. In agreement with recent observations (Bessadat *et al.*, 2017, Ayad *et al.*, 2018), the presence of *A. linariae* and *A. grandis* in Algeria, traditionally considered to be restricted, respectively, to either tomato or potato, was also confirmed on these two plant hosts. *Alternaria linariae* was found on potato and *A. grandis* on tomato in the coastal regions of Algiers and Tipaza, where these two crops often exist side by side or follow each other in the same plot during the seasons. *Alternaria grandis* has not been isolated from the Saharan regions that are characterized by warm and arid periods during each year. Analyses of nucleotidic polymorphism at the RPB2 locus has shown that *A. protenta* was recorded for the first time on tomato in the coastal region of Mostaganem (northwest of Algeria), where the climate is very different from that of the South. This species has already been reported on potato in Belgium (Landschoot *et al.*, 2017), but also in a particular region of the south of Algeria (El Oued), where the temperatures are high and the potato crops are grown under overhead irrigation (Ayad *et al.*, 2017).

Cross inoculations carried out with isolates representing the four identified large-spored *Alternaria* species (*A. solani*, *A. grandis*, *A. linariae* and *A. protenta*) on potato and tomato, showed that aggressiveness toward the two host plants varied according to the isolates but not with respect to particular fungal species. Similar experiments carried out on detached leaves with *A. linariae* and *A. grandis* (Rodrigues *et al.*, 2010; Cardoso, 2014; Gannibal *et al.*, 2014) have previously shown that these two species were able to cause disease on both plants with more severe symptoms on their respective traditional hosts. This suggests that the parasitic specialization of large-spored *Alternaria* species on solanaceous crops should be reconsidered.

ACKNOWLEDGEMENTS

The authors thank the UMR 1345 IRHS, Université d'Angers-INRA-Agrocampus Ouest (France) for hosting D. Ayad, and the Ministry of Higher Education and Scientific Research of Algeria for awarding her scholarship.

LITERATURE CITED

Ayad D., Leclerc S., Hamon B., Kedad A., Bouznad Z., Simoneau P., 2017. First report of early blight caused by *Alternaria protenta* on potato in Algeria. *Plant Disease* 101: 836.

Ayad, D., Hamon B., Kedad A., Bouznad Z., Simoneau P., 2018. First report of early blight caused by *Alternaria linariae* on potato in Algeria. *Plant Disease* 102: 2651.

Bessadat, N., Hamon B., Henni J.E., Simoneau P., 2016. First report of tomato early blight caused by *Alternaria grandis* in Algeria. *Plant Disease* 100: 533.

Bessadat N., Berruyer R., Hamon B., Bataille-Simoneau N., Benichou S., … Simoneau P., 2017. *Alternaria* species associated with early blight epidemics on tomato and other *Solanaceae* crops in northwestern Algeria. *European Journal of Plant Pathology* 148: 181–197.

Cardoso, C.R., 2014. *Potato and tomato early blight: molecular identification of Alternaria species, Host range and epidemics.* PhD Thesis, Universidade Federal de Viçosa, Brasil, 90 pp.

CIP (International Potato Center), 2008. *Alternaria solani* and *Alternaria alternata* on potato and outdoor production of tomato. *Bulletin OEPP/EPPO* 38: 319–321.

Dereeper, A., Guignon V., Blanc G., Audic S., Buffet S., … Gascuel O., 2008. Phylogeny. Fr: robust phylogenetic analysis for the non-specialist. *Nucleic Acids Research* 1: 36.

Droby, S., Dinoor A., Prusky D., Barkaigolan R., 1984. Pathogenicity of *Alternaria alternata* on potato in Israel. *Phytopathology* 74: 537-542.

Gannibal, P.B., Orina A.S., Mironenko N.V., Levitin M.M., 2014. Differentiation of the closely related species, *Alternaria solani* and *A. tomatophila*, by molecular and morphological features and aggressiveness. *European Journal of Plant Pathology* 139: 609–623.

Galloway, B.T., 1891. The new potato disease. *Garden and Field*, Adelaide, Australia 16, 158.

Goodwin, D.C., Lee S.B., 1993. Microwave miniprep of total genomic DNA from fungi, plants, protists and animals for PCR. *Biotechniques* 15: 438–444.

Jones, L.R., 1893. The new potato disease or early blight. *Vermont Agricultural Experimental Station Bulletin* 6: 66–70.

Jones, L.R., Grout A.J., 1897. Notes on two species of *Alternaria*. *Bulletin of the Torrey Botanical Society* 24: 254–258.

Landschoot, S., Vandecasteele M., De Baets B., Höfte M., Audenaert K., Haesaert G., 2017. Identification of *A. arborescens*, *A. grandis*, and *A. protenta* as new members of the European *Alternaria* population on potato. *Fungal Biology* 121: 172-188.

Lawrence, D.P., Gannibal P.B, Peever T.L., Pryor B.M., 2013. The sections of Alternaria: Formalizing species-groups concepts. Mycologia, 105: 530–546.

Liu, Y.J., Whelen S., Hall B.D., 1999. Phylogenetic relationships among ascomycetes: evidence from an RNA polymerase II subunit. *Molecular Biology and Evolution* 16: 1799 -1808.

Lourenço, V. Jr., Moya A., González-Candelas F., Carbone I., Maffia L.A., Mizubuti E.S.G., 2009. Molecular diversity and evolutionary processes of *Alternaria solani* in Brazil inferred using genealogical and coalescent approaches. *Phytopathology* 99: 765–774.

Nolte, P., 2008. Brown Spot and Black Pit of Potato: The Other Early Blight. *American Vegetable Grower* 56: 32–33.

Odilbekov, F., 2015. *Resistance to Early Blight in Potato and Genetic Structure of the Pathogen Population in Southeast Sweden*. PhD Thesis, Swedish University of Agricultural Science, 46 pp.

Rodrigues, T.T.M.S., Berbee M.L., Simmons E.G., Cardoso C.R., Reis A., Maffia L.A., 2010. First report of *Alternaria tomatophila* and *A. grandis* causing early blight on tomato and potato in Brazil. *New Disease Reports* 22: 28.

Simmons E.G., 2000. *Alternaria* themes and variations (244-286) species on *Solanaceae*. *Mycotaxon* 75: 1-115.

Simmons, E.G., 2007. *Alternaria*: an identification manual. CBS Biodiversity Series 6. CBS Fungal Biodiversity Centre, Utrecht, Netherlands, 775 pp.

Sung, G.H., Sung J.M., Hywel-Jones N.L., Spatafora J.W., 2007. A multi-gene phylogeny of Cavicipitaceae (Ascomycota, Fungi): identification of localized incongruence using a combinational bootstrap approach. *Molecular Phylogenetics and Evolution* 44: 1204- 1223.

Tymon L.S., Peeverand T.L., Johnson D.A., 2016. Identification and enumeration of small-spored *Alternaria* species associated with potato in the U.S. Northwest. *Plant Disease* 100: 465-472.

Van der Waals, J.E., Korsten L., Aveling T.A.S., 2001. A review of early blight of potato. *African Plant Protection* 70: 91-102.

Van der Waals, J.E., Korsten L., Slippers B., 2004. Genetic diversity among *Alternaria solani* isolates from potatoes in South Africa. *Plant Disease* 88: 959-964.

Woudenberg, J.H.C., Truter M., Groenewald J.Z., Crous P.W., 2014. Large-spored *Alternaria* pathogens in section *Porri* disentangled. *Studies in Mycology* 79: 1–47.

Zhang, M., Li P.H., Gen Y.H., Guo Y.X., Wu H.Y., 2017. First report of *Alternaria macrospora* causing leaf spot of *Solanum lycopersicum* in China. *Plant Disease* 101: 1545.

11

Grapevine trunk disease in European and Mediterranean vineyards: occurrence, distribution and associated disease-affecting cultural factors

Lucia GUERIN-DUBRANA[1,*], Florence FONTAINE[2], Laura MUGNAI[3]

[1] *Université de Bordeaux, ISVV, UMR1065 Santé et Agroécologie du Vignoble, Bordeaux Sciences Agro, F-33175 Gradignan, France*
[2] *Structure Fédérative de Recherche (SFR) Condorcet-FR CNRS 3417, Université de Reims Champagne-Ardenne, Unité de Recherche Résistance Induite et Bioprotection des Plantes EA 4707, BP 1039, (Cedex 2), F-51687 Reims, France*
[3] *Dipartimento di Scienze e Tecnologie Agrarie, Alimentari, Ambientali e Forestali (DAGRI) - Sez. Patologia vegetale ed Entomologia, University of Florence, Piazzale delle Cascine 28, 50144 Florence, Italy*
*Corresponding author: lucia.guerin@agro-bordeaux.fr

Competing Interests: The Author(s) declare(s) no conflict of interest.

Editor: Alan J.L. Phillips.

Abstract. A survey was carried out in 22 European and some non-European Mediterranean countries in 2015 and 2016, to gain insights into the main fungal grapevine trunk diseases (GTDs) in Europe. Information was obtained from 105 viticulture regions, representing approx. four million ha of vineyards. Vintage and table wines are the main industry products, followed by nursery plants and table grapes. Diverse scion cultivars are grown, with 13 cultivars providing 50% of survey respondent production, and Riesling Italico, Chardonnay, Cabernet Sauvignon and Merlot predominating. The five most used rootstocks are SO4, 110R, K5BB, 41B and 1103P, together representing about 50% of survey respondent production. Despite the diversity of grapevine training, Guyot and cordon remain the most widespread methods, and grapevines are mainly hand-pruned. The use of pneumatic shears and mechanical pruning is becoming more common in some countries. The survey questionnaire proposed a simplified set of symptoms or diseases occurring in mature vineyards [apoplexy, esca complex leaf symptoms (including Grapevine Leaf Stripe Disease, GLSD), dead cordon, Eutypa related dieback, Botryosphaeria or unidentified agent dieback, Phomopsis cane and leaf spot]. Apoplexy and esca/GLSD, defining the esca complex, were the most frequent and increasing syndromes in almost all countries, except Israel and the United Kingdom, where these diseases are not present or not recorded in the survey. Dead cordon and Phomopsis cane and leaf spot were mentioned as occurring in a large number of regions but not frequently. Within individual countries, the profile of GTDs varied according to the region, and to pedo-climatic and production conditions. GTDs on young vines were uncommon, but remain a subject of concern in some of European countries with large nursery production. This overview on spread and relevance of GTDs in Europe and related aspects will be a useful starting point for policy makers and for collaborative research on factors contributing to the increasing disease incidence of GTDs.

Keywords. Esca complex, Botryosphaeriaceae, *Phomopsis*, *Eutypa*, dieback.

INTRODUCTION

Europe has the greatest concentration of vineyards in the world. In 2016, Continental European vineyards covered approx. 4 million ha, which is more than the half the world vineyard area (Fontaine *et al.*, 2016). Between 2009 and 2014, the average annual production of wine was 167 million hectolitres (hL), representing 65% of world production and 70% of global wine exports (https://ec.europa.eu/agriculture/wine/statistics_en). The world wine market, considered by the sum of exports for all producing countries, is 104.3 million hL, valued at 28.3 billion euro. These production data reflect the economic importance of the wine industry in Europe. European vineyards also embody strong cultural heritage, demonstrated by the diversity of grapevine cultivars and viticultural practices (Black and Ulin, 2013).

Increasing reports of damage caused by grapevine trunk diseases (GTDs) in the last 20 years have raised concerns about these diseases affecting the culturally and economically important grapevine industries. These diseases, causing decline and plant death, are one of the greatest threats to the wine industry (Fontaine *et al.*, 2016). Increased disease incidence and severity are commonly attributed to different factors, including expanded planted area and increased productivity in the 1990s, and changes in cultural practices (Surico *et al.*, 2004). The ban of sodium arsenite in the beginning of this century for disease management in the European Union, has also been suggested as a factor, but it is well known that the upsurge of symptoms was recorded throughout Europe much earlier, at end of the 1980s (Graniti *et al.*, 2000; Surico *et al.*, 2004; Bertsch *et al.*, 2013).

GTDs cause reduced yields, they decrease the longevity of vineyards and reduce wine quality (Calzarano *et al.*, 2004, 2018; Lorrain *et al.*, 2012). In France, for example, about 13% of French vineyards are affected by wood diseases (Grosman and Doublet, 2012). In Sicily, Sidoti *et al.* (2000) reported a case where 15% of young vines in a vineyard showed signs of decline. Decline of young vines associated with black foot agents has been reported in different countries (Rego *et al.*, 2000; Halleen *et al.*, 2006, 2007; Cardoso *et al.*, 2013; Dos Santos *et al.*, 2014; Carlucci *et al.*, 2017; Úrbez-Torres *et al.*, 2014; Langenhoven *et al.*, 2018). Consequently, large disease management efforts are required to limit the spread and damage of these diseases, and to maintain the high levels of quality and productivity of European viticulture and wine production, as a major industry in Europe (Gramaje and Di Marco 2015; Gramaje *et al.*, 2018; Mondello *et al.*, 2018a).

Establishment of GTD control methods must consider the present socio-economic situation and new challenges. The increase in vineyard area in Europe is probably due to new European Union regulations making it easier to plant new vineyards (Reg. EU 1308/2013, AA.VV., 2013). Climatic changes also permit expanded geographical distribution of vineyards. The boundary of European viticulture has shifted northwards with new viticulture regions in the United Kingdom and Belgium (Mozell and Thach, 2014). In southern European countries, adapted cultural practices, such as irrigation, are developing to compensate water stress (Songy *et al.*, 2019).

Knowledge of the distribution of GTDs and the main factors associated with their development is essential to predict their spread, and to improve disease management. Data on some GTDs is sparse and incomplete through national or regional surveys, and a comprehensive inventory of GTD occurrence and frequency in Europe and Mediterranean regions is urgently required. Therefore, the European Cooperation in Science and Technology (COST) Action FA1303 "Sustainable Control of Grapevine Trunk Disease" (Fontaine and Armengol, 2014) created a network of multidisciplinary scientific institutes and companies to improve understanding of GTDs. A specific objective of this COST Action was to develop an overview of the occurrence of the different GTDs in each of the main viticulture regions of the European Union and neighbouring Mediterranean countries.

GTDs include several diseases, all defined as syndromes associated with or caused by pathogenic fungi growing in woody tissues of the vine.

Esca disease, previously called "apoplexy" or "folletage", was the first GTD reported in many European and Mediterranean countries (Viala, 1926; Mugnai *et al.*, 1999; Larignon, 2016). Esca was the original name given to a grapevine wood symptom, white rot, associated with symptoms in the crowns, causing sudden vine death (apoplexy). Typical leaf stripe symptoms were later associated with the same wood degradation. Consequently, the esca disease, was extended to the whole set of these symptoms, and from them, was described as chronic or mild forms and severe or acute forms, also named apoplectic forms or apoplexy. The leaf stripe symptom was characterized by multiple banding discoloration surrounding dry, brittle, light brown or red-brown necrotic tissues bordered by narrow red or yellow borders. These external foliar symptom were proposed by Surico (2009) to be treated as the external expression of a probably tracheomycotic disease, i.e. a separate disease from internal stem white rot ("esca", as originally proposed by French researchers). Surico (2009) defined the concept of the "esca complex", comprising different diseases according to the stage of vine life and the types of symptoms. Among them, the "Grapevine Leaf Stripe Disease", corresponding to the external specific symptom "tiger stripe" not associated to the internal white rot (Edwards *et al.*, 2001), could be seen more clearly in young diseased vines than in older vines. When GLSD is associated with white rot in the same plant, the name of esca or esca proper is used. During the present study, external visible symptoms on vines with characteristic leaf stripe patterns and sudden wilting were considered under the names, respectively, esca/GLSD and apoplexy. Apoplexy corresponds to a sudden wilting of entire vines or of one vine arm, including drying, shrivelling and leaf fall, and this can be caused by many factors, included root rots, not all of which have been described. Apoplexy and leaf stripe symptoms can both occur from late spring to the end of the growing season.

The causes of the external symptoms in the esca complex remain to be fully clarified, and Koch's postulates for these symptoms have not been completed. Two Ascomycetes, *Phaeomoniella chlamydospora* and *Phaeoacremonium minimum* (syn. *P. aleophilum*), are the most common vascular pathogens in GLSD and related diseases (brown-wood streaking, Petri disease). Present in vines alone or in combination, these fungi produce localized necrosis in wood xylem, such as black spots, eventually surrounded by pink to brown wood discoloration, and produce a range of phytotoxins that can be translocated to the leaves (Andolfi *et al.*, 2011). On older vines in European and Mediterranean regions, beside the Ascomycete fungi, it is common to find lignin-degrading Basidiomycete fungi (usually *Fomitiporia mediterranea*, and occasionally other Basidiomycetes) responsible for white rot. Other species of Ascomycetes from the Botryosphaeriaceae or Diatrypaceae (i.e *Eutypa lata*) can also be isolated from wood necroses from vines exhibiting typical leaf stripe foliar symptoms (Bruez *et al.*, 2014) without direct relationships between their presence and the foliar symptoms.

Eutypa dieback, also called "eutypiosis" or "dying-arm disease" was first described on apricot, and then on grapevine in the 1970s (Carter, 1991). Eutypa dieback has been reported in numerous countries from European and Mediterranean vineyards (Péros and Berger, 2003, Berraf-Tebbal *et al.*, 2011). In spring, new vine shoots are stunted with short internodes, the leaves are small chlorotic with marginal necroses and sometimes have dead interveinal tissues with the development of small fruit clusters (Rolshausen *et al.*, 2014). Foliar symptoms are associated with sectorial wood necroses or wedge-shaped cankers in perennial vine organs. The causal agent of Eutypa dieback is mainly the Ascomycete fungus *Eutypa lata*. Since the early 2000s, many other species in the Diatrypaceae have been isolated from necroses associated with Eutypa dieback symptoms (Trouillas *et al.*, 2010; Luque *et al.*, 2012, Pitt *et al.*, 2013, Rolshausen *et al.*, 2014), but *E. lata* remains the most common and virulent fungus causing the typical necrosis and foliar symptoms (Trouillas and Gubler, 2010). This fungus enters vines through pruning wounds, and colonizes the xylem tissues. There it produces phytotoxic compounds that may be transported by transpiration stream towards the aerial vine parts, producing stunting of new shoots (Molyneux *et al.*, 2002).

Botryosphaeria dieback of grapevine includes different symptoms and involves various Botryosphaeriaceae species (Úrbez-Torrez, 2011). The main symptoms include wedge-shaped cankers, dark streaking of grapevine wood, elongated black lesions on the shoots (cane cankers), progressive bud-break failure and plant dieback, and fruit rot in some hosts. At least 21 Botryosphaeriaceae species have been associated with Botryosphaeria dieback of grapevines (Úrbez-Torres, 2011), including *Diplodia seriata, Neofusicoccum parvum* and *Botryosphaeria dothidea* which are the most frequently isolated fungi (Úrbez-Torres, 2011). These pathogens may be present in asymptomatic tissues as latent infections (Gonzalez and Tello, 2011; Bruez *et al.*, 2014). The transition from the endophytic latent infection to active pathogen colonization could be related to factors such as host water stress (Luque *et al.*, 2010). Since the early 2000s, research on Botryosphaeriaceae on grapevine increasingly showed that species in this family play important roles in decline of vines (Larignon *et al.*, 2015). An overview of this decline in Europe is needed to identify the risk factors associated with Botryosphaeriaceae as wood pathogens.

The final GTD included in the present study was Phomopsis cane and leaf spot caused by *Diaporthe ampelina*, often associated with other less virulent species, which can also induce grapevine canker. Vines with Phomopsis cane and leaf spot have characteristic black necrotic irregular-shaped lesions on the cane internodes, occurring in spring and winter, and bleaching and cane cankers and sometimes branch dieback (Ravaz and Verge, 1925; Bugaret, 1984). Rachis necrosis and brown, shrivelled berries on fruit bunches occur near to harvest. The disease, still frequently named "excoriose" in many European viticulture regions, is caused by *Diaporthe* spp., particularly *Diaporthe ampelina* (previously called *Phomopsis viticola*) (van Niekerk *et al.*, 2005; Guarnaccia *et al.*, 2018). In internal wood tissue, *Diaporthe* spp. may cause perennial cankers and vascular discoloration similar to that observed in Botryosphaeria and Eutypa diebacks (Úrbez-Torres *et al.*, 2013; Baránek *et al.*, 2018).

On young grapevines, Petri disease and black-foot are the most damaging wood diseases (Rego *et al.*, 2000; Agustí-Brisach and Armengol, 2013; Carlucci *et al.*, 2017). The foliar symptoms of both diseases are similar, including typical decline symptoms such as leaf chlorosis with necrotic margins, stunting, budbreak delay, decreased growth and vigour, and wilting of leaves or shoots. In longitudinal stem section, vines affected by Petri disease have brown or black streaking in the xylem vessels, often with necroses at the graft unions. Various fungi have been isolated from the altered wood tissues, including *P. chlamydospora* and, less frequently, various species of *Phaeoacremonium* (Mugnai *et al.*, 1999; Gubler *et al.*, 2015) and *Cadophora* (Gramaje *et al.*, 2010).

Black-foot also manifests as necrotic root lesions, rootstock wood necroses, and gradual decline leading to vine death. The fungi causing black-foot are mainly *Cylindrocarpon*-like species belonging to the genera *Ilyonectria, Neonectria, Campylocarpon* in the Nectriaceae (Agusti-Brisach and Armengol, 2013). Decline of young vines may also be caused by other fungi or nematodes, alone or in pathogen complexes, including Botryosphaeriaceae, *Phomopsis, Pythium, Phytophthora*, which have also been isolated from declining vines in young vineyards (Halleen *et al.*, 2007; Agustí-Brisach and Armengol, 2013).

The above descriptions of the main GTDs shows the diversity of syndromes and associated pathogenic agents that are involved. For each GTD, incidence and symptom profiles may greatly vary according to regional or local factors such as climate, soil and cultural practices.

The species composition of grapevine trunk pathogen communities may be affected by environmental factors. Climate plays an important role in pathogen distribution (Merrin *et al.*, 1995; van Nierkerk *et al.*, 2011). For instance, geographical distribution of Botryosphaeriaceae species is strongly linked to climatic conditions. *Lasiodiplodia theobromae* is tolerant to high temperatures, while some species, such as *Diplodia seriata*, are very adaptable to different temperatures (Úrbez-Torres, 2011; Songy *et al.*, 2019). In Europe and Mediterranean regions, various pedo-climatic conditions, "terroirs", are associated with a large number of local varieties of *Vitis vinifera* subsp. *vinifera*, cultivated for mainly for wine, but also for fresh fruit and juice production. Vine training and cultural practices are also diverse, and these result from long grapevine-growing traditions.

Considering the wide diversity of viticulturists, the present study addressed the following questions on GTDs: (1) What is the distribution of each GTD disease on young or mature vines in European and Mediterranean countries? (2) Can information be obtained on the most proliferating or economically important GTDs? and (3) Can GTD occurrence and frequency be explained by regional or local factors? To provide answers to these questions, a large survey was carried out using a qualitative questionnaire. The survey included sets of questions to gather information on the occurrence of GTDs, cultural practices and terroir characteristics.

MATERIALS AND METHODS

Survey questionnaire

The questionnaire was designed to gather information about the occurrence and frequency of the main symptoms associated to GTDs on mature and young vines, and data on grapevine production and agronomic practices that could be related to disease occurrence, from the main grape production regions of European and Mediterranean countries. The questionnaire comprised two parts to obtain information on: 1) GTDs occurring in mature and young vineyards at regional scale; 2) vineyard production and agronomic practices at regional scale; 3) regional climate type in the seasons influencing disease and pathogen cycles, i.e. spring and summer; and 4) reference person(s) who collected the data in each surveyed country/ region.

The questionnaire included open and closed questions. For occurrence of GTDs on adult vines, six types of syndromes were included: (1) apoplexy, (2) esca/ GLSD, "dead cordon", (i.e. the result of dieback caused by wood canker agents observed in vineyard. Dead cordon is an non-specific symptom that may be related to (3) Eutypa dieback; (4) Botryosphaeria dieback; and (5), when the causal agent of cankers and dieback was not identified "dead cordon agent not identified" was reported. A sixth possible syndrome was also listed as Phomopsis cane and leaf spot.

On young vines, three types of declining symptoms were proposed: (1) Petri disease, (2) black foot and (3) decline caused by undetected agent, physiological factor or other agents. For each described syndrome, a scale that rated the frequency of the syndrome was defined as: 0, not present or not recorded (with 'not recorded' indicated with a code); 1, present but not frequent; or 2, frequent. Three other closed questions on the perception of importance were included: Is the syndrome perceived to increase in the region? Is the syndrome perceived to be worrying in the region? What is the susceptibility of the main cultivars? The questionnaire also included knowledge about the amount of GTD investigation in the country: Did laboratory or extension services carry out GTD diagnoses and detection? If yes, is this at regional or national level? The second part of the survey included questions on the range of production types within region or country, including: vintage wine, table wine, table grapes, grapevine nursery. Information was also requested on the main scion and rootstock cultivars, and on the relevant cultural practices linked to GTDs, including details of the common vine training, trellising systems, and pruning methods applied at regional scale.

Questionnaire dissemination and analysis

The questionnaire, formalized as tables on Excel sheets, was distributed in 2015, by email to each National corresponding researcher, as members of the COST Action FA1303 group. The COST Action included all the European countries with relevant grapevine cultivation plus some representative Mediterranean countries. In each country a reference person selected representative grape cultivation areas through collaboration with the most competent respondents in the area, and co-ordinated the data gathering from each respondent. These groups included researchers, extension staff and field technicians. The returned questionnaires with missing data or non-adapted responses were sent again. Focus respondents were surveyed in order to obtain feedback on the response accuracy and to validate some questionnaire responses. All participant responses were gathered in a database filing system. The descriptive analysis of the data was set up using simple calculations of proportions, rates and variable distributions. The analysis of contextual data developed a general view of the characteristics of the surveyed viticulture regions.

RESULTS

General information on grape production and grapevine growing methods

The questionnaire responses were received from 22 countries, including 19 European countries, one Maghreb country (Algeria) and two Middle East countries (Lebanon and Israel) (Table 1). Data from 105 viticulture regions, corresponding to the main regions of the surveyed countries, were recorded. The number of regions by country varied from one (England) to 12 (Italy). The vineyard areas in the surveyed countries totaled about 4 million ha, and varied greatly according to country, from 1,000 ha (England) to approx. one million ha (Spain). Among the surveyed countries, Spain, Italy, France and Turkey represented the main producers, with approx. 75% of the total vineyard area. Among the different types of production, Quality Wine ["Quality Wines Produced in Specified Regions" (QWPSR)] and Table Wine (TW) represented, respectively, the first and second ranges in almost all the surveyed countries, for 87 and 79% of reports (Table 2). Only in Algeria and Turkey Table Grape (TG) was reported as the main production. Table grape production was the second or third type of production in eight countries when rated for relevance, totaling 33 regions (Tables 1 and 2). Dried Raisin (DR) production was not commonly mentioned within the surveyed countries, but represents a major grape production component in Algeria and Turkey. Significant nursery plant production was reported in 16 countries, and generally this was rated as the third most important sector (77% of reports). Nursery production was reported as the 1st or 2nd type of production in importance only in six regions of five countries: Bulgaria, France, Hungary, Italy and Romania.

Eighty-six different scion cultivars were reported in the questionnaire responses. Eighteen were reported from at least four regions in different countries, including 66% of the reports (Figure 1). However, Chardonnay, Cabernet Sauvignon and Merlot were the most frequently grown varieties in the surveyed regions. Table 3 lists the other varieties recorded, as major ones in particular regions. The diversity of rootstock varieties reported in the regions surveyed was much less than that for scion variety. Twenty-four rootstock varieties were mentioned by respondents as the main rootstock planted varieties, but only eight rootstock varieties were mentioned in 89% of the reports (Figure 2). These were dominated by S04, 110R and K5BB. Three main training methods were reported, as Guyot (42% of reports), Cordon (33%) and Gobelet (14%). Three other training methods reported were: Pergola, especially in Italy, Lebanon and Algeria; Chablis, in France; and Sylvoz, in Hungary, Italy and Germany. The respondents generally noted that grapevines were mostly hand-pruned, but that pneumatic pruning shears for hand pruning and mechanical pruning were becoming increasingly used in some countries.

Distribution and frequency of GTDs

In 18 of 22 countries, GTD diagnoses and detection were carried out by laboratory or extension services at national (in ten countries) or regional (eight countries) levels (Table 1).

Table 1. General characteristics of grapevine production in surveyed European and Mediterranean countries.

Country	Country code	No. of surveyed vine cultivation regions	Vineyards area (× 1,000 ha)[b]	Type of production (in order of importance)[a]	Wine production (× 1,000 hL)[b]	GTD investigation diagnosis and detection (yes or no), regional or national level
European Union Mediterranean Countries						
Austria	AT	4	45	QW/TW/N	1,999	No
Bulgaria	BG	5	63	QW-TW/TG/N	745	No
Croatia	HR	2	29	QW/TW/N	842	Yes, national
Czech Republic	CZ	3	17	QW/TW/N	536	Yes, regional
France	F	7	789	QW/TW/N	46,534	Yes, regional
Germany	DE	4	102	QW/TW/N	9,202	Yes, regional
Greece	EL	3	110	QW/TG/N	2,800	Yes, regional
Hungary	HU	6	62	QW/TW/N	2,427	Yes, regional
Italy	IT	12	690	QW/TW/TG/N	44,229	Yes, national
Montenegro	ME	2	9	QW/TW/N	161	Yes,national
Portugal	PT	5	224	QW/TW/N	6,206	Yes, regional
Romania	Ro	5	192	QW/TW/N/TG	3,750	Yes, national
Serbia	RS	4	54	TW/TG	2,332	No
Slovakia	SK	6	16	QW/TW/TG	258	Yes, national
Slovenia	SI	3	16	QW/TW/N	494	Yes, national
Spain	SP	7	975	QW/TW/N	39,494	Yes, national
Switzerland	S	6	15	QW/TW/N	934	Yes, national
United Kingdom	UK	1	2	QW/TW	47	Yes, national
Non-European Union Mediterranean Countries						
Algeria	AL	71	74	TG/R/QW/TW	507	No
Israel	Il	8	8	QW/TW/TG/N	246	Yes, national
Lebanon	Le	3	14	QW/TG/TW	80	Yes, national
Turkey	T	6	502	TG/R/QW/TW	615	Yes, regional
Total	22	105	4066	-	1644,483	-

[a] QW (Quality wine); TW (Table wine); TG (Table grape), R (raisin), N (Nursery)
[b] 2014 data in OIV, 2014.

Table 2. Estimates of the relevance of grapevine production types, based on frequency of reports from the all viticultural regions surveyed.

Production type range	Quality wine (QW)		Table wine (TW)		Table grape (TG)		Nursery plants (N)		Raisins (R)		Total for production range
	No. Reports[a]	%[b]	No. Reports	%	No. Reports	%	No. Reports	%	No. Reports	%	
1st Production type	80	87.1	14	17.3	6	15.4	3	5.1	3	75	106
2nd Production type	11	11.8	64	79	19	48.7	3	7.7	1	25	98
3rd or 4th Production type	1	1.1	3	3.7	14	35.9	35	89.4	0	0	53
Total for type of production	93		81		39		41		4		258

[a] Number of reports over 105 regions surveyed.
[b] Percentage over total No. of reports for each production type.

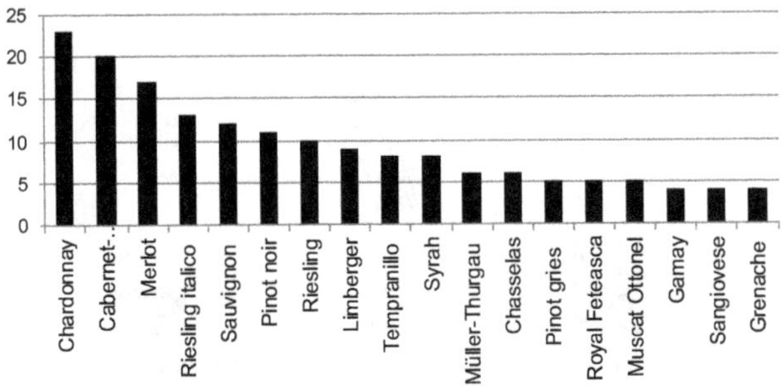

Figure 1. Numbers of reports of the main scion cultivars recorded in 105 surveyed regions.

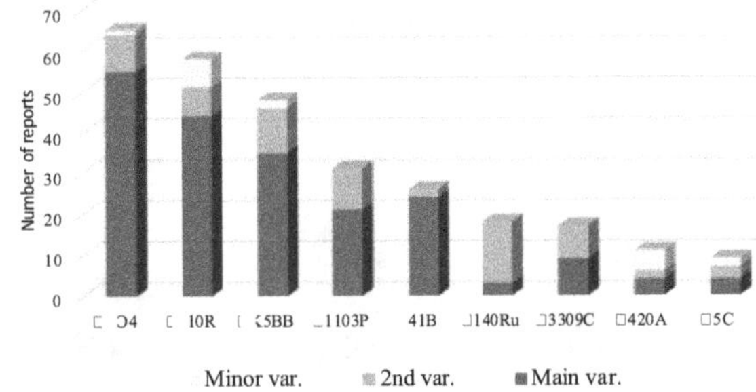

Figure 2. Number of reports of occurrence of the main rootstock varieties and their range in the regions where the survey was carried out.

Table 3. Grapevine scion varieties infrequently identified in survey questionnaire responses, with one to three mentions.

Scion variety	Report No.	Scion variety	Report No.
Furmint	3	Loureiro	1
Gamaret	3	Malvasia istriana	1
Grüne Veltliner	3	Mavrodaphni	1
Macabeu	3	Melon	1
Touriga Nacional	3	Meunier	1
Zweigelt	3	Montepulciano	1
Bobal	2	Montepulciano	1
Colombard	2	Montonico	1
Gewurztraminer	2	Muscat d'Alexandrie	1
Harslevelu	2	Negoska	1
Pinot blanc	2	Nuragus	1
Primitivo	2	Parellada	1
Yellow muscat	2	Passerina,	1
Agiorgitiko	1	Pecorino,	1
Airén	1	Plavac mali	1
Alfrocheiro preto	1	Portugais bleu	1
Aligoté	1	Prokupac	1
Alvarinho	1	Rebula	1
Arvine	1	Refosco	1
Baga	1	Rkatziteli	1
Bombino bianco	1	Saint-Laurent	1
Cabernet franc	1	Tamnjanika	1
Calabrese	1	Tinta Amarela	1
Carignan	1	Touriga Franca	1
Castelão	1	Trebbiano	1
Catarratto	1	Trebbiano romagnolo	1
Chenin	1	Treixadura	1
Dimiat	1	Ugni blanc	1
Fernão Pires	1	Vermentino	1
Garganega	1	Vernaccia	1
Glera	1	Vranac crni	1
Glera	1	Xarello	1
Grillo	1	Xinomavro	1
Jaen N	1	Zizak	1

GTDs on mature vines

Data on GTD occurrence in all the surveyed regions were combined. This revealed different frequency profiles for the different GTD syndromes reported in adult vines (Figure 3). The records of apoplexy and esca/GLSD showed similar profiles with a greater proportion of records for "frequent" occurrence , reaching 59% for apoplexy and 55.2% for esca/GLSD. These two syndromes were the most widespread and the only diseases reported in all the surveyed countries. They were present in most of the regions (apoplexy in 91% and esca/GLSD 86% over the 105 regions) (Table 4). In contrast, the frequency profiles of the three types of "dead cordon", caused by canker agents, Eutypa dieback, Botryosphaeria dieback, or "agent not identified", showed less occurrence, and were reported as frequent, respectively, in only 20, 30.5 and 27.6% of the regions. These diseases were not observed (or not reported), respectively, from 22, 27.6 and 22% of the regions. Phomopsis cane and leaf spot was present in 60.9% of the regions, but was usually not frequent.

The syndromes surveyed had different occurrences in different countries, although all syndromes highlighted in the questionnaire were widespread in Continental European and Mediterranean countries (Table 4). There was a similar frequency profile for both syndromes, apoplexy and esca/GLSD within a country, except in Hungary, where esca/GLSD is more frequent than apoplexy and in Romania and Serbia, where apoplexy is more frequent than esca/GLSD. Dead cordon, caused by Eutypa dieback, was always reported less frequently than apoplexy and esca/GLSD. It was actually reported as frequent only in 7 countries, even if it was present in all countries except Israel. Dead cordon caused by other pathogens (Botryosphaeriaceae or not identified agent) were frequently recorded in almost all countries, except from Austria, probably because of lack of observation.

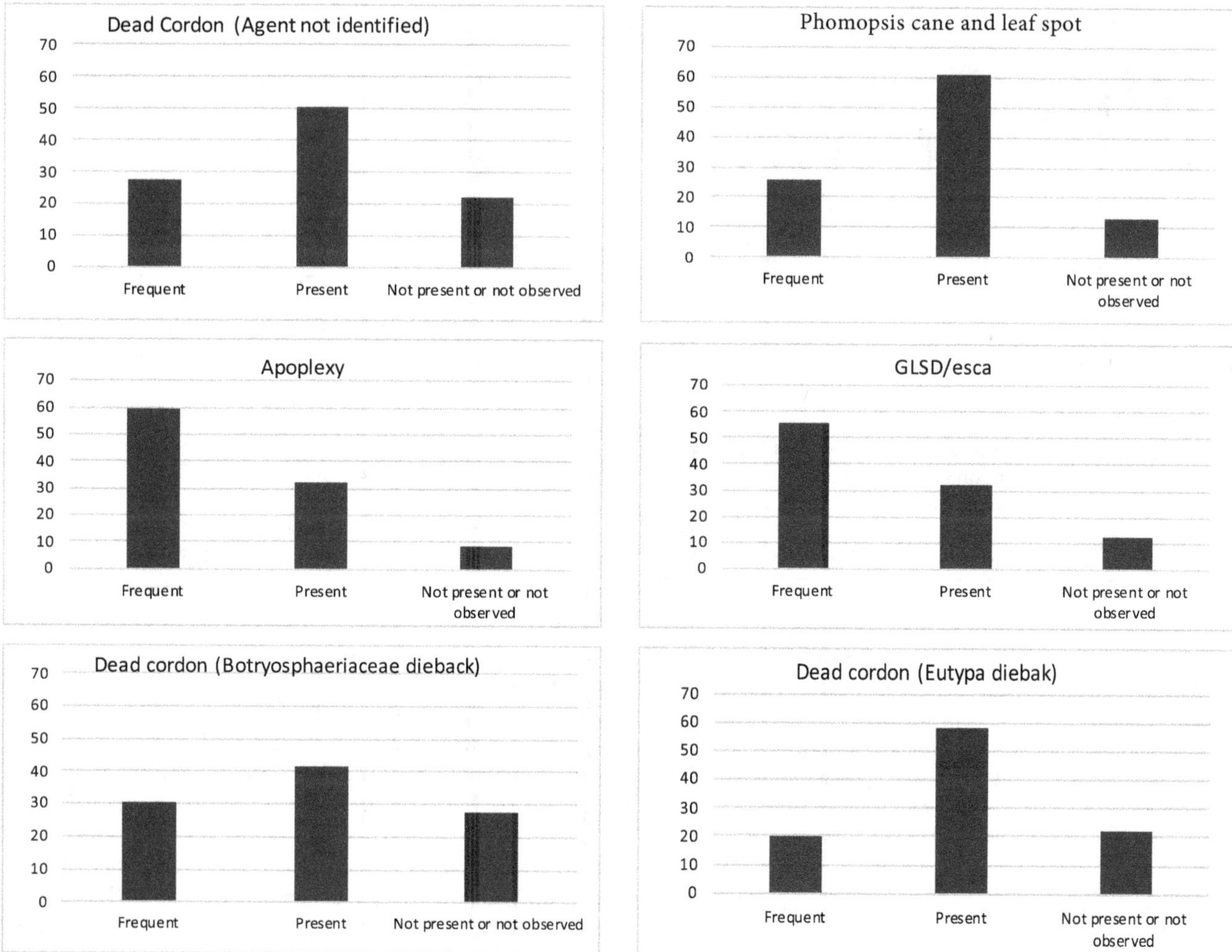

Figure 3. Reports (%) for all surveyed countries of the different GTDs on adult vines.

The level of occurrence of both of these syndromes varied according to the country and the region.

Cabernet Sauvignon was the most frequently recorded of the 18 main reported varieties (Figure 4) as the most susceptible variety to esca complex. Fourteen other varieties were cited at least once. Fewer reports concerned the susceptibility to the other GTDs. Some varieties like Ugni Blanc, Cabernet, Cabernet Sauvignon and Chasselas were reported as susceptible to Eutypa dieback in France and Germany. In Portugal, the varieties Touriga National, Loureiro, Alvarinho, Aragonez, Touringa Franca and Syrah were reported as susceptible to both esca complex disease and Botryosphaeria dieback. In Turkey, the variety Sultana Seedless was also noted as susceptible to these two disease syndromes.

GTDs on young vines

The occurrence of GTDs on young vines (Petri disease, black foot and general decline) is shown in Figure 5. The three forms on young vines were reported less frequently than the syndromes on adult vines. Decline diseases on young vines caused by Petri disease or unknown agent were mainly reported as "present but not frequent" (Petri disease, 44% and unknown agent, 61%). Petri disease was mentioned as frequent in nine regions from five countries (Hungary, Spain, Slovenia, Italy and Slovakia) (Table 5), and present in 14 countries, of the 22 surveyed countries. In seven other countries (Czech Republic, Romania, Austria, France, Lebanon, Algeria, Serbia), this decline was not observed (or not recorded). Black foot was mentioned as "frequent" only in two countries: Hungary and Spain, while it was noted "present", but not frequent, in nine other countries.

Table 4. Reported level of occurrence of Grapevine Trunk Diseases on adult vines in each surveyed European or Mediterranean country.

Country	No. of regions surveyed	Apoplexy/esca acute form						Esca/GLSD[a]						Dead cordon (Eutypa dieback)					
		Frequent		Present		Not present or not observed		Frequent		Present		Not present or not observed		Frequent		Present		Not present or not observed	
		No.	%	No.	%	No.	%	No.	%	No.	%	No.	%	No.	%	No.	%	No.	%
Algeria	6	3	50	0	0	3[b]	50	3	50	0	0	3[b]	50	3	50	0	0	3[b]	50
Austria	4	4	100	0	0	0	0	4	100	0	0	0	0	4	100	0	0	0	0
Bulgaria	5	0	0	5	100	0	0	1	20	4	80	0	0	0	0	5	100	0	0
Croatia	2	2	100	0	0	0	0	2	100	0	0	0	0	2	100	0	0	0	0
Czech Republic	3	1	33.3	2	66.7	0	0	0	0	3	100	0	0	0	0	3	100	0	0
England	1	0	0	1	100	0	0	0	0	1	100	0	0	0	0	1	100	0	0
France	7	6	85.71	1	14.29	0	0	7	100	0	0	0	0	1	14.29	6	85.71	0	0
Germany	4	3	75	0	0	1[b]	25	3	75	0	0	1[b]	25	0	0	3	75	1[b]	25
Greece	3	0	0	3	100	0	0	0	0	3	100	0	0	3	100	0	0	0	0
Hungary	6	2	33.3	4	66.7	0	0	5	83.3	1	16.7	0	0	0	0	5	83.3	1	16.7
Israel	5	0	0	5	100	0	0	0	0	5	100	0	0	0	0	0	0	5	100
Italy	12	10	83.3	2	16.7	0	0	11	91.7	1	8.3	0	0	0	0	11	91.7	1	8.3
Lebanon	3	3	100	0	0	0	0	3	100	0	0	0	0	0	0	3	100	0	0
Montenegro	2	1	50	1	50	0	0	2	100	0	0	0	0	0	0	1	50	1	50
Portugal	5	5	100	0	0	0	0	3	60	2	40	0	0	0	0	1	20	4	80
Romania	5	5	100	0	0	0	0	0	0	5	100	0	0	3	60	0	0	2	40
Serbia	5	4	80	1	20	0	0	0	0	1	20	4	80	5	100	0	0	0	0
Slovakia	6	0	0	6	100	0	0	0	0	6	100	0	0	0	0	6	100	0	0
Slovenia	3	3	100	0	0	0	0	3	100	0	0	0	0	0	0	3	100	0	0
Spain	6	6	100	0	0	0	0	6	100	0	0	0	0	0	0	6	100	0	0
Switzerland	6	4	66.7	2	33.3	0	0	4	66.7	2	33.3	0	0	0	0	6	100	0	0
Turkey	6	0	0	1	16.7	5[b]	83.3	1	16.7	0	0	5[b]	83.3	0	0	1	16.7	5[b]	83.3
Total	105	62		34		9		58		34		13		21		61		23	
%	100	59		32.4		8.6		55.2		32.4		12.4		20		58.1		21.9	

(Continued)

Table 4. (Continued).

Country	No. of regions surveyed	Dead cordon (Botryosphaeria dieback)						Dead cordon (agent not identified)						Phomopsis cane and leaf spot					
		Frequent		Present		Not present or not observed		Frequent		Present		Not present or not observed		Frequent		Present		Not present or not observed	
		No.	%	No.	%	No.	%	No.	%	No.	%	No.	%	No.	%	No.	%	No.	%
Algeria	6	1	16.7	0	0	5[b]	83.3	3	50	0	0	3[c]	50	0	0	3	50	3[b]	50
Austria	4	0	0	0	0	4	100	0	0	0	0	4	100	0	0	4	100	0	0
Bulgaria	5	1	20	1	20	3	60	2	40	3	60	0	0	0	0	5	100	0	0
Croatia	2	2	100	0	0	0	0	0	0	2	100	0	0	2	100	0	0	0	0
Czech Republic	3	0	0	3	100	0	0	3	100	0	0	0	0	0	0	3	100	0	0
England	1	0	0	1	100	0	0	0	0	1	100	0	0	1	100	0	0	0	0
France	7	1	14.3	1	14.3	5	71.4	2	28.6	4	57.1	1	14.3	0	0	7	100	0	0
Germany	4	0	0	3	75	1[b]	25	0	0	3	75	1[b]	25	0	0	3	75	1[b]	25
Greece	3	0	0	3	100	0	0	0	0	3	100	0	0	3	100	0	0	0	0
Hungary	6	3	50	2	33.3	1	16.7	3	50	3	50	0	0	0	0	4	66.7	2	33.3
Israel	5	0	0	5	100	0	0	0	0	5	100	0	0	0	0	5	100	0	0
Italy	12	6	50	3	25	3	25	5	41.7	5	41.7	2	16.	3	25	9	75	0	0
Lebanon	3	3	100	0	0	0	0	3	100	0	0	0	0	0	0	2	66.7	1	33.3
Montenegro	2	1	50	0	0	1	50	1	50	1	50	0	0	1	50	1	50	0	0
Portugal	5	3	60	2	40	0	0	0	0	1	20	4	80	1	20	4	80	0	0
Romania	5	0	0	4	80	1	20	3	60	0	0	2	40	5	100	0	0	0	0
Serbia	5	5	100	0	0	0	0	1	20	4	80	0	0	4	80	1	20	0	0
Slovakia	6	0	0	6	100	0	0	0	0	6	100	0	0	0	0	6	100	0	0
Slovenia	3	0	0	3	100	0	0	0	0	3	100	0	0	0	0	3	100	0	0
Spain	6	5	83.3	1	16.7	0	0	3	50	2	33.3	1	16.7	1	16.7	3	50	2	33.3
Switzerland	6	0	0	6	100	0	0	0	0	6	100	0	0	5	83.3	1	16.7	0	0
Turkey	6	1	16.7	0	0	5[b]	83.3	0	0	1	16.	5[b]	83.3	1	16.7	0	0	5[b]	83.3
Total	105	32		44		29		29		53		23		27		64		14	
%	100	30.5		41.9		27.6		27.6		50.5		21.9		25.7		60.9		13.3	

[a] Grapevine Leaf Stripe Disease.
[b] Not surveyed region.

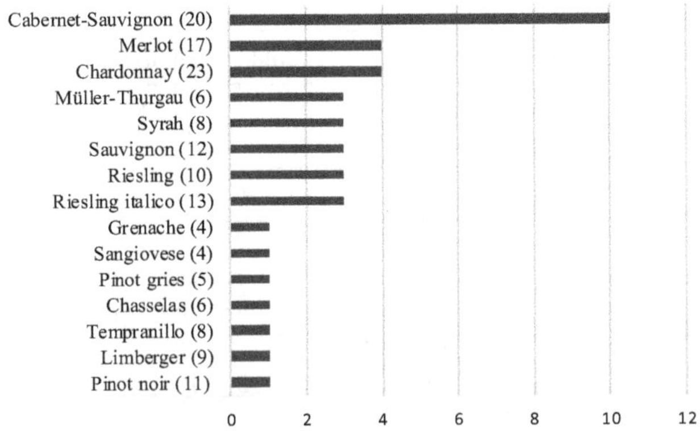

Figure 4 Numbers of survey response reports on cultivar susceptibility to esca complex (Total report number in brackets).

Focus on five sample countries

In order to give more details in specific situations, GTD distribution was analysed within five countries: Hungary, Bulgaria, Portugal, France and Italy.

Hungary

The profile of GTDs on adult vines varied according to the wine region (Table 6A). Apoplexy was recorded as "frequent and increasing" in two regions (Pécsi and Villány), characterized by dry spring and dry and hot summer weather, which may correspond to a sub-Mediterranean climate. The GTD profile in the Balat wine region differed from the other regions, with few recorded diseases or low levels of occurrence of GTD symptoms, such as apoplexy, esca/GLSD and dead cordon, in comparison with the five other regions. The Balaton area is characterised by the proximity of Lake Balaton and volcanic soils that create a peculiar environment. In the Egri region, the respondent reported frequent and increasing esca/GLSD and dead cordon caused by Botryosphaeria dieback. Dead cordon related to Eutypa dieback and Phomopsis cane and leaf spot were mentioned less frequently. They were recorded as "occasionally present" in three regions and "not recorded" in five regions.

Bulgaria

Contrasting GTDs profiles on adult vines were observed according to region (Table 6B). Esca/GLSD was recorded as "frequent" in the only region of the "sub-Balkan" area, the Rose Valley region. All the other disease syndromes on mature vines were recorded as "present" or "frequent" in this region. In the Northern region, characterized by a wet and cold spring weather, dead cordon related to Botryosphaeria dieback or to an unidentified canker agent were recorded as "frequent". In contrast, only in the South-Western region were all GTDs on mature vines rarely or never observed.

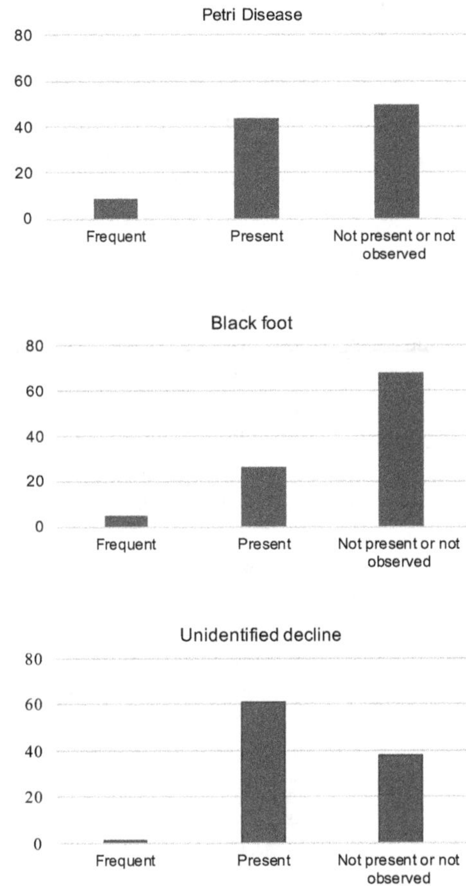

Figure 5. Proportions (%) of survey reports from all countries of different grapevine decline syndromes on young vines.

Portugal

Apoplexy, esca/GLSD and dead cordon related to Botryosphaeria dieback and Phomopsis cane and leaf spot were recorded as "frequent" in Portugal (Table 6C). In contrast, dead cordon related to Eutypa dieback was recorded as "absent". In Vinho verde, in the northern Portugal, apoplexy and Phomopsis cane and leaf spot were recorded as "frequent and worrying". The climate of that region is characterized by cool and wet springs and a cool summers. In Dao and Alentejo, apoplexy and esca/GLSD were recorded as "frequent and increasing"

Table 5. Reported levels of occurrence of GTD on young vines in surveyed European and Mediterranean countries.

Country	Region No.	Petri disease						Black foot[a]						Unidentified decline[b]					
		Frequent		Present		Not present or not observed		Frequent		Present		Not present or not observed		Frequent		Present		Not present or not observed	
		No.	%	No.	%	No.	%	No.	%	No.	%	No.	%	No.	%	No.	%	No.	%
Algeria	6	0	0	0	0	6[c]	100	0	0	0	0	6[c]	100	0	0	0	0	6[c]	100
Austria	4	0	0	0	0	4	100	0	0	0	0	4	100	0	0	4	100	0	0
Bulgaria	5	0	0	5	100	0	0	0	0	3	60	2	40	0	0	5	100	0	0
Croatia	2	0	0	2	100	0	0	0	0	0	0	2	100	0	0	0	0	2	100
Czech Repu-blic	3	0	0	0	0	3	100	0	0	0	0	3	100	0	0	3	100	0	0
England, UK	1	0	0	1	100	0	0	0	0	1	100	0	0	0	0	1	100	0	0
France	7	0	0	0	0	7	100	0	0	3	42.9	4	57.1	0	0	2	28.6	5	71.4
Germany	4	0	0	3	75	1[c]	25	0	0	3	75	1[c]	25	0	0	3	75	1[c]	25
Greece	3	0	0	3	100	0	0	0	0	3	100	0	0	0	0	3	100	0	0
Hungary	6	2	33.3	2	33.3	2	33.3	2	33.3	2	33.3	2	33.3	0	0	5	83.3	1	16.7
Israel	5	0	0	5	100	0	0	0	0	0	0	5	100	0	0	5	100	0	0
Italy	12	1	8.3	8	66.7	3	25	0	0	0	0	12	100	0	0	9	75	3	25
Lebanon	3	0	0	0	0	3	100	0	0	1	33.3	2	66.7	0	0	3	100	0	0
Montenegro	2	0	0	2	100	2	100	0	0	0	0	2	100	0	0	1	50	1	50
Portugal	5	0	0	4	80	1	20	0	0	5	100	0	0	1	20	4	80	0	0
Romania	5	0	0	0	0	5	100	0	0	0	0	5	100	0	0	1	20	4	80
Serbia	5	0	0	0	0	5	100	0	0	0	0	5	100	0	0	5	100	0	0
Slovakia	6	1	16.7	0	0	5	83.3	0	0	0	0	5	83.3	0	0	1	16.7	5	83.3
Slovenia	3	2	66.7	1	33.3	0	0	0	0	0	0	3	100	0	0	3	100	0	0
Spain	6	3	50	3	50	0	0	3	50	2	33.3	1	16.7	0	0	5	83.3	1	16.7
Switzerland	6	0	0	6	100	0	0	0	0	4	66.7	2	33.3	0	0	0	0	6	100
Turkey	6	0	0	1	16.7	5[c]	83.3	0	0	1	16.7	5[c]	83.3	0	0	1	16.7	5[c]	83.3
Total	105	9		46		52		5		28		71		1		64		40	
%	100	8.6		43.8		49.5		4.8		26.7		67.6		0.9		60.9		38.1	

[a] Ascertained presence of *Cylindrocarpon*-like spp.
[b] Undetected agent, physiological and other agents or cause excluded.
[c] Not surveyed regions.

and/or "worrying". The spring and summer climate in Dao is wet-warm and in Alentejo is dry-hot. Apoplexy, esca/GLSD and dead cordon related to Botryosphaeria dieback were recorded as "frequent and increasing" in Bairrada, located in the western Portugal, an area with mild climate with abundant rainfall (Vinho verde, Bairrada).

France

The seven surveyed regions showed contrasting profiles of GTD occurrence on mature vines (Table 6D). Apoplexy and esca /GLSD were reported as "frequent" in almost all the regions, and were considered as "increasing" and/or "worrying" in five of of seven regions. In contrast, Eutypa dieback and Phomopsis cane and leaf spot were recorded in all the regions but not as frequently occurring. Eutypa dieback was recorded as "frequent" in Charente, a region with Oceanic climate and where the susceptible cultivar Ugni Blanc (used for distilled wines) is widespread. In the Languedoc Roussillon area, GTDs were reported as "present" or "not observed", except for esca/GLSD. Dead cordon related to Botryosphaeria dieback was reported as "present" or "frequent" only in two regions, Champagne and Rhône.

Italy

In Italy, esca/GLSD and apoplexy have high and increasing importance in all regions (Table 6E). Eutypa dieback is also widespread, being reported as "present" in all regions except Sicily (probably linked to low

Table 6. Reported levels of occurrence of GTD on adult grapevines in surveyed regions in five selected countries: Hungary (A), Bulgaria (B), Portugal (C), France (D), Italy (E). i, increasing incidence; w, worrying presence; iw, increasing and worrying P, present; PO, present occasionally. Legend: light grey : present but not frequent; medium grey: frequent; dark grey: absent or not recorded.

A. Hungary

Spring climate[a]	Summer climate[b]	Region	Apoplexy	Esca /GLSD	Dead cordon (Eutypa dieback)	Dead cordon (Botryosphaeria dieback)	Dead cordon (causal agent not identified)	Phomopsis cane and leaf spot
uf	DHsu	Tokaj			PO			
WCsp	DHsu	Balat	PO	PO				
WWsp	DHsu	Egri		i	PO	i		PO
Dsp	DHsu	Pécsi	i				PO	
WWsp	DHsu	Neszmélyi						PO
Dsp	DHsu	Villány	i				PO	PO

B. Bulgaria

Spring climate[a]	Summer climate[b]	Region	Apoplexy	Esca /GLSD	Dead cordon (Eutypa dieback)	Dead cordon (Botryosphaeria dieback)	Dead cordon (causal agent not identified)	Phomopsis cane and leaf spot
WCsp	DHsu	Northern region (Danube plain)			PO			
WWsp	DHsu	Black sea region		PO	PO			PO
WWsp	DHsu	Sub-Balkan region (Rose valley)			PO	PO		
WWsp	DHsu	Southern region (Thracian Lowland)				never	PO	
WWsp	DHsu	South-west region (Struma River Valley)	PO	PO	PO	never	PO	PO

C. Portugal

Spring climate[a]	Summer climate[b]	Region	Apoplexy	Esca /GLSD	Dead cordon (Eutypa dieback)	Dead cordon (Botryosphaeria dieback)	Dead cordon (causal agent not identified)	Phomopsis cane and leaf spot
WWsp	DHsu	Dão	iw	i				PO
WWsp	Csu	Bairrada	i	i				PO
WCsp	Csu	Vinho Verde		PO				w
WCsp	DHsu	Douro			PO		w (PO)	
WWsp	DHsu	Alentejo	w	i				

D. France

Spring climate[a]	Summer climate[b]	Region	Apoplexy	Esca /GLSD	Dead cordon (Eutypa dieback)	Dead cordon (Botryosphaeria dieback)	Dead cordon (causal agent not identified)	Phomopsis cane and leaf spot
WCsp	Csu	Bourgogne	w					
WCsp	Csu	Champagne	i	iw				
WCsp	Csu	Pays de Loire	iw	iw			iw	
WCsp	DHsu	Charentes	i	i				
WCsp	DHsu	Bordeaux						
WWsp	DHsu	Languedoc-Roussillon						
WCsp	WHsu	Rhône	w	w				

(Continued)

Table 6. (Continued).

E. Italy

Spring climate[a]	Summer climate[b]	Region	Apoplexy	Esca /GLSD	Dead cordon (Eutypa dieback)	Dead cordon (Botryosphaeria dieback)	Dead cordon (causal agent not identified)	Phomopsis cane and leaf spot
WWsp	DHsu	Sardinia				i		
WWsp	Csu	Friuli Venezia Giulia	w	w			w	
Dsp	DHsu	Sicily		i				
WWsp	DHsu	Apulia		w		w		
WWsp	DHsu	Tuscany		w				
WWsp	WHS	Veneto		w				
WCsp	DHsu	Abruzzi	w	w				
WWsp	DHsu	Romagna						
WWsp	DHsu	Emilia		w				
WWsp	DHsu	Lombardy		w				
WCsp	DHsu	Piedmont						
WWsp	Csu	Trentino		w				

[a] Wet and cool spring (WCsp), wet warm spring (WWsp), dry spring (Dsp), uncertainty factor (uf).
[b] Dry and hot summer (DHsu),Wet and Hot summer (WHsu), cool summer (Csu).

amounts of rain, which is important for Eutypa inoculum spread. Eutypa dieback was responsible for limited damage. Cankers leading to dead cordon were also recorded in all regions, but were reported as "highly relevant" only in Sardinia, Friuli, Sicily, Apulia, Tuscany, Lombardy and Trentino. Phomopsis cane and leaf spot was also reported from all regions, was more widespread in Apulia, Veneto and Piedmont than elsewhere.

Increasing and/or "worrying" GTD occurrence

The GTDs were recorded as "increasing" and/or "worrying" by the correspondents in numerous countries in Europe and some of the Mediterranean area (Table 7). These diseases were reported as more important in old than in young vineyards, mainly due to the spread of apoplexy and esca complex/GLSD, respectively, in ten and 11 surveyed countries ("increasing" or/and "worrying" occurrence), particularly in the main grape-producing countries (France, Italy, Spain and Turkey). This represents about a third of recorded regions. Dead cordon related to Botryosphaeria dieback was the third disease problem, reported to be "increasing" and/or worrying in seven countries, and 16 of the 105 surveyed regions.

Dead cordon related to Eutypa dieback was recorded as "increasing" and/or "worrying" in ten regions of Serbia, Romania and Algeria. Dead cordon caused by unknown agents was "increasing" or "worrying" in five countries. Like Eutypa dieback, Phomopsis cane and leaf spot had more local importance, being reported as "increasing" or "worrying in eight regions within Romania, Portugal, Turkey and Montenegro.

The presence of decline diseases is perceived as a problem even in young vineyards: as "increasing" (three reports) or "increasingly worrying" (five reports) in some regions, particularly in Hungary and Spain.

On young vines, black foot was reported as "increasing" and/or "worrying" only in three regions of Spain, and Petri disease was considered as "increasing" in two regions of Spain and two regions of Hungary.

DISCUSSION

This study reports the first large-scale survey of GTDs in European and closeby Mediterranean countries, which aimed to determine the spread and severity of the main grapevine wood diseases, i.e. the disease syndromes associated with, or caused by, pathogenic fungi colonizing the woody tissues of grapevines. The diseases included in the survey are summarized in the Introduction. Some are included in "esca complex", i.e. a complex of interrelated diseases differently designated according to the stage of vine life when they occur and the type of symptom they cause.

The surveys carried out showed that apoplexy and esca/GLSD (i.e. often esca proper, as white rot in wood is frequently present with GLSD leaf symptoms) occurred in all the surveyed countries, and were

Table 7. Number of the surveyed regions in European and Mediterranean countries where alarming situations of GTD spread was reported.

Disease	Increasing OR worrying			Increasing AND worrying			Total	
	Surveyed Region		Country	Surveyed Region		Country	Surveyed Region	
	No.	%		No.	%		No.	%
Symptoms on adult vine								
Apoplexy	30	28.6	Hungary, Spain, Switzerland, Romania, Portugal, France, Italy, Algeria, Serbia, Germany	4	3.8	Portugal , France, Algeria	34	32.4
Esca /GLSD	29	27.6	Hungary, Portugal, France, Italy, Algeria,Turkey, Spain, Germany, Lebanon, Montenegro, Romania	11	10.5	Portugal, Czech, Spain, Austria	40	38.1
Dead cordon (Eutypa dieback)	8	7.6	Serbia, Romania	2	1.9	Algeria	10	9.5
Dead cordon (Botryosphaeria dieback)	14	13.3	Hungary, Romania, France, Italy, Spain, Serbia	2	1.9	Portugal, Spain	16	15.2
Dead cordon (causal agent not identified)	7	6.7	Algeria Romania Serbia, Portugal	2	1.9	Algeria, France	9	8.6
Phomopsis cane and leaf spot	8	7.6	Romania, Portugal, Turkey, Montenegro	0	0	-	8	7.6
Symptoms on young vine								
Petri disease	4	3.8	Hungary, Spain, Portugal	0	0	-	4	3.8
Black foot [a]	3	2.7	Spain, Portugal	0	0	-	3	2.7

[a] Ascertained presence of *Cylindrocarpon*-like spp.

generally recorded as the most frequently occurring GTDs. Both syndromes are related to esca complex, and they were also noted as increasing and worrying in many countries, in a large vineyard area in Europe. In mature vines, the other disease syndromes such as dead cordon related to *E. lata*, Botryosphaeriaceae or to unknown origin, and Phomopsis cane and leaf spot, were also recorded in almost all the surveyed countries. These diseases were often recorded as not frequent but increasing in several surveyed countries. In contrast, decline of young vine was less widespread and less important than decline of adult vines. Only a few viticulture regions were concerned about Petri disease or black foot, reporting it as frequently present only in a few countries.

Apoplexy and esca/GLSD are both very easily identifiable syndromes due to their characteristic symptoms, respectively, of sudden wilting and leaf stripe. The extensive occurrence of esca complex in the surveyed regions supports the endemic behaviour of this GTD, and its increasing presence, since the end of the 20th Century (Fischer and Kassemeyer, 2003; Surico, 2009). In contrast, if all responses on dead cordon syndromes are pooled, these occur as frequently as the esca complex. Among the diseases leading to dead cordon, Eutypa dieback was the most studied, and this disease is also easily detectable when foliar symptoms are visible (Munkvold *et al.*, 1994; Dubos, 1999). The situation is similar for Phomopsis cane and leaf spot, with typical symptoms on the canes (Úrbez-Torres *et al.*, 2013). Mixed infection makes it difficult to sharply separate the different dieback diseases. GTD diagnoses in vineyards may be difficult without specific foliar symptom expression, as is the case for Botryosphaeria dieback. In this survey, reports of Botryosphaeria dieback, probably based on previous etiological studies, were very likely to be underestimated due to the lack of specific surveys in many regions and countries. The significant number of reports "dead cordon" "agent unidentifed" supports this finding. Numerous pathogenic Botryosphaericeae species, presenting different life traits, may be involved in vine decline in European and Mediterranean countries (van Niekerk *et al.*, 2006; Bellée *et al.*, 2017). Their distribution and prevalence are related to many factors, such as climatic and agronomic influences. Some fungi are widespread in Europe, such as *Diplodia seriata*, reported in Spain (Armengol *et al.*, 2001; Martin and Cobos, 2007), Portugal (Phillips, 2002; Rego *et al.*, 2009), Hungary (Kovács *et al.*, 2017), Italy (Cristinzio, 1978; Burruano *et al.*, 2008; Spagnolo *et al.*, 2011, Carlucci *et al.*, 2009, 2015), France (Larignon *et al.*, 2001), Turkey (Akgül *et al.*, 2014), Germany (Fischer and Kassemeyer, 2003), Croatia (Kaliterna and Miličević, 2014), and Bulgaria (Nikolova, 2010). Similarly, *Neofusicoccum parvum* is reported from Croatia (Kaliterna *et al.*, 2013), in Turkey (Akgül *et al.* 2014), Portugal (Phillips, 2002), Spain (Luque *et al.*, 2009), Czech Republic (Baránek *et al.*, 2018) and Italy (Spagnolo *et al.*, 2011). Another virulent pathogen, *Lasiodiplodia theobromae*, was frequently isolated in Southern European and Mediterranean regions (El-Goorani and El Meleigi, 1972; Burruano *et al.*, 2008, Carlucci *et al.*, 2009), and recently was isolated from declining vines in France (Comont *et al.*, 2016).

The results of the survey showed that dead cordon related to Botryosphaeria dieback is increasing and worrying in more countries than those reporting Eutypa dieback. This could be related to the broad range of the Botryosphaeriaceae pathogenic species, characterised by various traits (endophytic, saprophytic, pathogenic) in numerous hosts. Furthermore, water stress has been reported to increase the severity of Botryosphaeria dieback on grapevine (van Nierkerk *et al.*, 2011; Amponsah *et al.*, 2014; Lawrence *et al.*, 2017). Incidence of disease caused by Botryosphaericeae in various plant hosts has increased in Europe during the last decades (Desprez-Loustau *et al.*, 2006; Slippers and Wingfield, 2007; Fabre *et al.*, 2011; Piškur *et al.*, 2011; Mehl *et al.*, 2014), and this may be related with climatic evolution towards more variable conditions, with increasing incidence and intensity of droughts over continental Europe and the Mediterranean region (Spinoni *et al.*, 2017). Further research on the group of pathogens implicated in grapevine decline is needed to better assess the distribution of the different pathogen species in European and Mediterranean countries, and to determine their life traits in relation with grapevine physiological status.

Results from this questionnaire also demonstrated occurrence of Phomopsis cane and leaf spot disease. This is not a wood disease and so not usually included among the Grapevine Trunk Diseases. Nevertheless, its main agents, *Diaporthe ampelina* and other *Diaporthe* species, are also recognized causes of cankers and dieback on grapevine (Úrbez-Torres *et al.*, 2013; Dissanayake *et al.*, 2015; Guarnaccia *et al.*, 2018). In the survey, the reports of Phomopsis cane and leaf spot, based on the typical easily visible symptoms on the basal internodes of canes, gave information on the disease spread in Europe. This highlights the need for accurate pathogen identification to analyse the involvement of *Diaporthe ampelina* (=*Phomopsis viticola*) and other *Diaporthe* species in perennial cankers and vine dieback in European vineyards.

Eutypa dieback was considered a menacing and spreading disease in Europe at the end of the 20th Century in Europe (Munkvold *et al.*, 1994; Dubos, 1999). However, the present survey has shown this disease is present, but not frequent, in a majority of regions. Increased knowledge of Eutypa dieback epidemiology and on efficient disease management (infected wood removal, wound protection and trunk renewal) may have contributed to decreased incidence of the disease (Dubos, 1999; Lecomte *et al.*, 2006). Nevertheless, Eutypa dieback is still present in 20% of the monitored region and was recorded as increasing or worrying in three countries, Serbia, Romania and Algeria.

In addition to the questions on the main GTDs diseases on vines, the survey also included an open question on other causes of decline observed in vineyards. Beside virus and phytoplasma diseases, that are widespread in Europe, *Verticillium* and *Fusarium* diebacks were also mentioned a few times.

The reports of increase of esca complex and other forms of decline on adult plants reported for numerous surveyed countries corroborated previous observations. Several hypotheses were advanced for this increase. In the 1960s–1970s there was a large increase in grapevine planted area, which often required intensive plant production methods, due to the large demand for of low-priced mass market grape products. This significant production led to increased vulnerability of vines to fungal agents of wood diseases (Surico *et al.*, 2004; Gramaje and Armengol, 2011). Domestic grape varieties well-adapted to local pedoclimatic conditions were still grown for inexpensive table or wine production, but were largely replaced by easily available international commercial varieties that better meet commercial needs. Cultural practices also changed, including the training systems, which have direct impacts on incidence of wood diseases incidence (Lecomte *et al.*, 2018), and use of mechanical pruning tools. Another factor leading to the increase in GTDs spread may be climatic change, which would affect plant disease prevalence and have impacts on the complex interactions between pathogens and hosts (Gregory *et al.*, 2009). Account should be taken of the increased awareness of these diseases, which leads to increased numbers of reports. The reported GTDs increasing in the surveyed viticulture regions matched the new wave of forest decline observed in European countries and around the world (Surico et al, 2004; Allen *et al.*, 2010). For instance, the results of this survey showed an increase in the apoplexy observed in different countries that might be related to the summer climatic conditions more favourable for expression of this syndrome, such as drought during summer coming after rain (Dubos, 1999, 2002; Surico *et al.*, 2000). Some studies have considered the relationships between GTDs and climate (Marchi *et al.*, 2006; van Niekerk *et al.* 2011; Calzarano *et al.*, 2018). The present survey results indicated the need for further study on these relationships, and Bois *et al.* (2017) proposed a methodology to address the relationships between grapevine pests and diseases and climate. The present survey has shown variation in GTDs occurrence among the regions within individual country. This variation could be explained by different climatic regimes. For instance, in Hungary, a higher level of occurrence for apoplexy was reported in two regions characterized by dry springs and hot/dry summers, compared to other regions of this country. Similarly, in Bulgaria, "Rose Valley" region showed a higher level of occurrence for esca complex/GLSD in comparison with other regions. This region is characterized by frequent rain during spring over a long period (pers. comm.), quite different from the south-eastern region (Struma), characterized by a continental Mediterranean climate, where GTDs were observed only occasionally. However, beyond these two examples, the data from this survey showed that climate did not always drive the occurrence of GTDs in a region.

Other factors, such as soil type, cultural practices, the rootstock and variety used and vine age may influence the host plant physiology, and hence the level of vulnerability to GTDs. Systematic monitoring and increased GTD research on epidemiology and host/pathogen interactions are needed to better predict the distribution and incidence of GTDs, in the context of climatic changes. Widespread monitoring has been instigated in many regions in Europe. For instance, in France, the National Observatory of GTDs shows large variations in disease incidence among particular grapevine plots (Bruez *et al.*, 2013).

The results of the present survey showed great diversity of grapevine varieties planted in the European and Mediterranean countries, probably reflecting only a small part of the diversity of cultivated varieties. However, this provides evidence of a long wine-producing tradition in these countries. The homogenization given by widespread planting of the same international varieties co-exists with increasing interest in renewal of vintage varieties that are well-adapted to local environments, and also new population variability selected to enlarge genetic backgrounds. In this survey, among the recorded varieties, the six most frequently cited (Cabernet Sauvignon, Merlot, Chardonnay, Riesling, Sauvignon and Pinot Noir) correspond to those commonly grown in the European vineyards (OIV, 2017). Among these varieties, Cabernet Sauvignon was the most reported as susceptible to esca. Cabernet Sauvignon has also been reported in the literature as susceptible to other GTDs, such as Botryosphaeria dieback (Larignon *et al.*, 2001, Quaglia *et al.*, 2009, Bruez *et al.*, 2013) and Eutypa dieback (Dubos, 1999). However, the cv. Merlot, also one of the main reported cultivars in the survey, was indicated as moderately susceptible, presenting less disease incidence in vineyards than Sauvignon blanc and Riesling (Murolo and Romanazzi, 2014), or than Cabernet Sauvignon (Christen *et al.*, 2007). From the National French GTD vineyard survey, Bruez *et al.* (2013) also showed that Merlot presented low esca complex incidence. This cultivar was considered to be tolerant to Eutypa dieback (Dubos, 1999). Regarding Chardonnay, Andreini *et al.* (2014) observed lower percentage of GLSD ("esca symptoms") on Chardonnay vines than on Cabernet Sauvignon or Trebbiano vines, and Bruez *et al.* (2013) reported different level of incidence according to the surveyed region. The number of studies on Chardonnay was less than those on Merlot and Cabernet Sauvignon. For the four other predominating cultivars, Riesling Italico, Pinot noir, Limberger and Riesling, few studies of their susceptibilities to GTDs have been reported. From a mature vineyards survey of 64 cultivars, Murolo and Romanazzi (2014) reported the greatest esca incidence for Riesling, and no symptoms present for Limberger. Pinot noir has been evaluated as moderately affected by Botryosphaeria dieback (Larignon *et al.*, 2001). Trebbiano and Sangiovese were reported as less susceptible to GLSD ("esca") than Cabernet Sauvignon (Andreini *et al.*, 2014), while Sangiovese showed lower percentage of esca symptomatic vines in comparison with Trebbiano in a separate study.

The behaviour of a particular cultivar may vary according to the environmental conditions. Li *et al.* (2017) showed for Cabernet Sauvignon vines of similar age that disease incidence varied from 0.10 to 11.7 in vineyards from one region. The cultivation of resistant or tolerant grapevine cultivars to diseases has been suggested as a major challenge for the 21st Century, to decrease pesticide use against foliar pathogens. Results of the present survey show it is also relevant to select cultivars less susceptible to GTDs. Evaluation of the different responses of the cultivars in particular environments remains essential.

The main varieties of rootstocks used for grafted vines reported in this study were SO4, 110R, K5BB and 1103P. It was generally accepted that rootstock mother plants can also be a source of infections by pathogenic fungi associated to Petri disease and esca complex. Gramaje *et al.* (2010) compared the response to infection of different rootstock varieties to Petri disease pathogens. They found that the rootstocks 140 Ru and 110R were the most greatly affected. Both were among the commonly reported rootstocks in our survey, and 110R was the second most frequently reported. Liminana *et al.* (2009) reported that the 16 year-old trunks of rootstocks 140Ru, 11R, Fercal and 101-14MG presented the most extensive necrosis. The susceptibility of rootstocks to pathogenic fungi associated to GTDs may be considered as well as the susceptibility of scion varieties conferred by the rootstock used. For instance, in an Italian vineyard, Marchi (2001) showed that vines cv. Trebbiano grafted on rootstocks 1103P, 420A or K5BB expressed more esca symptoms than those grafted on S04 or 140RU. In contrast, when two other varieties (cv. Fiano and Sauvignon blanc) were grafted on S04 they had greater esca incidence compared with the same cultivars grafted on 1103P (Murolo and Romanazzi, 2014). These authors explained their results by the greater drought resistance of 1103P than other rootstocks. Environmental conditions must be considered to explain the results, as well as the roles of other internal or external factors (Marchi *et al.*, 2006; Calzarano *et al.*, 2018) to explain these differences.

GTDs on young vines were reported in the survey with a lower occurrence than for GTD on adult plants. GTDs were rarely reported as "increasing" and/or "worrying" on young vines. Petri disease and black foot diseases were only reported to be increasing in Spain, and Petri disease, in Hungary. These diseases are characterized by general decline symptoms, but the diseases were reported only where where plant pathologists have made specific observations and diagnoses to identify the pathogens involved. Therefore, the lack of monitoring of young vineyards, resulting in an underestimation of the problems, cannot be discounted for this result. To our knowledge, this survey was the first to evaluate occurrence of decline diseases in young vines on a large geographical scale.

The large number of survey responses, from a total of 105 viticulture regions in 22 countries, allowed us to draw an overall picture of GTDs spread in Continental European and Mediterranean countries. Some countries with relevant grapevine production were missing from the survey including Tunisia, Morocco, and some Balkan and Central European countries (e.g. Poland). The survey data were based on correspondent declarative responses and not on formally recorded information. Differences between countries may be important: in 15 countries out of 22, GTD surveys or observatories were established at regional or national levels.

In other cases, the responses may be based on expert knowledge on the phytosanitary state, depending on personal perceptions. For some countries, the lack of information or correspondent response may have led to underestimation of occurrence of the disease syndromes. Therefore, the results presented in this paper need to be taken with some caution. Nevertheless, the survey data showed that almost all the surveyed countries (except four), had laboratories or extension services based on GTD survey or/and diagnostic/detection at regional or national levels. That information favoured validity of the collected data, even if the etiological complexity of dieback diseases requires more detailed investigations, such the recording of data from plots and etiological evidence provided by identification of the involved pathogenic fungi, especially for the dead cordon syndrome.

CONCLUSIONS

Taking account of the results of this survey with caution, all findings from this large-scale European consultation, extended to several Mediterranean countries and producers, highlighted the important roles of GTDs, particularly esca complex, in vine decline in European and Mediterranean grape-producing regions. Furthermore, the survey has emphasized the need to accurately determine the occurrence and distribution of the causes of grapevine decline, such as the involvement of Botryosphaericeae and *Diaporthe* species in dieback and vine death. Etiological and epidemiological studies should be conducted to identify and quantify the environmental and agronomic factors associated with GTDs. Also urgent is the need to share and disseminate all nursery and vineyard management practices that have proved useful for GTD prevention, or for limiting the damage resulting from these diseases. The recently established European Winetwork project was set up to increase knowledge transfer on current and innovative strategies for grapevine disease management in Europe (Mondello *et al.*, 2018b, www.winetwork.eu). This initiative needs to be maintained and promoted.

ACKNOWLEDGEMENTS

The authors acknowledge the contributions of all the participants in the European survey, for sharing information about GTDs and viticultural situations from their regions or countries. These people include: Akila Berraf (Algeria), Monika Riedl-Bauer, Harald Berger (Austria), Teodora Dzhabazova, Ivan Tsvetkov (Bulgaria), Marijan Bubola (Croatia), Miroslav Baránek, Ales Eichmeier (Czechia), Jim Newsome (England), Michel Badier (Pays De Loire, France), Bruno Doublet (France), Christel Chevrier (Languedoc Roussillon, France), Claire Grosjean (Bourgogne, France), Marie-Laure Panon (Champagne, Freance), Patrice Retaud (Charentes, France), Michael Fischer (Germany), Tzamos Sotiris, Epaminondas Paplomatas (Greece), Erzsébet Karaffa, Váczy Kálmán Zoltán (Hungary), Nedeljko Latinovic (Montenegro), David Ezra (Israel), Elia Choueiri (Lebanon), Cecilia Rego (Portugal), Liliana Tomoiaga (Romania), Sanja Živkovic (Serbia), Ervín Jankura (Slovakia), Denis Rusjan (Slovenia), Josep Armengol, Rebeca Cobos, David Gramaje, F. Javier Legorburu, Jordi Luque, Ramona M. Munoz (Spain), Olivier Viret (Switzerland) and Davut Soner Akgul (Turkey). We also acknowledge B. Benedetti (University of Firenze) and David Morais (UMR SAVE, INRA/Bordeaux Sciences Agro) for their assistance and Jean-Michel Boursiquot (Montpellier SupAgro) and Jean-Philippe Roby (Bordeaux Sciences Agro) for providing information on viticultural practices. This research is based upon work from COST Action FA1303 (Sustainable Control of grapevine trunk diseases), supported by COST (European Cooperation in Science and Technology).

LITERATURE CITED

AA.VV., 2013. The European Parliament and the Council of the European Union, *Regulation (EU) No 1308/2013 of the European Parliament and the Council of 17 December 2013 establishing a common organisation of the markets in agricultural products* and repealing Council Regulations (EEC) No 922/72, (EEC) No 234/79, (EC) No 1037/2001 and (EC) No 1234/2007, OJ L 347/671, 20.12.2013.

Agustí-Brisach C., Armengol J., 2013. Black-foot disease of grapevine: an update on taxonomy, epidemiology and management strategies. *Phytopathologia Mediterranea* 52: 245–261.

Akgül D.S., Savas Eskalen N.G.A., 2014. First report of wood canker caused by *Botryosphaeria dothidea, Diplodia seriata, Neofusicoccum parvum* and *Lasiodiplodia theobromae* on grapevine in Turkey. *Plant Disease* 98(4): 568–568.

Allen C.D., Macalady A.K., Chenchoum H., Bachelet D., McDowell N., ...Cobb N., 2010. A global overview of drought and heat-induced tree mortality reveals emerging climate change risks for forests. *Forest Ecology & Management* 259: 660–684

Amponsah N.T., Jones E.E., Ridgway H.J., Jaspers M.V., 2014. Factors affecting *Neofusicoccum luteum* infection and disease progression in grapevines. *Australasian Plant Pathology* 43(5): 547–556.

Andolfi L., Mugnai L., Luque J., Surico G., Cimmino A., Evidente A., 2011. Phytotoxins produced by fungi associated with grapevine trunk diseases. *Toxins* 3: 1569–1605.

Andreini L., Cardelli R., Bartolini S., Scalabrelli G., Viti R., 2014. Esca symptoms appearance in *Vitis vinifera* L.: influence of climate, pedo-climatic conditions and rootstock/cultivar combination. *Vitis* 53(1): 33–38.

Armengol J., Vicent A., Torne L., Garcia-Figueres F., Garcia-Jimenez J., 2001. Fungi associated with esca and grapevine declines in Spain: a three-year survey. *Phytopathologia Mediterranea* 40: S325- S329.

Baránek M., Armengol J., Holleinová J., Pečenka J., Calzarano F., Peňázová E., Vachůn M., Eichmeier A., 2018. Incidence of symptoms and fungal pathogens associated with grapevine trunk diseases in Czech vineyards: first example from a north-eastern European grape-growing region. *Phytopathologia Mediterranea* 57(3): 449–458.

Bellée A., Comont G., Nivault A., Abou-Mansour E., Coppin C., Dufour M.C., Corio-Costet M.F., 2017. Life traits of four Botryosphaeriaceae species and molecular responses of different grapevine cultivars or hybrids. *Plant Pathology* 66: 763–776.

Berraf-Tebbal A., Bouznad Z., Santos J.M., Coelho M.A., Péros J.P., Phillips A.J., 2011. *Phaeoacremonium* species associated with Eutypa dieback and esca of grapevines in Algeria. *Phytopathologia Mediterranea* 50: S86–S97.

Bertsch C., Ramirez-Suero M., Magnin-Robert M., Larignon P., Chong J., Spagnolo A., Clément C., Fontaine F., 2013. Grapevine trunk diseases: complex and still poorly understood. *Plant Pathology*, 62(2): 243-365.

Black R.E., Ulin R.C. (eds.), 2013. *Wine and culture: Vineyard to glass*. Bloomsbury (Academic), New York, USA, 336 pp.

Bois B., Zito S., Calonnec A., 2017. Climate vs grapevine pests and diseases worldwide: the first results of a global survey. *OENO one* 51(2): 133–139.

Bruez E., Grosman J., Doublet B., Lecomte P., Bertsch C., Fontaine F., Guérin-Dubrana L., Rey P., 2013. The

French Grapevine Trunk Wood Diseases Survey: overview of the data obtained over the last decade. *Phytopathologia Mediterranea* 52(2): 262–275.

Bruez E., Vallance J., Gerbore J., Lecomte P., Da Costa J.P., Guérin-Dubrana L., Rey P., 2014. Analyses of the Temporal Dynamics of Fungal Communities Colonizing the Healthy Wood Tissues of Esca Leaf-Symptomatic and Asymptomatic Vines. *PloS one* 9(5): e95928.

Bugaret Y., 1984. *L'excoriose de la Vigne. Recherche sur* Phomopsis viticola *Sacc. Nouvelles possibilités de lutte.* Thèse, Université Bordeaux I, 163 pp.

Burruano S., Mondello V., Conigliaro G., Alfonzo A., Spagnolo A., Mugnai L., 2008. Grapevine decline in Italy caused by *Lasiodiplodia theobromae*. *Phytopathologia Mediterranea* 47(2): 132–136.

Calzarano F., S. Di Marco, 2018. Further evidence that calcium, magnesium and seaweed mixtures reduce grapevine leaf stripe symptoms and increase grape yield. *Phytopathologia Mediterranea* 57: 459–471.

Calzarano F., Seghetti L., Del Carlo M., Cichelli A., 2004. Effect of esca on the quality of berries, musts and wines. *Phytopathologia Mediterranea* 43: 125–135.

Calzarano F., F. Osti, M. Baránek, S. Di Marco, 2018. Rainfall and temperature influence expression of foliar symptoms of grapevine leaf stripe disease (esca complex) in vineyards. *Phytopathologia Mediterranea* 57: 488–505.

Cardoso M., Diniz I., Cabral A., Rego C., Oliveira H., 2013. Unveiling inoculum sources of black foot pathogens in a commercial grapevine nursery. *Phytopathologia Mediterranea* 52: 298–312.

Carlucci A., Lops F., Raimondo M.L., Gentile V., Mucci M., Frisullo S., 2009. The Botryosphaeriaspecies from vineyards of Apulia. *Phytopathologia Mediterranea* 48: 180.

Carlucci A., Cibelli F., Lops F., Raimondo M.L., 2015. Characterization of Botryosphaeriaceae species as causal agents of trunk diseases on grapevines. *Plant disease* 99(12): 1678–1688.

Carlucci A., Lops F., Mostert L., Halleen F., Raimondo M.L., 2017. Occurence of fungi causing black foot on young grapevines and nursery rootstock plants in Italy. *Phytopathologia Mediterranea* 56: 10–39.

Carter M.V., 1991. The status of *Eutypa lata* a Pathogen. *Phytopathology Paper* 32: 59 pp.

Christen D., Schonmann S., Jermini M., Strasser R. J., Defago G., 2007. Characterization and early detection of grapevine (*Vitis vinifera*) stress responses to esca disease by in situ chlorophyll fluorescence and comparison with drought stress. *Environmental and Experimental Botany* 60(3): 504–514.

Comont G, Mayet V., Corio-Costet M.F., 2016. First report of *Lasiodiplodia viticola, Spencermartinsia viticola* and *Diplodia intermedia* associated with *Vitis vinifera* grapevine decline in French vineyards. *Plant Disease* 100 (11): 23–28.

Cristinzio G., 1978. Gravi attacchi di *Botryosphaeria obtusa* su vite in provincia di Isernia. *Informatore Fitopatologico* 6: 21–23.

Desprez-Loustau M.L., Marçais B., Nageleisen L.M., Piou D., Vannini A., 2006. Interactive effects of drought and pathogens in forest trees. *Annals of Forest Science* 63(6): 597–612.

Dissanayake A.J., Liu M., Zhang W., Udayanga D., Chukeatirote E., Li X.H., Yan J.Y., Hyde K.D., 2015. Morphological and molecular characterisation of *Diaporthe* species associated with grapevine trunk disease in China. *Fungal Biology* 119: 283–294.

Dos Santos R. F., Blume E., Muniz M. F. B., Harakawa R., da Ressurreição Garrido, L., Rego C., 2014. Characterization of *Campylocarpon pseudofasciculare* associated with black foot of grapevine in southern Brazil. *Phytopathologia Mediterranea* 53: 406–415.

Dubos B., 1999. *Maladies Cryptogamiques de la Vigne*. Editions F.ret, Bordeaux, France, 176 pp.

Dubos B., 2002. Le syndrome de l'esca. In: *Maladies Cryptogamiques de la Vigne*, 2nd ed. Féret, Bordeaux, 127–136.

Edwards J., Marchi G., Pascoe I.G., 2001. Young esca in Australia. *Phytopathologia Mediterranea* 40: 303–310.

El-Goorani M.A., El Meleigi M.A.,1972. Dieback of grapevine by *Botryodiplodia theobromae* Pat. in Egypt. *Phytopathologia Mediterranea* 11: 210–211.

Fabre B., Ioos R., Piou D., Marçais B., 2011. Is the emergence of Dothistroma needle blight of pine in France caused by the cryptic species *Dothistroma pini*? *Phytopathology* 102: 47–54.

Fischer M., Kassemeyer H.H., 2003. Fungi associated with Esca disease of grapevine in Germany. *Vitis* 42: 109–116.

Fontaine F., Armengol J., 2014. Sustainable control of grapevine trunk diseases. *Phytopathologia Mediterranea* 53: 584 (abstract).

Fontaine F., Gramaje D., Armengol J., Smart R., Nagy Z.A., … Corio–Costet M.F., 2016. *Grapevine Trunk Diseases. A review.* ©OIV publications, 1st Edition (Vol. May). Paris.

Gonzalez V., Tello M.L., 2011. The endophytic mycota associated with *Vitis vinifera* in central Spain. *Fungal diversity* 47(1): 29–42.

Gramaje D., Armengol J., 2011. Fungal trunk pathogens in the grapevine propagation process: potential inoculum sources, detection, identification, and management strategies. *Plant Disease* 95(9): 1040–1055.

Gramaje D., Di Marco S., 2015. Identifying practices likely to have impacts on grapevine trunk disease infections: a European nursery survey. *Phytopathologia Mediterranea* 54: 313–324.

Gramaje D., García-Jiménez J., Armengol J., 2010. Field evaluation of grapevine rootstocks inoculated with fungi associated with petri disease and esca. *American Journal Enology Viticulture* 61: 512–520.

Gramaje D., Úrbez-Torres J.R., Sosnowski M.R., 2018. Managing grapevine trunk diseases with respect to etiology and epidemiology: current strategies and future prospects. *Plant Disease* 102: 12–39.

Graniti A., Surico G., Mugnai L., 2000. Esca of grapevine: a disease complex or a complex of diseases? *Phytopathologia Mediterranea* 39: 16–20.

Gregory P.J., Johnson S.N., Newton A.C., Ingram J.S., 2009. Integrating pests and pathogens into the climate change/food security debate. *Journal of Experimental Botany* 60(10): 2827–2838.

Grosman J., Doublet B., 2012. Maladies du bois de la vigne. Synthèse des dispositifs d'observation au vignoble, de l'observatoire 2003–2008 au réseau d'épidémio-surveillance actuel. *Phytoma – La Défense des Végétaux* 651: 31–35.

Guarnaccia V., Groenewald J.Z., Woodhall J., Armengol J., Cinelli T., ... Crous P.W., 2018. Diaporthe diversity and pathogenicity revealed from a broad survey of grapevine diseases in Europe. *Persoonia: Molecular Phylogeny and Evolution of Fungi* 40: 135–153.

Gubler W.D., Mugnai L., Surico G., 2015. Esca, Petri and Grapevine leaf stripe disease. Pages 52–56 in: *Compendium of Grape Diseases Disorders and Pests*, 2nd Ed. (Wilcox W.F., Gubler W.D., Uyemoto J.K., eds), APS Press, St Paul, MN, USA.

Halleen F., Fourie P.H., Crous P.W., 2006. A review of black foot disease of grapevine. *Phytopathologia Mediterranea* 45: S55–S67.

Halleen F., Mostert L., Crous P.W., 2007. Pathogenicity testing of lesser-known vascular fungi of grapevines. *Australasian Plant Pathology* 36: 277–285.

Kaliterna J., Milicevic T., Bencic D., Duralija B., 2013. First report of *Neofusicoccum parvum* associated with grapevine trunk diseases in Croatia. *Plant disease* 97(12): 1656–1656.

Kaliterna J., Miličević T., 2014. Incidence and distribution of fungi from Diaporthaceae and Botryosphaeriaceae on grapevine in Croatia. *11th Conference of the European Foundation for Plant Pathology – Healthy plants – healthy people*. Krakov, Poland.

Kovács C., Balling P., Bihari Z., Nagy A., Sándor E., 2017. Incidence of grapevine trunk diseases is influenced by soil, topology and vineyard age, but not by *Diplodia seriata* infection rate in the Tokaj Wine Region, Hungary. *Phytoparasitica* 45(1): 21–32.

Langenhoven S., Halleen F., Spies C. F., Stempien E., Mostert L., 2018. Detection and quantification of black foot and crown and root rot pathogens in grapevine nursery soils in the Western Cape of South Africa. *Phytopathologia Mediterranea* 57: 519–537.

Larignon P., Fulchic R., Ceré L., Dubos B., 2001. Observation on black dead arm in French vineyards. *Phytopathologia Mediterranea* 40: 336–342.

Larignon P., Spagnolo A., Bertsch C., Fontaine F., 2015. First report of Young decline grapevine caused by *Neofusicoccum parvum* in France. *Plant disease* 99(12): 1859.

Larignon P., 2016. *Maladies Cryptogamiques du Bois de la Vigne: Symptomatologie et Agents Pathogènes*. http://www.vignevin.com, 2ème édition, janvier 2016.

Lawrence D.P., Hand F.P., Gubler W.D., Trouillas F.P., 2017. Botryosphaeriaceae species associated with dieback and canker disease of bay laurel in northern California with the description of *Dothiorella californica* sp. nov. *Fungal biology* 121(4): 347–360.

Lecomte P., Louvet G., Vacher B., Guilbaud P., 2006. Survival of fungi associated with grapevine decline in pruned wood after composting. *Phytopathologia Mediterranea* 45: S127–S130.

Lecomte P., Diarra B., Carbonneau A., Patrice Rey Chevrier C., 2018. Esca of grapevine and training practices in France: results of a 10-year survey. *Phytopathologia Mediterranea* 57: 472–487.

Li S., Bonneu F., Chadoeuf J., Picart D., Gégout Petit A., Guérin-Dubrana L., 2017. Spatial and Temporal Pattern Analyses of Esca Grapevine Disease in Vineyards in France. *Phytopathology* 107(1): 59–69.

Liminana J.M., Pacreau G., Boureau F., Ménard E., David S., Himonnet C., Fermaud M., Goutouly J.P., Lecomte P., Dumot V., 2009. Inner necrosis grapevine rootstock mother plants in the Cognac area (Charentes, France). *Phytopathologia Mediterranea* 48: 92–100.

Lorrain B., Ky I., Pasquier G., Jourdes M., Guerin-Dubrana L., Gény L., Rey P., Donèche B., Teissedre P.L., 2012. Effect of Esca disease on the phenolic and sensory attributes of Cabernet Sauvignon grapes, musts and wines. *Australian Journal Grape Wine Research* 18: 64–72.

Luque J., Martos S., Aroca A., Raposo R., Garcia-Figueres F., 2009. Symptoms and fungi associated with declining mature grapevine plants in northeast Spain. *Journal of Plant Pathology* 91: 381–390.

Luque J., Martos S., Garcia-Figueres F., 2010. Effects of water stress and inoculation with *Eutypa lata* and

Neofusicoccum parvum on young grapevine plants. *Phytopathologia Mediterranea* 49: 120–121(abstract).

Luque J., Garcia-Figueres F., Legorburu F.J., Muruamendiara A., Armengol J., Trouillas F.P., 2012. Species of Diatrypaceae associated with grapevine trunk diseases in Eastern Spain. *Phytopathologia Mediterranea* 51: 528–540.

Marchi G., 2001. Susceptibility to esca of various grapevine (*Vitis vinifera*) cultivars grafted on different rootstocks in a vineyard in the province of Siena (Italy). *Phytopathologia Mediterranea* 40: 27–36.

Marchi G., Peduto F., Mugnai L., Di Marco S., Calzarano F., Surico G., 2006. Some observations on the relationship of manifest and hidden esca to rainfall. *Phytopathologia Mediterranea* 45: 117–126.

Martin T.M., Cobos R., 2007. Identification of fungi associated with grapevine decline in Castilla y Leon (Spain). *Phytopathologia Mediterranea* 46: 18–25.

Mehl J.W.M., Slippers B., Roux J., Wingfield M.J., 2014. Botryosphaeriaceae associated with dieback of Schizolobium parahyba trees in South Africa. *Mycologia* 103: 534–553.

Merrin S.J., Nair N.G., Tarran J., 1995. Variation in *Phomopsis* recorded on grapevine in Australia and its taxonomic and biological implications. *Australasian Plant Pathology* 24: 44–56.

Molyneux R.J., Mahoney N., Bayman P., Wong R.Y., Meyer K., Irelan N., 2002. Eutypa dieback in grapevines: differential production of acetylenic phenol metabolites by strains of *Eutypa lata*. *Journal of Agricultural and Food Chemistry* 50: 1393–1399.

Mondello V., Songy A., Battiston E., Pinto C., Coppin C., … Fontaine F., 2018a. Grapevine trunk diseases: a review of fifteen years of trials for their control with chemicals and biocontrol agents. *Plant Disease* 102: 1189–1217.

Mondello V., Larignon P., Armengol J., Kortekamp A., Vaczy K., Prezman F., Serrano E., Rego C., Mugnai L., Fontaine F., 2018b. Management of grapevine trunk diseases: knowledge transfer, current strategies and innovative strategies adopted in Europe. *Phytopathologia Mediterranea* 57: 369–383

Mozell M.R., Thach L., 2014. The impact of climate change on the global wine industry: Challenge and solutions. *Wine Economics and Policy* 3: 81–89.

Mugnai L., Graniti A., Surico G., 1999. Esca (black measles) and brown wood-streaking: two old and elusive diseases of grapevines. *Plant Disease* 83: 404–417.

Munkvold G.P., Duthie J.A., Marois J.J. 1994. Reduction in yield and vegetative growth of grapevines due to Eutypa dieback. *Phytopathology* 84:186–192.

Murolo S., Romanazzi G., 2014. Effects of grapevine cultivar, rootstock and clone on esca disease. *Australasian Plant Pathologyyl* 43: 215–221.

Nikolova E., 2010. Еска По Лозата - Симптоми, Причинители И Средства За Борба. (Esca disease of grapevine – Symptoms, causal organisms and methods for control). Растениеъдни Науки 47: 195–205.

Péros J.P., Berger G., 2003. Diversity within natural progenesis of the grapevine dieback fungus *Eutypa lata*. *Current Genetics* 36: 301–309.

OIV, 2017. Focus OIV 2017. Distribution of the world's grapevine varieties. ISBN: 979-10-91799-89-8OIV - International organization of vine and wine, Paris, France.

OIV, 2014. OIV World Vitivinicultural Statistics 2013-2014. International Organisation of Vine and Wine, ISBN 979-10-91799-94-2, http://www.oiv.int/public/medias/6292/oiv-world-vitivinicultural-statistics-2013-2014-en.pdf

Phillips A.J.L., 2000. Excoriose, cane blight and related diseases of grapevines: a taxonomic review of the pathogens. *Phytopathologia Mediterranea* 39: 341–356

Phillips A.J.L., 2002. Botryosphaeria species associated with diseases of grapevines in Portugal. *Phytopathologia Mediterranea* 41: 3–18.

Piškur B., Pavlic D., Slippers B., Ogris N., Maresi G., Wingfield M.J., Jurc, D., 2011. Diversity and pathogenicity of Botryosphaeriaceae on declining *Ostrya carpinifolia* in Slovenia and Italy following extreme weather conditions. *European Journal of Forest Research* 130(2): 235–249.

Pitt W.M., Huang R., Steel C.C., Savocchia S., 2013. Pathogenicity and epidemiology of Botryosphaeriaceae species isolated from grapevines in Australia. *Australasian Plant Pathology* 42: 573–582.

Quaglia M., Covarelli L., Zazzarini A., 2009. Epidemiological survey on esca disease in Umbria, central Italy. *Phytopathologia Mediterranea* 48: 84–91.

Ravaz L., Verge G., 1925. Sur une maladie de la vigne, l'Excoriose. *Comptes rendus des Séances de l'Académie des Sciences*, Paris, France. 180, 313–315.

Rego M.C., Oliveira H., Carvalho A., Phillips A., 2000. Involvement of *Phaeoacremonium* spp. and *Cylindrocarpon destructans* with grapevine decline in Portugal. *Phytopathologia Mediterranea* 39: 76–79.

Rego C., Vaz A., Nascimento T., Cabral A., Oliveira H., 2009. Diseases incited by Botryosphaeriaceae fungi in Portuguese vineyards. *Phytopathologia Mediterranea* 48: 181 (abstract).

Rolshausen, P.E., Baumgartner K., Travadon R., Fujiyoshi P., Pouzoulet J., Wilcox W.F., 2014. Identification of Eutypa spp. causing Eutypa dieback of grapevine in eastern North America. *Plant Disease* 98 (4): 483–491.

Sidoti A., Buonocore E., Serges T., Mugnai L., 2000. Decline of young grapevines associated with *Phaeoacremonium chlamydosporum* in Sicily (Italy). *Phytopathologia Mediterranea* 39: 87–91.

Slippers B., Wingfield M.J., 2007. Botryosphaeriaceae as endophytes and latent pathogens of woody plants: diversity, ecology and impact. *Fungal Biology Review* 21: 90–106.

Songy A., Fernandez O., Clément C., Larignon P., Fontaine F., 2019. Grapevine trunk diseases under thermal and water stresses. *Planta*, https://doi.org/10.1007/s00425-019-03111-8.

Spagnolo A., Marchi G., Peduto F., Phillips A.J., Surico G., 2011. Detection of Botryosphaeriaceae species within grapevine woody tissues by nested PCR, with particular emphasis on the *Neofusicoccum parvum/N. ribis* complex. *European Journal of Plant Pathology* 129(3): 485–500.

Spinoni J., Nauman G., Vogt J.V., 2017. Pan-European seasonal trends and recent changes of drought frequency and severity. *Global and Planetary Change* 148: 113–130.

Surico G., 2009. Towards a redefinition of the diseases within the esca complex of grapevine. *Phytopathologia Mediterranea* 48: 5–10.

Surico G., Marchi G., Ferrandino F.J., Braccini P., Mugnai L., 2000. Analysis of the spatial spread of esca in some Tuscan vineyards (Italy). *Phytopathologia Mediterranea* 39: 211–224.

Surico G., Bandinelli R., Braccini P., Di Marco S., Marchi G., Mugnai L., Parrini C., 2004. On the factors that may have influenced the esca epidemic in Tuscany in the eighties. *Phytopathologia Mediterranea* 43: 136–143.

Trouillas F.P., Gubler W.D., 2010. Pathogenicity of Diatrypaceae species in grapevines in California. *Plant Disease* 94: 867–872.

Trouillas F.P., Úrbez-Torres J.R., Gubler W.D., 2010. Diversity of diatrypaceous fungi associated with grapevine canker diseases in California. *Mycologia* 102: 319–336.

Úrbez-Torres J. R., 2011. The status of Botryosphaeriaceae species infecting grapevines. *Phytopathologia Mediterranea* S50: 5–45.

Úrbez-Torres J.R., Peduto Hand F., Smith R.J., Gubler W.D., 2013. Phomopsis dieback: a grapevine trunk disease caused by *Phomopsis viticola* in California. *Plant Disease* 12: 1571–1579.

Úrbez-Torres J.R., Haag P., Bowen P., O'Gorman D.T., 2014. Grapevine Trunk Diseases in British Columbia: incidence and characterization of the fungal pathogens associated with black foot disease of grapevine. *Plant Disease* 98: 469–482.

van Niekerk J.M., Groenewald J.Z., D.F. Farr, Fourié P.H., Halleen F., Crous P.W., 2005. Reassessment of *Phomopsis* species on grapevines. *Australasian Plant Pathology* 34: 27–39.

van Niekerk J.M., Fourié P.H., Halleen F., Crous P.W., 2006. Botryosphaeria spp. As grapevine trunk disease pathogens. *Phytopathologia Mediterranea* 45: S43–S54.

van Niekerk J.M., Bester W., Halleen F., Crous P.W., Fourié P.H., 2011. The distribution and symptomatology of grapevine trunk disease pathogens are influenced by climate. *Phytopathologia Mediterranea* 50: 98–111.

Viala P., 1926. Recherches sur les maladies de la vigne: Esca. *Annales des Epiphyties* 12: 1–108.

12

Resistance of wild barley (*Hordeum spontaneum*) and barley landraces to leaf stripe (*Drechslera graminea*)

Arzu ÇELİKOĞUZ

Ankara University, Faculty of Agriculture, Department of Plant Protection, 06110, Ankara, Turkey
E mail: acelik@agri.ankara.edu.tr

Competing Interests: The Author(s) declare(s) no conflict of interest.

Editor: Andy Tekauz, Cereal Research Centre, Winnipeg, MB, Canada.

Summary. Wild barley (*Hordeum spontaneum*) and barley landraces are important sources of genetic variation for disease resistance. Thirty wild barley (*H. spontaneum*) genotypes and 30 barley landraces were evaluated for susceptibility to two *Drechslera graminea* isolates. Virulence differences were observed between the isolates, while the responses of the host genotypes to the isolates also varied. Of the *H. spontaneum* genotypes, 23% and 63%, respectively, were resistant to the Yozgat *D. graminea* isolate, and Eskişehir *D. graminea* isolates. On the other hand, 43% and 90% of the barley landraces were resistant to Yozgat and Eskişehir *D. graminea* isolates, respectively. *Hordeum spontaneum* genotypes 13, 24, 27, 29, 54, 86, and 91 exhibited resistance to both *D. graminea* isolates, while genotypes 14 and 32 showed intermediate reactions to the Yozgat isolate and resistant reactions to the Eskişehir isolate. Barley landraces 21, 37, 38, 39, 40, 73, 98, 128, 139, 153, 159,167, and 171 showed resistant reactions, and barley landrace 8 showed an intermediate reaction to both isolates. Barley landraces 3, 20, 24, 71, 101, 103, 104 and 160 exhibited intermediate responses to the Yozgat isolate and a resistant response to the Eskişehir isolate. Using resistant barley genotypes would reduce the need for pesticides for control of leaf stripe, and be an environmentally preferred strategy for disease control. The disease resistance present in wild barley and barley landraces are important for expanding the genetic basis of cultivated barley (*H. vulgare*). The resistant and intermediate genotypes identified in this study could be used as resistance sources in barley breeding, or landraces could be used directly for commercial barley production.

Keywords. Disease resistance, *Pyrenophora graminea, Hordeum vulgare.*

INTRODUCTION

Barley (*Hordeum vulgare* L.) is the second most cultivated cereal crop, after wheat, in Turkey, constituting 22% of the cereal production area. In this country, 7.1 million tons of barley are produced per year, with an average yield of 293 kg ha^{-1} (Tuik, 2017). It is believed that approx. 10,000 years ago, the first area where barley was cultivated was in the Fertile Crescent Region, located between the Mediterranean and Arab peninsulas, and bordered by the Tigris and Euphrates valleys (Harlan and Zohary, 1966; Nesbitt, 1995; Willcox, 1995; Ladizinski, 1998). Throughout history, this region has been considered as one of the richest centres of plant diversity (Zohary and Hopf, 2000). Turkey is uniquely situated in terms of plant genetic diversity, as it has an abundance of plant species and endemism due to a combination of geomorphologic, topographic, and seasonal diversity (Fao, 2015). Turkey is located at the intersection point between Mediterranean and Near East gene centres, and this area is one of the most significant genetic centres for barley (Kün, 1996).

Barley landraces (*H. vulgare* L. subsp. *vulgare*) are heterogeneous plant species grown by farmers and are populations exposed to natural and artificial selection (Brown, 2000), and barley landraces are genetically closer to modern varieties compared to wild barley (Thomas *et al.*, 1998). Local barley varieties are the main sources of seed used in regions of low annual rainfall and where 'traditional' agriculture is practiced (Ceccarelli and Grando, 2000). Wild barley (*H. spontaneum* C. Koch) is accepted as the progenitor of cultivated barley (*H. vulgare*), and its habitat is in the Fertile Crescent Region. This plant is indigenous to the area between the south and southeast of Turkey and the area between North Africa and southwest Asia (Harlan and Zohary, 1966; Nevo, 1992; Von Bothmer *et al.*, 1995). *Hordeum spontaneum* is often found in secondary habitats such as Mediterranean scrub lands or roadsides (Zohary and Hopf, 2000).

Wild barley genotypes and barley landraces are important resources for genetic variation, as they are highly adapted to abiotic and biotic stresses and can therefore be cultivated under unfavorable conditions (Allard and Bradshaw, 1964; Yitbarek *et al.*, 1998; Ellis *et al.*, 2000; Ceccarelli and Grando, 2000; Karakaya *et al.*, 2016a). This genetic variation provides a potential source of disease resistance alleles for breeding programmes (Allard and Bradshaw, 1964; Ceccarelli, 1996). Turkey is one of the most important genetic centres for barley, as landraces are widely planted and wild barley genotypes grow under natural conditions (Helbaek, 1969; Kün, 1996; Pourkheirandish and Komatsuda, 2007; Karakaya *et al.*, 2016a; Ergün *et al.*, 2017).

The causal agent of barley leaf stripe is the fungus *Drechslera graminea* (Rabenh. ex Schlecht.) Shoemaker (=*Helminthosporium gramineum* Rabh.) (teleomorph: *Pyrenophora graminea* (S. Ito & Kurib.). This fungus is a single-cycle, seed borne pathogen that causes reductions in barley yields and quality throughout world cereal production areas. The pathogen survives as mycelium within host pericarps and grows into developing seedlings via the coleorhizae when the barley seeds germinate (Platenkamp, 1976). Subsequently, the pathogen grows systemically in developing host plants (Çetinsoy, 1995; Mathre, 1997; Aktaş 2001). The first symptoms of disease occur as yellow stripes on seedling leaves, and these progress to chlorotic and necrotic stripes areas along the leaves. As results of the disease, sterile spikes and stunting also occur in affected plants (Tekauz and Chiko, 1980; Zad *et al.*, 2002). Severe infections result in drying out and premature death of plants (Mathre, 1997). Yield losses due to leaf stripe have been reported from various countries (Porta-Puglia *et al.*, 1986; Arabi *et al.*, 2004). The disease is present in Turkish barley fields, causing yield losses between 3% and 15% (Mamluk *et al.*, 1997). In 2012 and 2013, it was found that 40% of the surveyed barley fields in Central Anatolia were affected by leaf stripe (Karakaya *et al.*, 2016b). Kavak (2004) emphasized that in addition to yield losses, quarantine issues can be important, because the pathogen is readily seed transmitted. While barley leaf stripe can be controlled through the use of seed treatment fungicides, growing of resistant varieties would minimize the need for pesticides and be an economic and environmentally friendly method for controlling the disease.

Research has shown a diversity of morphological characters and virulence levels for populations of *D. graminea* (Gatti *et al.*, 1992; Jawhar and Arabi, 2006; Karakaya *et al.*, 2017). McDonald and Linde (2002) emphasized that pathogen populations that vary genetically can quickly evolve and overcome plant resistance. Significant virulence diversity and the possible results of a shrinking genetic basis of cultivated barley have been studied by plant pathologists and plant breeders (Jensen, 1988; Ulus and Karakaya, 2007; Çelik *et al.*, 2016). Plant breeders need sustainable sources of disease resistance for effective long-term disease control.

In this study, 30 barley landraces and 30 wild barley (*H. spontaneum*) genotypes were selected from a collection maintained by the Gene Bank of the Central Research Institute for Field Crops located in Ankara, Turkey. These host germplasm lines were assessed for their resistance to leaf stripe using two isolates of *D. graminea*.

MATERIALS AND METHODS

*Barley landraces, wild barley (*H. spontaneum*) genotypes and* Drechslera graminea *isolates*

Thirty *H. spontaneum* genotypes and 30 barley landraces were obtained from the Gene Bank of the Field Crops Central Research Institute, in Ankara, Turkey.

These plant lines had been collected from different regions of Turkey, and seeds of these genotypes had been multiplied from single spikes and maintained in the Gene Bank.

The virulent Yozgat isolate of *D. graminea* and the moderately virulent Eskişehir isolate were compared. The virulence of these isolates was previously determined by Çelik *et al.* (2016) and Karakaya *et al.* (2017). The isolates were maintained at the Mycology Laboratory of Ankara University, Faculty of Agriculture, Department of Plant Protection, Turkey. Barley cultivars 'Çumra 2001' and 'Larende' were used, respectively, as resistant and susceptible controls (Çelik *et al.*, 2016). The disease responses of these 30 landraces and 30 wild barley genotypes used in the present study against *Pyrenophora teres* f. *teres*, *P. teres* f. *maculata* and *Rhynchosporium commune* had been determined in previous studies (Çelik Oğuz *et al.*, 2017b; Çelik Oğuz *et al.*, 2019; Azamparsa *et al.*, 2019).

Treatments and disease evaluations

The sandwich method described by Mohammad and Mahmood (1974) was used to inoculate the barley landraces and wild barley genotypes. Seeds were surface sterilized with a 1% NaOCI solution for 3 min and then rinsed with sterile water. Cultures of *D. graminea* were grown on potato dextrose agar medium in Petri dishes at 22 ± 2°C for 10 d. Fifteen seeds of each host landrace or wild genotype were placed on the surface of half of a *D. graminea* culture followed by folding the other half of the culture over the seeds under sterile conditions. Cultures folded as 'sandwiches' were kept at 22°C for 4 d. After germination, seeds were incubated at 4°C for an additional 5 d. Three replications were used for each host line and isolate combination. Following treatment, the incubated seeds were taken from the sandwich cultures using sterile forceps and planted in pots containing growth medium (soil, sand, and animal manure at 3:1:1 w:w:w). The resulting plants were grown under greenhouse conditions of 15 ± 2°C at night, 22 ± 2°C during the day, using a 15/9 h light/dark regime. Pots were arranged on a greenhouse bench in a completely randomized fashion. Disease ratings were taken at 45 and 60 d after planting of inoculated seeds and were recorded separately. The responses of the plants to the two *D. graminea* isolates were evaluated using the scale developed by Tekauz (1983). Scale values were: 1 = infection < 5% (Resistant, R); 2 = infection 5–17% (Intermediate, I); and 3 = infection > 17% (Susceptible, S).

Data analyses

The percentage of leaf barley stripe was calculated using the following equation (Dumalasova *et al.*, 2014).

% Disease incidence = No. of infected plants ÷ Total no. of plants ×100.

Separate analyses of variances were performed for disease assessed at 45 and 60 d after planting for isolate, genotype and isolate*genotype interaction effects (Tables 1-4).

Biplot analysis was performed for the isolates and disease percentage values to assess the disease responses of each genotype tested to the two *D. graminea* isolates used (MSTAT, Michigan State University).

RESULTS

At 45 d after planting of inoculated seeds, there were 20 wild barley (*H. spontaneum*) genotypes showing susceptible reactions to the virulent Yozgat isolate of *D. graminea*, three genotypes with intermediate reactions and seven genotypes with resistant reactions to this isolate. At 60 days after planting, 21 genotypes were susceptible, two were intermediate and seven were resistant to the Yozgat isolate. Five genotypes were susceptible to

Table 1. Analysis of variance for the resistance of 30 *Hordeum spontaneum* genotypes 45 d after planting following seed inoculation with two isolates of *Drechslera graminea.*

Source	DF	SS	MS	F	P
H. spontaneum genotypes	29	66151.3	2281.1	1105.86	<0.001
Isolates	1	46217.7	46217.7	22406.17	<0.001
*H. spontaneum**isolates	29	33414.9	1152.2	558.60	<0.001
Error	120	247.5	2.1		
Total	179	146031.4			

Table 2. Analysis of variance for the resistance of 30 *Hordeum spontaneum* genotypes 60 d after planting following seed inoculation with two isolates of *Drechslera graminea.*

Source	DF	SS	MS	F	P
H. spontaneum genotypes	29	72401.8	2496.6	1842.90	<0.001
Isolates	1	52473.2	52473.2	38733.54	<0.001
*H. spontaneum**Isolates	29	36521.1	1259.3	929.60	<0.001
Error	120	162.6	1.4		
Total	179	161558.6			

Table 3. Analysis of variance for the resistance of 30 barley landraces 45 d after planting following seed inoculation with two isolates of *Drechslera graminea.*

Source	DF	SS	MS	F	P
Barley landraces	29	13938.64	480.64	706.65	<0.001
Isolates	1	5261.77	5261.77	7736.00	<0.001
Barley landraces*Isolates	29	7961.87	274.55	403.65	<0.001
Error	120	81.62	0.68		
Total	179	27243.90			

Table 4. Analysis of variance for the resistance of 30 barley landraces 60 d after planting following seed inoculation with two isolates of *Drechslera graminea.*

Source	DF	SS	MS	F	P
Barley landraces	29	13551.12	467.28	687.01	<0.001
Isolates	1	6646.66	6646.66	9772.10	<0.001
Barley landraces*Isolates	29	7878.26	271.66	399.41	<0.001
Error	120	81.62	0.68		
Total	179	28157.66			

the moderately virulent Eskişehir isolate, six genotypes had intermediate reactions, and 19 were resistant to this isolate. The numbers of genotypes showing susceptible, intermediate, and resistant reactions to the Eskişehir isolate remained unchanged at 60 d after planting (Table 5).

At 45 d after planting, seven barley landraces exhibited susceptible reactions to the Yozgat isolate of *D. graminea*, eight landraces showed intermediate reactions, and 15 ladraces were resistant to the isolate. At 60 d after planting, eight barley landraces were susceptible to the Yozgat isolate, nine landraces were intermediate, and 13 landraces showed resistant reactions to this isolate. At 45 d after planting, three landraces showed intermediate reactions and 27 landraces showed resistant reactions to the Eskişehir isolate. At the 60 day assessment, the reactions of the genotypes to the Eskişehir isolate were the same as those assessed at 45 d (Table 6).

The susceptible control barley cultivar 'Larende' exhibited susceptible reactions to both *D. graminea* isolates, and the resistant control cultivar 'Çumra 2001' was resistant to the two isolates.

Separate analyses of variance revealed statistically significant ($P < 0.01$) differences among the *H. sponatenum* genotypes and barley landraces and between the two *D. graminea* isolates, both at 45 and 60 d after planting of inoculated seeds. Significant ($P < 0.01$) isolate*genotype interactions were also detected (Tables 1-4).

Disease resistance evaluations require clear understanding of host/pathogen interactions. Visual analyses of these interactions are possible with biplot analyses. Low Component 1 negative values, and Component 2 values close to zero in biplots clearly illustrate the resistance of genotypes to disease (Yan and Falk 2002). In the biplot analyses, *H. spontaneum* genotypes 13, 24, 27, 29, 54, 86, 91, and the resistant control cultivar 'Çumra 2001' were grouped together, and representing the most resistant genotypes of those studied (Figure 1). The wild barley genotypes 32 and 14 showed intermediate responses to the Yozgat isolate, but they were resistant to the Eskişehir isolate. These two genotypes were closest to the point where the resistant genotypes were placed. Genotypes 1, 52, 62, 107, and the susceptible control cultivar 'Larende' which showed a susceptible reaction to both isolates, were between the two isolates in the biplot. The wild barley genotype 4, which was susceptible to both *D. graminea* isolates, was closer to the Eskişehir isolate biplot line because it was more susceptible to the Eskişehir isolate than to the Yozgat isolate (Table 1, Figure 1).

Barley landraces 39, 21, 38, 139, 98, 40, 159, 73, 167, 171, 37, 128, 153 and the resistant control cultivar 'Çumra 2001' were the genotypes that were most resistant to *D. graminea*. These landraces exhibited resistant reactions to both *D. graminea* isolates, and they were all at the same point on the biplot graph (Figure 2). Landraces 160, 24, 103, 20, 101, 104, 71, and 3 exhibited intermediate reactions to the Yozgat isolate and resistant reactions to the Eskişehir isolate. No genotypes were susceptible to both pathogen isolates, except for the susceptible control cultivar 'Larende'. Barley landraces 148 and 74 were susceptible to the Yozgat isolate and exhibited intermediate responses to the Eskişehir isolate (Figure 2).

DISCUSSION

The present study is the first evaluation of resistance to leaf stripe for these 30 barley landraces and 30 wild barley genotypes. Differences in host reactions to inoculation with *D. graminea* were detected. Virulence differences between two isolates of the pathogen were also evident. Overall, the barley landraces were more resistant to *D. graminea* than the *H. spontaneum* genotypes examined in this study. Other reports from Turkey and elsewhere have also shown variable levels of resistance in barley to *D. graminea*. Mueller *et al.* (2003) carried out a study using 612 barley accessions, and determined that they exhibited different reactions to natural infections by *D. graminea* under organic agriculture conditions.

Table 5. Reactions of 30 wild barley (*Hordeum spontaneum*) genotypes following inoculation with two isolates of *Drechslera graminea*. For disease values, the scale of Tekauz (1983) was used.

	D. graminea, Yozgat isolate					*D. graminea*, Eskişehir isolate			
Hordeum spontaneum genotype	45 d after planting		60 d after planting		*Hordeum spontaneum* genotype	45 d after planting		60 d after planting	
	Mean disease percent	Scale value	Mean disease percent	Scale value		Mean disease percent	Scale value	Mean disease percent	Scale value
1	100	3 (S)	100	3 (S)	1	22.2	3 (S)	22.2	3 (S)
4	25	3 (S)	25	3 (S)	4	33.3	3 (S)	33.3	3 (S)
5	16.6	2 (I)	33.3	3 (S)	5	0	1 (R)	0	1 (R)
6	25	3 (S)	25	3 (S)	6	0	1 (R)	0	1 (R)
8	40	3 (S)	40	3 (S)	8	0	1 (R)	0	1 (R)
9	66.6	3 (S)	83.3	3 (S)	9	12.5	2 (I)	12.5	2 (I)
13	0	1 (R)	0	1 (R)	13	0	1 (R)	0	1 (R)
14	14.2	2 (I)	14.2	2 (I)	14	0	1 (R)	0	1 (R)
16	55.5	3 (S)	55.5	3 (S)	16	0	1 (R)	0	1 (R)
24	0	1 (R)	0	1 (R)	24	0	1 (R)	0	1 (R)
27	0	1 (R)	0	1 (R)	27	0	1 (R)	0	1 (R)
29	0	1 (R)	0	1 (R)	29	0	1 (R)	0	1 (R)
32	12.5	2 (I)	12.5	2 (I)	32	0	1 (R)	0	1 (R)
33	28.5	3 (S)	28.5	3 (S)	33	0	1 (R)	0	1 (R)
38	66.6	3 (S)	66.6	3 (S)	38	0	1 (R)	0	1 (R)
44	33.3	3 (S)	33.3	3 (S)	44	16.6	2 (I)	16.6	2 (I)
45	71.4	3 (S)	71.4	3 (S)	45	11.1	2 (I)	11.1	2 (I)
49	42.8	3 (S)	42.8	3 (S)	49	14.2	2 (I)	14.2	2 (I)
52	71.4	3 (S)	100	3 (S)	52	28.5	3 (S)	28.5	3 (S)
54	0	1 (R)	0	1 (R)	54	0	1 (R)	0	1 (R)
62	75	3 (S)	75	3 (S)	62	50	3 (S)	50	3 (S)
66	50	3 (S)	50	3 (S)	66	0	1 (R)	0	1 (R)
70	40	3 (S)	40	3 (S)	70	0	1 (R)	0	1 (R)
76	71.4	3 (S)	71.4	3 (S)	76	0	1 (R)	0	1 (R)
80	87.5	3 (S)	87.5	3 (S)	80	11.1	2 (I)	11.1	2 (I)
86	0	1 (R)	0	1 (R)	86	0	1 (R)	0	1 (R)
91	0	1 (R)	0	1 (R)	91	0	1 (R)	0	1 (R)
93	75	3 (S)	75	3 (S)	93	10	2 (I)	10	2 (I)
99	75	3 (S)	75	3 (S)	99	0	1 (R)	0	1 (R)
107	57.1	3 (S)	57.1	3 (S)	107	28.5	3 (S)	28.5	3 (S)
Larende	80	3 (S)	80	3 (S)	Larende	60	3 (S)	60	3 (S)
Çumra 2001	0	1 (R)	0	1 (R)	Çumra 2001	0	1 (R)	0	1 (R)
	40.01*		41.09*			7.93*		9.31*	

*Significant at $P < 0.01$ (Tables 1 and 2).

More than 30% of the accessions were resistant to *D. graminea*. In the same study, a small group of accessions was selected and tested using the sandwich inoculation method for reactions to two aggressive *P. graminea* isolates. They found that the accessions BGRC 5592, HOR 333, HOR 11475, and OU J362 showed resistant reactions. Similarly, the sandwich method was applied in the present study, and we determined that 23% of the wild barley genotypes and 43% of the barley landraces were resistant to both isolates of *D. graminea*.

Arabi *et al.* (2004) tested ten widely cultivated barley varieties against a virulent *D. graminea* isolate (Sy3) in southern Syria. Differential reactions were observed among the varieties, and as the level of disease increased, there were decreases in crop yield, kernel weight, and plant biomass. It has also been reported

Table 6. Reactions of 30 barley landraces following inoculation with two isolates of *Drechslera graminea*. For disease valuaes, the scale of Tekauz (1983) was used.

	D. graminea, Yozgat isolate					*D. graminea*, Eskişehir isolate			
	45 d after planting		60 d after planting			45 d after planting		60 d after planting	
Barley landrace	Mean disease percent	Scale value	Mean disease percent	Scale value	Barley landrace	Mean disease percent	Scale value	Mean disease percent	Scale value
3	12.5	2 (I)	12.5	2 (I)	3	0	1 (R)	0	1 (R)
8	16	2 (I)	16	2 (I)	8	12.5	2 (I)	12.5	2 (I)
12	20	3 (S)	20	3 (S)	12	0	1 (R)	0	1 (R)
18	22.2	3 (S)	22.2	3 (S)	18	0	1 (R)	0	1 (R)
20	0	1 (R)	16.6	2 (I)	20	0	1 (R)	0	1 (R)
21	0	1 (R)	0	1 (R)	21	0	1 (R)	0	1 (R)
22	12.5	2 (I)	25	3 (S)	22	0	1 (R)	0	1 (R)
24	14.2	2 (I)	14.2	2 (I)	24	0	1 (R)	0	1 (R)
37	0	1 (R)	0	1 (R)	37	0	1 (R)	0	1 (R)
38	0	1 (R)	0	1 (R)	38	0	1 (R)	0	1 (R)
39	0	1 (R)	0	1 (R)	39	0	1 (R)	0	1 (R)
40	0	1 (R)	0	1 (R)	40	0	1 (R)	0	1 (R)
71	16.6	2 (I)	16.6	2 (I)	71	0	1 (R)	0	1 (R)
73	0	1 (R)	0	1 (R)	73	0	1 (R)	0	1 (R)
74	50	3 (S)	50	3 (S)	74	14.2	2 (I)	14.2	2 (I)
83	37.5	3 (S)	37.5	3 (S)	83	0	1 (R)	0	1 (R)
90	37.5	3 (S)	37.5	3 (S)	90	0	1 (R)	0	1 (R)
98	0	1 (R)	0	1 (R)	98	0	1 (R)	0	1 (R)
101	16.6	2 (I)	16.6	2 (I)	101	0	1 (R)	0	1 (R)
103	12.5	2 (I)	12.5	2 (I)	103	0	1 (R)	0	1 (R)
104	16.6	2 (I)	16.6	2 (I)	104	0	1 (R)	0	1 (R)
128	0	1 (R)	0	1 (R)	128	0	1 (R)	0	1 (R)
139	0	1 (R)	0	1 (R)	139	0	1 (R)	0	1 (R)
148	50	3 (S)	50	3 (S)	148	11.1	2 (I)	11.1	2 (I)
153	0	1 (R)	0	1 (R)	153	0	1 (R)	0	1 (R)
159	0	1 (R)	0	1 (R)	159	0	1 (R)	0	1 (R)
160	0	1 (R)	11.1	2 (I)	160	0	1 (R)	0	1 (R)
162	28.5	3 (S)	28.5	3 (S)	162	0	1 (R)	0	1 (R)
167	0	1 (R)	0	1 (R)	167	0	1 (R)	0	1 (R)
171	0	1 (R)	0	1 (R)	171	0	1 (R)	0	1 (R)
Larende	66.6	3 (S)	66.6	3 (S)	Larende	33.3	3 (S)	33.3	3 (S)
Çumra 2001	0	1 (R)	0	1 (R)	Çumra 2001	0	1 (R)	0	1 (R)
	12.10*		14.68*			1.26*		2.22*	

*Significant at $P < 0.01$ (Tables 3 and 4).

that a decrease in plant biomass affects vital activities of plants such as photosynthesis and respiration (Mathre, 1997). In Turkey, the reactions of 1,216 barley lines to barley leaf stripe were assessed and it was found that 25 lines were resistant and eight were intermediate to resistant to the disease (Albustan *et al.*, 1999). Ulus and Karakaya (2007) assessed the resistance of 15 widely used barley varieties to five *D. graminea* isolates, and determined that the barley cultivars 'Çumra 2001' and 'Yerçil 147' were resistant to all five isolates, and that the isolate Dg3 was the most virulent. Bayraktar and Akan (2012) reported that barley cultivars 'Durusu', 'Balkan 96 (Igri)', 'Çumra 2001' and 'Anadolu 98' were resistant to the 13 *D. graminea* isolates they tested, and that isolate 1003 was the most virulent.

Çelik *et al.* (2016) evaluated the reactions of three barley cultivars and 20 barley landraces to ten *D. graminea* isolates, and found that one barley landrace

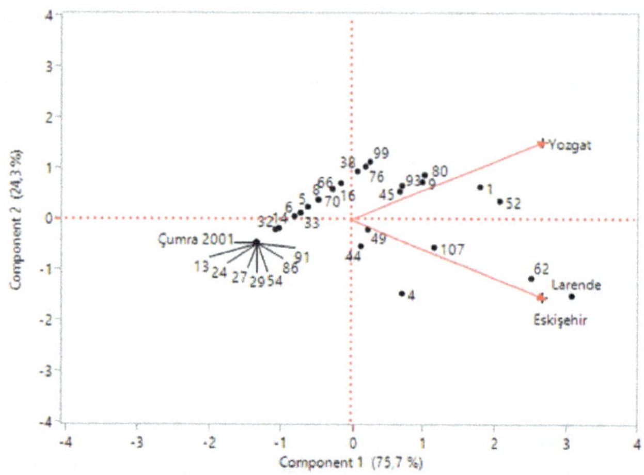

Figure 1. Biplot based on PCA analysis of the mean disease incidences of wild barley (*H. spontaneum*) genotypes inoculated with two isolates of *Drechslera graminea*.

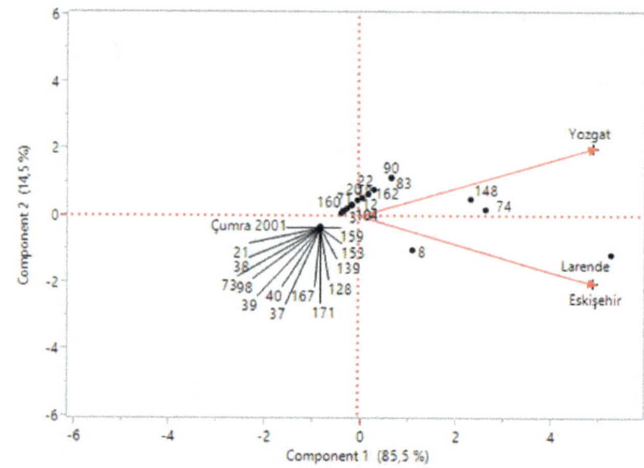

Figure 2. Biplot based on PCA analysis of the mean disease incidences of barley landraces inoculated with two isolates of *Drechslera graminea*.

was resistant to eight *D. graminea* isolates and had intermediate reactions to the other two isolates. The barley variety 'Çumra 2001' was resistant to all ten isolates. These authors also reported virulence differences among the isolates. In their study, average disease incidence for the Yozgat isolate was 40.2%, and for the Eskişehir isolates was 15.4%. Çelik Oğuz *et al.* (2017a), also found virulence differences among *D. graminea* isolates, and that only one of the 23 hulless barley lines they tested was resistant to three isolates of the pathogen. Karakaya *et al.* (2017) tested the same three isolates on 25 Iranian barley landraces and found similar virulence differences among the isolates. In their study, no Iranian barley landraces were resistant to all three isolates.

Turkey is a major genetic centre for cultivated and wild barleys, and there are barley genotypes in this country that are resistant to different abiotic and biotic stresses (Vavilov, 1951; Kün, 1996; Afanasenko *et al.*, 2000; Jakob *et al.*, 2014; Çelik *et al.*, 2016; Karakaya *et al.*, 2016a). Barley landraces and wild barley genotypes show great variation in agronomic traits as well as reaction to biotic stress factors. Resistance to different diseases has been reported in barley landraces and wild barley (*H. spontaneum*) genotypes (Azamparsa *et al.*, 2019; Karakaya *et al.*, 2017; Çelik and Karakaya, 2017; Çelik Oğuz *et al.*, 2017b; Çelik Oğuz *et al.*, 2019). Resistance among barley genotypes originating from the Middle East has been reported, with Anatolian landraces being superior compared to those from other origins, in terms of yield, drought, and disease tolerance (Chakrabarti, 1968; Khan and Boyd, 1969; Gökgöl, 1969).

In the present study, *H. spontaneum* genotypes 13, 24, 27, 29, 54, 86, and 91 were resistant to two *D. graminea* isolates. Genotypes 24, 27, and 54 were also resistant to virulent isolates of both forms of *Pyrenophora teres* (Çelik Oğuz *et al.*, 2019). *Hordeum spontaneum* genotype 13 was found to be resistant to virulent isolates of *P. teres* f. *maculata* (which causes the spot form of net blotch), while *H. spontaneum* genotype 29 was resistant to virulent isolates of *P. teres* f. *teres* (which causes the net form of net blotch) (Çelik Oğuz *et al.*, 2019). In addition, all seven of these wild barley genotypes showed resistance to up to four of six *Rhynchosporium commune* isolates (Azamparsa *et al.*, 2019).

Barley landraces 21, 37, 38, 39, 40, 73, 98, 128, 139, 153, 159, 167, and 171 were resistant to both *D. graminea* isolates examined in the present study. Among these landraces, landrace 40 was also resistant to virulent isolates of both forms of *P. teres* (Çelik Oğuz *et al.*, 2017b). Barley landraces 98, 167, and 171 were found to be resistant to virulent isolates of *P. teres* f. *maculata* (causing the spot form of net blotch), while barley landraces 21 and 153 were resistant to virulent isolates of *P. teres* f. *teres* (causing the net form of net blotch) (Çelik Oğuz *et al.*, 2017b). In addition, barley landraces 21, 38, 40, 218, 139, 153 and 167 showed resistance to up to three of six *R. commune* isolates (Azamparsa *et al.*, 2019).

The disease resistance of different *H. spontaneum* and barley landraces to other diseases has also been determined in other studies. For example, Kopahnke (1998) evaluated the reactions of wild barley and barley landraces to *P. teres* and found that 143 genotypes exhibited resistant reactions to all isolates tested. Jana and Bailey (1995) found resistance among the *H. spontaneum* genotypes and cultivated barley landraces obtained from Turkey and Jordan to the fungal pathogens *P. teres*

f. *maculata*, *P. teres* f. *teres*, and *Cochliobolus sativus*. The percentage of *H. spontaneum* accessions resistant to these pathogens (10.5%) was greater compared to that of the cultivated accessions (1.3%). Fetch *et al.* (2003) determined the reactions of 116 *H. spontaneum* genotypes originating from Jordan and Israel to six fungal pathogens. They showed that 98% of the genotypes from Jordan and 77% of the genotypes from Israel were resistant to Septoria leaf blotch, 70% and 90%, respectively from the two countries, were resistant to leaf rust, 72% and 78%, respectively, were resistant to net blotch, 58% and 70%, respectively, were resistant to powdery mildew, 53% and 46%, respectively, were resistant to spot blotch, and 2% and 26%, respectively from the two countries, were resistant to stem rust.

Wild barley (*H. spontaneum*) has greater genetic variation than cultivated barley (Saghai-Maroof *et al.*, 1994; Provan *et al.*, 1999; Nevo, 2004), and it is possible to crossbreed *H. spontaneum* with cultured barley (*H. vulgare*). Useful traits including disease resistance can be transferred to cultivated barley from *H. spontaneum* (Çelik and Karakaya, 2017), so wild barley is a significant potential genetic source for barley genetic improvement. Wild barley populations in the Middle East also possess considerable genetic variation (Nevo, 1992). It has been suggested that *H. spontaneum* genotypes should be preserved under *in situ* and *ex situ* conditions for barley improvement programmes, including those selecting for enhanced disease resistance (Nevo, 1992; Ceccarelli and Grando, 2000; Nevo, 2012). *Hordeum spontaneum* genotypes may show different resistance reactions based on their origins, and resistance genes can vary depending on geographic conditions (Sato and Takeda, 1997). *Hordeum spontaneum* populations from the Fertile Crescent Region, including parts of the Levant (eastern Mediterranean, including Turkey and Israel) and Iran are genetically variable for adaptation capability and population sustainability (Nevo, 2004; Jakob *et al.*, 2014). In the present study, *H. spontaneum* genotypes resistant to *D. graminea* isolates were observed. Seven and two of the *H. spontaneum* genotypes showed resistant reactions to the Yozgat isolate and two genotypes were of intermediate resistance. On the other hand, 19 of *H. spontaneum* genotypes were resistant to the Eskişehir isolate while six genotypes were of intermediate resistance to this isolate. The heterogenous nature of wild barley (*H. spontaneum*) resistance to diseases has been reported previously (Çelik and Karakaya, 2017; Karakaya *et al.*, 2016a). In a survey carried out in 2015, a total of 40 *H. spontaneum* populations in their natural habitat were examined, and it was determined that nine of these were disease-free. In these fields, the fungal pathogens *R. commune*, *Blumeria graminis* f. sp. *hordei*, *D. teres* f. *teres*, *D. teres* f. *maculata*, *Ustilago nigra*, *U. nuda*, *Puccinia hordei*, and *D. graminea* were identified (Karakaya *et al.*, 2016a).

The region between the south of the Fertile Crescent and the Himalayan mountains was the first area in which barley was domesticated (Azhaguvel and Komatsuda, 2007; Morrell and Clegg, 2007; Saisho and Purugganan, 2007). In recent years, barley varieties and yields in the area of Fertile Crescent have been under serious threat because of climate change and environmental pollution originating from human activities. Barley landraces provide gene resources that can be used to decrease the negative impacts of climate change (Mzid *et al.*, 2016). It is known that wild barleys and barley landraces have wide variation in terms of resistance to diseases (Simmonds, 1987; Çelik and Karakaya, 2017; Çelik Oğuz *et al.*, 2017b; Azamparsa *et al.*, 2019). In conventional agriculture systems, barley leaf stripe is controlled through treating seed with fungicides. However, European Union regulations state that under certified organic production practices, barley leaf stripe can only be controlled using hot water treatments. But, these may not always be fully effective. Barley stripe, which is very important under organic agriculture conditions, is prevalent in Northern Germany due to the cool and humid climatic conditions found there (Mueller *et al.*, 2003). Genetic resistance can be transferred from wild relatives to cultivated crops to decrease the use of chemicals (Laurei *et al.*, 1992).

In summary, barley leaf stripe is an important disease that can cause significant yield losses when no disease management practices are utilized. In the present study, new sources of resistance to *D. graminea* have been identified. The wild barley genotypes and barley landraces identified here could be used in plant breeding programmes to develop leaf stripe resistant genotypes, which would be ecosystem-friendly and also enhance farmer profitability.

ACKNOWLEDGEMENTS

Aziz Karakaya, Sinan Aydoğan and Özlem Akbaş are thanked for their help in this study. The Gene Bank of Central Research Institute for Field Crops provided seed of the germplasm lines used.

LITERATURE CITED

Afanasenko O.S., Makarova I.G., Zubkovich A.A., 2000. Inheritance of resistance to different *Pyrenophora teres* Dreschs. strains in barley accession CI 5791.

In: *Abstracts, Proceedings of 8th International Barley Genetics Symposium*, 22–27 October, Adelaide, South Australia, No, 2.26, 73–75.

Aktaş H., 2001. Önemli hububat hastalıkları ve sürvey yöntemleri. T.C. Tarım ve Köyişleri Bakanlığı Tarımsal Araştırma Genel Müdürlüğü Bitki Sağlığı Araştırmaları Daire Baskanlığı, Ankara.

Albustan S., Çetin L., Düşünceli F., Tosun H., Akar T., 1999. Orta Anadolu Bölgesi için TARM tarafından oluşturulan 1998 yılı arpa nörserilerinin yaprak lekesi (*Rhynchosporium secalis* (Oud.) J. J. Davis) ve arpa çizgili yaprak lekesi (*Pyrenophora graminea*) hastalıklarına karşı dayanıklılık bakımından değerlendirilmesi. In: *Hububat Sempozyumu* (H. Ekiz, ed.), 8-11 Haziran 1999, Konya, s,700–704.

Allard R.W., Bradshaw A.D., 1964. Implications of genotype-environment interaction in applied plant breeding. *Crop Science* 4: 503–508.

Arabi M.I.E., Jawhar M., Al-Safadi B., MirAli N., 2004. Yield response of barley to leaf stripe (*Pyrenophora graminea*) under experimental conditions in southern Syria. *Journal of Phytopathology* 152: 519–523.

Azamparsa M.R., Karakaya A., Ergün N., Sayim İ., Duran R.M., Özbek K., 2019. Identification of barley landraces and wild barley (*Hordeum spontaneum*) genotypes resistant to *Rhynchosporium commune*. *Tarım Bilimleri Dergisi* 25: 530–535.

Azhaguvel P., Komatsuda T., 2007. A phylogenetic analysis based on nucleotide sequence of a marker linked to the brittle rachis locus indicates a diphyletic origin of barley. *Annals of Botany* 100: 1009–1015.

Bayraktar H., Akan K., 2012. Genetic characterization of *Pyrenophora graminea* isolates and the reactions of some barley cultivars to leaf stripe disease under greenhouse conditions. *Turkish Journal of Agriculture and Forestry* 36:329–339.

Brown A.H.D., 2000. The genetic structure of crop landraces and the challenge to conserve them *in situ* on farms. In: *Genes in the field, On -farm conservation of crop diversity* (S.B. Brush, ed.) IPGRI/IDRC/Lewis Publishers. Boca Raton, London, New York, Washington, D.C, 29–48.

Ceccarelli S., Grando S., 2000. Barley landraces from the Fertile Crescent. A lesson for plant breeders. In: *Genes in the field, On -farm conservation of crop diversity* (S.B. Brush, ed.) IPGRI/IDRC/Lewis Publishers. Boca Raton, London, New York, Washington, D.C, 51–76.

Ceccarelli S., 1996. Adaptation to low/high input cultivation. *Euphytica* 92: 203–214.

Chakrabarti N.K., 1968. Some effects of ultraviolet radiation on resistance of barley to net blotch and spot blotch. *Phytopathology* 58: 467–471.

Çelik Oğuz A., Karakaya A., Ergün N., 2017a. Determination of the reactions of some Turkish hulless barley lines to *Drechslera graminea*. *Works of the Faculty of Agriculture and Food Sciences University of Sarajevo* LXII, 67 (2):196–202.

Çelik Oğuz A., Karakaya A., Ergün N., Sayim İ., 2017b. Turkish barley landraces resistant to net and spot forms of *Pyrenophora teres*. *Phytopathologia Mediterranea* 56: 217–223.

Çelik Oğuz A., Karakaya A., Murat Duran R., Özbek K., 2019. Identification of *Hordeum spontaneum* genotypes resistant to net blotch disease. *Tarım Bilimleri Dergisi* 25:115–122.

Çelik E., Karakaya A., 2017. Yabani arpa (*Hordeum spontaneum*) ve hastalıklara dayanıklılık. *Mustafa Kemal Üniversitesi Ziraat Fakültesi Dergisi* 22: 65–86.

Çelik Y., Karakaya A., Çelik Oğuz A., Mert Z., Akan K., ... Sayim İ., 2016. Determination of the reactions of some barley landraces and cultivars to *Drechslera graminea*. *Mediterranean Agricultural Sciences* 29: 43–47.

Çetinsoy S., 1995 Importance and control of barley leaf blight in Turkey. *Rachis* 14: 25–26.

Dumalasova V., Leiova-Svobodova L., Bartos P., 2014. Common bunt resistance of Czech and European winter wheat cultivars and breeder lines. *Czech Journal of Genetics and Plant Breeding* 50: 201–207.

Ellis R.P., Forster B. P., Robinson D., Handley L. L., Gordon D. C., ... Powell W., 2000. Wild barley: a source of genes for crop improvement in the 21st century? *Journal of Experimental Botany* 51: 342, 9–17.

Ergün N., Aydoğan S., Sayim İ., Karakaya A., Çelik Oğuz A., 2017. Arpa (*Hordeum vulgare* L.) köy çeşitlerinde tane verimi ve bazı tarımsal özelliklerin incelenmesi. *Tarla Bitkileri Merkez Araştırma Enstitüsü Dergisi* 26: 180–189.

FAO., 2015. *Wheat Landraces in Farmers' Fields in Turkey*: National Survey, Collection, and Conservation, 2009–2014, by Mustafa Kan, Murat Küçükçongar, Mesut Keser, Alexey Morgounov, Hafiz Muminjanov, Fatih Özdemir, Calvin Qualset.

Fetch Jr. T.G., Steffenson B. J.,Nevo E., 2003. Diversity and sources of multiple disease resistance in *Hordeum spontaneum*. *Plant Disease* 87: 1439–1448.

Gatti A., Rizza F., Delogu G., Terzi V., Porta-Puglia A., Vannacci G., 1992. Physiological and biochemical variability in a population of *Drechslera graminea*. *Journal of Genetics and Breeding* 46: 179–186.

Gökgöl M., 1969. *Serin iklim hububatı ziraatı ve ıslahı (buğday, çavdar, arpa ve yulaf)*. Tarım Bakanlığı Ziraat İşleri Genel Müdürlüğü. 407 s. Özaydın Matbaası, İstanbul.

Harlan J.R., Zohary D., 1966. Distribution of wild wheat and barley. *Science* 153: 1074–1080.

Helbaek H., 1969. Plant-collecting, dry-farming and irrigation agriculture in prehistoric Deh Luran. In: *Prehistory and human ecology of the Deh Luran plain. An early village sequence from Khuzistan, Iran* (F. Hole, K.V. Flannery, J.A. Neely, eds.), University of Michigan. Memoirs of the Museum of Anthropology, USA, 383–426.

Jakob S.S., Rödder D., Engler J.O., Shaaf S., Özkan H., … Kilian B., 2014. Evolutionary history of wild barley (*Hordeum vulgare* subsp. *spontaneum*) analyzed using multilocus sequence data and paleodistribution modeling. *Genome Biology and Evolution* 6: 685–702.

Jana S., Bailey K.L., 1995. Responses of wild and cultivated barley from West Asia to net blotch and spot blotch. *Crop Science* 35: 242–246.

Jawhar M., Arabi M.I.E., 2006. Genetic variability among *Pyrenophora graminea* isolates. *Australasian Plant Pathology* 35: 279–281.

Jensen J., 1988. Coordinator's report: Chromosome 5. *Barley Genetics Newsletter* 18: 61–63.

Karakaya A., Mert Z., Çelik Oğuz A., Ertaş M. N., Karagöz A., 2016a. Determination of the diseases occurring on naturally growing wild barley (*Hordeum spontaneum*) field populations. *Works of the Faculty of Agriculture and Food Sciences, University of Sarajevo* LXI, 66 (1): 291–295.

Karakaya A., Mert Z., Çelik Oğuz A., Çetin L., 2016b. Distribution of barley stripe disease in Central Anatolia, Turkey. *Selcuk Journal of Agriculture and Food Sciences* 30: 59–61.

Karakaya A., Çelik Oğuz A., Rahimi A., 2017. Response of Iranian barley landraces to *Drechslera graminea. Works of the Faculty of Agriculture and Food Sciences University of Sarajevo* LXII, 67 (2): 225–230.

Khan T.N., Boyd W.J.R., 1969. Physiologic specialization in *Drechslera teres. Australian Journal of Biological Sciences* 22: 1229–1235.

Kavak H., 2004. *Pyrenophora graminea* in fields sown-spring barley Angora in arid district of Turkey. *Pakistan Journal of Biological Sciences* 7: 1225–1228.

Kopahnke D., 1998. Evalution of barley for resistance to *Drechslera teres* (Sacc.) Shoem. *Beiträge zur Züchtungsforschung-Bundesansalt für Züchtungsforschung an Kulturpflanzen* 4: 1-3.

Kün E., 1996. *Tahıllar-1 (Serin İklim Tahılları)*. Ankara Üniversitesi Ziraat Fakültesi Yayınları, Yayın No:1451, Ankara.

Ladizinski G., 1998. How many tough-rachis mutants gave rise to domesticated barley? *Genetic Resources and Crop Evolution* 45: 411–414.

Laurei D.A., Snape J.W., Gale M.D., 1992. DNA marker techniques for genetic analysis in barley. In: *Barley: Genetics, Biochemistry, Molecular Biology and Biotechnology* (P.R, Shewry, ed.), The Alden Press, Ltd, Oxford. 115–132.

Mamluk O.F., Çetin L., Braun H.J., Bolat N., Bertschinger L., …Düsünceli F., 1997. Current status of wheat and barley disease in the Central Anatolia Plateau of Turkey. *Phytopathologia Mediterranea* 36: 167–181.

Mathre D.E., 1997. *Compendium of Barley Diseases*. Second edition. St Paul, Minnesota, American Phytopathological Society Press, 90 pp.

McDonald B.A., Linde C., 2002. Pathogen population genetics, evolutionary potential, and durable resistance. *Annual Review of Phytopathology* 40: 349–379.

Mohammad A., Mahmood M., 1974. Inoculation techniques in Helminthosporium stripe of barley. *Plant Disease Reporter* 58: 32–34.

Morrell P.L., Clegg M.T., 2007. Genetic evidence for a second domestication of barley (*Hordeum vulgare*) east of the Fertile Crescent. *Proceedings of the National Academy of Sciences of the United States of America* 104: 3289–3294.

Mueller K. J., Valè G., Enneking D., 2003. Selection of resistant spring barley accessions after natural infection with leaf stripe (*Pyrenophora graminea*) under organic farming conditions in Germany and by sandwich test. *Journal of Plant Pathology* 85: 9-14.

Mzid R., Chibani F., Ayed R.B., Hanana M., Breidi J., … Chalak L., 2016. Genetic diversity in barley landraces (*Hordeum vulgare* L. subsp. *vulgare*) originated from Crescent Fertile region as detected by seed storage proteins. *Journal of Genetics* 95: 733–739.

Nesbitt M., 1995. Clues to agricultural origins in the northern Fertile Crescent. *Diversity* 11: 142–143.

Nevo E., 1992. Origin, evolution, population genetics and resources for breeding of wild barley, *Hordeum spontaneum*, in the Fertile Crescent. In: *Barley: genetics, biochemistry, molecular biology and biotechnology* (P. R. Shewry, ed.) C.A.B. International. UK, 19–43 pp.

Nevo E., 2004. Population genetic structure of wild barley and wheat in the Near East Fertile Crescent: Regional and local adaptive patterns. In: *Cereal Genomics* (P.K. Gupta, R.K Varshney, eds.) Springer, Dordrecht, 135–165.

Nevo E., 2012 Evolution of wild barley and barley improvement. In: *Advance in Barley Sciences* (C. Li, G. Zhang, X. Liu, J. Eglinton, eds.), Proceedings of 11th Int. Barley Genetics Symposium. Zhejiang University Press- Springer, 1-16.

Platenkamp R., 1976. Investigations on the infection pathway of *Drechslera graminea* in germinating barley. *Review of Plant Pathology* 56: 319–320.

Porta-Puglia A., Delogu G., Vanacci G., 1986. *Pyrenophora graminea* on winter barley seed: effect on disease incidence and yield loss. *Journal of Phytopathology* 117: 26–33.

Pourkheirandish M., Komatsuda T., 2007. The importance of barley genetics and domestication in a global perspective. *Annals of Botany* 100: 999–1008.

Provan J., Russell Booth J.R., Powell A.W., 1999. Polymorphic chloroplast simple sequence repeat primers for systematic and population studies in the genus *Hordeum. Molecular Ecology* 8: 505–511.

Saghai Maroof M.A., Biyashev R.M., Yang G.P., Zhang Q., Allard R.W., 1994. Extraordinarily polymorphic microsatellite DNA in barley: species diversity, chromosomal locations and population dynamics. *Proceedings of the National Academy of Sciences of the United States of America* 91: 5466–5470.

Saisho D., Purugganan M.D., 2007. Molecular phylogeny of domesticated barley traces expansion of agriculture in the Old World. *Genetics* 177: 1765–1776.

Sato K., Takeda K., 1997. Net blotch resistance in wild species of *Hordeum. Euphytica* 95: 179–185.

Simmonds N.W. (ed.)., 1987. Principles of crop improvement. Longman, New York, NY.

Tekauz A., 1983. Reaction of Canadian barley cultivars to *Pyrenophora graminea*, the incitant of leaf barley stripe. *Canadian Journal of Plant Pathology* 5: 294- 301.

Tekauz A., Chiko A.W., 1980. Leaf stripe of barley caused by *Pyrenophora graminea*: occurrence in Canada and comparisons with barley stripe mosaic. *Canadian Journal of Plant Pathology* 2: 152–158.

Thomas W.T.B., Baird E., Fuller J.D., Lawrence P., Young G.R., … Powell W., 1998. Identification of a QTL decreasing yield in barley linked to Mlo powdery mildew resistance. *Molecular Breeding* 4: 381–393.

Tuik., 2017. https://biruni.tuik.gov.tr/medas/?kn=92&locale=tr- Access date: 22.11.2018

Ulus C., Karakaya A., 2007. Assessment of the seedling reactions of some Turkish barley cultivars to barley stripe. *Tarım Bilimleri Dergisi* 13: 409–412.

Vavilov N.I., 1951. *The origin, variation, immunity and breeding of cultivated plants*, (translated from the Russian by K. S. Chester). Chronica Botanica; New York: Stechert-Hafner.

Von Bothmer R., Jacobson N., Baden C., Jorgensen R.B., Linde-Laursen I., 1995. *An ecogeographical study of the genus Hordeum*. Second edition. Rome: International Plant Genetic Resources Institute (IPGRI), Food and Agriculture Organization (FAO).

Willcox G., 1995. Archeobotanists sleuth out origins of agriculture from early Neolithic sites in the Eastern Mediterranean. *Diversity* 11: 141–142.

Yan W., Falk D.E., 2002. Biplot analysis of host-by-pathogen data. *Plant Disease* 86: 1396–1401.

Yitbarek S., Berhane L., Fikadu A., Van Leur J.A.G., Grando S., Ceccarelli, S,, 1998. Variation in Ethiopian barley landrace populations for resistance to barley leaf scald and net blotch. *Plant Breeding* 117: 419–423.

Zad J., Aghakhani M., Etebarian R., Okhovat M., 2002. Barley leaf stripe disease. *Meded Rijksuniv Gent Fak Landbouwkd Toegep Biol Wet* 67:279–281.

Zohary D., Hopf M., 2000. *Domestication of plants in the old world: the origin and spread of cultivated plants in West Asia, Europe, and the Nile Valley*. Third edition. Oxford University Press, New York.

13

Endophytic fungal communities of ancient wheat varieties

Giulia CASINI[1,2], Thaer YASEEN[3], Ahmed ABDELFATTAH[4,5*], Franco SANTORO[6], Leonardo VARVARO[1], Sandro DRAGO[2], Leonardo SCHENA[4]

[1] *Department of Agriculture and Forestry Science, University of Tuscia, Via S. Camillo de Lellis s.n.c., 01100 Viterbo (VT), Italy*
[2] *Enbiotech S.r.l., via Quarto dei mille, 6, 90144 Palermo (PA), Italy*
[3] *Regional Office for the Near East and North Africa Region (RNE), Food and Agriculture Organization of the United Nations (FAO), Cairo, Egypt*
[4] *Dipartimento di Agraria, Università Mediterranea di Reggio Calabria, Località Feo di Vito, 89122 Reggio Calabria (RC), Italy*
[5] *Department of Ecology, Environment and Plant Sciences, Stockholm University, SE-106 91, Stockholm, Sweden*
[6] *CIHEAM - Istituto Agronomico Mediterraneo, Via Ceglie 9, 70010 Valenzano (BA), Italy*
* Corresponding author: ahmed.abdelfattah@su.se

Competing Interests: The Author(s) declare(s) no conflict of interest.

Editor: Rasoul Zare, Iranian Research Institute of Plant Protection, Tehran, Iran.

Abstract. The fungal community composition and structure of two ancient tetraploid wheat varieties, native to the Sicilian territory of Italy, Perciasacchi (winter wheat) and Tumminia (spring wheat) were investigated using High Throughput Sequencing (HTS). This showed a predominance of Ascomycetes and Basidiomycetes including *Alternaria, Fusarium, Mycosphaerella, Filobasidium, Cystofilobasidium, Cryptococcus, Leucosporidium, Dioszegia, Puccinia, Sporobolomyces, Aureobasidium, Cladosporium, Holtermanniella* and *Gibberella*. Principal Coordinates Analysis (PCoA) and Linear discriminant analysis Effect Size (LEfSe) showed that *Aureobasidium, Leucosporidium* and *Puccinia* differentiated between the two wheat varieties. In addition, the microbial association analysis suggested that some endophytic taxa play important roles within the wheat fungal community. Genera such as *Cryptococcus* and *Cystofilobasidium* were shown to have consistent antagonistic activity against *Gibberella* spp., while, *Acremonium* and a group of unidentified ascomycetes had mutual exclusion relationships with *Puccinia*. Since both *Gibberella* and *Puccinia* contain several economically important pathogens of wheat, the detected fungal interactions may indicate microbial-mediated resistance in these wheat varieties.

Keywords. TS, fungal diversity, microbial ecology, microbiome, domestication.

INTRODUCTION

Plant-associated microorganisms, collectively referred to as the plant microbiota, are known to influence plant physiological development, and response to environmental changes by conferring stress tolerance, disease resistance, and affecting interactions with competitors and predators

(Lugtenberg and Kamilova 2009; Berendsen *et al.*, 2012). Endophytes are microorganisms that can be isolated from surface-disinfested plant tissues or extracted from within plants, and that do not harm plant growth (Hallmann *et al.*, 1997). Endophytes have gained importance in basic and translational science due to their potential roles as biocontrol agents (Dorworth and Callan 1996; Massart *et al.*, 2015; Abdelfattah *et al.*, 2018).

Wheat (*Triticum* sp.) is a global staple food crop adapted to a wide variety of environmental conditions, including marginal areas (Salmon and Clark, 1913). In Italy, production of durum wheat (*Triticum durum*) accounts for more than 50% of the total European production (European Commission, Eurostat and DG Agriculture and Rural Development). Production of this crop is mainly concentrated in the hot and dry southern regions of Italy, including Sicily (https://gain.fas.usda.gov, http://dati.istat.it). Recently, local ancient wheat varieties have acquired increased importance due to their peculiar organoleptic properties, which has fueled expanding craft milling and bakery enterprises (Shewry and Hey, 2015; Jankielsohn and Miles, 2017). For example, in Sicily, the importance of local ancient tetraploid wheat landraces is increasing. This is due to the quality of their flour, their low nutritional requirements, and their ability to grow in dry environmental conditions typical of the Sicilian region. This has sparked scientific curiosity towards the molecular determinants of such adaptation capabilities.

The possibility of finding microbial strains capable of improving wheat productivity and tolerance to biotic and abiotic factors has motivated scientific studies in the last decade (Coombs and Franco, 2003; Larran *et al.*, 2007; Velazquez-Sepulveda *et al.*, 2012; Hubbard *et al.*, 2014). Advances in sequencing and computational applications, such as metabarcoding (Taberlet *et al.*, 2012), have enabled detailed characterization of microbial diversity (Turner *et al.*, 2013; Ofek-Lalzar *et al.*, 2014). Most studies have focused on the bacteria while little is known about fungi. It is becoming increasingly clear that fungal microbiota play critical roles in plant growth, development and stress tolerance (Shendure and Ji, 2008; Nicolaisen *et al.*, 2014; Abdelfattah *et al.*, 2018).

It remains unclear whether locally-adapted wheat varieties are capable of recruiting distinct microbiota, and, if so, to what extent this differential recruitment is influenced by environment and the host genotype. As a first step towards deciphering the contribution of wheat microbiota to crop adaptation and yield, we here report molecular characterization of the composition of the fungal endophytic communities of two locally adapted Sicilian tetraploid wheat varieties, the winter type Perciasacchi and the spring type Tumminia, grown in two agricultural fields.

MATERIALS AND METHODS

Experimental design

This study was carried out in two fields (Field 1 and Field 2), of almost 1 ha each, located in the Madonie area of the Sicilian inland. Field 1 (37.7813410 N, 14.2852990 E), was located at 850 m above sea level and was previously used for legume culture. Field 2 (37.740245 N, 14.239368 E) was located at 800 m above sea level and was uncultivated for almost 5 years. One half of both fields was sown with Perciasacchi in the second week of November 2015, while the other half was sown with Tumminia in the first week of February 2016. The fields and seed were not subjected to any treatments before or after sowing. Based on the field topography, 23 sampling plots were selected in Field 1 and 27 were selected in Field 2.

Plant sampling, surface sterilization and sample preparation

During the heading phase of both varieties, a total of 50 samples were collected from the selected sampling plots. Each sample consisted of ten plants including their roots. Samples were transported to the laboratory and stored at 4°C before processing. The aerial parts of each plant were cut using sterile scissors and a stem portion, of approx. 15 cm above the crown, was kept for sap extraction. Stem and root samples were surface-sterilized using 5% sodium hypochlorite (NaClO), and then rinsed in sterile water. Sap extraction was carried out with the use of a new method, the CIHEAM-IAMB patented 'Method for the extraction of sap from plant material and apparatus for carrying out the method' (https://patents.google.com/patent/WO2017017555A1/en). This method of extracting plant sap from vessels and xylem using the pressure of a syringe has the advantage of reducing plant components in the final extract that can inhibit enzyme activity, adversely affecting the results of diagnosis. Sap extract was obtained by inserting 1 mL of phosphate-buffered saline (PBS) solution into one terminal of the plant stem using a syringe. PBS with the sap extracts was collected from the other terminal of the plant stem in a sterile 1.5 mL capacity tube, which was later used for plating and DNA extraction. Twenty-eight samples were randomly chosen, seven from each half of each field (14 from each wheat variety), for high throughput sequencing (HTS).

Culture conditions for endophytic fungi and DNA extraction

Endophytic fungi were cultured by aseptically transferring 100 μL of each wheat sap extract to 9 cm diam. Petri dishes containing semi-selective Nutrient Yeast Dextrose Agar (NYDA; containing 10 g L^{-1} glucose, 5 g L^{-1} yeast extract, 8 g L^{-1} nutrient broth, 18 g L^{-1} agar, 250 mg L^{-1} streptomycin sulphate and 250 mg L^{-1} ampicillin) (Janisiewicz and Roitman, 1988). The inoculated dishes were incubated at 26°C for 48-72 h. After incubation, a small portion of the growing mycelium was transferred to new Potato Dextrose Agar (PDA). A small portion of the subsequently grown mycelium was transferred from each plate into a unique sterile 1.5 mL capacity tube, and stored at -20°C. DNA extraction was carried out from 400 μL of wheat sap extract and the mixed fungal mycelium, using the NucleoSpin® Plant II extraction kit (Macherey-Nagel) following the manufacturer's instructions and doubling the pre-lysis and lysis incubation periods.

Metabarcoding analyses

Samples for metabarcoding were grouped into the following four categories including seven biological replicates: P1 (Perciasacchi grown in Field 1), T1 (Tumminia grown in Field 1), P2 (Perciasacchi grown in Field 2) and T2 (Tumminia grown in Field 2).

Amplifications of the fungal ribosomal Internal Transcribed Spacer 2 (ITS2) region were performed using the forward primer ITS86F (Turenne *et al.*, 1999) and mix reverse primers ITS4-Mix 1, ITS4-Mix 2, ITS4-Mix 3, and ITS4-Mix 4 (Tedersoo *et al.*, 2014; 2015). Amplifications of the ITS2 region were performed using KAPA HiFi Hot Start ReadyMix kit (KAPA Biosystems), under a temperature profile of 95°C for 3 min, 35 cycles at 98°C for 20 sec, 56°C for 15 sec, 72°C for 30 sec, followed by an elongation step of 72°C for 1 min. PCR products were visualized on 2% agarose gel in order to verify the successful amplification and the absence of contamination. PCR purification was performed using Agencourt AMPure XP beads kit and following the user manual instructions (Beckman Coulter). Amplicon indexing was carried out using Nextera XT v2 Index Kit (Illumina) and followed by a second PCR purification as previously described. Amplification products were quantified by fluorimetry using Qubit (Invitrogen) and pooled in equimolar concentrations before sequencing reactions in MiSeq (Illumina) according to the manufacturer guidelines (support.illumina.com). Datasets generated during this study were deposited and are available at the National Center for Biotechnology Information (NCBI), Sequence Read Archive (SRA), under the accession number PRJNA449228 (www.ncbi.nlm.nih.gov/bioproject/ PRJNA449228).

Sequencing data analyses

Raw reads were trimmed using Trimmomatic v. 0.32 (Bolger *et al.*, 2014), using a sliding window of six bases, Q-score average of ≥ 20, and a minimum sequence length of 150 bp. The paired-end reads were then assembled using PANDAseq Assembler (Masella *et al.*, 2012), setting a minimum sequence length value of 150 bp and a minimum overlapping value of 20 bp. Reads were then checked and filtered for chimeric sequences using VSEARCH v.1.11.1 (Rognes *et al.*, 2016) and the UNITE dynamic database released on November 20, 2016 (https://unite.ut.ee). The same database was also used for creating Operational Taxonomic Units (OTUs), with a similarity threshold of 99% and for taxonomy assignments with BLAST method (Altschul *et al.*, 1990) as implemented in QIIME v. 1.9.1 (Caporaso *et al.*, 2010). These taxonomic units were collapsed to describe the fungal community at the genus, family, order, class, and phylum levels.

Downstream analyses

The downstream analyses were conducted using QIIME 1.9.1 pipeline (Caporaso *et al.*, 2010) as described by (Abdelfattah *et al.*, 2017). The OTU table was normalized by rarefaction to an even depth of 57,623 for cultured mycelium and 42,448 for sap extract samples. This was to reduce sample heterogeneity as well as to keep samples with acceptable numbers of sequences to be used in statistical and taxonomic analyses. Fungal richness and abundance were calculated through alpha-diversity analyses determined by Shannon's Diversity, Simpson, Chao1, and Observed OTUs indices. The diversity results were then compared using a nonparametric two-sample t-test, and the *P*-values were calculated through 999 Monte Carlo permutations. Beta diversity analysis was performed using Bray Curtis dissimilarity metrics and the results were used to conduct Principal Coordinates Analysis (PCoA). PCoA graphs were also implemented with the taxonomic information and plotted on a 3D graph using EMPeror (Vázquez-Baeza *et al.*, 2013). Beta diversity results were used to compare groups of samples (fields, wheat varieties and sample categories) using Permanova analyses. Differentially abundant taxa between the two fields, the wheat varieties and the four sample categories were detected using the Linear discriminant analysis Effect Size (LEfSe) (Segata *et al.*, 2011), setting *P*-value thresholds for the factorial Kruskal-Wallis test and the pairwise Wilcoxon test of 0.05, and the logarithmic Effect Size (LDA) cut-off >2.

Fungal association network

Inferred fungal associations (co-occurrence and mutual exclusion) within each wheat variety were computed using the CoNet (v1.1.1. beta) plugin within Cytoscape (v3.6.1). The associations of OTUs present in at least 20 samples were identified using an ensemble of correlation metrics (Spearman and Pearson coefficients) and distance metrics (Bray–Curtis and Kullback–Leibler dissimilarity measures). For each association metric and each edge, 100 renormalized permutation and bootstrap scores were generated following the ReBoot procedure (Faust *et al.*, 2012). The measure-specific *P*-values from multiple association metrics were merged using the Simes method (Sarkar and Chang, 1997), and false-discovery rate corrections were performed using Benjamini–Hochberg multiple testing correction (Benjamini and Hochberg, 1995). Only 1,000 top- and 1,000 bottom-ranking edges from each association measure were kept in the network analysis, and only edges supported by at least two of the four association metrics were retained in the final network inference of associations among taxa.

Table 1. Sequencing results obtained from analyses of DNA extracted from cultured mycelium isolated from four wheat variety and field categories: P1 = Perciasacchi from Field 1; T1 = Tumminia from Field 1; P2 = Perciasacchi from Field 2; and T2 = Tumminia from Field 2.

Category	No. of reads	Total OTUs	Rarefied OTUs[a]
P1	83,331	927	927
T1	127,221	960	960
P2	57,623	615	615
T2	84,962	753	753

[a]Rarefaction depth 57,623

Table 2. Sequencing and alpha diversity results from analyses of DNA extracted from wheat sap extracted from four wheat variety and field categories: **P1** = Perciasacchi from Field 1; **T1** = Tumminia from Field 1; **P2** = Perciasacchi from Field 2; and **T2** = Tumminia from Field 2.

Category	Sequencing results			Alpha diversity metrics			
	No. of reads	Total OTUs	Rarefied OTUs[a]	Simpson	Shannon	Chao1	Observed Species
P1	459,150	5,454	3,181	0.95	5.62	1249.40	651.52
T1	299,769	2,646	1,805	0.96	5.67	1252.82	555.92
P2	566,026	4,047	2,815	0.92	4.61	1072.07	507.20
T2	949,383	6,612	3,799	0.93	5.15	908.93	498.35

[a] Rarefaction depth 42,448

RESULTS

Sequencing data processing

HTS generated a total of 1,509,410 unpaired reads from fungal mycelium and 4,492,327 unpaired reads from wheat sap extract. After trimming, pairing, quality and chimera filtering, 687,093 high-quality fungal sequences were retained from mycelium and 2,330,956 from sap extract. The resulting sequences were then assigned to 2,943 OTUs for mycelium samples and 6,885 OTUs for sap samples. A summary of the sequencing results is shown in Table 1 and Table 2.

Cultured endophytic fungi from wheat

The sequencing of the ITS2 region of the endophytic fungi isolated in culture plates provided the taxonomic composition of the cultivable endophytic fungi (Figure 1). A total of 12 species from ten genera were identified. The number of genera per sample varied from four to nine. Ascomycota was the representative phylum of all the cultured genera, where *Mycosphaerella*, *Acremonium*, *Cladosporium*, *Aureobasidium*, and *Fusarium* were shared between all the assessed samples. Other genera were detected only in some sample categories: *Fusarium* and *Alternaria* from P1, T1 and T2, *Microdochium* from P1 and T1, and *Monographella* from T1 and T2).

Endophytic fungal community composition

The taxonomic assignment of the OTUs obtained from the HTS of wheat sap samples elucidated the composition of the endophytic fungal community among the four sample categories (P1, P2, T1 and T2). Overall, 28 OTUs were identified, belonging to 26 fungal genera (Figure 2). However, the relative abundance (RA) of the detected fungal taxa showed some differences between the sample categories, at the phylum and class levels (Figure 3). For example, members of Ascomycota were predominant in P1 (60% RA) and T2 (68% RA). In contrast, Basidiomycota were the most abundant in P2 (39%

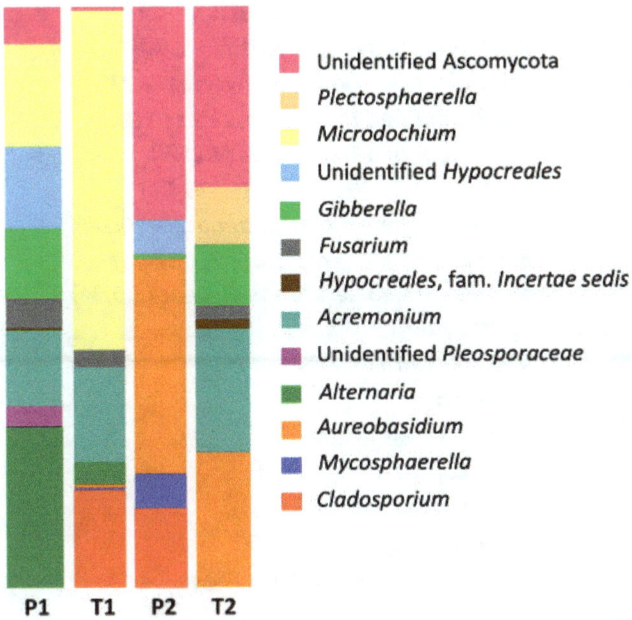

Figure 1. Taxonomic composition of cultured endophytic fungi. Genera distributions of endophytic fungi grown in cultures and analyzed using HTS, among four wheat variety and field categories: P1 = Perciasacchi from Field 1; T1 = Tumminia from Field 1; P2 = Perciasacchi from Field 2; and T2 = Tumminia from /Field 2.

RA) and T1 (59% RA). Field 2 gave the greatest number of unidentified fungal sequences, in particular in the T2 category. Dothideomycetes and Tremellomycetes were the main representative fungal classes in all the samples, with an average RA of 30%, followed by Sordariomycetes (7.5%), Agaricomycetes (2.5%), Microbotryomycetes (4.4%), Pucciniomycetes (3.4%), Leotiomycetes (1.0%), Ustilaginomycotina *incertae-sedis* (0.4%), and Taphrinomycetes (0.2%). At the genus level, *Alternaria*, *Mycosphaerella*, *Cladosporium*, *Filobasidium*, *Holtermanniella*, *Cystofilobasidium* and *Cryptococcus* were detected in all the sample categories.

Richness and diversity of endophytic fungi

The number of detected fungal OTUs per sample varied from 2,646 to 6,612. Alpha diversity indices showed that samples collected from Field 1 had greater fungal diversity compared to those from Field 2 (Table 2). The two-sample t-test based on Shannon index revealed a significant difference between the two fields ($P = 0.02$). However, there was no significant differences, either between the same variety grown in different fields or between the two varieties grown within the same field.

The PCoA plots, showing beta-diversity results, showed that samples collected from the two fields tended to segregate separately. In addition, the two wheat varieties clustered into different groups within each field (Figure 4). In agreement with PCoA and alpha diversity results, statistical comparisons using non-parametric Permanova tests showed significant variation of the fungal communities between the two fields ($P = 0.001$). In addition, the fungal communities were significantly different between the two wheat varieties ($P = 0.044$). A significant difference was also detected where the same varieties were grown in different fields ($P = 0.002$ for Field 1 and 0.005 for Field 2).

Some of the detected taxa had significantly different RAs in the investigated fields and varieties. In particular, *Mycosphaerella*, *Dioszegia*, *Filobasidium*, *Protomyces* and *Alternaria* varied significantly between the two fields (Figure 5), while *Aureobasidium*, *Leucosporidium* and *Puccinia* varied between the two varieties (Figure 6).

Fungal interactions

The co-occurrence and mutual exclusion of specific OTUs were analyzed for each wheat variety. The resulting networks, after statistical calculations and removal of unstable edges/links, were characterized, for Perciasacchi by 132 nodes (OTUs) linked with 420 edges with a clustering coefficient of 0.606, and for Tumminia, by 134 nodes (OTUs) linked with 583 edges and a clustering coefficient of 0.731 (Figure 7). Overall, the interactions between fungal phylotypes were characterized by a greater number of co-occurrences (401 in Perciasacchi and 510 in Tumminia) compared to mutual exclusions (19 in Perciasacchi and 73 in Tumminia). In both varieties, *Cryptococcus*, *Cystofilobasidium* and *Holtermanniella* were the dominant genera, with consistent co-occurrence interactions between each other as well as within each genus. *Cryptococcus* and *Cystofilobasidium* had mutual exclusion relationships with *Gibberella* species. On the other hand, *Gibberella*, *Acremonium* and a group of unidentified ascomycetes had a mutual exclusion interaction with *Puccinia*. *Cryptococcus* excluded *Sporobolomyces* only in Perciasacchi.

DISCUSSION

This study has demonstrated the limits of culture-dependent methods for characterization of the endophytic fungal communities. Metabarcoding analyses from crude sap extract allowed the detection of 26 genera belonging to Ascomycota and Basidiomycota, while

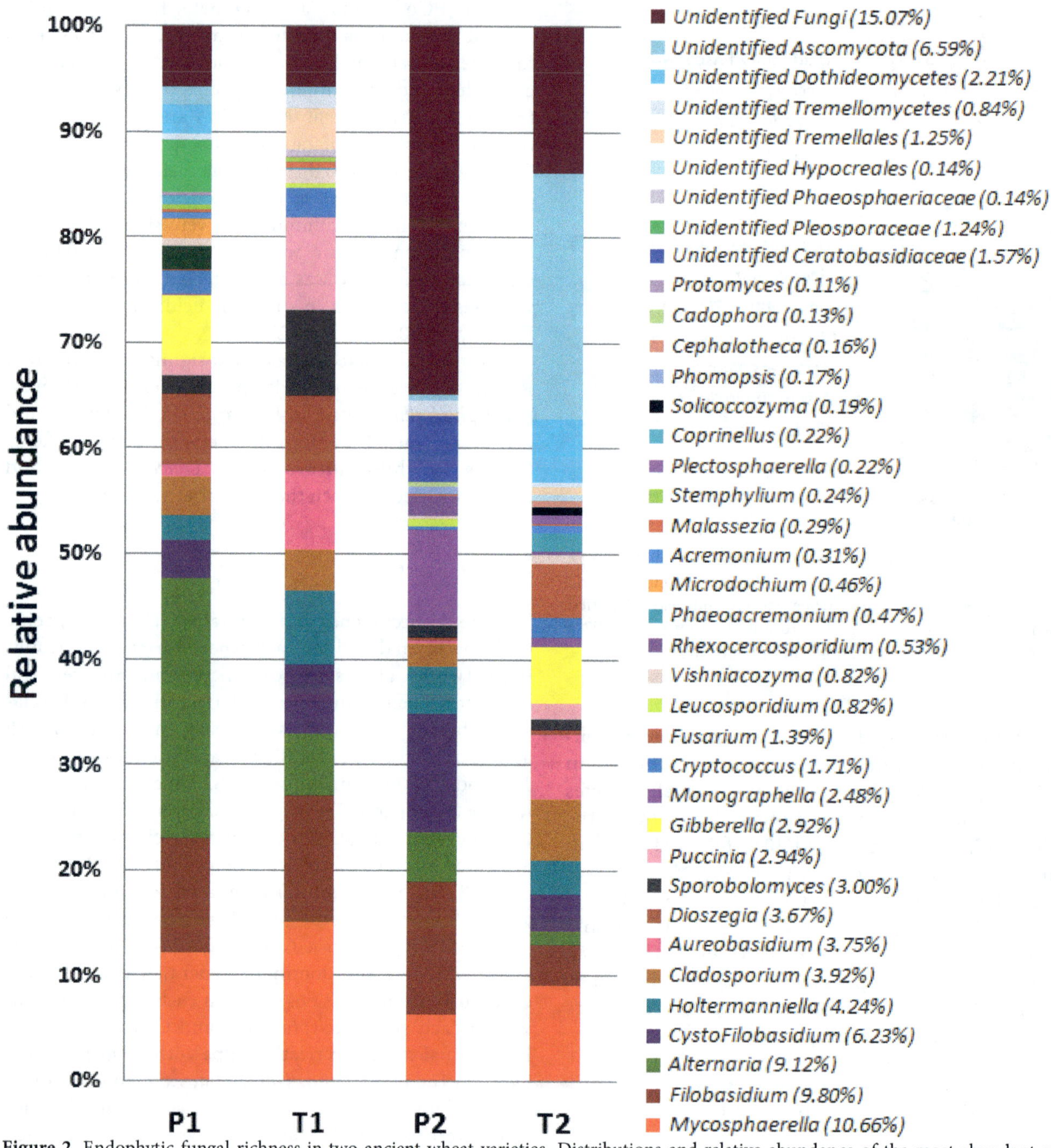

Figure 2. Endophytic fungal richness in two ancient wheat varieties. Distributions and relative abundance of the most abundant genera detected among four sample categories, from analyses of wheat sap samples. Average relative abundance for each genus is reported in parentheses. P1 = Perciasacchi from Field 1; T1 = Tumminia from Field 1; P2 = Perciasacchi fromField 2; and T2 = Tumminia from Field 2. Mean values of relative abundance are indicated in the parenthesis.

fungal isolations allowed the detection of only ten genera belonging to Ascomycota. Since isolations were carried out using one culture medium under one condition, a broader range of culturing conditions may be required to capture greater proportion of wheat endophytes. Nevertheless, nine out of 12 fungal species grown in plate cultures were also detected in the wheat sap extracts, therefore a good correspondence was obtained between the fungal taxa identified by the two methods. The variability in detection of the two methods could be due to

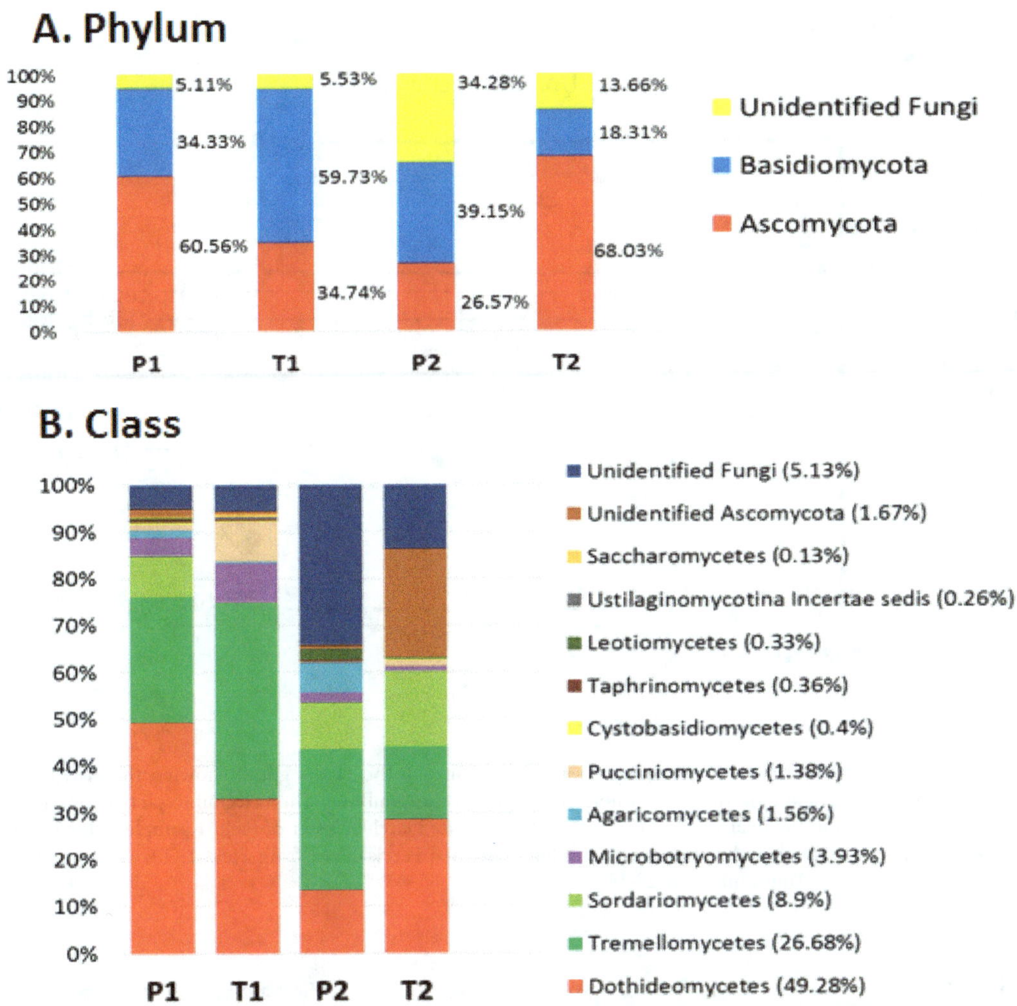

Figure 3. Structure and richness of the wheat endophytic fungal communities. Richness of fungal phyla (A) and classes (B) resulting from downstream analyses of the HTS performed on DNA extracted from wheat sap. Average relative abundance for each class is reported in parentheses. Sample categories are: P1 = Perciasacchi from Field 1; T1 = Tumminia from Field 1; P2 = Perciasacchi from Field 2; and T2 = Tumminia from Field 2.

bias of the culturing technique, which may have favored some taxa over others. Alternatively, the choice of primers for HTS analyses may have impact on the numbers of detected taxa (Tedersoo *et al.*, 2015). Moreover, it is important to highlight that the sap extract obtained with the patented method was successful in fungal isolation and molecular analyses.

The variations observed in the endophytic fungal communities between each wheat variety grown in the two fields could be explained by the different locations and agronomic conditions at the two field sites. Field 1 had been to crop rotation with legumes, while Field 2 was not cultivated for five previous years. Previous studies have reported that geographical location as well as crop management practices can affect the composition of soil microbial communities and, as consequence, the composition of endophytic microbial communities (Göre and Bucak, 2007; Sapkota *et al.*, 2017; Soman *et al.*, 2017).

The fungal communities were different between the two wheat varieties. This is not surprising since host genotype is considered to be a major factor determining the composition of endophyte communities (Sapkota *et al.*, 2015). These results highlight the importance of investigating microbial diversity of ancient crop varieties as possible sources of beneficial microorganisms.

Alternaria, *Cladosporium*, *Sporobolomyces*, *Dioszegia* and *Cryptococcus* are reported to be ubiquitous, and they have been detected in the present study as well as in the phyllospheres and grain of several commercial wheat

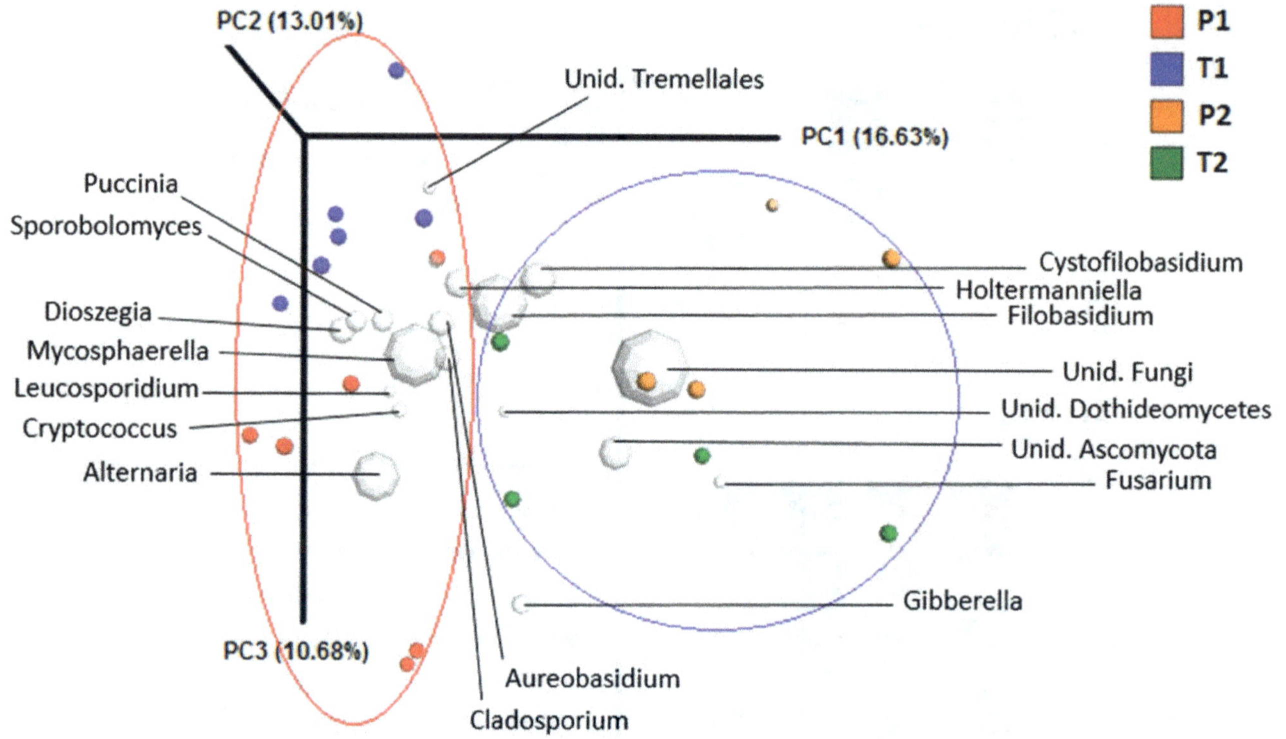

Figure 4. Principal Coordinates Analysis (PCoA). Integration of taxonomy in the PCoA graph showing the distribution of the 18 most abundant endophytic fungal genera among the four sample categories (white symbols). The taxonomic plot weight and the distance from the sample plots are proportional to the relative abundance of each genus. Red circle: cluster of Field 1 samples; blue circle: cluster of Field 2 samples. The different coloured symbols indicate the sample categories: P1 = Perciasacchi from Field 1; T1 = Tumminia from Field 1; P2 = Perciasacchi from Field 2; and T2 = Tumminia from Field 2.

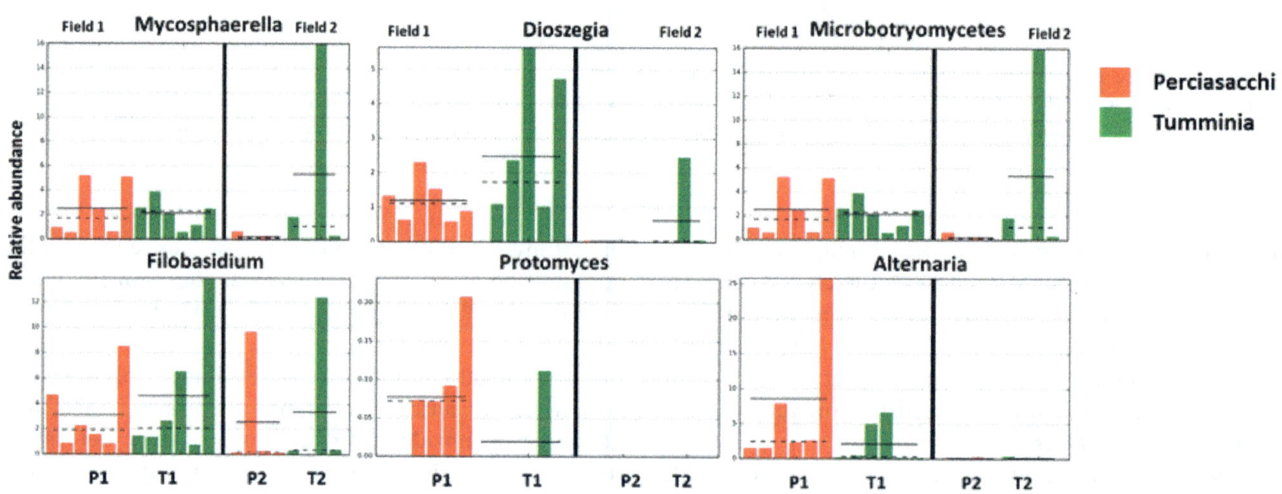

Figure 5. Differentially abundant genera between two ancient wheat varieties and two fields. Histograms indicate significant differences calculated by LEfSe analyses (*Pa*< 0.05 and LDA cut-off >2). P1 = Perciasacchi from Field 1; T1 = Tumminia from Field 1; P2 = Perciasacchi from Field 2; and T2 = Tumminia from Field 2. Category means (straight line) and medians (dotted line) are indicated.

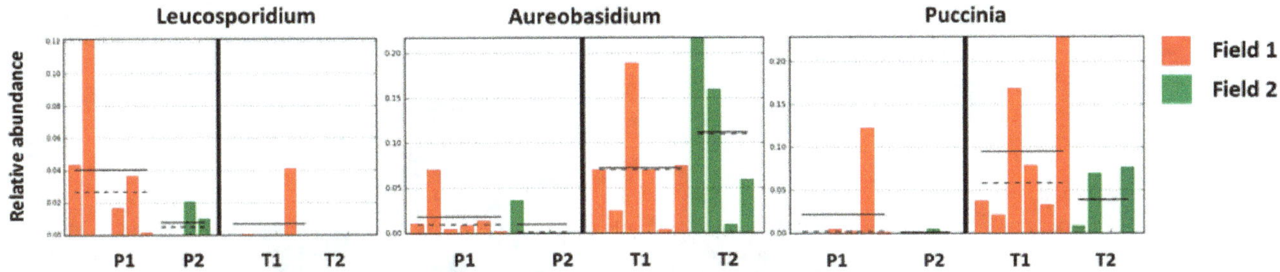

Figure 6. Discriminant fungal genera between two ancient wheat varieties. Differential genera between variety Perciasacchi (red) and Tumminia (green), resulting from LEfSe analysis ($P < 0.05$, LDA cut-off > 2). Histograms represent differential taxa distributions between the two wheat varieties and among the four sample categories: P1 = Perciasacchi from Field 1; T1 = Tumminia from Field 1; P2 = Perciasacchi from Field 2; and T2 = Tumminia from Field 2). Category means (straight line) and medians (dotted line) are indicated.

varieties (Nicolaisen *et al.*, 2014; Sapkota *et al.*, 2015). On the other hand, genera such as *Filobasidium*, *Holtermanniella* and *Cystofilobasidium* were detected in Perciasacchi and Tumminia, but have not been detected in the phyllospheres of modern cultivars. Conversely, *Pyrenophora*, *Epicoccum*, *Phoma* and *Sphaeosphaeria* belonging to core OTUs in the phyllospheres of modern wheat cultivars (Nicolaisen *et al.*, 2014) were not found in the present study. Differences between ancient and modern cultivars may be the consequence of different breeding histories as well as cultivar selection and/or growing conditions. Further research comparing varieties with contrasting pedigrees is necessary to firmly define taxa shared, or differentiating between, wheat genotypes, as previously reported for other crop species (Liu *et al.*, 2018).

Despite the presence of important fungal pathogens, no disease symptoms were observed in the collected samples from both fields. The presence of non-pathogenic fungi and plant resistant genes may account for this observation. However, an important contribution from beneficial fungi for maintaining plant health is likely. The analysis of the association networks suggested that key endophytes played roles within the wheat fungal communities detected. For instance, Basidiomycete endophytes such as *Cryptococcus* and *Cystofilobasidium* were the predominant genera and may have had consistent antagonistic activity against *Gibberella* species. Similarly, *Acremonium* and a group of unidentified ascomycetes mutually excluded *Puccinia*. Since both *Gibberella* and *Puccinia* contain several pathogens that cause economically important diseases in wheat, the detected interactions may indicate microbial-mediated resistance in wheat varieties. Furthermore, *Holtermanniella* and *Cystofilobasidium* were detected in the two ancient wheat varieties, and have not been reported to colonize modern wheat cultivars. This indicates possible negative impacts of breeding and domestication on native beneficial microbiomes.

In conclusion, the results from this study have expanded knowledge of the endophytic fungal communities associated with ancient wheat varieties, and have allowed formulation of hypotheses on the roles of the fungi in host plants. Further studies, including comparison between ancient and modern wheat microbiota, are required to clarify the roles of specific fungi, and to facilitate their exploitation as alternative means for improving plant health and resilience.

ACKNOWLEDGMENTS

This research was funded by The Italian Ministry of Education, University and Research with the grant *Innovazione di prodotto e di processo nelle filiere dei prodotti da forno e dolciari* (PON03PE_00090_01). Special thanks is given to farms La Mandralisca Azienda Agricola, Gangi (PA) and Azienda Agricola Aldo Gallina, Santa Venera (EN) for allowing the use of parts of their fields to carry out this study. The authors are grateful to Senga Robertson-Albertyn and Davide Bulgarelli (University of Dundee, United Kingdom) for English revision and the critical comments of this paper.

LITERATURE CITED

Abdelfattah, A., Cacciola S. O., Mosca S., Zappia R., Schena L., 2017. Analysis of the fungal diversity in citrus leaves with greasy spot disease symptoms. *Microbial Ecology* 73: 739–749.

Abdelfattah, A., Malacrinò A., Wisniewski M., Cacciola S. O., Schena L., 2018. Metabarcoding: A powerful tool to investigate microbial communities and shape future plant protection strategies. *Biological Control* 120: 1–10.

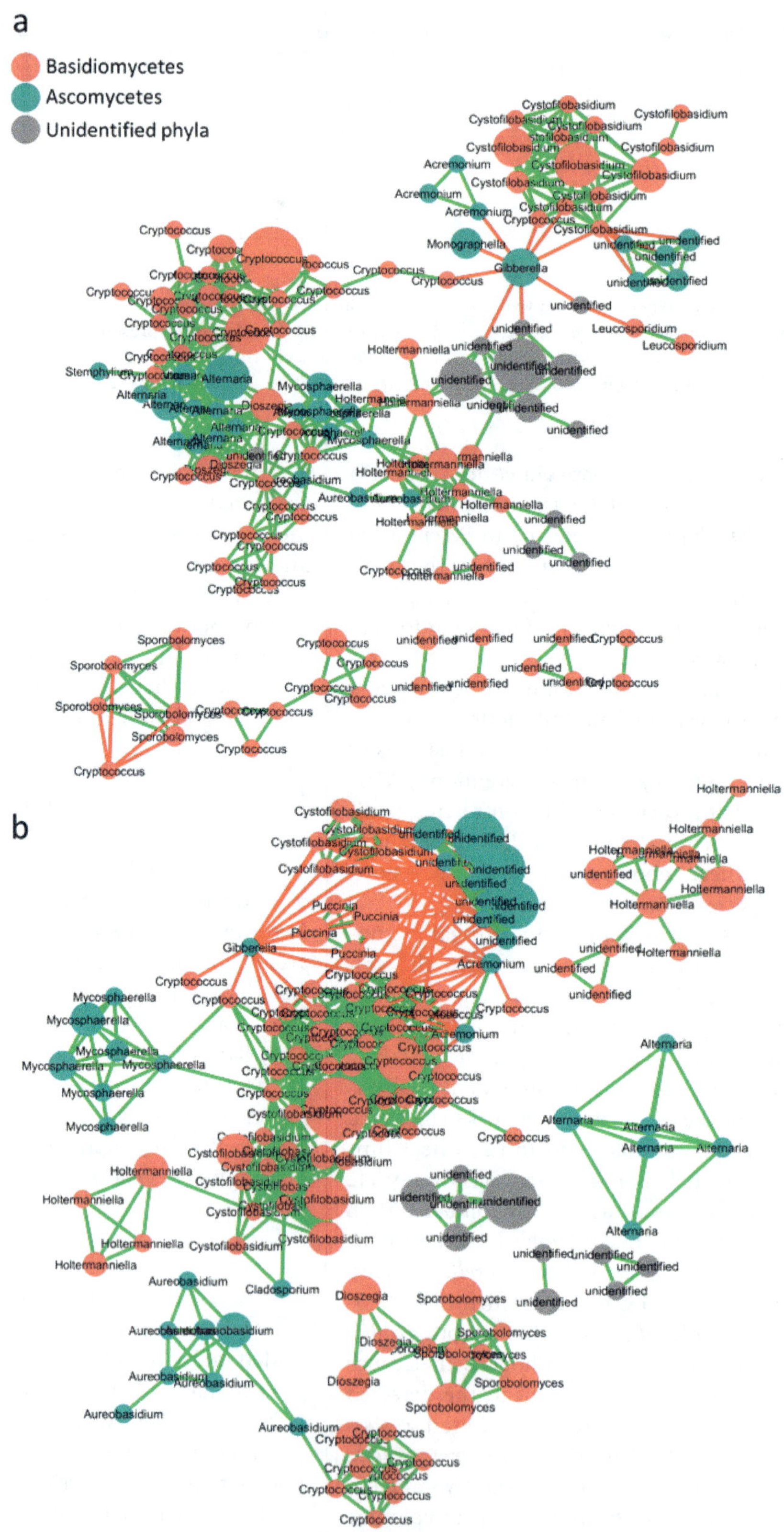

Figure 7. Microbial association network showing the interactions (co-occurrence and mutual exclusion) represented by green links for wheat variety Perciasacchi (a) and red links for Tumminia (b) The size of the nodes indicates OTU abundance. Node colours differentiate fungal phyla. Interactions were calculated by CoNet and visualised in Cytoscape 3.6.

Altschul, S. F., Gish W., Miller W., Myers E. W., Lipman D. J., 1990. Basic local alignment search tool. *Journal of Molecular Biology* 215: 403–410.

Benjamini, Y., Hochberg Y., 1995. Controlling the false discovery rate: a practical and powerful approach to multiple testing. *Journal of the Royal Statistical Society*. Series B (Methodological): 289–300.

Berendsen, R. L., Pieterse C. M., Bakker P. A., 2012. The rhizosphere microbiome and plant health. *Trends in Plant Science* 17: 478–486.

Bolger, A. M., Lohse M., Usadel B., 2014. Trimmomatic: a flexible trimmer for Illumina sequence data. *Bioinformatics* 30: 2114–2120.

Caporaso, J. G., Kuczynski J., Stombaugh J., Bittinger K., Bushman F. D., ... Knight R., 2010. QIIME allows analysis of high-throughput community sequencing data. *Nature Methods* 7: 335–336.

Coombs, J. T., Franco C. M. M., 2003. Isolation and Identification of Actinobacteria from Surface-Sterilized Wheat Roots. *Applied and Environmental Microbiology* 69: 5603–5608.

Dorworth, C. E., Callan B. E., 1996. Manipulation of endophytic fungi to promote their utility as vegetation biocontrol agents. In: *Endophytic Fungi in Grasses and Woody Plants* (S. C. Reddin and L. M. Carris eds), APS press., St. Paul: 209–216.

Faust, K., Sathirapongsasuti J. F., Izard J., Segata N., Gevers D., ... Huttenhower C., 2012. Microbial Co-occurrence Relationships in the Human Microbiome. *PLoS Computational Biology* 8: e1002606.

Göre, M. E., C. Bucak, 2007. Geographical and seasonal influences on the distribution of fungal endophytes in Laurus nobilis. *Forest Pathology* 37: 281–288.

Hallmann, J., Quadt-Hallmann A., Mahaffee W. F., Kloepper J. W., 1997. Bacterial endophytes in agricultural crops. *Canadian Journal of Microbiology* 43: 895–914.

Hubbard, M., Germida J. J., Vujanovic V., 2014. Fungal endophytes enhance wheat heat and drought tolerance in terms of grain yield and second-generation seed viability. *Journal of Applied Microbiology* 116: 109–122.

Janisiewicz, W., Roitman J., 1988. Biological control of blue mold and gray mold on apple and pear with Pseudomonas cepacia. *Phytopathology* 78: 1697–1700.

Jankielsohn, A., Miles C., 2017. How do older wheat cultivars compare to modern wheat cultivars currently on the market in South Africa? *Journal of Horticultural Science and Research* 1: 42–47.

Larran, S., Perelló A., Simón M. R., Moreno V., 2007. The endophytic fungi from wheat (*Triticum aestivum* L.). *World Journal of Microbiology & Biotechnology* 23: 565–572.

Liu, J., Abdelfattah A., Norelli J., Burchard E., Schena L., ... Wisniewski M., 2018. Apple endophytic microbiota of different rootstock/scion combinations suggests a genotype-specific influence. *Microbiome* 6: 18.

Lugtenberg, B., Kamilova F., 2009. Plant-growth-promoting rhizobacteria. *Annual Review of Microbiology* 63: 541–556.

Masella, A. P., Bartram A. K., Truszkowski J. M., Brown D. G., Neufeld J. D., 2012. PANDAseq: paired-end assembler for Illumina sequences. *BMC Bioinformatics* 13: 31.

Massart, S., Martinez-Medina M., Jijakli M. H., 2015. Biological control in the microbiome era: Challenges and opportunities. *Biological Control* 89: 98–108.

Nicolaisen, M., Justesen A. F., Knorr K., Wang J., Pinnschmidt H. O., 2014. Fungal communities in wheat grain show significant co-existence patterns among species. *Fungal Ecology* 11: 145–153.

Ofek-Lalzar, M., Sela N., Goldman-Voronov M., Green S. J., Hadar Y., Minz D., 2014. Niche and host-associated functional signatures of the root surface microbiome. *Nature Communications* 5: 4950.

Rognes, T., Flouri T., Nichols B., Quince C., Mahé F., 2016. VSEARCH: a versatile open source tool for metagenomics. *PeerJ* 4: e2584.

Salmon, C., Clark A. J., 1913. Durum Wheat. *Farmers' Bulletin*, Ed. Washington Government Printing Office 534.

Sapkota, R., Jørgensen L. N., Nicolaisen M., 2017. Spatiotemporal Variation and Networks in the Mycobiome of the Wheat Canopy. *Frontiers in Plant Science* 8: 1357.

Sapkota, R., Knorr K., Jørgensen L. N., O'Hanlon K. A., Nicolaisen M., 2015. Host genotype is an important determinant of the cereal phyllosphere mycobiome. *New Phytologist* 207: 1134–1144.

Sarkar, S. K., Chang C.-K., 1997. The Simes method for multiple hypothesis testing with positively dependent test statistics. *Journal of the American Statistical Association* 92: 1601–1608.

Segata, N., Izard J., Waldron L., Gevers D., Miropolsky L., ... Huttenhower C., 2011. Metagenomic biomarker discovery and explanation. *Genome Biology* 12: R60–R60.

Shendure, J., Ji H., 2008. Next-generation DNA sequencing. *Nature Biotechnology* 26: 1135–1145.

Shewry, P. R., Hey S., 2015. Do "ancient" wheat species differ from modern bread wheat in their contents of bioactive components? *Journal of Cereal Science* 65: 236–243.

Soman, C., Li D., Wander M. M., Kent A. D., 2017. Long-term fertilizer and crop-rotation treatments differentially affect soil bacterial community structure. *Plant and Soil* 413: 145–159.

Taberlet, P., Coissac E., Pompanon F., Brochmann C., Willerslev E., 2012. Towards next-generation biodiversity assessment using DNA metabarcoding. *Molecular Ecology* 21: 2045–2050.

Tedersoo, L., Anslan S., Bahram M., Põlme S., Riit T., … Abarenkov K., 2015. Shotgun metagenomes and multiple primer pair-barcode combinations of amplicons reveal biases in metabarcoding analyses of fungi. *MycoKeys* 10: 1–43.

Tedersoo, L., Bahram M., Põlme S., Kõljalg U., Yorou N. S., … Abarenkov K., 2014. Global diversity and geography of soil fungi. *Science* 346: 1256688.

Turenne, C. Y., Sanche S. E., Hoban D. J., Karlowsky J. A., Kabani A. M., 1999. Rapid Identification of Fungi by Using the ITS2 Genetic Region and an Automated Fluorescent Capillary Electrophoresis System. *Journal of Clinical Microbiology* 37: 1846–1851.

Turner, T. R., James E. K., Poole P. S., 2013. The plant microbiome. *Genome Biology* 14, 209.

Vázquez-Baeza, Y., Pirrung M., Gonzalez A., Knight R., 2013. EMPeror: a tool for visualizing high-throughput microbial community data. *GigaScience* 2: 16–16.

Velazquez-Sepulveda, I., Orozco-Mosqueda M., Prieto-Barajas C., Santoyo G., 2012. Bacterial diversity associated with the rhizosphere of wheat plants (*Triticum aestivum*): Toward a metagenomic analysis. *Phyton International Journal Experimental Botany* 81: 81–87.

14

Incidence of Fusarium foot and root rot of cereals under conservation agriculture in north west Tunisia

Samira CHEKALI[1,*], Samia GARGOURI[2], Moncef BEN HAMMOUDA[3], Hatem Cheikh M'HAMED[4], Bouzid NASRAOUI[5]

[1] *Laboratoire de Protection des Végétaux, Institut National de Recherche Agronomique de Tunisie, Pôle Régional de Recherche-Développement Agricole du Nord-Ouest Semi-Aride au Kef, Université de Carthage, BP 221, 7100, Le Kef, Tunisia*
[2] *Laboratoire de Protection des Végétaux, Institut National de la Recherche Agronomique de Tunisie, Université de Carthage, Rue Hédi Karray, 2049, Ariana, Tunisia*
[3] *Laboratoire de Physiologie de la Production des Céréales, Ecole Supérieure d'Agriculture du Kef, Université de Jandouba, 7119 Le Kef, Tunisia*
[4] *Laboratoire de Recherche de Science et Techniques Agronomiques, Institut National de Recherche Agronomique de Tunisie, Université de Carthage, Rue Hédi Karray, 2049, Ariana, Tunisia*
[5] *Laboratoire de Recherche des Bioagresseurs et Protection Intégrée en Agriculture, Institut National Agronomique de Tunisie, Université de Carthage, 43 Avenue Charles Nicolle, 1082, Tunis-Mahrajène, Tunisia*
*Corresponding author: samirachekali@gmail.com

Competing Interests: The Author(s) declare(s) no conflict of interest.

Editor: Antonio Moretti, Consiglio Nazionale delle Ricerche, CNR, Bari, Italy.

Abstract. Conservation agriculture, based on direct drilling for crop establishment, has emerged in Tunisia since 1999/2000 as an alternative to conventional agriculture based on conventional drilling. The main objective of this approach is to ensure yield stability from crops and replenishment of soil organic matter. Previous research has demonstrated increased risks from pathogens favoured by mulching. The impacts of direct drilling on yields, and Fusarium foot and root rot of durum wheat, oat and barley, were studied over three successive growing seasons (2009/10, 2010/11, 2011/12) in northwest Tunisia. Disease incidence was estimated by the recovery frequency of *Fusarium* spp. isolates from stem bases and roots of plants of the three cereals. In addition, disease severity was assessed by occurrence of whiteheads that senesced prematurely, and the browning intensity on durum wheat stem bases. Grain yields were recorded at the ripening stages. *Fusarium culmorum* and *F. pseudograminearum* were isolated from the three cereals, with *F. culmorum* being the dominant pathogen. Direct drilling increased the incidence (60%) of these pathogens on stem bases and roots during the warmer seasons of 2009/10 and 2010/11, but less infection (37%) compared to conventional drilling was observed in the wetter season of 2011/12. Direct drilling increased the percentage of whiteheads of durum wheat (73%), but did not affect disease severity, which was estimated by the discolouration on stem bases and grain yield.

Keywords. Direct drilling, conventional drilling, Fusarium foot and root rot, cereals, grain yield.

INTRODUCTION

Conservation agriculture technology is increasingly relevant for addressing the needs of resource-poor farmers, and the challenges of resource degradation, sustainability, food insecurity, poverty alleviation, climate change, labour shortages and high energy costs (Kassam *et al.*, 2012). Direct drilling (hereafter abbreviated as DD) for crop establishment is practiced on about 111 million ha worldwide, and the proportional of adoption by farmers is 47% in South America, 38% in the United States and Canada, 12% in Australia,4% in New Zealand, and 4% in the rest of the world, including Europe, Asia and Africa (Derpsch *et al.*, 2010). In North Africa, conservation agriculture systems have been promoted, particularly in Morocco and Tunisia. In Morocco, 4,000 ha of DD have been reported, despite long-term research on DD farming initiated in the early 1980s (Mrabet, 2012). In Tunisia, the promotion and development of these approaches was farmer-centered, and the area under DD increased from 27 ha in 1999 to nearly 6,000 ha in 2007 and 8,000 ha in 2008 (FAO, 2011). DD is now applied on 12,000 ha, which cover 0.8 % of cereal agricultural land, an area distributed among 200 farmers and operated by 102 direct drilling machines (ICARDA, 2016).

Under DD, the soil is left undisturbed from harvest to planting, except for narrow strips where the seeds are planted and the fertilizer is applied (Schillinger *et al.*, 1999). Despite the benefits, DD may constitute potential risk for disease development since leaving infected plant residues on soil surfaces contributes to increasing pathogen inoculum. This may influence the incidence of diseases caused by pathogens which survive on crop debris. These include the complexes of species causing foot and root rot (Smiley *et al.*, 2005). This pathogen complex includes *Fusarium pseudograminearum* (teleomorph *Giberella coronicola*), *F. culmorum*, *F. avenaceum* (teleomorph *Giberella avenacea*), *Bipolaris sorokiniana* (teleomorph *Cochliobolus sativus*), and *Microdochium nivale* (teleomorph *Monographella nivalis*). Under adopted conventional drilling (CD) in Tunisia, *F. culmorum* followed by *F. pseudograminearum* are the main pathogens isolated from cereals in the semi-arid regions (Gargouri, *et al.* 2001; Boughalleb *et al.* 2008). Both of these fungi survive in crop residues, but with different biologies. *F. culmorum* is can grow at lower temperatures than *F. pseudograminearum* (Doohan *et al.*, 2003), and persists as hyphae within stubble residues and as chlamydospores in soil (Burgess, 2011).

These pathogens produce lesions on the coleoptiles, roots, and sub-crown internodes of affected plants, and cause browning of the stem bases. Damage to cereals is often unnoticed until white heads appear shortly before crop maturity, or as shriveled grain that is noted during harvest (Papendick and Cook 1974; Burgess *et al.*, 2001). Infection is favoured by wet conditions shortly after sowing of crops, and disease severity increases and yield reductions become significant when infected plants are under water stress and/or exposed to high temperatures late in the growing season (Cook, 1981; Paulitz *et al.*, 2002.; Smiley *et al.*, 2005; Chekali *et al.*, 2011). In Tunisia, losses in grain yields as much as 25% occur during the dry seasons (Chekali *et al.*, 2013). Crop rotation is a major control method for these pathogens, reducing disease incidence in legume / fallow / cereals cropping systems (Chekali *et al.*, 2016).

Reports of effects of DD on development of cereal foot and root rot diseases have been varied among regions. For DD adopted in Tunisia, there is no information on these diseases. The objective of the present study was to evaluate the impacts of DD on Fusarium foot and root rot development and yields of durum wheat, oat and barley subjected to different cropping sequences over 3 years in northwest Tunisia.

MATERIALS AND METHODS

Site description

A field trial was conducted during the 2009/10, 2010/11 and 2011/12 growing seasons, to compare DD to CD. The trial was located at the Experimental Station of Institut National des Grandes Cultures(INGC) located at Boussalem (36°36'35" N, 8°58'17" E), at an altitude of 124 m. The area has a warm temperate Mediterranean climate. The soil at the trial site contained 54% clay, 31% silt and 15% sand. The winter is the rainy season. According to Köppen(1936), the climate is classified as hot-summer Mediterranean climate with annual mean precipitation of 500–600 mm. This precipitation varied between 400 and 700 mm during the three growing seasons of this study (Table 1).

Experimental design and field lay-out

The trial was set up as a Split Plot Design. Four blocks were divided into two Main Plots, corresponding to CD and DD. These were each divided into five Sub Plots, each of 360 m^2 (30 m × 12 m), of the following rotations scenarios: durum wheat after oat, oat after durum wheat, durum wheat after barley, barley after durum wheat or durum wheat after faba bean. Prior to sowing, the soil was harrowed, and durum wheat var.

Table 1. Monthly rainfall (mm) recorded during the experimental period. The critical period for pathogen infection and disease development is highlighted in gray.

Growing seasons	Sep.	Oct.	Nov.	Dec.	Jan.	Feb.	Mar.	Apr.	May.	Jun.	Jul.	Aug.	Total
2009/10	104.5	38.5	29.0	37.0	44.0	32.5	70.0	41.5	28.0	0.0	0.0	0.0	425.0
2010/11	14.5	64.0	88.0	42.0	55.0	102.0	42.0	420.0	26.5	49.0	4.0	0.0	529.0
2011/12	28.0	228.0	85.0	89.5	50.0	105.5	60.5	74.5	5.5	35.0	7.0	5.0	751.0

Table 2. Minimum and maximum temperatures recorded during the experimental period. The critical period for pathogen infection and disease development is highlighted in gray.

Growing seasons		Sep.	Oct.	Nov.	Dec.	Jan.	Feb.	Mar.	Apr.	May.	Jun.	Jul.	Aug.
2009/10	Max.	27.9	24.2	22.5	14.7	17.3	17.8	16.9	22.5	27.1	30.8	35.0	35.2
	Min.	17.4	12.6	7.5	7.6	5.0	5.1	6.4	11.4	13.0	15.9	20.8	21.1
2010/11	Max.	28.1	26.1	20.0	18.1	17.2	14.5	18.0	23.1	23.8	32.0	37.3	37.4
	Min.	17.4	13.0	10.5	4.9	5.2	4.5	7.8	9.0	13.5	15.7	20.4	19.7
2011/12	Max.	33.6	22.0	20.1	15.1	14.2	11.4	17.4	22.7	26.4	36.5	37.8	37.8
	Min.	18.3	13.2	12.0	7.0	5.3	4.7	8.1	10.6	11.2	19.1	21.7	20.3

'Rezzek', a local line of oat, barley var. 'Rihane'or fababean var. 'Sabeur' were sown. Ammonium nitrate fertilizer was added at different plant growth stages. Weeds, diseases (powdery mildew, Septoria blotch, rust) and insects (aphids) were controlled throughout the trial. Grain harvests were in June each season (Table 3).

Pathogen isolation and disease incidence

Forty mature plants of each of the three cereals were removed from individual plots using a zsampling pattern in each plot, and the plants were then thoroughly washed with tap water. Sections (each 2 mm length) of stem bases and roots were surface sterilized for 2 min in 6% NaClO and then for 10 s in 70% alcohol, followed by rinsing in sterile distilled water. These tissue pieces were then dried, and placed on 25% potato dextrose agar (¼ PDA) containing 100 mg L^{-1} streptomycin sulfate. Resulting fungal cultures were incubated for 5 d (12 h light/12 h dark) at 25°C. *Fusarium* spp. hyphae were transferred on carnation leaf agar (CLA) to favour conidium production, and were morphologically identified according to Leslie and Summerell (2006). *F. culmorum* and *F. pseudograminearum* identifications were confirmed by molecular techniques using the protocol of Möller *et al.* (1992) for DNA extraction. Primers and protocols used for PCR amplification of *F. culmorum* were those of Schilling and Geiger (1996), and for *F. pseudograminearum* those of Aoki and O'Donnell (1999). The incidence of infections was determined by the frequency of isolation of each *Fusarium* sp. from stem bases and roots.

Disease severity

Disease severity was assessed only for durum wheat, by determining prematurely senescing inflorescences (whiteheads) and the brownish discolourations on the stem bases. At the anthesis stage (Feekes 10.51), (late April), whiteheads were estimated visually three times in each treatment, and expressed as percent of total heads per plot. The brown colour that appeared between the first and the second internodes of stems was evaluated using a 0-3 scale (0 = no symptoms, 1 = light browning at the first internode (0-25%), 2 = clear browning extending to the second internode (25-50%), or 3 = dark browning extending to the second internode (75-100%)). Ratings were converted to severity indices using the following formula:

$$\text{Severity index} = \frac{((0)\times(0)+(1)\times(1)+(2)\times(2)+(3)\times(3))}{\text{Total number of observed plants}}$$

Yield estimates

Grain yields from different crop rotations were measured during each growing season.

Table 3. Technical itinerary adopted during the experimental period.

Growing season	2009/10		2010/11		2011/12	
	CD	DD	CD	DD	CD	DD
Soil management (from October to November)						
Harrowing number (0–15 cm deep)	3	0	2	0	2	0
Sowing rates (kg ha^{-1})						
Durum wheat	160	160	160	160	160	160
Oat	150	150	150	150	150	150
Barley	120	120	120	120	120	120
Fababean	120	120	120	120	120	120
Fertilizer applications (kg ha^{-1}) Ammonium nitrate (NH_4NO_3)						
- *Seedling stage (Feekes 1)*						
Durum wheat	120	120	100	100	100	100
Oat	100	100	100	100	100	100
Barley	100	100	100	100	100	100
- *Stem elongation stages (Feekes 7)*						
Durum wheat	100	100	130	130	100	100
Oat	100	100	130	130	100	100
Barley	100	100	130	130	100	100
Pesticide treatments (L ha^{-1})						
Herbicide:						
Before sowing - Glyphosate (as Round up+)	0	2	0	2	0	2
- *Seedling stage (Feekes 1)*						
- Durum wheat: Fénoxaprop-Ethyl, Iodosulfuron-Méthyl Sodium and Méfenpyr-diethyl	1	1	1	1	1	1
- Oat: 2-4-D-Acide, di Métosulam	1	1	1	1	1	1
- Barley: Pinoxadan, Cloquintocet-mexyl	1–0.8	1–0.8	1–0.8	1–0.8	1–0.8	1
- Fababean: Bentazone	1.25	1.25	1.25	1.25	1.25	1.25
Fungicide						
- Epoxiconazole	1	1	1	1	-	-
Insecticide						
- Deltamethrine	1	1	1	1	-	-

Statistical analyses

Data were analyzed using ANOVA to test the significance of main effects with four replicates, growing season, sowing method (DD *vs* CD) and crop sequences as main factors, using SPSS Statistics version 20 software published by IBM Crop 2011. Means comparisons was performed using Student's LSD tests (P = 0.05 or 0.01).

RESULTS

Identification of pathogens

During the three growing seasons, morphological identification of isolates recovered from 4,800 stem base fragments and 4,800 root fragments of the three cereals revealed that *F. culmorum* was isolated from 39.3% of stem bases and 16.2% from roots, and *F. pseudogramiearum was isolated from* 2.4% of stem bases and 0.9% of roots. Molecular identification using the specific primers OPT18 R and OPT18 F amplified 470 bp of *F. culmorum* fragments, and Fp1-1 and Fp 1-2 amplified 520 bp of *F. pseudograminearum* fragments.

Disease incidence

Data analyses revealed that growing season conditions significantly affected the incidence of infection by both *F. culmorum* (P = 0.02) and *F. pseudograminearum* (P < 0.01) on stem bases of the cereals. In addition,

Table 4. ANOVA of the effects of growing season (GS), sowing method (DD *vs* CD) and crop sequences (CS) on incidence of *Fusarium culmorum* (*Fc*) or *F. pseudograminearum* (*Fpg*) infections of durum wheat, oat or barley stem bases and roots during the experimental period.

SV	DF	*Fc* Stem base		*Fc* Roots		*Fpg* Stem base		*Fpg* Roots	
		F value	Pr> F	F value	Pr> F	F value	Pr> F	F value	Pr> F
GS	2	3.910	0.024	0.007	0.993	11.53	0.000	2.050	0.136
DD *vs* CD	1	33.740	0.000	15.130	0.000	5.540	0.021	0.070	0.782
CS	4	0.930	0.446	1.220	0.309	2.620	0.042	0.680	0.602
CS × DT	2	0.610	0.545	0.790	0.457	5.977	0.004	0.370	0.964
GS × CS	8	1.500	0.170	0.870	0.543	2.301	0.029	0.480	0.864
DT × CS	4	1.990	0.120	0.630	0.638	3.533	0.011	0.430	0.786

there was a statistically significant incidence difference between the DD and CD treatments for *F. culmorum* on stem bases and roots (both $P < 0.01$) and for *F. pseudograminearum* only on stem bases ($P = 0.02$). However, no statistically significant effects were detected for the crop sequence effects, except for the incidence of *F. pseudograminearum* recovered from the stem bases ($P = 0.04$), and between the different tested factors (Table 4). Hence, combined data analyses were carried out for all of the crop sequences.

DD increased the incidence infection of *F. culmorum* on stem bases of durum wheat, oat and barley combined, compared to CD during the trial period. This effect was greater in the two dryer seasons, at 60%; ($P < 0.01$) in 2009/10, and 58% ($P < 0.01$) in 2010/11, than in the wetter season of 2011/12 (32%: $P < 0.01$). On roots, the incidence of infection by this pathogen (60%) was increased ($P < 0.05$) by DD only in 2010/11 (Figure 1).

Fusarium pseudograminearum was present in isolation only in the two growing seasons 2009/10 and 2010/11. Despite the low incidence on stem bases of the three cereals, DD significantly increased incidence of this pathogen (80%; $P < 0.02$) during the 2009/10 growing season, compared to CD (Figure 2).

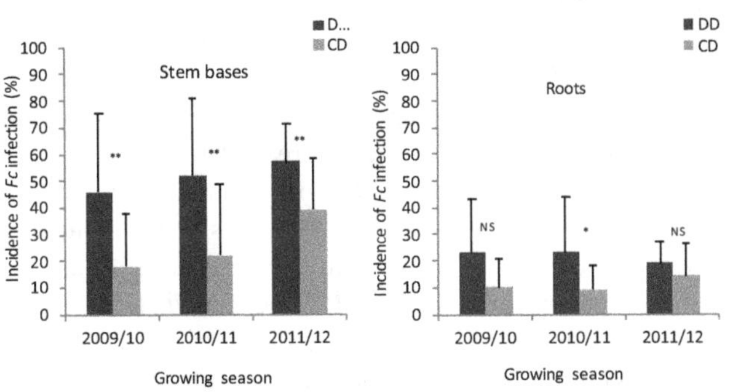

Figure 1. Mean incidence of *Fusarium culmorum* infections on stem bases and roots of durum wheat, oat or barley as affected by different sowing methods (DD *vs* CD) during three growing seasons. Error bars are 2 × standard deviations. NS: No significant difference (P> 0.05). *: significant difference ($P < 0.05$). **: highly significant difference ($P < 0.01$) according to Student's Test.

Disease severity on durum wheat

DD affected occurrence of whiteheads only in the 2010/11 growing season. This drilling method greatly increased the percentage of durum wheat whiteheads (97%; $P < 0.01$; Figure 3), but had no effect on disease severity estimated by discoloration on stem bases.

Durum wheat yields, under DD vs CD and Fusarium foot and root rot

The average of grain yields of durum wheat estimated over the three growing seasons were not significantly different between DD and CD. The yields were estimated 2,700 kg ha^{-1} from DD, and 2,600 kg ha^{-1}from CD, in the 2009/10 and 2010/11 growing seasons. However, in the 2011/12 growing season, the durum wheat yields were 4,760 kg ha^{-1} from DD and 4,660 kg ha^{-1}from CD. No statistically significant interaction was detected between DD or CD and infection incidence or disease severity, or for occurrence of whiteheads.

DISCUSSION

Conservation agriculture and crop residue management may have impacts on development of soilborne pathogens and the diseases they cause, in medium or long term periods. In this study durum wheat, oat and barley were subjected to different crop rotations and DD or CD crop establishment methods, and assessed for dif-

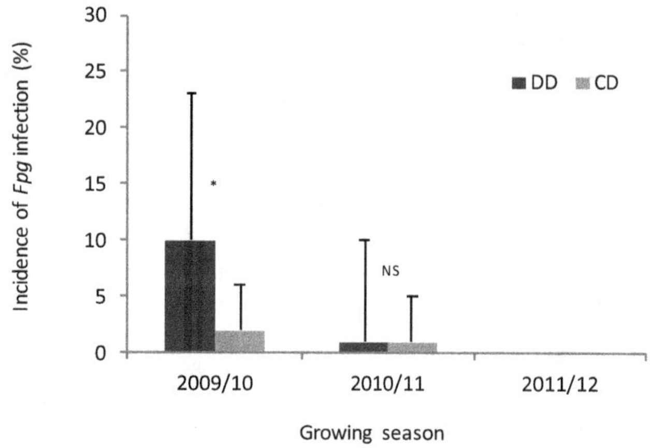

Figure 2. Mean incidence of *Fusarium pseudograminearum* stem base and root infections resulting from different cereal sowing methods (DD *vs* CD) for durum wheat, oat of barley during three growing seasons. Error bars are 2 × standard deviations. NS: No significant difference ($P> 0.05$). *: significant difference ($P < 0.05$).

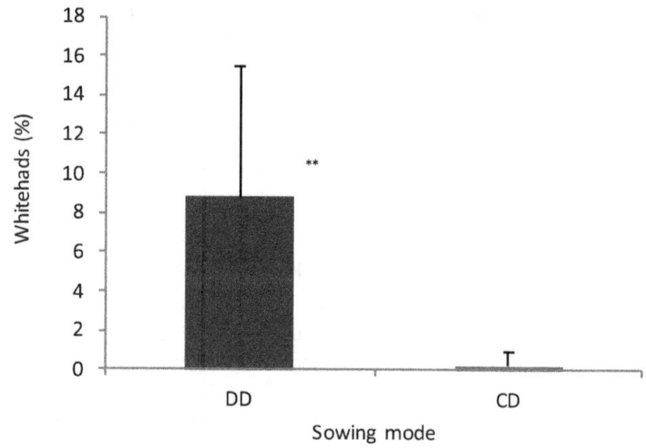

Figure 3. Mean proportions of whiteheads in durum wheat for different sowing methods (DD vs CD) during the 2010/11 growing season. Error bars are 2 × standard deviations. **: highly significant difference ($P < 0.01$) according to Student Test.

ferences in root and crown rot diseases caused by *Fusarium* spp. *Fculmorum* and *F. pseudograminearum* were frequently isolated from stem bases and roots of the three cereals, These results are in agreement with those of Gargouri (2003) and Boughalleb *et al.* (2008), who reported the dominance of *F. culmorum* in Tunisia.

The presence and distribution of each of these species varied in the three experimental years. Both pathogens were isolated as a complex during dry growing seasons (2009/10, 2010/11), while in the wet season (2011/12) only *F. culmorum* was isolated. Similar results were obtained by Lipps and Deep (1991), and Steinkellner and Langer (2004) in Austria. Precipitation and temperature occurring during the experimental period probably influenced the development of each species. Rainfall over the three growing seasons was 425 mm in 2009/10, 529 mm in 2010/11 and 751 mm in 2011/12, with considerable fluctuation during each season. Precipitation over the infection periods (January-February) was almost the same in 2010/11 and 2011/12, but less in 2009/10, and during the disease development period (March-April) was slightly greater in 2009/10 and 2011/12 than in 2010/11 (Table 1). In parallel, the mean recorded temperatures from infection to disease development (January-April) in 2009/10 (11.3–14.3°C) and 2010/11 (10.4–14.5°C) were greater than those in 2011/12 (8.9–14.7°C) (Table 2). The relatively low temperatures in January (9.8°C) and February (8°C) of 2012 (Table 2), in comparison with the other two seasons, probably favoured the development only of *F. culmorum*. These results indicate that this pathogen developed at low temperatures, as reported by Doohan *et al.* (2003), Gargouri (2003), Pitt and Hocking (2009) and Poole (2010). This indicates that *F. culmorum* is more adapted to cold regions than *F. pseudograminearum*, which develops better when it is warm. Some authors have suggested that the relationship between the two *Fusarium* species and DD depends on physical, nutritional and microbiological characteristics of soils. Smiley and Patterson (1996) demonstrated that diseases caused by these pathogens increased with amount of soil surface crop residues that occur, and they are directly correlated with amounts of soil organic nitrogen and carbon. However, under DD, more soil water is conserved than for CD, which could reduce the incidence of cereal foot rot, especially if caused by *F. culmorum* which is enhanced by plant water stress (Papendick and Cook, 1974). On the other hand, *F. pseudograminearum* becomes a serious problem on wheat both in low and high precipitation zones of the Pacific northwest of the United States of America, and was more dependent than *F. culmorum* on crop residues for survival (Paulitz*et al.*, 2002).

The present research showed that after only one year of the experiment, DD increased the incidence of infection by the two fungi, mainly on stem bases of the three different cereals. This effect was consistent with those reported by Windels and Wiersma (1992), Smiley and Patterson (1996) and Bailey *et al.* (2001); and those obtained by Cook (2001), Paulitz*et al.* (2002)and Schroeder and Paulitz (2006). These authors suggested that the main reasons for the increasing range and prevalence of root diseases on wheat and barley in the United States of America and many other cereal-growing areas of the world are the increased frequency of cereals in the crop rotations, and the use of minimum or no-tillage. Other authors have expressed opposite views. Papendick and Cook (1974) explained that under DD, more water is conserved in soil, leading to reduced incidence of cereal foot rot, especially if caused by *F. culmorum*, which is enhanced by plant water stress. Govaertset *al.* (2006) suggested that DD may decrease the effects of the disease by improving soil quality, water retention and microbial activity. In our research conditions, we suggest that both high temperature and water stress at flowering time both had major effects, increasing the incidence of the disease.

The severity of disease estimated by whiteheads appearing prematurely on durum wheat was greatly increased (97%) by DD in 2010/11, which is not in agreement with the results of Wildermuth *et al.* (1997). They found fewer whiteheads (4.3%) after DD,and more (19.3%) after reduced drilling and CD (12.2%) and *F. pseudograminearum* infection. Others have suggested that DD reduces the formation of whiteheads by inducing humid crop canopy atmospheres. Results from the present study can be explained by insufficient moisture at soil level, which was probably caused by high evapotranspiration in the high temperature recorded in the region.

DD did not increase disease severity, estimated by the brown discolourations on durum wheat stem bases. In addition, the grain yields were not affected.

CONCLUSIONS

Results of this study have shown that *F. culmorum* followed by *F. pseudograminearum* were the main pathogens recovered from stem bases and roots of the three cereals examined. The results also showed that for the three cereal types, DD increased incidence of *F. culmorum* infection by 58-60% in dry seasons, and by 32% in a wetter season. However, incidence of *F. pseudograminearum* infection increased by 80% in dry seasons under DD. In contrast to several other reports, DD increased disease severity by 97%, as expressed by whiteheads appearing in durum wheat prior to maturity. This may be attributed to the high temperatures and water stress at the experimental site.

There was no difference between DD and CD for disease severity, estimated by the discolouration of cereal stem bases.

In conservation agriculture, for cereal production, inputs are reduced, soil quality improves, and grain yields get close to, or slightly greater than, those from conventional agriculture. This is because there is more available water for the crop growth where conservation agriculture is applied. In the present study, no significant differences in grain yields from the three cereals were observed between DD and CD.

Given the high incidence of *F. culmorum* infections on durum wheat, oat and barley under DD observed in this study, further research is required at more sites over long periods, and in different farming systems. Furthermore, these studies should also consider host variety/species effects on diseases caused by *Fusarium* spp.

ACKNOWLEDGEMENTS

We thank S. Arfaoui and M.A. Hanachi of INGC Experimental Station, and S. Aloui-Rezgui of Ecole Supérieure d'Agriculture du Kef for their assistance in this research.

LITERATURE CITED

Aoki T., O'Donnell K., 1999. Morphological and molecular characterization of *Fusarium pseudograminearum* sp. nov., formerly recognized as the Group 1 population of *F. graminearum*. *Mycologia* 91: 597–609.

Bailey K.L., Gossen B.D., Lafond G.P., Watson P.R., Derksen D.A., 2001. Effect of tillage and crop rotation on root and foliar diseases of wheat and pea in Saskatchewan from 1991 to 1998: Univariate and multivariate analysis. *Canadian Journal Plant Science* 81: 789–803.

Boughalleb N., Souli M., Oueled Sghair H., 2008. Occurrence and geographic Tunisian distribution of wheat Fusarium head blight and Fusarium root rot in Jendouba's area of Tunisia. *African Journal of Plant Science Biotechnology* 2: 32–26.

Burgess L.W., Backhouse D.B, Summerell A, 2001. Crown rot of wheat. 5; In: *Fusarium Paul E Nelson Memorial Symposium* (B.A. Summerell, I.F. Leslie, D. Backhouse, W.L. Bryden, L.W. Burgess, Ed.), The American Phytopathological Society Press, St. Paul, MN, USA, 271–295.

Burgess L.W., 2011. Foot, crown, and root rots of cereals insidious diseases of global importance: A Tunisian perspective. *Tunisian Journal of Plant Protection* 6 (Guest Editorial).

Chekali S., Gargouri S., Paulitz T., Nicol J.M., Rezgui M., Nasraoui B., 2011. Effects of *Fusarium culmorum* and water stress on durum wheat in Tunisia. *Crop Protection* 30: 718–725.

Chekali S., Gargouri S., Berraies S., Gharbi M.S., Nicol M.J., Nasraoui B., 2013. Impact of Fusarium foot and root rot on yield of cereals in Tunisia. *Tunisian Journal of Plant Protection* 8: 75–86.

Chekali S., GargouriS, Rezgui M, Paulitz T, Nasraoui B., 2016. Impacts of previous crops on Fusarium foot and root rot, and on yields of durum wheat in North West Tunisia. *Phytopathologia Mediterranea* 55: 253–261.

Cook R.J., 1981. Fusarium diseases of wheat and other small grains in North America. In: *Fusarium: Diseases, Biology and Taxonomy* (P.E. Nelson, T.A. Toussoun, R.J Cook, Ed.), The Pennsylvania State University Press, University Park., USA, 39–52.

Cook RJ., 2001. Management of wheat and barley root diseases in modern farming systems. *Australasian Plant Pathology* 30:119–126.

Derpsch R., Friedrich T., Kassam A., Hongwen L., 2010. Current status of adoption of no-till farming in the world and some of its main benefits. *International Journal of Agriculture & Biological Engineering* 1:1-25.

Doohan F.M., Brennan J., Cooke B.M., 2003. Influence of climatic factors on *Fusarium* species pathogenic to cereals. *European Journal of Plant Pathology* 109: 755–768.

FAO, 2011. Conservation Agriculture. Website at: http://www.fao.org/ag/ca/6c.html.

Gargouri S., 2003. *Evaluation de l'incidence de la pourriture du pied et étude de la diversité génétique de la structure des populations des espèces de Fusarium associées à la maladie.* PhD Thesis, Département de Biologie, Faculté des Sciences de Tunis, Université de Tunis El Manar, Tunisie, 108 pp.

Gargouri S., Hajlaoui M.R., Guermech A., Marrakchi M., 2001. Identification des espèces fongiques associées à la pourriture du pied du blé et étude de leur répartition selon les étages bioclimatiques en Tunisie. *Bulletin OEPP/EPPO Bulletin* 31: 499–503.

Govaerts B., Mezzalama M., Sayre K.D., Crossa J., Nicol J.M., Deckers J., 2006. Long-term consequences of tillage, residue management, and crop rotation on maize/wheat root rot and nematode populations in subtropical highlands. *Applied Soil Ecology* 32: 305–315.

IBM Crop. Released, 2011. IBM Statistics for Windows, Version 20.0. Armonk, NY: IBM Corp.

ICARDA., 2016. Conservation Agriculture: strengthening crop production in marginal areas. https://www.icarda.org/update/conservation-agriculture-strengthening-crop-production-marginal-areas#sthash.J35pMNnM.dpbs.

Kassam A., Friedrich T., Derpschc R., Lahmard R., Mrabet R., 2012. Conservation agriculture in the dry Mediterranean climate. *Field Crop Research* 132: 7-17

Köppen W., 1936. Das geographische System der Klimate. In: *Handbuch der Klimato - logie* (W. Köppen, R . Geiger , ed.), Gebrüder Borntraeger, Berlin, 1-44.

Leslie J.F., Summerell B.A., 2006. The Fusarium Laboratory Manual. *Blackwell Publishing*, 388 pp.http://eu.wiley.com/WileyCDA/WileyTitle/productCd-0813819199.html

Lipps P.E., Deep I.W., 1991. Influence of tillage and crop rotation on yield, stalk rot, and recovery of *Fusarium* and *Trichoderma* spp. from corn. *Plant Disease* 75: 828–833.

Moller E.M., Bahnweg G., Sandermann H., Geiger H.H., 1992. A simple and efficient protocol for isolation of high molecular weight DNA from filamentous fungi, fruit bodies, and infected plant tissues. *Nucleic Acids Research* 22: 6115–6116.

Mrabet R., Moussadek R., Fadlaoui A., Van Ranst E., 2012. Conservation agriculture in dry areas of Morocco. *Field Research* 132: 84–94.

Papendick R.I., Cook R.J., 1974. Plant water stress and development of Fusarium foot rot in wheat subjected to different cultural practices. *Phytopathology* 64: 358–363.

Pautlitz T.C., Smiley R.W., Cook R.J., 2002. Insights into the prevalence and management of soil borne cereal pathogens under direct seeding in Pacific Northwest, USA. *Canadian Journal of Plant Pathology* 24: 416–428.

Pitt J.I., Hocking A.D., 2009. *Fungi and Food Spoilage.* 3rd ed. Vol. II, London: Blackie Academic and Professional, 536 p.

Poole G.J., 2010. *Identification of quantitative Trait Loci (QTL) for resistance to Fusarium crown rot in wheat and survey of* Fusarium pseudograminearum *and* F. culmorum *in the Pacific Northwest of the U.S.* PhD thesis, Department of Plant Pathology, Washington State University, Washington State, USA, 278 pp.

Schilling A.G., Moller E.M., Geiger H.H., 1996. Polymerase chain reaction-based assays for species-specific detection of *Fusarium culmorum, F. graminearum,* and *F. avenaceum. Phytopathology* 86: 515–522.

Schillinger W.F., Cook R.J, Papendick R.I., 1999. Increased dry land cropping intensity with no-till barley. *Agronomy Journal* 91:744–752.

Schroeder K.L., Paulitz T.C, 2006. Root diseases of wheat and barley during the transition from conventional tillage to direct seeding. *Plant Disease* 90: 1247–1253.

Smiley R.W., Patterson L.M., 1996. Pathogenic fungi as associated with Fusarium foot rot of winter wheat in the semi-arid Pacific Northwest USA. *Plant Disease* 80: 944–949.

Smiley R.W., Gourlie J.A., Easley G.S., Patterson L.M., Whittaker R.G., 2005. Crop damage estimates for crown rot of wheat and barley in the Pacific Northwest. *Plant Disease* 89: 595–604.

Steinkellner S., Langer I., 2004. Impact of tillage on the incidence of *Fusarium* spp. In soil. *Plant and soil* 267: 13–22.

Wildermuth G.B., Thomas G.A, Radford B.J, McNamara R.B., Kelly A., 1997. Crown rot and common root rot in wheat grown under different tillage and stubble treatments in southern Queensland. *Australia Soil & Tillage Research* 44: 211–224.

Windels C.E., Wiersma J.V., 1992. Incidence of *Bipolaris* and *Fusarium* on sub crown internodes of spring barley and wheat grown in continuous conservation tillage. *Phytopathology* 82: 699-705.

15

Biological control of clementine branch canker, caused by *Phytophthora citrophthora*

Mohamed ZOUAOUI[1], Badiâa ESSGHAIER[1], Maroua WESLATI[1], Marwa SMIRI[1], Mohamed Rabeh HAJLAOUI[2], Najla SADFI-ZOUAOUI[1,*]

[1] *Laboratoire de Mycologie, Pathologies et Biomarqueurs LR16ES05, Faculté des Sciences de Tunis, Université de Tunis El Manar 2092, Tunis, Tunisie*
[2] *Laboratoire de Biotechnologie Appliquée à l'Agriculture, INRA Tunisie, Université de Carthage, 2049 Ariana, Tunisie*
*Corresponding author: najla.sadfi@fst.utm.tn

Competing Interests: The Author(s) declare(s) no conflict of interest.

Editor: Matteo Garbelotto, University of California, Berkeley CA, USA.

Summary. Significant losses of clementine trees (*Citrus* × *clementina*) due to *Phytophthora* branch canker were observed in the Cap Bon Peninsula in northeastern Tunisia. This disease is caused by *Phytophthora citrophthora*. The low efficiency of available cultural and host resistance disease management methods, and potential harmful impacts of pesticide strategies, lead to a search of new control alternatives. This study investigated potential biocontrol agents. The halophilic bacterial strains *Bacillus pumilus* M3-16 and *Halomonas elongata* L80, previously selected for their antifungal activity, were assessed in a greenhouse trial. In addition, 69 endophytic bacteria were isolated from citrus roots and assessed for their antagonistic activities. Five isolates were selected because they showed strong growth inhibition of *P. citrophthora* in dual culture tests. Based on phenotypic characteristics, biochemical tests and sequence analyses of the 16S region of rDNA, the bacteria were identified as *Bacillus pumilus* (S19), *Bacillus amyloliquefaciens* (S24), *Bacillus siamensis* (S54) *Paenibacillus polymyxa* (S31), and *Pseudomonas veronii* (S40). These metabolized different carbon sources, and possessed antibiotic genes, produced siderophores and phytohormones, and solubilized phosphates. In greenhouse trials, two endophytic strains (S24, S31) and two extremophile strains (M3-16, L80) were also assessed, these bacteria significantly reduced ($P < 0.0001$) necrotic lesions on host plants, by 35% for strain S24 of *Bacillus amyloliquefaciens* to 51% for strain L80 of *Halomonas elongata*. This is the first report of *Halomonas* sp. Controlling a *Phytophthora* sp. The strains described in this study offer a foundation for developing an efficient biofungicide.

Keywords. Branch canker, biocontrol, *Bacillus*, *Halomonas elongata*.

INTRODUCTION

Citrus fruits are an economically important crop in Tunisia, which produces approx. 350,000 tons of citrus fruits annually (Mahjbi *et al.*, 2016) on over 21,000 ha (FAO, 2013). However, intensification of cropping techniques, exchange of plant material, and competition between citrus producing countries have contributed to the emergence of disease epidemics such as Citrus greening, Citrus canker and tristeza (Cambra*et al.*, 2000; Gottwald *et al.*,2002; Bove, 2006). Among the pathogenic fungi infecting citrus, *Phytophthora* is one of the most virulent, causing damping off, root rot, brown rot, and gummosis (Graham *et al.*, 2003). Although ten species have been reported to be pathogenic toward citrus trees around the world, three cause the most serious disease, stem gummosis, as well as root and fruit rot. These are *Phytophthora citrophthora*, *P. nicotianae* (syn. *P. parasitica*), and *P. palmivora* (Erwin *et al.*, 1996). These three *Phytophthora* species have distinct temporal and climatic requirements, so their relative distributions and importance vary in different production areas (Matheron *et al.*, 1997). *Phytophthora nicotianae* and *P. palmivora* are major causes of citrus diseases in the United States of America, whereas *P. citrophthora* predominantly causes disease in the Mediterranean Basin (Graham *et al.*, 1998; Cacciola*et al.*, 2008; Khanchouch *et al.*, 2017).

Clementine production has expanded in the last few years to reach second in citrus production in Tunisia, after the 'Maltaise' cultivar, and third in citrus exports (GIFruits, 2018). This progress comes after the introduction of new varieties such as 'Marisol' and 'Hernandina' via the FAO program in 1996 from Corsica, France.

Since 2015, in the Cap-Bon Peninsula in northeastern Tunisia, a region which represents 70% of the country's citrus fruit production, a new serious disease has appeared, which was found to be caused by *Phytophthora citrophthora* (Zouaoui *et al.*, 2016). This syndrome was reported in Spain in 2008 (Alvarez *et al.*, 2008a) and in South Africa in 2010 (Schutte *et al.*, 2010), and is characterized by host gum exudation, browning of the crust, cracking, and cankers. Reasons for emergence of this disease are still unclear, but could include several hypotheses including climate change, the introduction of susceptible varieties, the appearance of a virulent strain of *P. citrophthora* (Cohen *et al.*, 2003) and the roles of snails and ants as dispersal vectors of *P. citrophthora* (Alvarez *et al.*, 2009).

The use of chemical pesticides is decreasing, due to potential negative side effects, including contamination of soil and ground water, health risks to humans, and development of pesticide resistance in pathogens (Conacher and Mes, 1993; Jepson *et al.*, 2014). Biocontrol, particularly the suppression of plant pathogens by antagonistic bacteria, is a potential alternative.

Endophytes are plant-associated microorganisms that live in plant tissues without causing any detrimental effects to their hosts (Kloepper *et al.*, 2006; Ryan *et al.*, 2008). In many studies, endophytic bacteria such as *Pseudomonas* and *Bacillus* have provided effective ways to manage plant diseases, and improve plant growth (Lee *et al.*, 2008; Khabbaz *et al.*, 2015). These bacteria commonly produce modified or unmodified peptides, simple heterocyclic (phenazine) and aliphatic compounds, hydrogen cyanide (HCN), siderophores, volatile compounds, proteolytic enzymes such as cellulase and β-1, 3-glucanase (Susi *et al.*, 2011; Radhakrishnan *et al.*, 2017; Biessy and Filion, 2018). *Bacillus* spp. have several advantages such as their ubiquity, spore production, heat resistance, and antibiotic production. *Pseudomonas* spp. are excellent competitors towards soil fungal and bacterial microflora. *Pseudomonas* bacteria have the ability to use plant exudates as nutrients (Lugtenberg *et al.*, 2002; Espinosa-Urgel 2004), and to produce chelating ferric ions and a wide variety of secondary metabolites, and induce systemic resistance (ISR) in plants. All these characteristics are essential for effective biocontrol.

Previous studies have shown the efficiency of halotolerant and moderately halophilic bacteria isolated from different Tunisian Sebkhas (shallow salt lakes) to control several phytopathogenic fungi, including *Fusarium sambucinum* causing dry rot of potato and gray mold in strawberry and tomato fruits (Sadfi *et al.*, 2001). The halophilic bacteria are known to produce a wide range of extracellular antibiotics and volatile compounds, as well as antifungal enzymes such as protease, chitinase and glucanase (Sadfi-Zouaoui *et al.*, 2008a; Essghaier *et al.*, 2009b).

The objectives of the present study were to: (i) isolate and evaluate endophytic bacteria from citrus roots for control of Citrus branch canker, (ii) identify plant growth-promoting bacteria (PGPB) traits of antagonism and antibiotic and phytohormone production, and the degradation of different carbon sources, and (iii) assess the abilities of selected halolerant and endophytic bacteria to reduce development of host necrosis caused by *P. citrophthora* under greenhouse conditions.

MATERIALS AND METHODS

Isolation of citrus root endophytic bacteria

Endophytic bacteria were isolated from citrus roots. Ten different citrus cultivars, all grafted on sour orange, were randomly collected from Cap Bon Peninsula. Sampled roots were surface disinfected with 1% sodium hypochlorite solution for 10 min. The external portion of each root (approx. 5 mm from the margin) was removed with a sterile scalpel, and the root tissue was triturated in a sterile porcelain mortar in 10 mM phosphate buffer (pH 7.2). The root extracts were spread on tryptic soy agar, (TSA, Biolife) and then incubated at 28°C for 48

h. Resulting bacterial isolates were initially screened for antagonistic properties, and grouped by morphological characteristics (Haque *et al.*, 2016).

Antagonistic bacteria

Two halophilic antifungal bacterial strains from Tunisian Sebkhas, *Bacillus pumilus* isolate number M3-16, and *Halomonas elongata* L80, were used in this study. The morphological, physiological, and molecular characteristics of these strains were previously described and their 16S rDNA sequences have been deposited in the GenBank database under the accession numbers EU435355 for M3-16 and EU435356 for L80 (Essghaier *et al.*, 2009a).

Pathogen inoculum

The strain *Phytophthora citrophthora* (E1P1) was previously reported as the causal agent of clementine trunk and branch canker in Tunisia (Zouaoui *et al.*, 2016). The ITS1/ITS4 rDNA sequences of this strain have been deposited in the GenBank database under the accession number KX269827.

Dual culture tests for antagonism

Antagonistic bacteria were selected by the co-culture test. For each test, a mycelium plug (5 mm diam.) of *P. citrophthora* was transplanted at 2.5 cm from each side of the bacterial strip located in the middle of each Petri dish. These tests were carried out on potato dextrose agar, and the incubation was for 7 d at 28°C. The percentage inhibition of *P. citrophthora* was calculated according to the formula developed by Whipps (1987): I = (R 1 -R 2) / R1 * 100. R1 was the radial distance of the pathogen colony measured from the centre to the side of the Petri plate, and R2 was the radial distance of the colony from the centre to the bacterial strip (Sadfi-Zouaoui *et al.*, 2008a).

Morphological characterization of bacterium strains

The effective endophytic bacteria were phenotypically characterized according to morphological, physiological, and chemical analyses, based on the following tests: colony and cell morphology, motility, Gram staining, pigmentation, and catalase and oxidase tests (Trotel-Aziz *et al.*, 2008). In addition, biochemical features of the highly antagonistic bacteria were investigated using API 20NE and ZYM strips (bioMérieux), following the manufacturer's instructions.

Molecular characterization of strains

To extract the DNA, bacteria from 24 h tryptic soy broth (TSB, Biolife) cultures were processed by centrifugation at 8000 rpm for 10 min. The bacterial pellets were then suspended in 500 μL of TE solution (10 mM Tris-HCl, 1 mM EDTA, pH 8) and treated with 17 μL of lysozyme (30 mg mL^{-1}) for 30 min at 37°C. Six μL of proteinase K (20 mg mL^{-1}) and 40 μL of 10% SDS were then added to this suspension which was incubated at 37°C. The solution was homogenized thoroughly after the addition of 100 μL of 5M NaCl and 80 μL of CTAB / NaCl (10% 0.7M) and incubated at 65°C for 10 min. An equal volume of chloroform/isoamyl alcohol (24/1) solution was then added. The aqueous phase obtained after centrifugation for 20 min at 12000 rpm was transferred to a new tube, to which 2 volumes of isopropanol was added. After incubation at -20°C for 1 h, the preparation was centrifuged for 5 min at 13000 rpm. The precipitated DNA was washed with 70% ethanol, dried and then resuspended in 50 μL of TE solution (Sambrook *et al.*, 1989). The DNA obtained was stored at -20°C.

Amplification of the 16 rDNA was carried out using PCR, with the universal primers 27f (5′- AGA GTT TGA TYM TGG CTC AG-3′) and 1492r (5′-TAC CTT GTT AYG ACT T-3′) (Reysenbach *et al.*, 1992). The PCR profile was initial denaturation at 96°C for 3 min followed by 30 annealing cycles at 57°C for 30 s, extension at 72°C for 2 min, and denaturation at 96°C for 30 s, and an extension cycle of 72°C for 7 min. The PCR products were cleaned using the PCR purification kit (Promega) and sent for sequencing. The sequences were compared using the BLAST program (http://www.ncbi.nlm.nih.gov/BLAST/) for identification of the isolates. A neighbour-joining phylogenetic tree was produced by MEGA v7.0 (Kumar *et al.*, 2016).

Modes of action of antagonistic bacteria

Cellulase production

Endoglucanase activity of the strains was determined as described by Miller (1959), based on the amount of sugar released during the hydrolysis of cellulose. This is determined colorimetrically, using detection by dinitrosalicylic acid 3 (DNS). The protocol for each sample was as follows: 200 μL of supernatant was added to 200 μL of the substrate 1% carboxymethyl cellulose (CMC) and 200 µL of sodium citrate buffer (pH 4.8), then the reaction mixture was incubated at 50°C for 30 min. The reaction was stopped by the addition of 800µL of DNS. The mixture was then heated to 100°C for 15 min and then cooled in an ice bath. The nil controls each consisted of 200µL of buffer plus 800µL of DNS. Absorbance was determined at 540 nm, and converted to the concentration of reducing sugars using a standard curve created by different concentrations of glucose as standard. One unit of enzyme activity was defined as the amount of enzyme that released 1 µM of reducing sugars per 30 min (Assareh *et al.*, 2012).

Siderophore production

A medium containing chromium azurol S (CAS) was used to detect the secretion of siderophores (Schwyn and Neilands, 1987). The principle is that the culture medium is initially blue, due to the iron-hexadecyl-trimethyl-ammonium complex (iron/CAS/HDTMA complex), which turns orange when the iron is displaced by siderophore produced by the microorganism. This competition for iron will favour the siderophore forming orange ferri-siderophores. After 14 d of culture incubation at 28°C, red to orange halos form around bacterial colonies. Visual examination ensured the ability of microorganisms to produce siderophores. The strain colonies were classified for growth as + (0 to 10 mm), ++ (10 to 20 mm), or +++ (20 to 30 mm).

Phosphate solubilization

Phosphate solubilization was detected with the method of Islama *et al.* (2007), using the National Botanical Research Institute's phosphate (NBRIP) growth medium. This contained (per litre) 10 g glucose, 5 g $Ca_3(PO_4)_2$, 5g $MgCl_2·6H_2O$, 0.25 g $MgSO_4·7H_2O$, 0.2 g KCl, 0.1 g $(NH_4)_2SO_4$, and 15 g agar. pH 7.0. Inoculated plates were incubated at 24±2°C. The solubilization zones around bacterial colonies were assessed 7 d after inoculation, by subtracting the diameters of bacterial colonies from the total zone diameters (dp). The strains colonies were classified as + (5 mm <dp> 15 mm), ++ (15 <dp> 20 mm), or +++ (dp>20 mm).

Indole acetic acid (IAA) production

For each sample, 100 µL of fresh culture was inoculated into 10 mL of TSB medium and incubated at 28°C for 5 d. with agitation. The cultures were each centrifuged at 10000 rpm for 5 min, 1 mL of the supernatant was then removed, 2 mL of Salkowskire agent and 100 µL of orthophosphoric acid (10 mM) were added, and the cultures were then incubated for 30 min at 25°C. Observation of pink color indicated production of IAA (Ahmad *et al.*, 2008; Tarnawski *et al.*, 2006). Quantity of IAA was determined by absorbance at 530 nm, converted to concentration of IAA (μgmL^{-1}) using a standard IAA curve.

PCR detection of antibiotic biosynthesis genes

The five selected bacterial strains were screened for production of lipopeptide antibiotics, using specific primers that amplify genes from fengycin (FENDF, GGCCCGTTCTCTAAATCCAT; FENDR, GTCATGCTGACGAGAGCAAA), bacillomycin (BMYBF, GAATCCCGTTGTTCTCCAAA; BMYBR, GCGGGTATTGAATGCTTGTT), bacilysin (BACF, CAGCTCATGGGAATGCTTTT; BACR, CTCGGTCCTGAAGGGACAAG), Surfactin (SRFAF, TCGGGACAGGAAGACATCAT; SRFAR, CCACTCAAACGGATAATCCTGA) and iturin (ITUD1F, GATGCGATCTCCTTGGATGT; ITUD1R, ATCGTCATGTGCTGCTTGAG). The PCR amplifications were performed in 15 µL reaction mixtures each containing 0.06 µL (5 U μL^{-1}) *Taq* DNA polymerase, 1 µL (20 ng μL^{-1}) of DNA template, 1.5 µL of 10× PCR buffer, 0.45 µL (50 mmol L^{-1}) of $MgCl_2$, 1 µL (10 mmol L^{-1}) of dNTPs, 0.3 µL (10 µmol L^{-1}) of each primer, 0.75 µL (10 mg L^{-1}) of bovine serum albumin (BSA), and 9.64 µLof filter-sterilised Milli-Q water (Mora *et al.*, 2011; Khabbaz *et al.*, 2015).

Biocontrol screening in planta

Pathogen inoculum

Phytophthora citrophthora (strain E1P1) was grown on V8 juice agar (JV8A) in Petri plates for sporangium production. The strain was incubated for 2 d in the dark at 24°C. The Petri plates were then filled with the Chen-Zentmyer's Salt Solution (containing, per L of distilled water: 1.64 g (Ca $(NO_3)_2$, 0.05 g KNO_3, 0.48 g $MgSO_4$, and 1 mL of chelate iron solution (13.05 g L^{-1}EDTA; 7.5 g L^{-1} KOH; 24.9 g $L^{-1}FeSO_4$; pH.7), and incubated under continuous fluorescent light at 24°C. Sporangia were produced within 1 to 2 days.

Bacterium strains

Four bacteria were selected for *in-planta* tests. Two halophilic strains (M3-16, L80) and two endophytic bacteria (S24 and S31) were used. They were grown for 48 h on TSA, supplemented with 5% NaCl for the two halophilic strains. After 48 h, colonies were scraped from the agar surfaces in Petri dishes, and were diluted in sterilized saline solution (1% NaCl). Bacterium concentrations were determined by dilution plating on TSA, and adjusted to 10^8 colony forming units (CFU) mL^{-1} for inoculations.

Experimental design

One hundred 2-year-old clementine plants ('Hernandina'), grafted on 'Carrizo' citrange rootstock, were grown in a partially controlled greenhouse at the Tunisian National Institute of Agricultural Research. The experiment was arranged in a completely randomized design. Five treatments were applied, each on 20 plants with two inoculation points for each plant. These were: untreated plant inoculated with *P. citrophthora* (E1P1), d, (ii) plant inoculated with antagonist *B. amyloliquefaciens* (S24) and *P. citrophthora*, (iii) plant inoculated with the antagonist *P. polymyxa* (S31) and *P. citrophthora*, (iv) plant inoculated with the halophilic antagonist *H. elongata* (L80) and *P. citrophthora*, and (v) plant inoculated with the halophilic antagonist *B. pumilus* (M3-16) and *P. citrophthora*.

Co-inoculations

For each plant, two cuts of length 5 to 10 mm were made with a scalpelon the bark of the scion, at 15 to 20 cm from each other. Each wound was immediately drop-inoculated with 100 µL of suspension of the bacterial strains (10^8 CFU mL^{-1}). The liquid was allowed to dry for 20–40 min at air temperature, and the incision was filled with a JV8A plug containing sporangia of *P. citrophthora* E1P1 placed in direct contact with the stem cambium. The wound was then wrapped with foil, moistened, and sealed with a strip of paraffin film to prevent desiccation.

Pathogenicity assessments and data analyses

The results were assessed 10 d after inoculation. Bark of the inoculated area on each plant was scraped with a scalpel, and margins of lesions were measured with a ruler. Statistical analyses for these data were performed using R statistical software version 3.4.0 (R Core Team, 2017). Normal distributions of the data were verified by the Shapiro-Wilk test before being subject to the non-parametric Kruskal-Wallis tests. Additionally, average reduction in lesion size [% reduction = 100 − (average lesion size for treatment × 100)/average lesion size for experimental control)] was calculated for the treatments that displayed statistically significant differences from the non-treated controls (Alvarez *et al.*, 2008b).

RESULTS

Isolation, characterization and identification of antagonistic bacteria

Among the 69 bacteria isolated from citrus roots, five isolates (S19, S24, S31, S40 and S54) were shown to be the most effective for inhibiting growth of *P. citrophthora* in the dual culture tests (Figure 1). Growth inhibition percentage (GI%) ranged from 67% for S31 to 74% for isolate S19. The GI% for both S24 and S54 was 69%, and for S40 was 62% (Figure 2).

Isolates S19, S24, S31, and S54 were identified as belonging to *Bacillus* and S40 to *Pseudomonas* on the basis of several phenotypic features (Table 1). The colony morphology of the *Bacillus* isolates was circular with cream pigmentation, the isolates had rod cell shape and were Gram positive, while the *Pseudomonas* isolate had pale yellow colonies, coccoid cell shape and was Gram negative (Table 1). According to API 20NE strip (bioMérieux) tests, positive reactions were observed among all the isolates for nitrate reduction, esculin and gelatin hydrolyses. Negative reactions were observed for L-tryptophane, D-glucose, L-arginine, urease, capric acid, adipic acid and phenylacetic acid. All other reactions were different among the isolates (Table 1). For

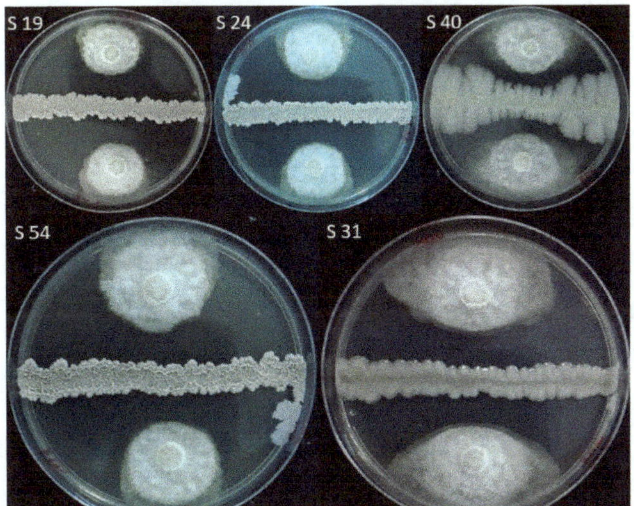

Figure 1. Dual culture showing antagonistic activities of the endophytic bacteria (strains S19, S24, S31, S40 and S54) toward *Phytophthora citrophthora*.

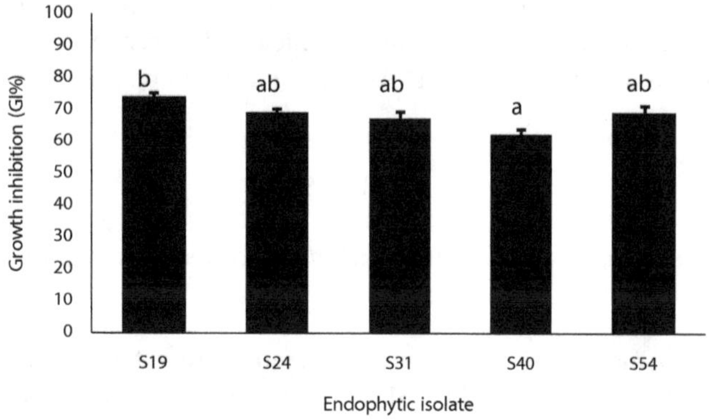

Figure 2. Growth inhibition rates of the endophytic bacteria toward *Phytophthora citrophthora*. Growth inhibition values followed by different letters are significantly different (Tukey Test, $P< 0.05$, n = 6, (± standard errors)).

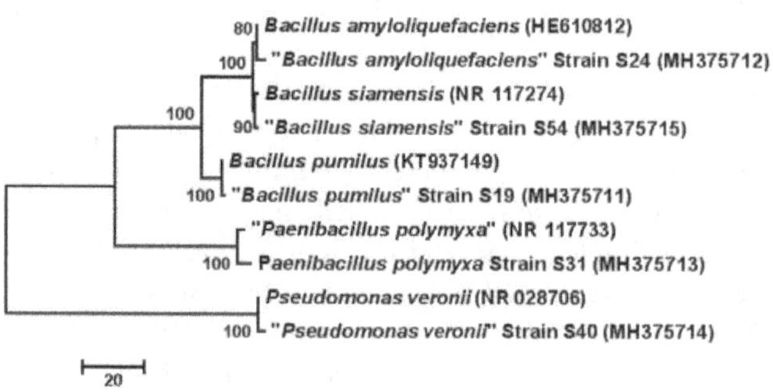

Figure 3. Neighbour-joining tree based on 16S rDNA sequences showing the phylogenetic relationship between strains S19, S24, S31, S40 and S54. Bootstrap values (expressed as percentages of 1000 replications) >70% are given at the nodes.

API ZYM strip tests, positive reactions were observed for alcaline phosphatase, esterase (C4), esterase lipase (C8), acid phosphatase and naphthol-AS-BI-phosphohydrolase. Positive reactions for α-chymotrypsin were only observed for the isolates S31 and S40, and positive reaction for α-galactosidase occurred only for the isolate S19. All other reactions were negative (Table 1).

BLAST analyses of the 16SrDNA sequences showed that there was 99% homology of isolate S19 with *Bacillus pumilus* L14 (KT937149), isolate S24 with *Bacillus amyloliquefaciens* BD18C2-B16 (HE610812), isolate S31 with *Paenibacillus polymyxa* DSM13, S40 with *Pseudomonas veronii* CIP104663 (NR_028706), and isolate S54 with *Bacillus siamensis*PD-A10 (NR_117274). The 16S rDNA sequences were deposited in the GenBank database, under the accession numbers MH375711 for S19, MH375712 for S24, MH375713 for S31, MH375714 for S40 and MH375715 for S54 (Figure 3).

Table 1. Phenotypic characterization of selected antagonistic bacteria[a].

Characteristic	S19	S24	S31	S40	S54
Colony morphology	Circular	Circular	Circular	Circular	Circular
Cellshape	Rod	Rod	Rod	coccoid	Rod
Mobility	+	+	+	+	+
Pigmentation	Cream	Cream	Cream	Cream	Cream
Gram strain	+	+	+	-	+
Catalase	+	+	+	+	+
Oxidase	+	-	-	+	+
Nitrate reduction	+	+	+	+	+
L-tryptophane	-	-	-	-	-
D-glucose	-	-	-	-	-
L-arginine	-	-	-	-	-
Urease	-	-	-	-	-
Esculinhydrolysis	+	+	+	+	+
Gelatinhydrolysis	+	+	+	+	+
4-nitrophenyl-βD-galactopyranoside	-	-	+	-	-
L-arabinose	+	+	+	+	-
D-mannose	+	+	-	+	+
D-mannitol	+	+	-	+	+
N-acetyl-glucosamine	+	+	-	-	+
D-maltose	+	+	-	+	+
Potassium gluconate	+	-	+	+	+
Capricacid	-	-	-	-	-
Adipicacid	-	-	-	-	-
Malicacid	+	+	-	+	+
Trisodium	+	-	-	+	+
Phenylacetic	-	-	-	-	-
Alcaline phosphatase	+	+	+	+	+
Esterase (C4)	+	+	+	+	+
Lipase (C8)	+	+	+	+	+
Lipase (C 14)	-	-	-	-	-
Leucine arylamidase	-	-	-	-	-
Valine arylamidase	-	-	-	-	-
Cystine arylamidase	-	-	-	-	-
Trypsin	-	-	-	-	-
α-chymotrypsin	-	-	+	+	-
Acid phosphatase	+	+	+	+	+
Naphthol-AS-BI-phosphohydrolase	+	+	+	+	+
α-galactosidase	+	-	-	-	-
ß-galactosidase	-	-	-	-	-
ß-glucuronidase	-	-	-	-	-
α-glucosidase	-	-	-	-	-
ß-glucosidase	-	-	-	-	-
N-acetyl-ß-glucosaminidase	-	-	-	-	-
α-mannosidase	-	-	-	-	-
α-fucosidase	-	-	-	-	-

[a]S19, S24, S31 and S54: Strains of *Bacillus* sp.; S40: Strain of *Pseudomonas* sp.
(+) Positive reaction ; (-) negative reaction.

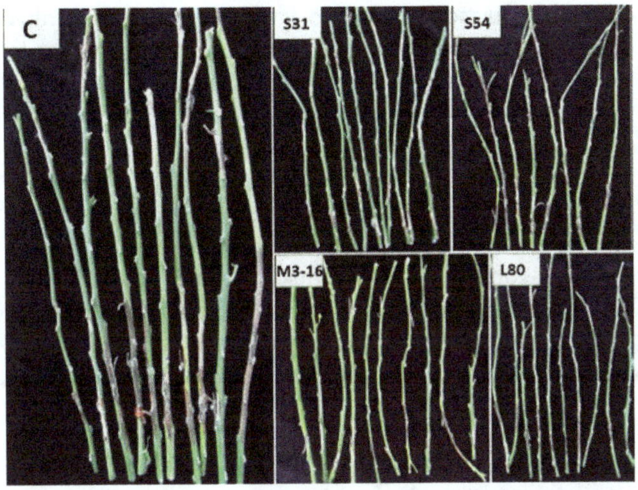

Figure 4. Suppression of branch canker on clementine mandarin for the trial conducted under partly controlled greenhouse conditions. Host branches were cut and leaves removed for improved view of the symptoms. C: Plants inoculated with *Phytophthora citrophthora* (Pc) without bacterial treatment. S31 and S24 indicate plants treated with endophytic bacteria, and M3-16 and L80 indicate those treated with halophilic bacteria.

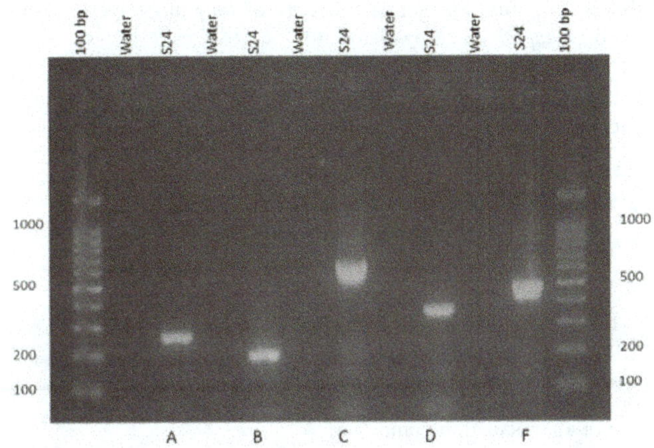

Figure 5. PCR amplification products of antibiotic biosynthetic genes in *Bacillus amyloliquefaciens* (S24), with primers (A) fengycin (FENDF/R) 269 bp; (B) surfactin (SRFAF/R) 201 bp; (C) iturin A (ITUD1F/R) 647 bp; (D) bacillomycin (BMYBF/R) 370 bp; and (E) bacilysin (BCAF/R) 498 bp

Modes of action of antagonistic bacteria

Production of antibiotic genes

The five lipopeptides genes Fengycin, Surfactin, Iturin A, Bacillomycin and Bacilysin were detected in all the *Bacillus* strains (*B. pumilus*S19 and S24, *B. amyloliquefaciens*, and *B. siamensis* (S54). However, the Bacillomycin gene was not detected in the *P. polymyxa* strain (S31) (Table 2; Figure 5).

Cellulase production

The *P. veronii* strain (S40) was the most efficient for production of cellulase (2.64 U mL^{-1}), followed by the three strains of *Bacillus* with S19 producing 1.62 U mL^{-1}, S24 producing 2.13 U mL^{-1}and S54 producing 2.51 U mL^{-1}. Cellulase production from *Paenibacillus polymyxa* strain S31 was 1.21 UmL^{-1} (Table 2).

Siderophore production

All the strains produced siderophores. Strain S24 produced the smallest halo (+) from 1 to 10 mm, and the strains S31 and S54 produced the largest halo zones (+++) of 20 to 30 mm (Table 2).

Table 2. Modes of action of selected antagonistic bacteria in dual culture assays with *Phytophthora citrophthora.*

Strain	Hydrolytic enzyme[a]					PGPB traits[b]			Antibiotic molecular screening[a]				
	Lip	Amy	Ure	Prot	Cell (UmL^{-1})[c]	Sid	Phos	Aux	Fen	Bcin	Bsin	Surf	Itu A
S19	-	+	-	-	1.62±0.12ab	+	++	-	+	+	+	+	+
S24	+	+	-	+	2.13±0.05ab	+	++	-	+	+	+	+	+
S31	-	+	-	+	1.21±0.13a	+++	+	+	+	-	+	+	+
S40	-	+	-	-	2.64±0.06b	+	+++	-	-	-	-	-	-
S54	+	+	-	+	2.51±0.02ab	+++	++	-	+	+	+	+	+

[a] (+) Positive reaction (-) and negative reactions for: Lip , Lipase; Amy, Amylase; Ure, Urease; Prot, protease; Cell, cellulase; Sid, Siderophore production; Phos, phosphate solubilization; Aux, Auxin production; Fn, Fengycin; Bcin, Bacillomycin; Bsin, Bacilysin; Surf,: Surfactin; Itu A, Iturin A.
[b] PGPB: Plant growth-promoting bacteria.+, Growth ranging from 1 to 10 mm; ++, growth ranging from 10 to 20 mm; +++, growth ranging from 20 to 30 mm.
[c]Mean cellulase activity (U mL^{-1}) followed by different letters are significantly different ($P < 0.05$, n = 6; Tukey tests) (± standard errors).

Table 3. *In planta* antagonistic activity of four bioactive bacteria toward *Phytophthora citrophthora.*

Treatments with bioactive bacteria[a]	Mean lesion length(cm)[b]	Average reduction in lesion size (%)[c]
Control	4.9±0.47a[d]	
S24	3.2±0.42bc	35
S31	2.9±0.19b	41
L80	2.4±0.20b	51
M3-16	3.1±0.22c	37

[a] Treatments by bioactive bacteria applied at 10^8 UFC mL^{-1}.
[b] Mean lesion length 10 d after inoculation with *P. citrophthora*. Mean of 40 inoculation points.
[c] Average reduction in lesion size [% reduction = 100 – (Av. lesion size treatment × 100)/Av. lesion size control] was calculated for the treatments that displayed statistically significant differences from the non treated controls.
[d] Means followed by the same letter are not significantly different ($P < 0.05$, n = 40) according to Dunn tests.

Phosphate solubilization

The three *Bacillus* strains produced clear phosphate solubilized zones, ranging from 15 to 20 mm after 9 d. The *Ps. veronii* strain S40 produced the largest zone of more than 20 mm. *Paenibacillus polymyxa* (S31) was the weakest strain for this character, with zones less than 15 mm (Table 2).

Auxin production

Only the strain *Paenibacillus polymyxa* S31 produced IAA (18.5 μg mL^{-1},Table 2).

Suppression of branch canker on clementine mandarin plants

The endophytic and halophilic bacterium strains all reduced ($P<0.0001$; Kuskalwallis tests) the size of necrotic lesions caused by *P. citrophthora* on clementine plants 10 dafter inoculations, compared to the non-treated plants (Figure 4). The strain L80 of *Halomonas elongata* was the most effective for lesion reduction, giving 51% reduction of mean lesion size (Table 3). The other bacteria reduced lesion sizes, proportional reductions not exceeding 41%.

DISCUSSION

The Tunisian citrus industry has flourished in recent years, and this has been partly due to increased clementine production. However, trunk and branch canker caused by *P. citrophthora* represents a serious threat to productive clementine cultivation. Biological control of this new disease has not been previously investigated. In order to establish an eco-friendly strategy for management of *Phytophthora*, 69 isolates of bacteria from citrus roots were identified and screened for antagonistic activity against *P. citrophthora*. Five strains had strong *in vitro* antagonistic effects against *P. citrophthora*. Molecular characterization allowed us to identify these strains, three of which were *Bacillus*, one was *Paenibacillus*, and the fifth was *Pseudomonas*. Köberl *et al.* (2011) found that *Bacillus* and *Paenibacillus* represent 96% of the antagonists towards phytopathogens in agriculture. Size of the inhibition zone in dual culture was the first criterion used to select the antagonist candidates. The use of PDA as a rich nutrient medium could exclude competition as a mode of action. No physical contact between the isolates and the pathogen in the dual culture tests could be related to the production of antifungal metabolites that inhibit mycelium growth (Montealegre *et al.*, 2003; Lee *et al.*, 2008). The growth inhibition proportion (GI%) of the selected bacteria varied from 74% for *B. pumilus* S19 to 62% for *Ps. veronii* S40. Among those most antagonistic bacteria, *B. amyloliquefaciens* and *P. polymyxa* were selected using a preliminary *in vivo* trial (results not shown) to study their ability to suppress branch canker under greenhouse conditions. In addition, two halophilic bacteria isolated from Tunisian sebkhas were selected. These were *Bacillus pumilus* M3-16 and *Halomonas elongata* L80. These halophilic bacteria produced extracellular antifungal enzymes such as chitinase, glucanase, and protease, and were characterized by tolerance and stability in the presence of extreme conditions (pH, temperature and salts) compared to other antifungal enzymes reported in the literature (Essghaier *et al.*, 2009b, 2010, 2012). Other reports have shown that halophilic bacteria with inability to produce inhibition zones in solid medium co-culture tests, have also demonstrated promising disease inhibition in *in vivo* tests. *Bacillus thuringiensis* strains unable to form inhibition zones in dual cultures were very effective *in vivo* on potato tubers (Sadfi *et al.*, 2001). *Bacillus subtilis* effectively inhibited grey mold on wounded tomatoes but was ineffective *in vitro* on PDA medium (Sadfi-Zouaoui *et al.*, 2008b).

The endophytic bacteria isolated from citrus roots demonstrated abilities to produce cell wall-degrading enzymes such as endoglucanase and protease, which are important for breaking through plant cell walls. Reinhold-Hureket *al.* (2006) confirmed that endoglucanases are essential for *Azoarcus* sp. to colonize rice roots. Many other studies have confirmed that bacterial endophytes are mainly recruited from soil via rhizosphere and root systems, to reach xylem and phloem vessels and colonize plants intra- and extra-cellularly (Liu *et al.*, 2017). Little is known about endophyte colonization of citrus tissues. Lacava *et al.*(2007) studied the colonization of *Citrus sinensis* by endophytic bacteria, choosing the model of a *K. pneumoniae* strain labelled with GFP genes. They concluded that the endophytic bacterium strain colonized xylem vessels of *C. roseus* branches and roots.

The greenhouse trial of the present study revealed that the endophytic *B. amyloliquefaciens* and the halophilic *B. pumilus* produced similar results, reducing necroses in citrus plants by, respectively, 35 and 37%. Several other studies have demonstrated the ability of *B. amyloliquefaciens* to control *in planta* infections by *Phytophthora* spp. (Chung *et al.*, 2005; Anandhakumar *et al.*, 2008; Li *et al.*, 2014; Zhang *et al.*, 2016). Anandhakumar *et al.* (2008) showed that *B. amyloliquefaciens* reduced red core and crown rot diseases of strawberry in a greenhouse trial, exhibiting a similar level of disease control (up to 59%) as the chemical fungicide Aliette®. The second most effective bacterium in the present study was *P. polymyxa,* which reduced necrosis by 41%. Recent publications have also indicated the potential of *P. polymyxa* as a bio-pesticide (Grady *et al.*, 2016; Weselowski*et al.*, 2016; Luo *et al.*, 2018). However, it is the ability of this bacterium to inhibit *Phytophthora* diseases of citrus has not been previously explored.

The greatest *in vitro* inhibition of *P. citrophthora* was obtained with *H. elongata*, with 51% reduction of host necrosis. This is the first record of potential for *H. elongata*as a biocontrol agent against *Phytophthora* sp. In order to study the PGPB traits of this bacterium, direct and indirect mechanisms to promote citrus growth were evaluated. Direct mechanisms include increased availability of plant nutrients (biofertilization), from phosphate solubilization and azote fixation. According to the biochemical test 'Gallerie API 20 NE', all the five strains were able to reduce atmospheric azote (N_2) to nitrite (NO_2^-) and then to nitrate (NO_3^-). In addition, the *P. polymyxa* S31 strain could synthesize IAA enhances plant cell growth and proliferation (Grady *et al.*, 2016). The indirect mechanism was demonstrated by the abilities of all the candidate bioactive bacteria to produce siderophores, lytic enzymes such as lipase, protease and cellulase, and to possess most of the lipopeptide genes. Interactions between lipopeptides can become synergistic to enhance their respective activities (Romero *et al.*, 2007; Malfanova *et al.*, 2012; Li *et al.*, 2014). Previous studies have highlighted the role of *Bacillus* lipopeptides in biological control of plant diseases (Ongena and Jaques, 2008; Mora *et al.*, 2015; Paraszkiewicz *et al.*, 2017). Beside their involvement in antifungal and antimicrobial activities, some bacteria also facilitate root colonization and modulate plant immunity (Ongena and Jaques, 2008).

CONCLUSIONS

This study has demonstrated potential to overcome the new serious disease in clementine orchards by adopting an ecological biocontrol approach. After screening endophytic bacteria from citrus roots, five bacteria displayed strong antagonistic traits as *in vitro* pathogen inhibition rates, metabolization of different carbon sources, PGPB traits, and antibiotic production. In the greenhouse trial, the strains L80 of *H. elongata* and S31 of *P. polymyxa* showed the greatest potential as candidates for control the disease. This is the first biological control study of clementine branch canker. The bacterial strains S31 and L80 offer a good basis for developing efficient biofungicides, and should be further investigated under field conditions.

ACKNOWLEDGEMENTS

This research was supported by funds from the PRF Project (No. PRF14CTA-BIOTEC02) and by the Ministry of Higher Education and Scientific Research of Tunisia (LR16ES05).

LITERATURE CITED

Ahmad F., Ahmad I., Khan M.S., 2008. Screening of free-living rhizospheric bacteria for their multiple plant growth promoting activities. *Microbiological Research* 163: 173–181.

Alvarez L.A., Vicent A., De la Roca E., Bascón J., Abad-Campos P., Armengol J., García-Jiménez J.,2008a. Branch cankers on citrus trees in Spain caused by *Phytophthora citrophthora. Plant Pathology* 57: 84–91.

Alvarez L.A., Vicent A., Solar J.M., De la Roca E., Bascon J., García-Jiménez J., 2008b. Comparison of application methods of systemic fungicides to suppress branch cankers in clementine trees caused by *Phytophthora citrophthora. Plant Disease* 92(9): 1357–1363.

Alvarez L.A., Gramaje D., Abad-Campos P., García-Jiménez J., 2009. Role of the *Helix aspersa* snail as a vector of *Phytophthora citrophthora* causing branch cankers on clementine trees in Spain. *Plant Pathology* 58: 956–963.

Anandhakumar J., Zeller W., 2008. Biological control of red stele (*Phytophthora fragariae* var. *fragariae*) and crown rot (*P. cactorum*) disease of strawberry with rhizobacteria. *Journal of Plant Diseases and Protection* 115: 49–56.

Assareh R., Shahbani Zahiri H., Akbari Noghabi K., Aminzadeh S., Bakhshi Khaniki G., 2012. Characterization of the newly isolated *Geobacillus* sp. T1, the efficient cellulase-producer on untreated barley and wheat straws. *Bioresource Technology* 120: 99–105.

Biessy A., Filion M., 2018. Phenazines in plant-beneficial *Pseudomonas* spp.: biosynthesis, regulation, function and genomics. *Environmental Microbiology* 20: 3905–3917.

Bove J.M., 2006. Huanglongbing: A destructive, newly-emerging, century-old disease of citrus. *Journal of Plant Pathology* 88: 7–37.

Cacciola S.O., Magnano di San Lio G., 2008. Management of citrus diseases caused by *Phytophthora* spp. In:*Integrated Management of Diseases Caused by Fungi, Phytoplasma and Bacteria* (A. Ciancio and K. G. Mukerji ed.), Springer Netherlands, Dordrecht, Netherlands, 61–84.

Cambra M., Gorris M.T., Marroquín C., Román M.P., Olmos A., Martínez M.C., de Mendoza A.H., López A., Navarro L., 2000. Incidence and epidemiology of Citrus tristeza virus in the Valencian Community of Spain. *Virus Research* 71: 85–95.

Chung S., Kim S.D., 2005. Biological control of phytopathogenic fungi by *Bacillus amyloliquefaciens* 7079; Suppression rates are better than popular chemical fungicides. *Journal of Microbiology and Biotechnology* 15: 1011–1021.

Cohen S., Allasia V., Venard P., Notter S., Verniere C., Panabieres F., 2003. Intraspecific variation in *Phytophthora citrophthora* from citrus trees in Eastern Corsica. *European Journal of Plant Pathology* 109(8): 791–805.

Conacher H.B.S., Mes J., 1993. Assessment of human exposure to chemical contaminants in foods. *Food Additives & Contaminants* 10: 5–15.

Erwin D.C., Ribeiro O.K., 1996. *Phytophthora Diseases Worldwide.American Phytopathological Society*, St. Paul, MN, 562 pp.

Espinosa-Urgel M., 2004. Plant-associated *Pseudomonas* populations: molecular biology, DNA dynamics, and gene transfer. *Plasmid* 52: 139–150.

Essghaier B., Bejji M., Jijakli H., Boudabous A., Sadfi-Zouaoui N., 2009b. High salt-tolerant protease from a potential biocontrol agent *Bacillus pumilus* M3-16. *Annals of Microbiology* 59: 553.

Essghaier B., Fardeau M.L., Cayol J.L., Hajlaoui M.R., Boudabous A., Jijakli H., Sadfi-Zouaoui N., 2009a. Biological control of grey mould in strawberry fruits by halophilic bacteria. *Journal of Applied Microbiology* 106: 833–846.

Essghaier B., Hedi A., Bejji M., Jijakli H., Boudabous A., Sadfi-Zouaoui N., 2012. Characterization of a novel chitinase from a moderately halophilic bacterium, *Virgibacillus marismortui* strain M3-23. *Annals of Microbiology* 62: 835–841.

Essghaier B., Rouaissi M., Boudabous A., Jijakli H., Sadfi-Zouaoui N., 2010. Production and partial characterization of chitinase from a halotolerant *Planococcus rifitoensis* strain M2-26. *World Journal of Microbiology and Biotechnology* 26: 977–984.

FAO, 2013. The Food and Agriculture Organization Corporate Statistical Database.Rome: Food and agriculture Organisation of United Nations (FAO). Available at: http://www.fao.org/faostat/en/#data/QC. Accessed June 9, 2017.

GIFruits, 2018. Export Agrumes 2018(T). Available at:http://gifruits.com/?p=2262&lang=fr

Gottwald T.R., Graham J.H., Schubert T.S., 2002. Citrus canker: the pathogen and its impact. Plant Health Progress. DOI: 10.1094/PHP-2002-0812-01-RV. http://www.plantmanagementwork.

Grady E.N., MacDonald J., Liu L., Richman A., Yuan Z.C., 2016. Current knowledge and perspectives of *Paenibacillus*: a review. *Microbial Cell Factories* 15: 203.

Graham J.H., Bright D.B., McCoy C.W., 2003. *Phytophthora*-Diaprepes Weevil Complex: *Phytophthora* spp. relationship with citrus rootstocks. *Plant Disease* 87: 85–90.

Graham J.H., Timmer L.W., Drouillard D.L.,Peever T.L., 1998. Characterization of *Phytophthora* spp. causing outbreaks of citrus brown rot in Florida. *Phytopathology* 88: 724–729.

Haque M.A., Yun H.D.,Cho K.M., 2016. Diversity of indigenous endophytic bacteria associated with the roots of Chinese cabbage (*Brassica campestris* L.) cultivars and their antagonism towards pathogens. *Journal of Microbiology* 54: 353–363.

Ippolito A., Nigro F., Lima G., 1997. Influence of the scion on the susceptibility of sour orange rootstock to *Phytophthora* gummosis and root rot. *Phytopathologia Mediterranea* 36: 81–86.

Islama M.T., Deora A., Hashidoko Y., Rahman A., Ito T., Tahara S., 2007. Isolation and identification of potential phosphate solubilizing bacteria from the rhizoplane of *Oryza sativa* L. cv. BR29 of Bangladesh. *Zeitschrift Fur Naturforschung C* 62: 103–110.

Jepson P.C., Guzy M., Blaustein K., Sow M., Sarr M., Mineau P., Kegley S., 2014. Measuring pesticide ecological and health risks in West African agriculture to establish an enabling environment for sustainable intensification. *Philosophical Transactions of the Royal Society B Biological Sciences* 369: 20130491. DOI: 10.1098/rstb.2013.0491

Khabbaz S.E., Zhang L., Cáceres L.A., Sumarah M., Wang A.,Abbasi P.A., 2015. Characterisation of antagonistic *Bacillus* and *Pseudomonas* strains for biocontrol potential and suppression of damping-off and root rot diseases. *Annals of Applied Biology* 166: 456–471.

Khanchouch K., Pane A., Chriki A., Cacciola S.O., 2017. Major and emerging fungal diseases of citrus in the Mediterranean Region. In: *Citrus Pathology. InTechOpen*. DOI: 10.5772/66943.

Kloepper J.W., Ryu C.M., 2006. Bacterial endophytes as elicitors of induced systemic resistance. In: *Microbial Root Endophytes* (B.J.E. Schulz, C.J.C. Boyle, T.N. Sieber, ed.), Springer Berlin Heidelberg, Berlin, Germany: 33–52.

Köberl M., Müller H., Ramadan E.M., Berg G., 2011. Desert farming benefits from microbial potential in arid soils and promotes diversity and plant health. *Plos One* 6: e24452.

KumarS., Stecher G., Tamura K., 2016. MEGA7: Molecular evolutionary genetics analysis version 7.0 for bigger datasets. *Molecular Biology and Evolution* 33: 1870–1874.

Lacava P. T., Araujo W. L., Azevedo J. L., 2007. Evaluation of endophytic colonization of *Citrus sinensis* and *Catharanthus roseus* seedlings by endophytic bacteria. *Journal of Microbiology* 45: 11–14.

Laville E., 1984. La gommose à *Phytophthora* des clémentiniers en Corse. In *Arboriculture Fruitière*: 32–35.

Lee K.J., Kamala-Kannan S., Sub H.S., Seong C.K.,Lee G.W., 2008. Biological control of *Phytophthora* blight in red pepper (*Capsicum annuum* L.) using *Bacillus subtilis. World Journal of Microbiology and Biotechnology* 24: 1139–1145.

Li B., Li Q., Xu Z., Zhang N., Shen Q., Zhang R., 2014. Responses of beneficial *Bacillus amyloliquefaciens* SQR9 to different soilborne fungal pathogens through the alteration of antifungal compounds production. *Frontiers in Microbiology* 5: 636.

Liu H., Carvalhais L.C., Crawford M., Singh E., Dennis P.G., Pieterse C.M.J., Schenk P.M., 2017. Inner plant values: diversity, colonization and benefits from endophytic bacteria. *Frontiers in Microbiology* 8: 2552.

Lugtenberg B.J.J., Chin-A-Woeng T.F.C., Bloemberg G.V., 2002. Microbe–plant interactions: principles and mechanisms. *Antonie van Leeuwenhoek* 81: 373–383.

Luo Y., Cheng Y., Yi J., Zhang Z., Luo Q., Zhang D., Li Y., 2018. Complete genome sequence of industrial biocontrol strain *Paenibacillus polymyxa* HY96-2 and further analysis of its biocontrol mechanism. *Frontiers in Microbiology* 9: 1520.

Mahjbi, A., Oueslati A., Baraket G., Salhi-Hannachi A., Zehdi Azouzi S., 2016. Assessment of genetic diversity of Tunisian orange, *Citrus sinensis* (L.) Osbeck using microsatellite (SSR) markers. *Genetics and Molecular Research* 15(2). DOI: 10.4238/gmr.15026564

Malfanova N., Franzil L., Lugtenberg B., Chebotar V., Ongena M., 2012. Cyclic lipopeptide profile of the plant-beneficial endophytic bacterium *Bacillus subtilis* HC8. *Archives of Microbiology* 194: 893–899.

Matheron M.E., Porchas M., Matejka J.C., 1997. Distribution and seasonal population dynamics of *Phytophthora citrophthora* and *P. parasitica* in Arizona citrus orchards and effect of fungicides on tree health. *Plant Disease* 81: 1384–1390.

Miller G.L., 1959. Use of dinitrosalicylic acid reagent for determination of reducing sugar. *Analytical Chemistry* 31: 426–428.

Montealegre R.J., Rodrigo R., Luz María P., Rodrigo H., Polyana S., Ximena A.B., 2003. Selection of bioantagonistic bacteria to be used in biological control of *Rhizoctonia solani* in tomato. *Electronic Journal of Biotechnology* 6(2). DOI: 10.4067/S0717-34582003000200006

Mora I., Cabrefiga J., Montesinos E., 2011. Antimicrobial peptide genes in *Bacillus* strains from plant environments. *International Microbiology* 14: 213–223.

Mora I., Cabrefiga J., Montesinos E., 2015. Cyclic lipopeptide biosynthetic genes and products, and inhibitory activity of plant-associated *Bacillus* against phytopathogenic bacteria. *Plos One* 10:e0127738.

Ongena M., Jacques P., 2008. *Bacillus* lipopeptides: versatile weapons for plant disease biocontrol. *Trends in Microbiology* 16: 115–125.

Paraszkiewicz K., Bernat P., Siewiera P., Moryl M., Paszt L.S., Trzcinski P., Jalowiecki L., Plaza G., 2017. Agricultural potential of rhizospheric *Bacillus subtilis* strains exhibiting varied efficiency of surfactin production. *Scientia Horticulturae* 225: 802–809.

Radhakrishnan R., Hashem A., Abd Allah E.F., 2017. *Bacillus*: A biological tool for crop improvement through bio-molecular changes in adverse environments. *Frontiers in Physiology* 8: 667.

Reinhold-Hurek B., Maes T., Gemmer S., Van Montagu M., Hurek T., 2006. An endoglucanase is involved in infection of rice roots by the not-cellulose-metabolizing endophyte Azoarcus sp. strain BH72. *Molecular Plant-Microbe Interaction* 19: 181–188.

Reysenbach A.L., Giver L.J., Wickham G.S., Pace N.R.,

1992. Differential amplification of rRNA genes by polymerase chain reaction. *Applied and Environmental Microbiology* 58: 3417–3418.

R Core Team, 2017. *R: A Language and Environment for Statistical Computing*. R Foundation for Statistical Computing, Vienna, Austria. URL https://www.R-project.org/.

Ryan R.P., Germaine K., Franks A., Ryan D.J., Dowling D.N., 2008. Bacterial endophytes: recent developments and applications. *FEMS Microbiology Letters* 278: 1–9.

Romero D., de Vicente A., Rakotoaly R.H., Dufour S.E., Veening J.W., Arrebola E., Cazorla F.M., Kuipers O.P., Paquot M., Perez-Garcia A., 2007. The iturin and fengycin families of lipopeptides are key factors in antagonism of *Bacillus subtilis* toward *Podosphaera fusca*. *Molecular Plant Microbe Interactions* 20: 430–440.

Sadfi N., Chérif M., Fliss I., Boudabbous A., Antoun H., 2001. Evaluation of bacterial isolates from salty soils and *Bacillus thuringiensis* strains for the biocontrol of *Fusarium* dry rot of potato tubers. *Journal of Plant Pathology* 83: 101–118.

Sadfi-Zouaoui N., Hannachi I., Andurand D., Essghaier B., Boudabous A., Nicot P., 2008b. Biological control of *Botrytis cinerea* on stem wounds with moderately halophilic bacteria. *World Journal of Microbiology and Biotechnology* 24: 2871–2877.

Sadfi-Zouaoui N., Essghaier B., Hajlaoui M.R., Fardeau M.L., Cayaol J.L., Ollivier B., Boudabous A., 2008a. Ability of moderately halophilic bacteria to control grey mould disease on tomato fruits. *Journal of Phytopathology* 156: 42–52.

Sambrook J., Fritsch E.F., Maniatis T., 1989. *Molecular Cloning: a Laboratory Manual*. Cold Spring Harbor Laboratory Press, Cold Spring Harbor, NY, USA.

Schutte G.C., Botha W.J., 2010. *Phytophthora citrophthora* trunk and branch canker on Clementine mandarins in the Western Cape province of South Africa. *South African Journal of Plant and Soil* 27: 215–220.

Schwyn B., Neilands J.B., 1987. Universal chemical assay for the detection and determination of siderophores. *Analytical Biochemistry* 160: 47–56.

Susi P., Aktuganov G., Himanen J., Korpela T., 2011. Biological control of wood decay against fungal infection. *Journal of Environmental Management* 92: 1681–1689.

Tarnawski S.,Hamelin J., Jossi M., Aragno M., Fromin N., 2006. Phenotypic structure of *Pseudomonas* populations is altered under elevated pCO_2 in the rhizosphere of perennial grasses. *Soil Biology and Biochemistry* 38: 1193–1201.

Trotel-Aziz P., Couderchet M., Biagianti S., Aziz A., 2008. Characterization of new bacterial biocontrol agents *Acinetobacter*, *Bacillus*, *Pantoea* and *Pseudomonas* spp. mediating grapevine resistance against *Botrytis cinerea*. *Environmental and Experimental Botany* 64: 21–32.

Weselowski B., Nathoo N., Eastman A.W., MacDonald J.,Yuan Z.C., 2016. Isolation, identification and characterization of *Paenibacillus polymyxa* CR1 with potentials for biopesticide, biofertilization, biomass degradation and biofuel production. *BMC Microbiology* 16: 244.

Whipps J.M., 1987. Effect of media on growth and interactionsbetween a range of soil-borne glasshouse pathogens and antagonistic fungi. *New Phytologist* 107: 127–142.

Zhang M., Li J., Shen A., Tan S., Yan Z., Yu Y., Xue Z., Tan T., Zeng L., 2016. Isolation and identification of *Bacillus amyloliquefaciens* IBFCBF-1 with potential for biological control of *Phytophthora* blight and growth promotion of pepper. *Journal of Phytopathology* 164: 1012–1021.

Zheng F.C., Ward E., 1998. Variation within and between *Phytophthora* Species from rubber and citrus trees in china, determined by polymerase chain reaction Using RAPDs. *Journal of Phytopathology* 146: 103–109.

Zouaoui M., Dhieb C., Ben Abdelali N., Hajlaoui M.R., Sadfi Zouaoui N., 2016. First report of clementine trunk and branch canker caused by *Phytophthora citrophthora* in Tunisia. *New Disease Reports* 34, 29. DOI: 10.5197/j.2044-0588.2016.034.029

16

Genetic variability, chemotype distribution, and aggressiveness of *Fusarium culmorum* on durum wheat in Tunisia

Safa OUFENSOU[1,2,*], Virgilio BALMAS[1], Barbara SCHERM[1], Domenico RAU[1], Salvatore CAMIOLO[1], Vanda Assunta PROTA[1], Mossadok BEN-ATTIA[2], Samia GARGOURI[3], Matias PASQUALI[4], Safia EL-BOK[6], Quirico MIGHELI[1,5]

[1] *Dipartimento di Agraria, Università degli Studi di Sassari, Via E. De Nicola 9, 07100 Sassari, Italy*

[2] *Laboratoire de Bio-surveillance de l'Environnement, Faculté des Sciences de Bizerte, Route de Tunis, 7021Zarzouna, Université de Carthage, Tunisia*

[3] *Laboratoire de Protection des Végétaux, Institut National de Recherche Agronomique de Tunisie, Rue Hédi Karray, 2049 Ariana, Tunisia*

[4] *DeFENS-Department of Food Environmental and Nutritional Sciences, Università degli Studi di Milano, Via Celoria,2, 20133 Milano, Italy*

[5] *Unità di ricerca-Istituto Nazionale di Biostrutture e Biosistemi, Università degli Studi di Sassari, 07100 Sassari, Italy*

[6] *Laboratoire de Biodiversité, Biotechnologies et Changements Climatiques, Faculté des Sciences de Tunis, Université Tunis-El Manar, 2092 Tunis, Tunisia*

*Corresponding author: safaoufensou@gmail.com

Competing Interests: The Author(s) declare(s) no conflict of interest.

Editor: Antonio Moretti, Consiglio Nazionale delle Ricerche, Bari, Italy.

Abstract. *Fusarium culmorum* is the most commonly reported root rot pathogen in Tunisian durum wheat. Isolates of the pathogen from four durum wheat growing areas in the north of Tunisia were analyzed for their chemotypes. Two chemotypes were detected at unequal abundance (96% of 3-ADON and 4% of NIV). Distribution of a SNP mutation located at the position 34 bp after the first exon of the EF-1α partial sequence was analysed, to verify whether the haplotype was specifically associated to Fusarium root rot. A and T haplotypes were homogeneously distributed in three different Tunisian regions (Mateur, Beja and Bousalem) but not for the region of Bizerte, from which greatest number of A haplotype strains were detected. The isolates were tested for their virulence under glasshouse conditions, and a mean of 91% of crown and root infection was observed. Chemotype influenced virulence, but there was no significant influence of the geographical origin or haplotype on virulence. The distribution of three inter simple sequence repeats (ISSR) was examined, to better understand the structure of *F. culmorum* populations in Tunisia. A total of 27 fragments were obtained with eight polymorphic bands. Cluster analysis showed a high level of similarity between isolates. Analysis of molecular variance confirmed that there was little genetic differentiation among *F. culmorum* strains from different locations.

Keywords. *Fusarium* crown and root rots, population structure, trichothecenes.

INTRODUCTION

Durum wheat is extensively grown in Tunisia, covering over 40% of the cereal-producing areas (Fakhfakh *et al.*, 2011). *Fusarium culmorum* is reported to be among the most prevalent pathogens responsible of foot and root rot (FRR) of durum wheat and other small grain cereal crops (Burgess *et al.*, 2001; Smiley *et al.*, 2005; Wagacha and Muthomi, 2007; Scherm *et al.*, 2013). FRR is particularly severe in areas affected by water stress, that are usually found in Tunisia (Gargouri *et al.*, 2001), in Southern Italy (Balmas *et al.*, 2006) and in Turkey (Tunali *et al.*, 2006), Iran (Eslahi, 2012) and Syria (El-Khalifeh *et al.*, 2009). *Fusarium culmorum* infects host plants at the initial growing stages, causing rotting of root and crown tissues. In some cases, lesions and browning of the coleoptiles or seedling death may occur. When *F. culmorum* infects wheat plants at later growing stages, brown spots on basal internodes can be observed. Under high humidity conditions, reddish-pink discolourations of the nodes appear due to the development of sporulating pathogen mycelium (Scherm *et al.*, 2013; Balmas *et al.*, 2015).

Fusarium culmorum causes serious problems since it causes yield reductions and mycotoxin accumulation. The pathogen produces type B trichothecenes (Smiley *et al.*, 2005; Miedaner *et al.*, 2008; Obanor *et al.*, 2010; Pasquali *et al.*, 2016). The compounds are harmful to humans and animals, resulting in cancer development and other generally irreversible effects (Bennett and Klich, 2003; Pestka and Smolinski, 2005). Type B trichothecenes are also considered as virulence factors through inhibition of host plant defence mechanisms (Wagacha and Muthomi, 2007; Scherm *et al.*, 2013). Increased virulence of *F. culmorum* strains is associated to their ability to contaminate plant tissues with high doses of trichothecenes, although this correlation has not always been confirmed (Gang *et al.*, 1998; Scherm *et al.*, 2011), or these compounds may have minor effects ith they are translocated from the basal portions of host plants (Winter *et al.*, 2013).

For the most important mycotoxins, the maximum permitted levels have been set, for grains and cereal-derived products used for human or livestock consumption (EC, 2006). However, in some cases, high levels of toxin and the possible interaction between concomitantly occurring mycotoxins, represent toxicological risks, and food safety is seriously hampered (Balmas *et al.*, 2015).

For *F. culmorum*, chemotypes have been recognized within the type B trichothecene mycotoxins (Pasquali *et al.*, 2016). Distinct chemotypes are recognized according to their production of deoxynivalenol (DON) and related derivatives or nivalenol (NIV) (Scherm *et al.*, 2013). Discrimination of DON and NIV may provide insight into the toxigenic potential of *F. culmorum* strains. To our knowledge, all the *F. culmorum* strains from wheat in Italy belong to 3-ADON (Quarta *et al.*, 2005; Covarelli *et al.*, 2014) apart from two NIV strains found in two Italian regions, Tuscany and Emilia-Romagna (Prodi *et al.*, 2010), and one isolate that was characterized as NIV in Sardinia (Balmas *et al.*, 2015). Search for chemotypes associated with FFR throughout the Middle East demonstrated that 100% of *F. culmorum* strains in Turkey belong to the 3-ADON chemotype (Yörük and Albayrak, 2012), while in Syria 55% of the strains were 3-ADON and 45% were NIV chemotypes (Alkadri *et al.*, 2013). In other surveys (Yörük and Albayrak 2012; Alkadri *et al.*, 2013; Mert-Türk and Gencer, 2013; Motallebi *et al.*, 2015) dominance of the *F. culmorum* 3-ADON chemotype was further highlighted.

The DON chemotype is most widely found (Scherm *et al.*, 2013), whereas NIV producers are less frequent in many European countries (Bakan *et al.*, 2001; Jennings *et al.*, 2004), and in Tunisia (Kammoun *et al.*, 2010; Rebib *et al.*, 2014), Turkey (Yörük and Albayrak, 2012) and the United States of America (Mirocha *et al.*, 1994).

The nucleotide sequence of the translation elongation factor 1-α (EF1-α) gene, encoding a part of a highly conserved ubiquitous protein involved in translation, was first used in fungi in *Fusarium* (O'Donnell *et al.*, 1998). As a single-locus identification tool, EF1-α shows a high level of sequence polymorphism among related species, hence it was considered a useful genetic region for phylogenetic and taxonomic studies, allowing reliable identification as an alternative to rDNA or β-tubulin (O'Donnell, 2000; Roger *et al.*, 1999).

Knowledge of chemotype distribution within *F. culmorum* populations originating from several agro-ecological areas in Tunisia would provide useful information on strain fitness in the field, representing a reliable resource for the development of effective disease control strategies (Strange and Scott, 2005).

The goals of the present study were: 1) to characterize a representative *F. culmorum* population isolated from FRR-affected durum wheat plants grown in different agro-ecological areas of Tunisia; 2) to assess genetic variability and population structure of Tunisian *F. culmorum* strains, based on EF1-α sequence polymorphism and ISSR markers; 3) to assess the virulence of *F. culmorum* strains on wheat plants to verify the hypothesis that a specific haplotype is associated with FRR; and 4) to examine the distribution of genetic chemotypes and gather information about the potential toxigenicity of the fungal population that might contaminate durum wheat in Tunisia.

MATERIALS AND METHODS

Sampling, fungal isolation and isolate storage

FRR distribution on durum wheat was monitored during the 2015 growing season. A total of 88 fields were investigated in two climatic regions (sub humid and higher semi-arid) in Northern Tunisia (Bortoli *et al.*, 1969). These regions include more than 90% of the Tunisian wheat production areas (Figure 1). Sixty-eight fields were from the sub-humid region (Bizerte, Mateur and Beja), and 20 fields were from the higher semi-arid region (Bousalem). The fields, separated by approx. 10 km, were randomly selected. For each field in which FRR symptoms were observed, 20 plants were randomly collected along diagonal transects in different field zones. Plant samples were transferred in paper bags to the laboratory and stored at 4°C until analysed.

All fungal strains were obtained from the basal stems of diseased durum wheat plants. Fungal isolation was carried out according to Balmas *et al.* (2015), and monospore cultures were prepared as described by Burgess *et al.* (1994). All monospore strains collected were identified based on morphological traits, as described by Burgess *et al.* (1994).

For further analyses, all strains were stored at -80°C in 15% glycerol in the *Fusarium* collection of the Dipartimento di Agraria, University of Sassari, Italy.

Genomic DNA extraction and molecular characterization

Mycelia of fungal strains were each collected with a sterile spatula from PDA plates, after 5 d of incubation at 25°C in darkness. Genomic DNA was extracted from each isolate according to Aljanabi and Martinez (1997), and stored at 4°C. For each isolate, a partial sequence of the translation elongation factor 1α (EF1-α) was amplified (Balmas *et al.*, 2015), in a total of 50 μL reaction mixture containing: 10-25 μg of DNA template, 1.5 mM $MgCl_2$, 0.2 mM dNTPs, 0.5 μM of each of the primers TEF1 ATGGGTAAGGA(A/G)GACAAGAC and TEF2 GGA(G/A)GTACCAGT(G/C)ATCATGTT, 2 units *Taq* polymerase (Invitrogen). The PCR programme included one cycle at 98°C for 2 min, 35 cycles at 98°C for 15 s, 60°C for 15 s and 72°C for 30 s, followed by a final extension at 72°C for 5 min. PCR products were purified (PureLink^{TM} Quick PCR Purification Kit, Invitrogen) following the manufacturer's instructions, and amplicon

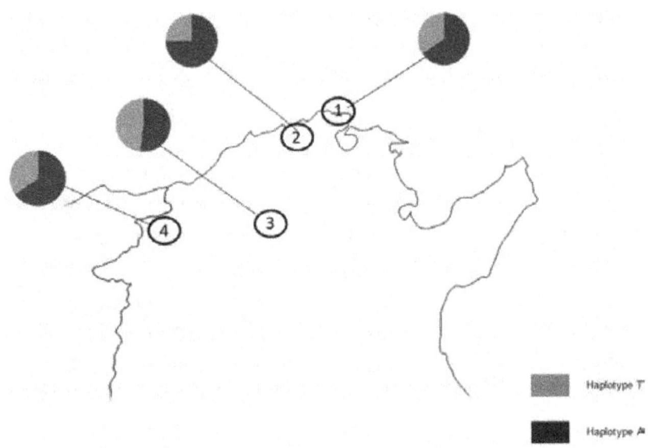

Figure 1. Map indicating sample sites in Tunisia where Fusarium isolates associated with root rot of wheat were obtained. Proportions (%) of haplotypes are indicated. Localities of Bizerte (1), Mateur (2), Beja (3) and Bousalem (4) are also indicated.

concentration was estimated with a fluorometer (Qubit^{TM}, Invitrogen), for sequencing optimisation. Sequencing was performed with 3500 Genetic Analyser (Life Technologies). For each isolate, both forward and reverse strands were sequenced and aligned using the multiple alignment program ClustalOmega (http://www.ebi.ac.uk/Tools/msa/clustalo/). To confirm morphological identification, EF1-α partial sequences were compared with consensus sequences available in *Fusarium*-ID and GenBank databases (O'Donnell *et al.* 2012). Based on the polymorphic nucleotide (T–A) located at position 34 after the first exon of the EF1-α partial sequence gene (Balmas *et al.* 2010, 2015), a Single Nucleotide Polymorphism (SNP) was targeted. All the sequences were then deposited in GenBank (Table 1).

To determine the genetic chemotype of each isolate, a *Tri12* multiplex PCR was carried out according to Ward *et al.* (2002) using two different primer sets, 12CON/12NF CATGAGCATGGTGATGTC/TCTCCTCGTTGTATCTGG for NIV discrimination and 12CON/12-3F CATGAGCATGGTGATGTC/CTTTGGCAAGCCCGTGCA for DON discrimination. PCR conditions were performed at 98°C (2 min) for one cycle; 98°C (10 s), 59°C (10 s), and 72°C (20 s) for 30 cycles; and a final cycle of 72°C for 5 min.

Genetic variability among the *F. culmorum* isolates was evaluated by an ISSR-PCR assay, using three different ISSR primers (Table 3). PCR conditions and annealing temperature were modified according to the primer sequence: 48°C for ISSR4 [$(GAGG)_3GG$; Albayrak *et al.* 2016]; and 55°C for both ISSR5 [$(AG)_9G$)] and ISSR6 [$(AC)_8YG$; Mishra *et al.* 2003]. PCR products were visualized on agarose gel (1.5% agarose) after electrophoretic

Table 1. Locations, chemotypes, Single nucleotide polymorphisims (SNP), mean aggressiveness scores and GenBank accession numbers for *Fusarium* isolates associated with Fusarium root rot of wheat in Tunisia.

Isolate	Location	Coordinates	Chemotype[a]	SNP[b]	Aggressiveness[c] Mean ± SD	EF-1α GenBank Accession No.
Fu-BI87	Bizerte	N37 12.423 E940.384	3-ADON	A	90.00 ± 31.6 [abc]	MF510992
Fu-BI84	Bizerte	N37 12.423 E940.384	3-ADON	T	77.50 ± 41.6 [abc]	MF510990
Fu-BI151	Bizerte	N37 13.964 E9 42.388	3-ADON	A	95.00 ± 15.81 [ab]	MF511031
Fu-BI169	Bizerte	N37 13.964 E9 42.388	3-ADON	T	50.00 ± 44.1 [cde]	MF511041
Fu-BI3	Bizerte	N36 36.677 E8 40.158	3-ADON	A	95.00 ± 15.81 [ab]	MF510936
Fu-BI36	Bizerte	N36 36.677 E8 40.158	3-ADON	A	100.0 ± 0.0 [a]	MF510958
Fu-BI135	Bizerte	N36 36.677 E8 40.158	3-ADON	A	77.50 ± 36.2 [abc]	MF511027
Fu-BI33	Bizerte	N37 14.416 E9 43.842	3-ADON	T	100.0 ± 0.0 [a]	MF510955
Fu-BI42	Bizerte	N37 14.416 E9 43.842	3-ADON	A	100.0 ± 0.0 [a]	MF510963
Fu-BI50	Bizerte	N37 14.416 E9 43.842	3-ADON	T	82.50 ± 33.4 [abc]	MF510969
Fu-BI54	Bizerte	N37 14.416 E9 43.842	3-ADON	T	85.00 ± 31.6 [abc]	MF510971
Fu-BI58	Bizerte	N37 14.416 E9 43.842	3-ADON	A	62.50 ± 37.7 [abcd]	MF510975
Fu-BI105	Bizerte	N37 14.416 E9 43.842	3-ADON	T	100.0 ± 0.0 [a]	MF511008
Fu-BI120	Bizerte	N37 14.416 E9 43.842	3-ADON	T	85.00 ± 31.6 [abc]	MF511017
Fu-BI123	Bizerte	N37 14.416 E9 43.842	3-ADON	A	100.0 ± 0.0 [a]	MF511019
Fu-BI13	Bizerte	N37 08.734 E9 46.500	3-ADON	A	55.00 ± 42.2 [bcde]	MF510943
Fu-BI15	Bizerte	N37 08.734 E9 46.500	NIV	A	65.00 ± 39.4 [abcd]	MF510944
Fu-BI4	Bizerte	N37 13.964 E9 42.390	3-ADON	A	92.50 ± 23.72 [ab]	MF510937
Fu-BI118	Bizerte	N37 13.964 E9 42.390	3-ADON	A	90.00 ± 24.15 [abc]	MF511015
Fu-BI67	Bizerte	N37 14.416 E9 43.842	3-ADON	A	87.50 ± 24.30 [abc]	MF510980
Fu-BI150	Bizerte	N37 14.416 E9 43.842	3-ADON	A	90.00 ± 31.6 [abc]	MF511030
Fu-BI170	Bizerte	N37 14.416 E9 43.842	3-ADON	A	100.0 ± 0.0 [a]	MF511042
Fu-BI183	Bizerte	N37 14.416 E9 43.842	3-ADON	A	100.0 ± 0.0 [a]	MF511046
Fu-BI6	Bizerte	N37 06.858 E9 46.944	NIV	T	77.50 ± 41.6 [abc]	MF431609
Fu-BI8	Bizerte	N37 06.858 E9 46.944	NIV	T	72.50 ± 39.9 [abc]	MF510939
Fu-BI28	Bizerte	N37 06.858 E9 46.944	NIV	T	75.00 ± 40.8 [abc]	MF510951
Fu-BI161	Bizerte	N37 07.868 E9 45.396	3-ADON	T	75.00 ± 28.87 [abc]	MF511039
Fu-BI159	Bizerte	N37 07.868 E9 45.396	3-ADON	A	100.0 ± 0.0 [a]	MF511037
Fu-BI152	Bizerte	N37 07.868 E9 45.396	3-ADON	A	90.00 ± 31.6 [abc]	MF511032
Fu-BI149	Bizerte	N37 07.868 E9 45.396	3-ADON	A	100.0 ± 0.0 [a]	MF511029
Fu-BI128	Bizerte	N37 07.868 E9 45.396	3-ADON	A	100.0 ± 0.0 [a]	MF511023
Fu-BI103	Bizerte	N37 07.868 E9 45.396	3-ADON	A	100.0 ± 0.0 [a]	MF511006
Fu-BI35	Bizerte	N37 07.868 E9 45.396	3-ADON	A	100.0 ± 0.0 [a]	MF510957
Fu-BI9	Bizerte	N37 10.539 E9 44.507	3-ADON	A	85.00 ± 33.7 [abc]	MF510940
Fu-BI10	Bizerte	N37 10.539 E9 44.507	3-ADON	T	100.0 ± 0.0 [a]	MF510941
Fu-BI19	Bizerte	N37 10.539 E9 44.507	3-ADON	A	100.0 ± 0.0 [a]	MF510946
Fu-BI22	Bizerte	N37 10.539 E9 44.507	3-ADON	A	100.0 ± 0.0 [a]	MF510948
Fu-BI25	Bizerte	N37 10.539 E9 44.507	3-ADON	A	100.0 ± 0.0 [a]	MF510949
Fu-BI27	Bizerte	N37 10.539 E9 44.507	3-ADON	A	100.0 ± 0.0 [a]	MF510950
Fu-BI30	Bizerte	N37 10.539 E9 44.507	3-ADON	A	87.50 ±27.00 [abc]	MF510952
Fu-BI76	Bizerte	N37 10.539 E9 44.507	3-ADON	T	100.0 ± 0.0 [a]	MF510984
Fu-BI90	Bizerte	N37 10.539 E9 44.507	3-ADON	T	95.00 ± 15.81 [ab]	MF510994
Fu-BI140	Bizerte	N37 10.539 E9 44.507	3-ADON	A	100.0 ± 0.0 [a]	MF511028
Fu-BI156	Bizerte	N37 10.539 E9 44.507	3-ADON	T	100.0 ± 0.0 [a]	MF511036
Fu-BI98	Mateur	N37 08.774 E9 57.402	3-ADON	A	100.0 ± 0.0 [a]	MF511001
Fu-BI117	Mateur	N37 08.774 E9 57.402	3-ADON	A	100.0 ± 0.0 [a]	MF511014
Fu-BI127	Mateur	N37 08.774 E9 57.402	3-ADON	A	100.0 ± 0.0 [a]	MF511022

(Continued)

Table 1. (Continued).

Isolate	Location	Coordinates	Chemotype[a]	SNP[b]	Aggressiveness[c] Mean ± SD	EF-1α GenBank Accession No.
Fu-MA1	Mateur	N37 08.862 E9 33.401	3-ADON	A	100.0 ± 0.0 [a]	MF431610
Fu-MA12	Mateur	N37 08.862 E9 33.401	3-ADON	T	100.0 ± 0.0 [a]	MF510942
Fu-BI21	Mateur	N37 08.862 E9 33.401	3-ADON	A	16.00 ±16.80 [e]	MF510947
Fu-BI97	Mateur	N37 08.862 E9 33.401	3-ADON	A	92.50 ± 23.72 [ab]	MF511000
Fu-BI100	Mateur	N37 08.862 E9 33.401	3-ADON	T	97.50 ± 7.91 [a]	MF511003
Fu-BI112	Mateur	N37 08.862 E9 33.401	3-ADON	A	77.50 ± 36.2 [abc]	MF511013
Fu-MA95	Mateur	N37 08.862 E9 33.400	3-ADON	A	95.00 ±15.81 [ab]	MF510998
Fu-MA101	Mateur	N37 08.862 E9 33.400	3-ADON	A	100.0 ± 0.0 [a]	MF511004
Fu-BI107	Mateur	N37 10.729 E9 45.649	3-ADON	T	100.0 ± 0.0 [a]	MF511010
Fu-BE31	Beja	N36 42.475 E9 10.687	3-ADON	A	95.00 ± 15.81 [ab]	MF510953
Fu-BE32	Beja	N36 42.475 E9 10.687	3-ADON	T	95.00 ± 15.81 [ab]	MF510954
Fu-BE48	Beja	N36 42.475 E9 10.687	3-ADON	A	80.00 ± 42.2 [abc]	MF510967
Fu-BE55	Beja	N36 42.475 E9 10.687	3-ADON	A	100.0 ± 0.0 [a]	MF510972
Fu-BE69	Beja	N36 42.475 E9 10.687	3-ADON	T	82.50 ± 37.4 [abc]	MF510981
Fu-BE78	Beja	N36 42.475 E9 10.687	3-ADON	T	95.00 ± 15.81 [ab]	MF510985
Fu-BE106	Beja	N36 42.475 E9 10.687	3-ADON	A	100.0 ± 0.0 [a]	MF511009
Fu-BE129	Beja	N36 42.475 E9 10.687	3-ADON	A	92.50 ± 16.87 [ab]	MF511024
Fu-BE131	Beja	N36 42.475 E9 10.687	3-ADON	T	67.50 ± 44.2 [abcd]	MF511025
Fu-BE132	Beja	N36 42.475 E9 10.687	3-ADON	T	100.0 ± 0.0 [a]	MF511026
Fu-BE154	Beja	N36 42.475 E9 10.687	3-ADON	T	95.00 ± 15.81 [ab]	MF511034
Fu-BE162	Beja	N36 42.475 E9 10.687	3-ADON	A	100.0 ± 0.0 [a]	MF511040
Fu-BE171	Beja	N36 42.475 E9 10.687	3-ADON	A	100.0 ± 0.0 [a]	MF511043
Fu-BE180	Beja	N36 42.475 E9 10.687	3-ADON	A	100.0 ± 0.0 [a]	MF511044
Fu-BE182	Beja	N36 42.475 E9 10.687	3-ADON	T	100.0 ± 0.0 [a]	MF511045
Fu-BE37	Beja	N36 38.159 E9 06.458	3-ADON	A	9750 ± 7.91 [a]	MF510959
Fu-BE38	Beja	N36 38.159 E9 06.458	3-ADON	A	100.0 ± 0.0 [a]	MF510960
Fu-BE41	Beja	N36 38.159 E9 06.458	3-ADON	A	100.0 ± 0.0 [a]	MF510962
Fu-BE52	Beja	N36 38.159 E9 06.458	3-ADON	A	80.00 ± 36.9 [abc]	MF510970
Fu-BE60	Beja	N36 38.159 E9 06.458	3-ADON	A	92.50 ± 23.72 [ab]	MF510977
Fu-BE63	Beja	N36 38.159 E9 06.458	3-ADON	T	92.50 ± 12.08 [ab]	MF510979
Fu-BE72	Beja	N36 38.159 E9 06.458	3-ADON	A	90.00 ± 31.6 [abc]	MF510982
Fu-BE75	Beja	N36 38.159 E9 06.458	3-ADON	T	97.50 ± 7.91 [a]	MF510983
Fu-BE83	Beja	N36 38.159 E9 06.458	3-ADON	T	100.0 ± 0.0 [a]	MF510989
Fu-BE104	Beja	N36 38.159 E9 06.458	3-ADON	T	92.50 ± 23.72 [ab]	MF511007
Fu-BE126	Beja	N36 38.159 E9 06.458	3-ADON	T	92.50 ± 16.87 [ab]	MF511021
Fu-BO44	Bousalem	N36 34.444 E8 54.849	3-ADON	T	100.0 ± 0.0 [a]	MF510964
Fu-BO46	Bousalem	N36 34.444 E8 54.849	3-ADON	A	90.00 ± 31.6 [abc]	MF510965
Fu-BO47	Bousalem	N36 34.444 E8 54.849	3-ADON	A	100.0 ± 0.0 [a]	MF510966
Fu-BO56	Bousalem	N36 34.444 E8 54.849	3-ADON	A	75.00 ± 31.18 [abc]	MF510973
Fu-BO57	Bousalem	N36 34.444 E8 54.849	3-ADON	A	85.00 ± 31.6 [abc]	MF510974
Fu-BO59	Bousalem	N36 34.444 E8 54.849	3-ADON	A	100.0 ± 0.0 [a]	MF510976
Fu-BO79	Bousalem	N36 34.444 E8 54.849	3-ADON	T	77.50 ± 27.51 [abc]	MF510986
Fu-BO80	Bousalem	N36 34.444 E8 54.849	3-ADON	T	100.0 ± 0.0 [a]	MF510987
Fu-BO82	Bousalem	N36 34.444 E8 54.849	3-ADON	A	100.0 ± 0.0 [a]	MF510988
Fu-BO85	Bousalem	N36 34.444 E8 54.849	3-ADON	A	100.0 ± 0.0 [a]	MF510991
Fu-BO91	Bousalem	N36 34.444 E8 54.849	3-ADON	A	85.00 ± 33.7 [abc]	MF510995
Fu-BO102	Bousalem	N36 34.444 E8 54.849	3-ADON	A	100.0 ± 0.0 [a]	MF511005
Fu-BO110	Bousalem	N36 34.444 E8 54.849	3-ADON	A	100.0 ± 0.0 [a]	MF511011

(Continued)

Table 1. (Continued).

Isolate	Location	Coordinates	Chemotype[a]	SNP[b]	Aggressiveness[c] Mean ± SD	EF-1α GenBank Accession No.
Fu-BO111	Bousalem	N36 34.444 E8 54.849	3-ADON	A	97.50 ± 7.91 [a]	MF511012
Fu-BO119	Bousalem	N36 34.444 E8 54.849	3-ADON	A	100.0 ± 0.0 [a]	MF511016
Fu-BO153	Bousalem	N36 34.444 E8 54.849	3-ADON	T	90.00 ± 31.6 [a]	MF511033
Fu-BO155	Bousalem	N36 34.444 E8 54.849	3-ADON	A	95.00 ± 15.81 [a]	MF511035
Fu-BO160	Bousalem	N36 34.444 E8 54.849	3-ADON	A	92.50 ± 23.72 [ab]	MF511038
Fu-BO2	Bousalem	N36 34.444 E8 54.900	3-ADON	T	90.00 ± 31.6 [abc]	MF510935
Fu-BO7	Bousalem	N36 34.444 E8 54.900	3-ADON	T	100.0 ± 0.0 [a]	MF510938
Fu-BO16	Bousalem	N36 34.444 E8 54.900	3-ADON	A	92.50 ± 23.72 [ab]	MF510945
Fu-BO34	Bousalem	N36 30.263 E8 47.347	3-ADON	T	95.00 ± 15.81 [ab]	MF510956
Fu-BO92	Bousalem	N36 37.668 E8 54.874	3-ADON	T	100.0 ± 0.0 [a]	MF510996

[a] 3-acetyldeoxynivalenol, 3-ADON; nivalenol, NIV.
[b] Single nucleotide polymorphism (SNP) at position 34 after first exon of EF-1α (Balmas *et al.* 2010).
[c] Data followed by the same letters are not significantly different from each other (Tukey test).

Table 2. Statistical significance of impacts of location, chemotype or haplotype on pathogenicity of *Fusarium* isolates associated with root rot of wheat.

Variable	Number	Degrees of freedom	P-value
Location		3	0.32
Bizerte	440		
Mateur	120		
Beja	250		
Bousalem	230		
Chemotype		1	0.00
3-ADON	1000		
NIV	40		
Haplotype		1	0.87
A	660		
T	380		

Table 3. ISSR primers used in this study.

Primer	Sequence	(P/M)[a]	Reference
ISS4	$(GAGG)_3GG$	(4/12)	Albayrak *et al.* (2016)
ISSR5	$(AG)_9G$	(0/5)	Mishra *et al.* (2003)
ISSR6	$(AC)_8YG$	(4/10)	Mishra *et al.* (2003)

[a]*P* number of polymorphic markers with polymorphism level above 2%; M monomorphic markers.

separation in 1 × TAE buffer of 1 h for ISSR5 and ISSR6, or 2 h for ISSR4. Fragment size was estimated by comparison with a 1 Kb Plus DNA Ladder (Invitrogen). Gel images were analysed using a Gel-Doc XR+ System (Bio-Rad). ISSR markers were visually scored as presence (1) or absence (0) of each band.

Pathogenicity experiments

For each *F. culmorum* isolate, agar plugs (8 mm diam.) covered with mycelium were cut from actively growing colony margins of 5-d-old cultures grown on PDA, and were used to inoculate durum wheat seeds. A single durum wheat seed (cv. Saragolla) was deposited on each mycelial plug and was planted in the centre of a plastic pot containing sterilized potting mix composed of equal proportions peat soil, sand and redwood soil. The pots were incubated on a bench in a glasshouse for 20 d, with day and night temperatures of 25-30 °C and 18-25 °C, and with daily irrigation. For each isolate, ten replicates were established. According to, After 3 weeks, disease severity index was assessed using an empirical scale of five disease severity classes of disease severity (Balmas *et al.*, 2006), where 0 = no browning; 1 = 1 to 25%; 2 = 50%; 3 = 75%; and 4 = no plant emergence). For each plant, stem browning were also assessed.

Statistical analyses

Disease severity data were analysed using Minitab version 17.1.0 software. *Post hoc* analyses (Tukey's HSD test of multiple comparisons) were subsequently performed considering 95% confidence level.

A phylogenetic tree was constructed from pairwise distance matrix by UPGMA, applying MEGA version 7.0 software. Bootstraps analysis with 1,000 replicates was carried out to estimate the statistical support for different tree branches, and the number on each branch represented the bootstrap value.

Analysis of molecular variance (AMOVA), provided by the Arlequin version 3.5.1.2 software, using 1,023 permutations, was used to calculate the variance within the *F. culmorum* collection, based on ISSR molecular markers.

RESULTS

Distribution of Fusarium culmorum isolates in Tunisia, and their molecular chemotyping

A total of 104 *F. culmorum* isolates were obtained from the basal stems of symptomatic durum wheat plants. The greatest number of isolates (44) was collected from the the sub-humid region of Bizerte.

Among the 104 *F. culmorum* isolates 100 (96%) were of the 3-ADON chemotype, and were obtained from all the Tunisian regions considered. Only four isolates were ascribed to the NIV chemotype, and all NIV chemotype isolates were collected from two fields of the sub-humid region of Bizerte

EF1-α haplotype distribution

The identification of both SNPs (A or T) of the EF1-α partial sequence of the *F. culmorum* isolates confirmed the presence of the two haplotypes within the *F. culmorum* population. Both haplotypes were homogenously and equally distributed in the region of Beja, whereas the A haplotype was dominant in the other three tested regions (Figure 1).

ISSR marker analysis

Based on three ISSR primers (ISSR4, ISSR5, and ISSR6), 27 bands were scored, and ranged from 0.39 to 2.6 Kb. The most informative primer was ISSR4 (12 amplicons). Bootstrapping gave values less than 50%. The cluster analysis with the UPGMA using pairwise genetic distances indicated high similarity between the *F. culmorum* isolates (Figure 2). Results from AMOVA provided an estimated pattern of population differentiation. All ISSR variation was distributed among isolates within populations (98.3% of the total variance). A Small proportion (1.7%) of the variability was explained by differences between populations and was not statistically significant ($P = 0.12$; Table 4). In addition, no statistically significant correlations were detected between cluster-

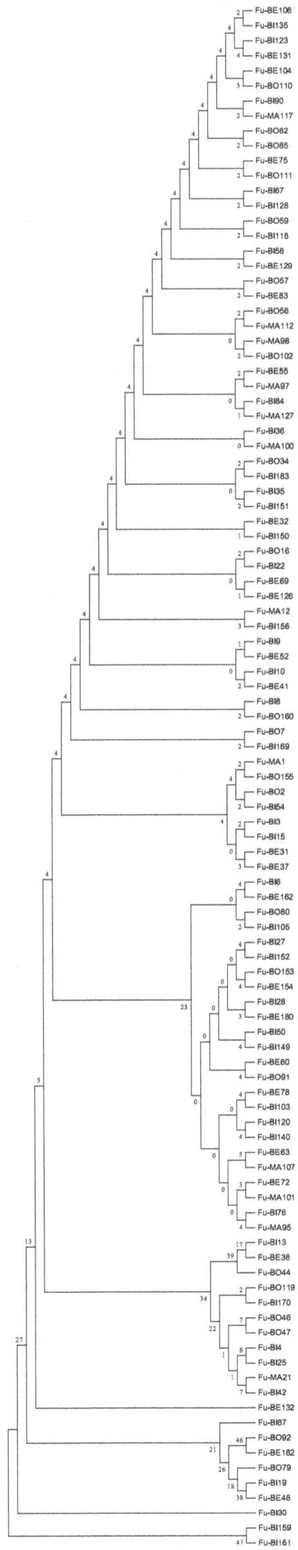

Figure 2 UPGMA analysis of *Fusarium culmorum* populations, based on combined ISSR data analyzed using using pairwise distances matrix. The numbers on the branches represent bootstrap support values.

Table 4 Analysis of molecular variance of ISSR for 104 isolates of *Fusarium culmorum* from four Tunisian populations.

Source of variation	$d.f^a$	SSD^b	Variance components	Percentage of variation
Among population	3	3.142	0.01282 Va	1.71
Within population	99	73.052	0.73790 Vb	98.29
Total	102	76.194	0.75072	

Va and FST : *P* (rand. value > obs. value) = 0.12121
P (rand. value = obs. value) = 0.00000
P-value = 0.12121±0.01052

[a] *df* degrees of freedom;
[b] *SSD* sums of squared deviations;
[c] *P* probability of obtaining equal or large value determined by1023 randomizations of the treatments.

ing analyses, virulence, chemotype and haplotype of the isolates.

Aggressiveness of Fusarium culmorum *isolates*

In the pathogenicity test, root symptoms caused by *F. culmorum* isolates were detected in almost all cases (91%), with severity values ranging from 16 to 100%. A highly significant effect of chemotype on isolate aggressiveness was observed, with the 3-ADON isolates being more aggressive than the NIV isolates ($P = 0.007$; Table 1). However, few NIV chemotype strains were analyzed, which did not allow robust statistical analysis. In contrast, no statistically significant effects were detected for geographical origin ($P = 0.324$) or haplotype ($P = 0.877$) on isolate aggressiveness.

DISCUSSION

Fusarium culmorum is reported as the predominant cause of FRR disease in Tunisia, since incidence of the pathogen is promoted by dry springs and irregular rainfall (Gargouri *et al.*, 2001). Pedo-climatic conditions in wheat-producing Tunisian regions are very similar to those in Sardinia (Balmas *et al.*, 2015), and in Turkey (Tunali *et al.*, 2008), where *F. culmorum* is the most common species causing FRR.

The present research aimed to investigate the genetic variability and the structure of *F. culmorum* populations associated with FRR on durum wheat in Tunisia. *F. culmorum* strains were collected from different wheat production areas, showing different climatic conditions. Two distinct chemotypes (3-ADON and NIV) were distinguished in the Tunisian populations of *F. culmorum*, the 3-ADON chemotype was the most common. This confirms previous reports by Rebib *et al.* (2014) on two fields showing FRR symptoms. The NIV chemotype was detected for the first time as associated to FRR in Tunisia, although this chemotype was found only in two fields in the Bizerte region.

Based on chemotype characterization of Italian *Fusarium* species, Covarelli *et al.* (2015) suggested that climatic conditions may strongly affect the occurrence of 3-ADON and 15-ADON, whereas NIV contamination may occur regardless of climatic conditions. Kammoun *et al.* (2010) reported that most isolates (98%) causing Fusarium head blight were 3-ADON producers, while 2% were NIV producers and originated from Bizerte, results which agree with those reported here. However, further investigations are required to better understand if specific agronomic or environmental conditions favour the presence of NIV-chemotype strains (Beyer *et al.*, 2014).

While previous reports from Tunisia have focused on FHB chemotyping (Bensassi *et al.*, 2009; Kammoun *et al.*, 2010), the present study represents the first geographic survey in different areas of Tunisia, and of the chemotypes of *F. culmorum* causing FRR on durum wheat. These results are in agreement with previous studies carried out in northern Mediterranean countries, including Italy, France, Portugal and Yugoslavia (Logrieco *et al.*, 2003; Pasquali and Migheli, 2014), as well as Germany and the United Kingdom (Tóth *et al.*, 2004; Jennings *et al.*, 2004). These reports confirm that the DON chemotype predominates among *F. culmorum* strains recovered from cereal grains.

Genetic chemotyping is an essential tool for characterizing *F. culmorum* populations causing root rot on wheat, but the presence of an amplification product reflects the possibility of a particular toxin being produced, whereas biosynthesis of the toxin remains to be confirmed by chemical analysis.

In our previous study of *F. culmorum* strains collected in the Sardinian region, association of haplotype A (EF1-α polymorphism) with FRR was highlighted (Balmas *et al.*, 2015). To verify this association, an *ad hoc* survey on FRR-causing isolates from Tunisia was carried out. In this study, both A and T haplotypes were observed within the Tunisian *F. culmorum* population. No significant association was found between A-haplotype and FRR. It is therefore evident that, at least in Tunisia, the presence of the two haplotypes is not associated with specific fitness or virulence advantage in FRR pathogens on wheat.

ISSR markers were used to assess the genetic variability of *F. culmorum* populations in Tunisia, aiming to verify the findings of Rebib *et al.* (2014) over a large geographic area. They suggested a high level of similarity among populations comparing two Tunisian fields. The present data confirmed that no clear trends were apparent in the distribution of the genetic variability with regard to the geographic origin within Tunisia. Our observations are in agreement with those reported by Albayrak *et al.* (2016), who showed similarity coefficients of 65.7-94.3% among *F. culmorum* isolates using ISSR4 and 41 other primers. The present report also confirms a previous study on the distribution of RAPD markers, suggesting that the low level of genetic differentiation among Tunisian populations of *F. culmorum* is mostly interpreted as the outcome of asexual reproduction in this pathogen (Gargouri *et al.* 2003). Similarly, Gargouri *et al.* (2003) indicated that no structuring had been observed at small or large geographic scales in this fungal species. Based on this assumption, the lack of a geographic structure in Tunisian populations of *F. culmorum* also suggests that spore dispersal probably occurs over a wide geographic area. These factors probably influence the level of genetic diversity within populations.

Using three markers, two of which (ISSR5 and ISSR6) were the same used in the present work, Mishra *et al.* (2003) found that 81% of the ISSR bands were polymorphic among *F. culmorum* populations: ISSR5 generated 28 ISSR distinct genotypes and ISSR6 produced 22 genotypes among 75 examined isolates. Moreover, the pattern of genetic diversity was largely associated to the geographical origin of the isolates (Mishra *et al.* 2003). In our study, no clear spatial clustering or relationships between variability and geographical regions were observed. Other genotyping assays have failed to show clear correlation between genetic variability and the geographic origin, mostly because the tested populations were from limited agro-ecological areas (Gargouri *et al.* 2003; Mishra *et al.* 2003).

The current survey provides a first insight into the genetic diversity of the *F. culmorum* population causing FRR in the main durum wheat growing regions of Tunisia. Further analyses with greater numbers of *F. culmorum* isolates from throughout Tunisia are warranted. Nonetheless, these preliminary data provide knowledge at the country scale on chemotyping and haplotyping, as an aid to ensure food safety monitoring, and for development of effective disease prevention and control strategies.

ACKNOWLEDGEMENTS

This research was supported by Fondazione di Sardegna and by Regione Sardegna ["Selection of sustainable phenolic inhibitors of trichothecene biosynthesis: in vitro studies and computational insights into the structure-activity relationship (SSPIT)"]; by Ministère Tunisien de l'Enseignement Supérieur et de la Recherche Scientifique [Faculté des Sciences de Bizerte, Laboratoire de Bio-surveillance de l'environnement (LR01/ES14)]. S.O. acknowledges financial support by SIPCAM Italia SpA.

LITERATURE CITED

Albayrak G., Yörük E., Gazdağli A., Sharifnabi B., 2016. Genetic diversity among *Fusarium graminearum* and *F. culmorum* isolates based on ISSR markers. *Archives of Biological Sciences* 68: 333–343.

Aljanabi S.M. Martinez I., 1997. Universal and rapid salt-extraction of high quality genomic DNA for PCR-based techniques. *Nucleic Acids Research* 22: 4692–4693.

Alkadri D., Nipoti P., Doll K., Karlovsky P., Prodi A., Pisi A., 2013. Study of fungal colonization of wheat kernels in Syria with a focus on Fusarium species. *International Journal of Molecular Science* 14: 5938–5951.

Bakan B, Pinson L., Cahagnier B., Melcion D., Sémon E., Richard-Molard D., 2001. Toxigenic potential of *Fusarium culmorum* strains isolated from French wheat. *Food Additives and Contaminants* 18: 998–1003.

Balmas V., Delogu G., Sposito S., Rau D., Migheli Q., 2006. Use of a complexation of tebuconazole with β-cyclodextrin for controlling foot and crown rot of durum wheat incited by *Fusarium culmorum*. *The Journal of Agricultural and Food Chemistry* 54: 480–484.

Balmas V., Migheli Q., Scherm B., Garau P., O'Donnell K., ... Geiser D.M., 2010. Multilocus phylogenetics show high levels of endemic fusaria inhabiting Sardinian soils (Tyrrhenian Islands). *Mycologia* 102: 803–812.

Balmas V., Scherm B., Marcello A., Beyer M., Hoffmann L., ... Pasquali M., 2015. *Fusarium* species and chemotypes associated with Fusarium head blight and Fusarium root rot on wheat in Sardinia. *Plant Pathology* 64: 972–979.

Bennett J.W., Klich M., 2003. Mycotoxins. *Clinical Microbiology Reviews* 16: 497–516

Bensassi F., Zaied C., Abid S., Hajlaoui M.R., Bacha H., 2009. Occurrence of deoxynivalenol in durum wheat in Tunisia. *Food Control* 21: 281–285.

Beyer M., Pogoda F., Pallez M., Lazic J., Hoffmann L., ... Pasquali M., 2014. Evidence for a reversible drought induced shift in the species composition of mycotoxin producing Fusarium head blight pathogens isolated from symptomatic wheat heads. *International Journal of Food Microbiology* 182–183: 51–56.

Bortoli L., Gounot M., Jacquinet C., 1969. Climatologie et Bioclimatologie de la Tunisie septentrionale. *Annales de l'INRA Tunisie* 42: 1-23.

Burgess L.W., Summerell B.A., Bullock S., Gott K.B., Backhouse D., 1994. *Laboratory Manual for Fusarium Research.* 3rd ed. *Fusarium* Research Laboratory, Department of Crop Sciences, University of Sydney and The Royal Botanic Gardens, Sydney, NSW, Australia, 136 pp.

Burgess L.W., Backhouse D., Summerell B.A., Swan L.J., 2001. Crown rot of wheat. In: *Fusarium: Paul E. Nelson Memorial Symposium* (B.A. Summerell, J.F. Leslie, D. Backhouse, W.L. Bryden, ed.), APS press, American Phytopathological Society. St. Paul, MN, USA, 271–294.

Covarelli L., Beccari G., Prodi A., Generotti S., Etruschi F., … Mañes J., 2015. *Fusarium* species, chemotype characterization and trichotechene contamination of durum and soft wheat in an area of central Italy. *Journal of the Science of Food and Agriculture* 95: 540–551.

European Commission, 2006. Commission Regulation (EC) No. 1881/2006 of 19/12/2006, setting maximum levels for certain contaminants in foodstuffs. *Official Journal of the European Union*, L364, 5-24.

El-Khalifeh M., El Ahmed A., Yabrak M., … Nachit M., 2006. Variation of cultural and morphological characteristics in *Fusarium* spp. pathogens of common root rot disease on wheat in Syria. *Arab Journal of Plant Protection* 24: 67–74.

Eslahi M.R., 2012. Fungi associated with root and crown rot of wheat in Khuzestan province, Iran. *Journal of Crop Protection* 1: 107–111.

Fakhfakh M.M., Yahyaoui A., Rezgui S., Elias E.M., Daaloul A., 2011. Identification and pathogenicity assessment of *Fusarium* spp. sampled from durum wheat fields in Tunisia. *African Journal of Biotechnology* 10: 6529–6539.

Gang G., Miedaner T., Schuhmacher U., Schollenberger M., Geiger H.H., 1998. Deoxynivalenol and nivalenol production by *Fusarium culmorum* isolates differing in aggressiveness toward winter rye. *Phytopathology* 88: 879–884.

Gargouri S., Hajlaoui M.R., Guermech A., Marrakchi M., 2001. Identification des espèces fongiques associées à la pourriture du pied du blé et étude de leur répartition selon les étages bioclimatiques en Tunisie. *Bulletin OEPP/EPPO Bulletin* 31: 499–503.

Gargouri S., Bernier L., Hajlaoui M.R., Marrakchi M., 2003. Genetic variability and population structure of the wheat foot rot fungus, *Fusarium culmorum*, in Tunisia. *European Journal of Plant Pathology* 109: 807–815.

Jennings P., Coates M.E., Turner J.A., Chandler E.A., Nicholson P., 2004. Determination of deoxynivalenol and nivalenol chemotypes of *Fusarium culmorum* isolates from England and Wales by PCR assay. *Plant Pathology* 53: 182–190.

Kammoun L.G., Gargouri S., Barreau C., Richard-Forget F., Hajlaoui M., 2010. Trichothecene chemotypes of *Fusarium culmorum* infecting wheat in Tunisia. *International Journal of Food Microbiology* 140: 84–89.

Logrieco A., Bottalico A., Mulé G., Moretti A., Perrone G., 2003. Epidemiology of toxigenic fungi and their associated mycotoxins for some Mediterranean crops. *European Journal of Plant Pathology* 109: 645–667.

Mert-Türk F., Gencen R., 2013. Distribution of the 3-AcDON, 15-AcDON, and NIV chemotypes of *Fusarium culmorum* in the North-West of Turkey. *Plant Protection Science* 49: 57–64.

Miedaner T., Cumagun C.J.R., Chakraborty S., 2008. Population genetics of three important head blight pathogens *Fusarium graminearum*, *F. pseudograminearum* and *F. culmorum*. *Plant Pathology Journal* 156: 129–139.

Mirocha C.J., Xie W., Xu Y., Wilcoxson R.D., Woodward R.P., … Behele G., 1994. Production of trichothecene mycotoxins by *Fusarium graminearum* and *Fusarium culmorum* or barley and wheat. *Mycopathologia* 128: 19–23.

Mishra P.K., Fox R.T.V., Culhan A., 2003. Inter-simple sequence repeats and aggressiveness analyses revealed high genetic diversity, recombination and long-range dispersal in *Fusarium culmorum*. *Annals of Applied Biology* 143: 291–301.

Motallebi P., Alkadri D., Pisi A., Nipoti P., Tonti S., … Prodi A., 2015. Pathogenicity and mycotoxin chemotypes of Iranian *Fusarium culmorum* isolates on durum wheat, and comparisons with Italian and Syrian isolates. *Phytopathologia Mediterranea* 54: 437–445.

Obanor F., Erginbas-Orakci G., Tunali B., Nicol J.M., Chakraborty S., 2010. *Fusarium culmorum* is a single phylogenetic species based on multilocus sequence analysis. *Fungal Biology* 114: 753–765.

O'Donnell K, Kistler H.C., Cigelnik E., Ploetz R.C, 1998. Multiple evolutionary origins of the fungus causing Panama disease of banana: concordant evidence from nuclear and mitochondrial gene genealogies. *Proceedings of the National Academy of Sciences* USA 95: 2044–2049.

O'Donnell K., 2000. Molecular phylogeny of the *Nectria haematococca-Fusarium solani* species complex. *Mycologia* 92: 919–938.

O'Donnell K., Humber R.A., Geiser D.M., Kang S., Park B., … Rehner S.A., 2012. Phylogenetic diversity of insecticolous fusaria inferred from multilocus DNA sequence data and their molecular identification via FUSARIUM-ID and Fusarium MLST. *Mycologia* 104: 427–445.

Pasquali M., Migheli Q., 2014. Genetic approaches to chemotype determination in type B-trichothecene producing *Fusaria*. *International Journal of Food Microbiology* 189: 164–182.

Pasquali M, Beyer M., Logrieco A., Audenaert K., Balmas V., … Vogelgsang S., 2016. A European Database of *Fusarium graminearum* and *F. culmorum* trichothecene genotypes. *Frontiers in Microbiology* 7: 1-11.

Pestka J.J., Smolinski A., 2005. Deoxynivalenol: toxicology and potential effects on humans. *Journal of Toxicology and Environmental Health Part B: Critical Reviews* 8: 39–69.

Prodi A., Salomoni D., Alkadri D., Tonti S., Nipoti P., … Pancaldi D., 2010. Presence of deoxynivalenol and nivalenol chemotypes of *Fusarium culmorum* isolated from durum wheat in some Italian regions. *Journal of Plant Pathology* 92 (Suppl.4): 96.

Quarta A., Mita G., Haidukowski M., Santino A., Mulè G., Visconti M., 2005. Assessment of trichothecene chemo- types of *Fusarium culmorum* occurring in Europe. *Food Additives and Contaminants* 22: 309–315.

Rebib H., Bouraoui H., Rouaissi M., Brygoo Y., Boudabbous A., … Sadfi-Zouaoui N., 2014. Genetic diversity assessed by SSR markers and chemotyping of *Fusarium culmorum* causal agent of foot and root rot of wheat collected from two different fields in Tunisia. *European Journal Plant Pathology* 139: 481–495.

Roger AJ., Sandblom O., Doolittle W.F., Philippe H., 1999. An evaluation of elongation factor 1 alpha as a phylogenetic marker for eukaryotes. *Molecular Biology and Evolution* 16: 218–233.

Scherm B., Orrù M., Balmas V., Spanu F., Azara E., … Migheli Q., 2011. Altered trichothecene biosynthesis in *TRI6*-silenced transformants of *Fusarium culmorum* influences the severity of crown and foot rot on durum wheat seedlings. *Molecular Plant Pathology* 12: 759–771.

Scherm B., Balmas V., Spanu F., Pani G., Delogu G., … Migheli Q., 2013. *Fusarium culmorum*: causal agent of foot and root rot and head blight on wheat. *Molecular Plant Pathology* 14: 323–341.

Smiley R.W., Gourlie J.A., Easley S.A., Patterson L.M., 2005. Pathogenicity of fungi associated with the wheat crown rot complex in Oregon and Washington. *Plant Disease* 89: 949–57.

Strange R.N., Scott P.R., 2005. Plant disease: a threat to global food security. *Annual Review of Phytopathology* 43: 83–116.

Tóth B., Mesterhàzy A., Nicholson P., Téren J., Varga J., 2004. Mycotoxin production and molecular variability of European and American isolates of *Fusarium culmorum*. *European Journal of Plant Pathology* 110: 587–599.

Tunali B., Büyük O., Erdurmus D., Ozseven I., Dermici A., 2006. Fusarium Head Blight and deoxynivalenol accumulation of wheat in Marmara region and reactions of wheat cultivars and lines to *F. graminearum* and *F. culmorum*. *Plant Pathology Journal* 5: 150–156.

Tunali B., Nicol J.M., Hodson D., Uckun Z., Büyük O., … Bağci S.A. (2008). Root and crown rot fungi associated with spring, facultative, and winter wheat in Turkey. *Plant Disease* 92: 1299–1306.

Wagacha J.M., Muthomi J.W., 2007. *Fusarium culmorum*: infection process, mechanisms of mycotoxin production and their role in pathogenesis in wheat. *Crop Protection* 26: 877–885.

Ward T.J., J.P. Bielawski, H.C. Kistler, E. Sullivan and K. O'Donnell, 2002. Ancestral polymorphism and adaptive evolution in the trichothecene mycotoxin gene cluster of phytopathogenic *Fusarium*. *Proceedings of the National Academy of Sciences* USA 99: 9278–9283.

Winter M., Koopmann B., Doll K., Karlovsky P., Kropf U., … Von Tiedemann A., 2013. Mechanisms regulating grain contamination with trichothecenes translocated from the stem base of wheat (*Triticum aestivum*) infected with *Fusarium culmorum*. *Phytopathology* 103: 682–689.

Yörük E., Albayrak G., 2012. Chemotyping of *Fusarium graminearum* and *F. culmorum* isolates from Turkey by PCR assay. *Mycopathologia* 173: 53–61.

17

Activity of extracts from three tropical plants towards fungi pathogenic to tomato (*Solanum lycopersicum*)

Pascal Noël MEKAM[1,2,*], Serena MARTINI[2], Julienne NGUEFACK[1], Davide TAGLIAZUCCHI[2], Ghislaine Ndonkeu MANGOUMOU[1], Emilio STEFANI[2,3,*]

[1] *Department of Biochemistry, Faculty of Science, University of Yaoundé 1, PO. BOX 812 Yaoundé, Cameroon*
[2] *Department of Life Sciences, University of Modena and Reggio Emilia, Via Amendola 2, 42122 Reggio Emilia, Italy*
[3] *University Centre for International Cooperation and Development (CUSCOS), via Università 4, 41121 Modena, Italy*
*Corresponding authors: noelmekam@gmail.com, emilio.stefani@unimore.it

Competing Interests: The Author(s) declare(s) no conflict of interest.

Editor: Jean-Michel Savoie, INRA Villenave d'Ornon, France.

Summary. Antifungal properties were assessed of water and ethanol extracts from the pan-tropical plants *Oxalis barrelieri* L., *Stachytarpheta cayennensis* L., and *Euphorbia hirta* L. against *Fusarium oxysporum* f. sp. *vasinfectum*, *Alternaria solani* Sorauer, and *Rhizoctonia solani* Kuhn. The plant extracts inhibited fungal growth *in vitro* at 1.25-20 mg mL^{-1}, and the degrees of inhibition increased in a dose-dependent manner. Ethanol extracts from the plants inhibited fungal growth by 80-100%, while water extracts showed less antifungal activity, with maximum growth inhibition of 62%. Growth inhibition from ethanol extracts was two- to three-fold greater than for water extracts at equivalent concentrations. Antifungal activity of the extracts varied with their content and composition of phenolics, flavonoids, tannins, and alkaloids. In greenhouse experiments, spraying tomato plants (*Solanum lycopersicum* L.) with ethanol extract from *E. hirta* at 2.5 mg mL^{-1} did not cause phytotoxicity, and increased plant size, when compared to untreated plants. Spraying *E. hirta* ethanol extract on tomato plants infected by *R. solani* reduced disease severity up to 80%, when compared to non-sprayed plants. These results demonstrate potential of leaf extracts from *E. hirta*, *O. barrelieri*, and *S. cayennensis* as biofungicides for the control of *R. solani*, *A. solani*, and *F. oxysporum*, which are among the most important causal agents of tomato diseases.

Keywords. *Euphorbia hirta*, *Oxalis barrelieri*, *Stachytarpheta cayennensis*, antifungal activity, plant growth promotion.

INTRODUCTION

In Cameroon, tomato is the most important vegetable crop, with production of over 1.18 million tons harvested from 92,626 hectares in 2016 (FAO, 2017). However, phytopathogenic fungi are responsible for the most prevalent diseases, including: late blight, caused by *Phytophthora infestans* (Mont.) de Bary; early blight, caused by *Alternaria* spp.; damping-off and seedling blights, caused by *Pythium* spp.; *Rhizoctonia solani* Kuhn and *Verticillium albo-atrum* Reinke & Berthold; and Fusarium wilt and root rot, caused by *Fusarium oxysporum* f. spp. (Jones *et al.*, 2014). Furthermore, bacterial wilt caused by *Ralstonia solanacearum* (Smith) Yabuuchi *et al.*, and bacterial canker caused by *Clavibacter michiganensis* subsp. *michiganensis* (Smith) Davis *et al.* are also reported to cause severe crop losses in epidemic years (Fontem *et al.*, 1999).

Most strategies for pest management rely on frequent application of synthetic chemical pesticides and copper compounds, to avoid yield and quality losses. These may have consequences for human health, the environment, and from development of resistant pathogen strains, due to continuous use of the same available chemicals (Ishii and Hollomon, 2015; Lucas *et al.*, 2015). For these reasons, it is important to identify new effective strategies for disease management that pose reduced risks to human health and the environment.

Biofungicides have emerged as the main alternative to conventional fungicides, and the application of plant-derived products is a potential choice in disease management (Ribera and Zuñiga, 2012; Pusztahelyi *et al.*, 2015). Tropical plants are rich sources of bioactive chemicals for the development of biopesticides as safe disease control agents, as reviewed by Suprapta (2016). Examples include papaya anthracnose (*Colletotrichum gloeosporioides*) that can be efficiently managed using extracts from tickberry (*Lantana camara, Verbenaceae*) (Ademe *et al.*, 2013); and grey mold (*Botrytis cinerea*) of blackcurrant controlled by extracts from hyssop (*Hyssopus officinalis, Lamiaceae*) and summer savory (*Satureja hortensis, Lamiaceae*) (Sesan *et al.*, 2015). Several plants are reported to produce diverse arrays of low molecular mass antimicrobial compounds, often called 'natural products'. These may also play important roles in plant physiology. Some of these compounds may provide antimicrobial potential and beneficial effects on plants, such as early seed germination, plant growth promotion, improved crop yield, and increased tolerance to abiotic and biotic stresses (Wink, 2010). The compounds can also enhance postharvest shelf-life of perishable products (Ji *et al.*, 2005; Yang *et al.*, 2011; Kharchoufi *et al.*, 2018; Scavo *et al.*, 2019).

Many local plants commonly present in several Cameroon provinces, including *Oxalis barrelieri* L., *Stachytarpheta cayennensis* L., and *Euphorbia hirta* L., have been gaining attention, based on their ethnobotanical uses, phytochemical and pharmacological properties, and their easiness to be cropped in tropical areas. *Euphorbia hirta* has been chemically studied and found to possess antifungal activity against *Fusarium moniliforme* Sheldon and *Phoma sorghina* Saccardo (Karanga *et al.*, 2017). *Oxalis barrelieri* plant extracts showed inhibitory effects on mycelium growth and conidia germination of *F. oxysporum* and *P. infestans* (Dakole *et al.*, 2016). Plant extract of *S. cayennensis* inhibited the growth of different bacteria (Okoye *et al.*, 2010).

Effects of plant extracts on fungal pathogens of tomato have not been studied. The present study aimed to: i) examine the antifungal activity of water and hydro-ethanolic extracts of *E. hirta*, *O. barrelieri* and *S. cayennensis* against three major phytopathogenic fungi affecting tomato (*i.e. R. solani*, *A. solani*, and *F. oxysporum*); ii) determine the phytochemical composition of these extracts; and iii) assess the growth promoting and protective effects of the *E. hirta* ethanol extract on tomato. Once antimicrobial activity is confirmed, such extracts or their components may have potential to be developed as innovative agrochemicals, for implementing sustainable pest management in organic and integrated tomato production.

MATERIALS AND METHODS

Plant material

Three pan-tropical plant species, *Oxalis barrelieri* L. (*Oxalidaceae*), *Stachytarpheta cayennensis* L. (*Verbenaceae*) and *Euphorbia hirta* L. (*Euphorbiaceae*) were collected from a local area (Central Region, Yaoundé-Mbankomo, Cameroon) (Figure 1). These species were chosen based on their previously studied phytochemical properties (Senthikumar, 2018). They were grown until the flowering stage and harvested. After harvesting, the plant material was shade-dried for 2 weeks, and the dried leaves were milled into powder.

Tomato seeds 'Leader F1', obtained from ISI Sementi, Fiorenzuola d'Arda, Italy, were sown and transplanted in a greenhouse into trays containing a commercial soil mix (Dueemme Marketing srl.). The greenhouse was maintained at a constant temperature of 27°C, with a 16 h photoperiod each day. Tomato plants were used for experiments 3 weeks after sowing, at the five to six true leaf stage.

Fungus strains

Fusarium oxysporum f. sp. *vasinfectum*, strain FUSITS04 and *Alternaria solani* Sorauer, both originat-

Figure 1. Pan-tropical plants examined in the present study: *Oxalis barrelieri* L. (A), *Stachytarpheta cayennensis* L. (B), and *Euphorbia hirta* L. (C). These plants are shown at the beginning of the flowering stage, during which they were harvested and dried.

ing from Cameroon, and *Rhizoctonia solani* Kuhn (courtesy of P. Nipoti, University of Bologna, Italy) were used during the experiments. These fungi were isolated from diseased tomato plants, and were maintained on 3.9% potato dextrose agar (PDA) at 4°C until used for experiments. *R. solani* was also used for greenhouse *in planta* experiments.

Preparation of crude plant extracts

Before extraction, leaf powders were defatted (1:6, w/v) by mixing in 600 mL of anhydrous n-hexane under continuous stirring at room temperature (21 ± 1°C) for 24 h. After filtration through fine cloth, each defatted plant residue was hexane-evaporated in a laminar flow cabinet. Extraction of the plant residues was achieved using either distilled water or 70% hydro-ethanol solution. Plant material was soaked (1:6, w/v) in 600 mL of distilled water or in 600 mL of ethanol solution under continuous stirring for 24 h, followed by filtration through Whatman No. 1 filter paper (11 µm pore size) and centrifugation at 5,200 × *g* for 10 min. The supernatants were then collected and the solvents evaporated overnight (12 h) in a ventilated oven at 50°C. The dried pellets obtained were designated as: WE*ox* (water extract of *O. barrelieri*), EE*ox* (ethanol extract of *O. barrelieri*), WE*st* (water extract of *S. cayennensis*, EE*st* (ethanol extract of *S. cayennensis*), WE*eu* (water extract of *E. hirta*), or EE*eu* (ethanol extract of *E. hirta*). The extracts were stored at 4°C until used.

Effects of plant extracts on fungus growth in vitro

Effects of different water and ethanol extracts on growth of fungi were assayed *in vitro* on PDA agar plates, amended with different concentrations of the plant extracts, using the supplemented agar method described by Rios *et al.* (1988).

Five increasing amounts of each plant extract, from 0.125 to 2 g, were added to flasks containing 100 mL of PDA medium before autoclaving (at 121°C for 15 min). This gave final concentrations of each extract in poured PDA plates of 1.25, 2.50, 5, 10, or 20 mg mL^{-1}. PDA plates without extract additions were used as experimental controls. Each agar plate was then inoculated with a 5 mm diam. mycelium plug taken from the margin of a 7-d-old culture, and kept in an incubator at 27°C. Growth was assessed after 7 d, by measuring two opposing diameters of the fungus colony. Growth inhibition relative to the controls was calculated according to the following equation:

$$\text{Growth inhibition (\%)} = \frac{D - d}{D} \times 100$$

where, D = colony diameter in the control PDA plate, and d = colony diameter in the amended PDA plate.

Each fungus-extract combination was replicated in five plates, and all experiments were independently repeated three times.

Greenhouse in planta *experiments*

Application of plant extract and fungus inoculation

To perform *in planta* experiments, the most active ethanol extract of *E. hirta* (EE*eu*) was chosen for assessment against *A. solani*. Prior to these experiments, a preliminary assay was carried out on a set of tomato plants to assess for possible phytotoxic effects of EE*eu*. Since no visible phytotoxic effects were observed in this assay after 1 week, an aqueous solution of EE*eu* was used as a spray onto the shoots canopy of tomato plants, at a concentration of 2.50 mg mL^{-1}.

Inoculum of *R. solani* was prepared from 7-d-old cultures grown on PDA by gently blending the mycelium with the agar to obtain an inoculum paste. Inoculation was carried out 48 h after EE*eu* applications, by transferring the tomato plants into new pots containing infested soil-perlite (3:1, v:v). The soil was infected by setting 2.5 g of *R. solani* inoculum paste at the bottom of a hole made at mid-depth in each pot (Logemann *et al.*, 1992). The tomato plants were carefully uprooted from their previous pots, partially cut at their root ends and transferred into the inoculated pots, so that the inoculum paste became in contact with the wounded roots. The pots were then arranged in a completely randomized design in a greenhouse maintained at 27°C, under a 16 h photoperiod each day and at 70% relative humidity, and were watered appropriately. The experimental design considered four experimental treatments: i) plants sprayed with EE*eu* (TE); ii) plants inoculated with *R. solani* (TR); iii) plants sprayed with EE*eu*, followed by inoculation with *R. solani* (TE+R); or iv) untreated tomato plants, as the experimental control (TC). Ten replicate plants were used for each treatment, experiments were repeated three times independently.

Evaluation of tomato plant growth promotion and disease

For each treatment, the heights of the treated plants were measured at 7, 14, and 21 d after extract application, to evaluate effects of the extract on plant growth.

Disease severity was recorded for TR and TE+R treatments at 5, 10, 15, and 20 d after inoculation. Severity was scored using a 0-5 scale, were: 0 = healthy plant, 1 = 1-10% of leaves with initial wilts, 2 = 11- 25% of leaves with wilts, 3 = 26-49% of leaves showing wilting and chlorosis, 4 = 50-74% of leaves showing pronounced wilting and development of necrotic areas, and 5 = whole leaves wilting. Severity was calculated according to the following equation:

$$\text{Disease severity (\%)} = \frac{\Sigma dn}{DN} \times 100$$

where, d = severity score; n = number of disease plants with the same severity score; N = total number of the examined plants, and D = the greatest severity score.

Disease reduction was calculated by comparing disease severity observed on plants using the following equation:

$$\text{Disease reduction (\%)} = \frac{S - s}{S} \times 100$$

where, S = disease severity on tomato inoculated with *R. solani* (treatment TR), and s = disease severity on tomato treated with EE*eu* then inoculated with *R. solani* (treatment TE+R).

Analyses of phytochemical contents and antioxidant potential

Preliminary analyses of crude plant extracts were performed to assess their phytochemical compositions. Each extract was analyzed for: i) total phenolic content using the Folin-Ciocalteu reagent, using the method of Singleton *et al.* (1999); ii) total flavonoid content, using the method of Zhishen *et al.* (1999); iii) total tannin content, using the method described by Verzelloni *et al.* (2010); iv) total alkaloid content, using bromocresol green reagent and the method described by Tabasum *et al.* (2016); v) total polysaccharides content using phenol-sulfuric acid after mild acid hydrolysis, using the method of Dubois *et al.* (1956); and vi) total protein content using Bradford's reagent, and Bradford's method (Bradford, 1976).

Antioxidant activity of the plant extracts was measured using the free radical cation 2,2-azinobis-3-ethyl benzothiazoline-6-sulfonic acid, using the method described by Re *et al.* (1999).

Statistical analyses

All data were statistically analyzed using the MaxStat Lite software, version 3.60 (available at: https://maxstat-lite.soft112.com/). Data were subjected to analysis of variance (ANOVA) and the means comparison carried out using Tukey's multiple range test at $P \leq 0.05$. Principal component analyses (PCA) were performed using the software package Solo, version 8.6.1, (Eigenvector Research, Inc. Manson, WA, USA), considering the analytical properties as variables.

RESULTS

Characteristics of plants extracts

The mean yields of the extracts, based on initial dry biomass of leaves, varied between 5 and 8% (w/w). More precisely, extraction productivity for the different plant extracts was: WE*ox*, 6.2%; EE*ox*, 5.1%; W*est*, 7.4%; EE*st*, 5.3%; WE*eu*, 8.2%; and EE*eu*, 5.6%.

Extracts differed in colour: water extracts were dark brown, and the ethanol extracts were grayish-green.

Antifungal activity of plant extracts

The addition of the plant extracts to PDA medium at all concentrations did not affect medium consistency, but only its colour. Growth of the three phytopathogenic fungi on PDA medium, without addition of plant extracts, reached the following mean colony diameters after 1 week: 8.50 cm for *R. solani*, 7.50 cm for *F. oxysporum* f. sp. *vasinfectum* and 7.38 cm for *A. solani*. Addition of the plant extracts to PDA medium inhibited fungus growth in all of the substrate/fungus combinations, and inhibition was related to extract concentration in the medium (Figure 2A). Increasing concentration of EE*eu* led to complete inhibition of *F. oxysporum*. The same concentration of WE*eu* also inhibited growth of *F. oxysporum*, but to a lesser extent than for EE*eu* (Figure 2B). The tested plant extracts consistently reduced mycelium growth of *R. solani*, *F. oxysporum*, and *A. solani*.

The ethanol extracts gave greater antifungal activity than the water extracts. As little as 1.25 mg mL^{-1} of ethanol extracts from the three plants were sufficient to inhibit fungus growth from 10 to 28% (depending on fungus). Between 10 to 20 mg mL^{-1} of ethanol extract, growth inhibition was 90–100% (Figures 3A, 4A, and 5A). Therefore, the ethanol extracts in sufficient concentration completely inhibited growth of all three fungi. At low doses, water extracts from all three plants also inhibited fungus growth by 2 to 28% (depending on fungus) but differently from the ethanol extracts. Growth inhibition from the water extracts was never as great as with the ethanol extracts (Figures 3B, 4B and 5B). This confirms the greater antifungal activity of the ethanol than water extracts.

Fungus growth inhibition was dose dependent for all the fungus/extract combinations, but the inhibitory effects of extracts varied for extracts from the different plants. In general, extracts from *E. hirta* gave greater inhibition than those from *S. cayennensis*, which were more active than the extracts from *O. barrelieri*.

An effect was observed on the morphology of *F. oxysporum* grown on PDA plates amended with WEs. Mycelium colour and texture were greatly modified on amended PDA plates, as shown in Figure 6. These changes in morphology were not apparent for the other two fungi growing on WEs-amended media.

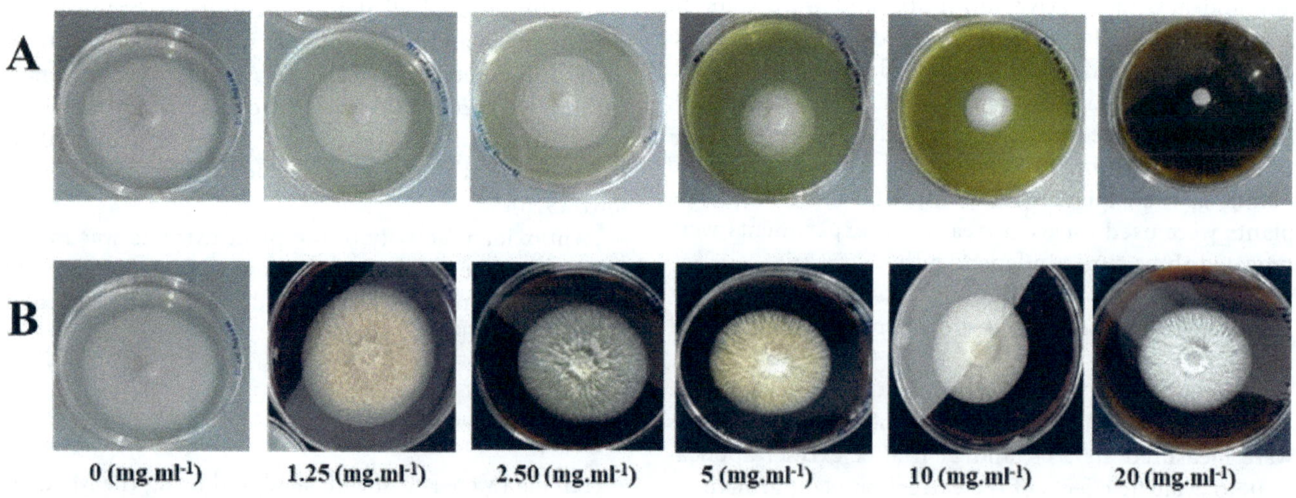

Figure 2. Inhibition of mycelium growth of *Fusarium oxysporum* f. sp. *vasinfectum* after 7 d on PDA supplemented with increasing concentrations of ethanol extract (A) and water extract (B) from *Euphorbia hirta*.

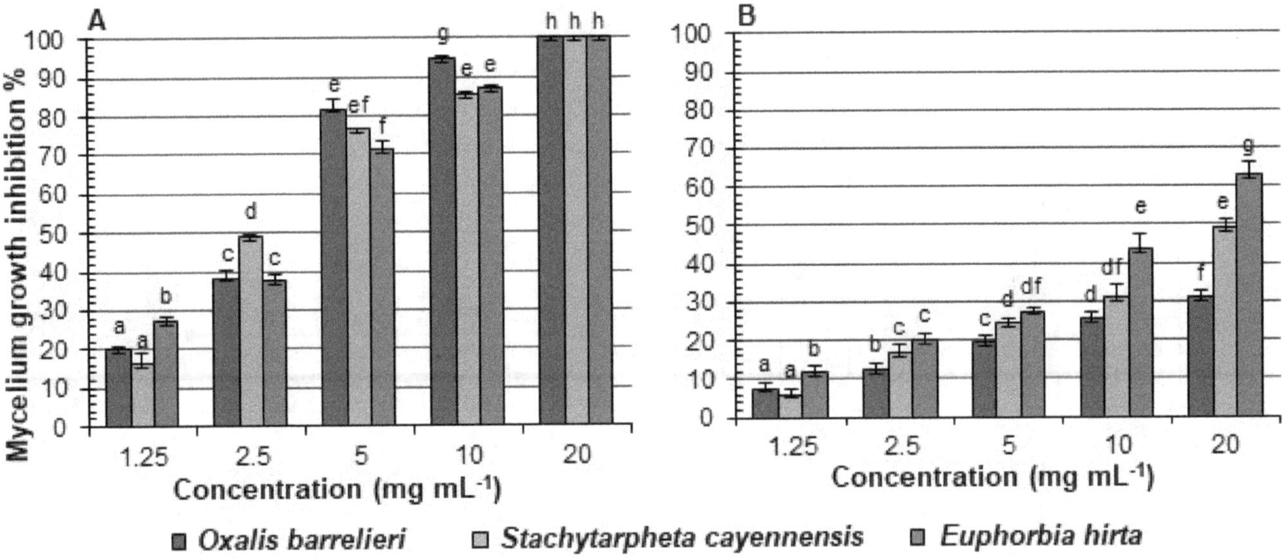

Figure 3. Mycelium growth inhibition (%) of *Fusarium oxysporum* f. sp. *vasinfectum* at 1.25, 2.50, 5, 10 or 20 mg mL^{-1} concentrations of the ethanol extracts (A) or water extracts (B) for extracts from three different plant species. Data are means ± standard deviation from three experiments, each with five replicates. At each extract concentration, the means accompanied by different letters are significantly different, according to ANOVA paired with Tukey's tests ($P \leq 0.05$).

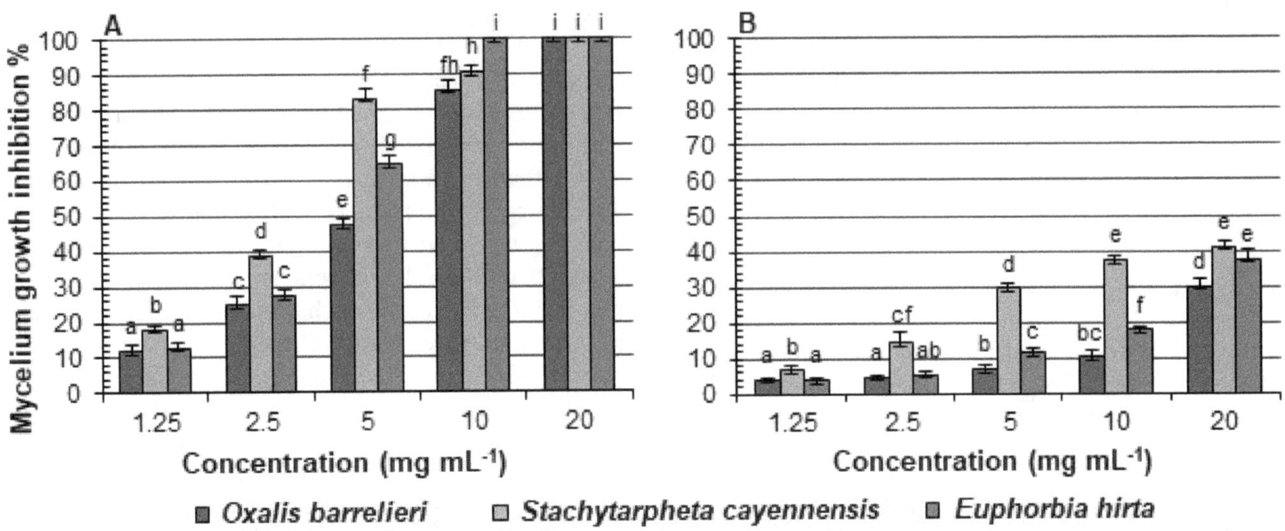

Figure 4. Mycelium growth inhibition of *Rhizoctonia solani* at 1.25, 2.50, 5, 10 and 20 mg mL^{-1} concentrations of the ethanol extracts (A) and water extracts (B) from three plant species. Data are means ± standard deviation from three experiments, each with five replicates. At each concentration, the means accompanied by different letters are significantly different, according to ANOVA paired with Tukey's tests ($P \leq 0.05$).

Effects of Euphorbia hirta *ethanol extract on tomato plant height*

The height of all tomato plants sprayed with EE*eu* (TE) was significantly increased after 7 d (Figure 7). Plant size increased during the following weeks and was greater ($P \leq 0.05$) in EE*eu*-treated plots, when compared to controls. Three weeks after spraying with EE*eu*, mean plant heights from the different extract treatments were: TE, 20.4 cm; TC, 14.5 cm; TE+R, 12.3 cm; and TR. 11.2 cm.

Disease reduction by Euphorbia hirta *ethanol extract*

Data of disease severity and reduction are summarized in Table 1. Tomato plants infected with *R.*

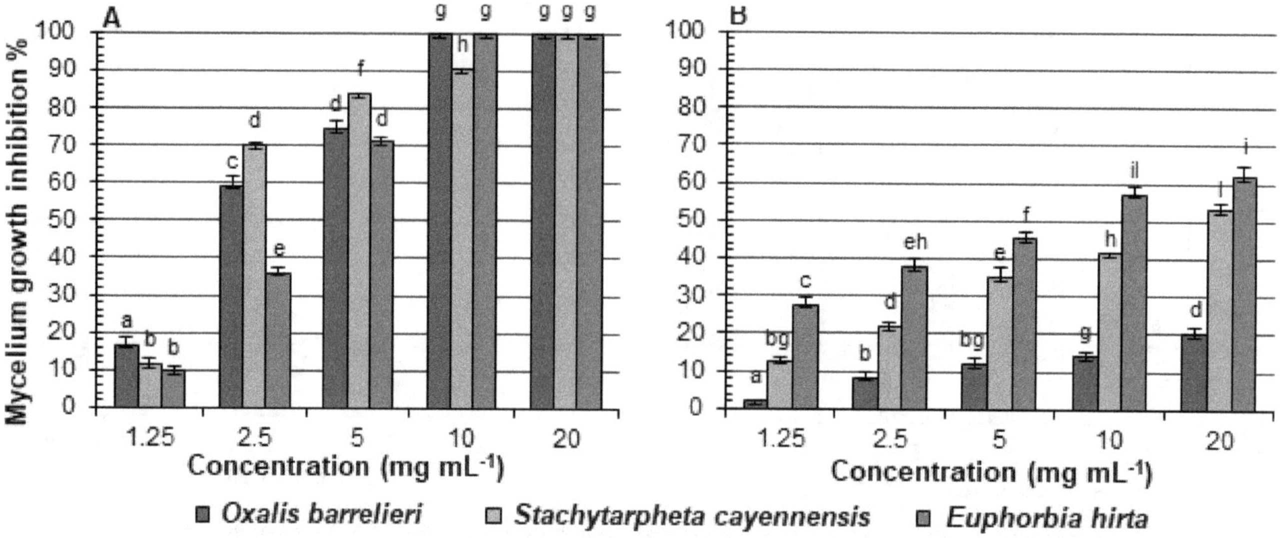

Figure 5. Mycelium growth inhibition of *Alternaria solani* at 1.25, 2.50, 5, 10 and 20 mg mL^{-1} concentrations of the ethanol extracts (A) and water extracts (B) from three plant species. Data are means ± standard deviation from three experiments, each with five replicates. At each concentration, the values accompanied by different letters are significantly different, according to ANOVA paired with Turkey's tests ($P \leq 0.05$).

solani (TR) showed greater disease severity, when compared to EE*eu*-treated plants inoculated with *R. solani* (TE+R). The final disease severity estimates gave 56.7 ± 5% from the TE+R treatment and 80 ± 2% from the TR treatment. This demonstrated that the *E. hirta* ethanol extract efficiently protected tomato plants from *R. solani* infections.

Phytochemical content and antioxidant capacity of the plant extracts

Means values of total phenolic, flavonoid, alkaloid, tannin, protein, and sugar contents and radical antioxidant activity of all the tested plant extracts are presented in Table 2. Except for the alkaloids, whose contents in plant extracts were very similar, the quantification of other chemical components was different for all tested plant extracts. EE*st* had the greatest amounts of phenolics (101.7 ± 11.8 mg gallic acid g^{-1}), flavonoids (33.5 ± 0.7 mg catechin g^{-1}), and tannins (7 ± 1.9 mg catechin g^{-1}). WE*ox* had the greatest amount of proteins (8.1 ± 0.8 mg BSA g^{-1}). The polysaccharide components were greater in water extracts compared to the ethanol extracts, except for EE*st* (179 ± 27.3 mg glucose g^{-1}) and WE*st* (94.8 ± 18.9 mg glucose g^{-1}).

Table 2 shows that WE*ox* and WE*st* were more effective in scavenging ABTS radical cation, with values, respectively, of 262.67 ± 41.48 and 250.33 ± 40.54 mg of L-ascorbic acid g^{-1}. Mean values for the other extracts were: for EE*st*, 168.33 ± 2.36; for EE*ox*, 157.22 ± 4.46; for WE*eu*, 112.53 ± 8.01, and for EE*eu*, 83 ± 26.40 mg of L-ascorbic acid g^{-1}.

The chemical nature of the extracts correlated with the extraction procedures. For instance, *O. barrelieri* was quite productive for total phenolics, when extracted with water (more than double the quantity of phenolics in WE than in EE), whereas *S. cayennensis* and *E. hirta* were much more productive when extracted with ethanol. *Euphorbia hirta* gave the maximum productivity using ethanol for all components, other than total polysaccharides. Total polysaccharides were more abundant in WEs of all three plants. This was expected, since polysaccharides are more soluble in water than in ethanol solutions (Guo *et al.*, 2017). Conversely, total alkaloids and total flavonoids were better extracted with ethanol than water from all three plants: in particular, *E. hirta* was very productive in total flavonoids when extracted with ethanol, compared to water extraction.

The cumulative percentage (74.7%) of the total variance explained by the first two principal components (PC1 and PC2) allowed designing the bidimensional plot represented in Figure 8. The distribution of extracts along PC1 and PC2 shows a clear distinction between ethanol and water extracts, respectively, positive and negative scores on PC1. In order to understand which variables accounted most for this distribution, they were added to the bidimensional plot. The water extracts had negative scores on PC 1 and were characterized by greater polysaccharide contents than the respective ethanol extracts, and the greatest

Figure 6. Morphological modification (colour, structure and texture) of *Fusarium oxysporum* f. sp. *vasinfectum* growing on modified PDA amended with water extract from *Oxalis barrelieri*. Left, PDA amended with 5 mg mL^{-1} of water extract; right, PDA without plant extract.

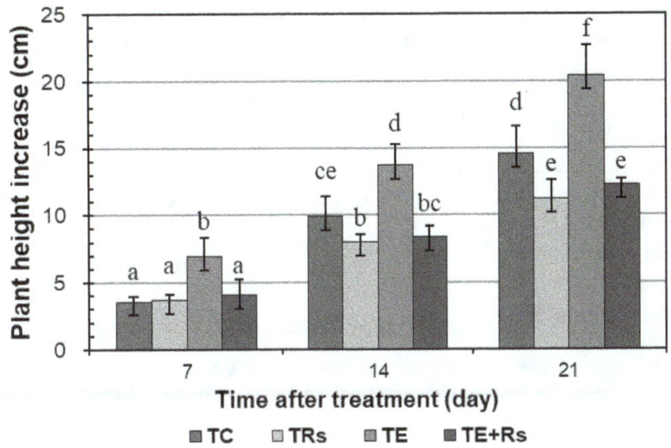

Figure 7. Height (cm) of tomato plants, with or without inoculation of *Rhizoctonia solani*, after treatment with *Euphorbia hirta* ethanol extract, for plants measured at 7, 14 or 21 days after treatment. TC: Negative control, sprayed with water; TRs: positive control, inoculated with *R. solani*; TE: Plants treated with *E. hirta* extract; TE+Rs: Plants treated with *E. hirta* extract and inoculated with *R. solani*. Values are means ± standard deviation, calculated for 30 plants per treatment. Values accompanied by different letters are significantly different, according to ANOVA paired with Tukey's tests ($P \leq 0.05$).

antioxidant activities (WE*ox* and WE*st*, with positive scores on PC 2). Otherwise, ethanol extracts, which were positively linked to PC 1, were characterized by the greatest phenolic, tannin and flavonoid contents (particularly EE*st* and EE*eu*), and were more effective for growth inhibition of the fungi than the respective water extracts, depicted by the positive scores and positive correlation on PC 1.

DISCUSSION

The potential of plant-derived molecules as effective compounds for management of plant pests and diseases raised considerable attention during the last 10-15 years (Reignault and Walters, 2007; Martinez, 2012). This has been particularly important in areas where chemical pesticides may have deleterious impacts on the sustainability of agricultural systems and on food safety (Tripathi and Dubey, 2004; Shuping and Eloff, 2017).

Table 1. Influence of *Euphorbia hirta* ethanol extract on fungal disease caused by *Rhizoctonia solani* on tomato plants.

Treatment	Disease index (%)			
	5 d	10 d	15 d	20 d
R. solani (T_{Rs})	16 ± 2.00^{b}	37.33 ± 3.06^{d}	60.67 ± 7.57^{e}	80 ± 2.00^{f}
Extract + *R. solani* (T_{E+Rs})	8 ± 2.00^{a}	24.67 ± 5.77^{c}	38.67 ± 6.11^{d}	56.67 ± 5.03^{e}
	Disease reduction (%)			
	50.53 ± 6.37	33.43 ± 17.74	35.47 ± 13.59	29.24 ± 4.52

Values are the means ± standard deviation, calculated for 30 plants per treatment.
Different letters indicate significant differences in disease indices, according to ANOVA paired with Tukey's tests ($P \leq 0.05$).

Table 2. Total phenolics, flavonoids, alkaloids, tannins, proteins, polysaccharides content, and antioxidant capacity of three tropical plant extracts from *Oxalis barrelieri*, *Stachytarpheta cayennensis* and *Euphorbia hirta*.

Components	*Oxalis barrelieri*		*Stachytarpheta cayennensis*		*Euphorbia hirta*	
	Ethanol extract	Water extract	Ethanol extract	Water extract	Ethanol extract	Water extract
Total phenolics (mg GAE g^{-1})	34.24 ± 2.65^{b}	70.19 ± 6.80^{d}	101.71 ± 11.84^{e}	49.99 ± 4.62^{c}	71.58 ± 2.34^{d}	24.99 ± 3.87^{a}
Total flavonoids (mg CE g^{-1})	7.52 ± 0.52^{b}	7.03 ± 0.35^{b}	33.54 ± 0.69^{e}	22.95 ± 1.6^{d}	15.16 ± 0.1^{c}	3.47 ± 0.68^{a}
Total tannins (mg CE g^{-1})	0.96 ± 0.44^{ab}	3.39 ± 0.04^{c}	6.96 ± 1.89^{d}	0.94 ± 0.29^{a}	5.00 ± 1.18^{d}	1.73 ± 0.38^{b}
Total alkaloids (mAbs g^{-1})	76.75 ± 21.75^{b}	39.55 ± 12.09^{a}	73.00 ± 12.73^{b}	70.10 ± 16.83^{b}	68.25 ± 11.67^{b}	41.70 ± 18.95^{ab}
Total proteins (mg BSA g^{-1})	4.14 ± 0.28^{b}	8.08 ± 0.82^{e}	5.28 ± 0.22^{c}	4.41 ± 0.27^{b}	6.32 ± 0.27^{d}	1.75 ± 0.11^{a}
Total polysaccharides (mg glucose g^{-1})	46.40 ± 19.24^{a}	83.12 ± 4.01^{b}	94.76 ± 18.88^{b}	179.27 ± 27.25^{c}	59.81 ± 5.07^{a}	84.65 ± 10.71^{b}
Antioxidant activity (mg L-ascorbic acid g^{-1})	157.22 ± 4.46^{c}	262.67 ± 41.48^{e}	168.33 ± 2.36^{d}	250.33 ± 40.54^{e}	112.53 ± 8.01^{b}	83.00 ± 26.4^{a}

Values means ± standard deviation of three replicates.
Uppercase letters in the same row indicate significant differences according to ANOVA paired with Tukey's tests ($P \leq 0.05$).

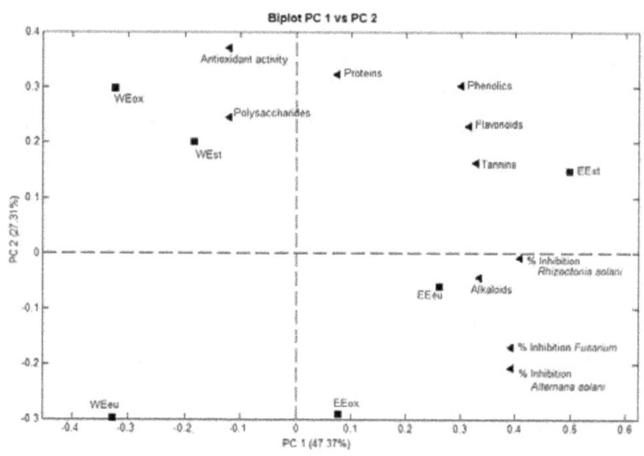

Figure 8. Main biplot (loadings and scores) obtained from principal component analysis for data from ethanol and water extract evaluations, along principal components 1 (PC1) and 2 (PC2). WE and EE indicate, respectively, values for water and ethanol extracts. ox: *Oxalis barrelieri* L., st: *Stachytarpheta cayennensis* L., and eu: *Euphorbia hirta* L. Symbol ▪ depicts the different extracts (scores), whereas ◂ represents the bioactivities and biochemical properties of the extracts (loadings).

In the present study, leaves from three pan-tropical plants, *O. barrelieri*, *S. cayennensis*, and *E. hirta*, were used as potential sources of bioactive molecules against three important phytopathogenic fungi, affecting tomato in Cameroon and worldwide. The three plant species were chosen as sources of possible bioactive/antimicrobial compounds, since they are very common as ruderal herbs in young fallows (Tchiengué, 2012). Additionally, preliminary ethnopharmacological studies showed that these species are relevant as medicinal plants (FAO, 1999).

Our results highlighted the potential fungicidal/fungistatic activity of these plant extracts *in vitro*. Antifungal effects were always more pronounced using hydro-ethanol extracts (EEs) than water extracts (WEs), and were concentration and extract-pathogen-interaction dependent. This confirms the results of Kotze and Eloff (2002), who reported that, in most cases, water extracts had a low antimicrobial efficacy. Therefore, for several plant species, the use of less polar solvents increases the extraction efficiency and concentration of antimicrobial molecules (Eloff, 1998), such as flavonoids and phenolics (Table 2). The complete inhibition of fungal growth was recorded using a concentration ranging from 10 to 20 mg mL^{-1} of EEs obtained from the three plants tested. The antifungal efficacy of such extracts increased linearly ($P \leq 0.05$) in relation to the concentration used, both for EEs and for WEs. This indicates the presence of antifungal molecules in both extracts. The preliminary phytochemical composition of WEs and EEs of the three plants species used in our experiments showed various concentrations of active biomolecules. The concentration of phenolics was greatest in EE*st* and EE*eu* and least in WE*eu*. Conversely, for *O. barrelieri*, more phenolics were found in WE*ox* than EE*ox*. This is possibly due to the diverse nature of phenolics produced by *O. barrelieri* compared, for instance, with *E. hirta* (Mekam *et al.*, 2019). Phenolics (or phenols) are a large class of compounds classified as simple phenols or polyphenols. They are found in all plants and consist of simple phenols, benzoic and cinnamic acids, coumarins, tannins, lignins, lignans and flavonoids (Khoddami *et al.*, 2013). Therefore, it is expected that different plants may produce phenolics in different quantities and of different chemical structures. Phenolics are frequently synthesized in plants in response (or as protection) to stress, such as pathogen infection, insect attack, UV radiation or wounding. Therefore, these compounds have important roles in plant defense (Mandal *et al.*, 2010). In most plant species, the key step in phenolic syntheses is the conversion of phenylalanine. This reaction, catalyzed by the phenylalanine ammonia-lyase enzymes, leads to the production of various hydroxycinnamic acids, benzoic acids and derivatives of polyphenols, which are important antifungal biomolecules often found in plants (Raymond Chia and Dykes, 2010).

As highlighted in Table 2, phenolic contents of the different plant extracts were: EE*st* > EE*eu* > WE*ox* > WE*st* > EE*ox* > WE*eu*. Nevertheless, their antifungal activities did not follow this order, since the EEs gave greater antifungal activity than WEs, when tested at equivalent concentrations. EE*eu* gave the greatest antifungal activity, completely inhibiting mycelium growth of *F. oxysporum* f. sp. *vasinfectum* and *A. solani*, at a concentration of 10 mg mL^{-1}. At the same concentration, WE*eu* exhibited moderate antifungal activity, inhibiting mycelium growth of *F. oxysporum* f. sp. *vasinfectum* by 18% and *A. solani* by 57%. In general, hydro-ethanol treatments extracted more (and possibly more diverse) antifungal compounds than water, and this may be related to the less polar nature of ethanol as an extractant, when compared to water. Considering the minimum effective concentration of EEs that completely inhibited fungal growth, the measured EE activity was twice to three times greater than for WEs. These results agree with those of other reports that have showed the efficiency of some hydro-ethanolic plant extracts have the greatest and widest range of *in vitro* activities, resulting in complete inhibition of fungal growth. Galani *et al.* (2013) reported that the EEs of *Ageratum conyzoides* and *Callistemon citrinus* completely inhibited *Phythophthora infestans* at 5,000 ppm, and that of *Ocimum gratissimum* at 10,000 ppm. Dakole *et al.* (2016) reported that the EEs of *Ageratum conyzoides* and *Callistemon citrinus* were the most active for inhibiting radial growth of *Phytophthora infestans. Cymbopogon citratus* and *Ocimum gratissimum* were the most active against radial growth and conidia germination of *Fusarium oxysporum* f. sp. *lycopersici* at a concentration of 6,250 μg mL^{-1}. This variability in antifungal activity of extracts suggested that a correlation exist between the nature of plant extracts and their concentrations of active phytochemicals. The amounts, biochemical nature and activity of phenolics (including flavonoids and tannins) and alkaloids contained in the present study extracts was dependent on: i) their solubility in water or ethanol; ii) the absence of inhibitors; iii) their synergism with polysaccharides and proteins present in the extracts; iv) the differences in modes of action; and, v) the structure and biology of the phytopathogenic fungi (Lapornik *et al.*, 2005). Therefore, the phenolics present in these plant extracts displayed important and direct inhibition of growth of three tomato pathogens and may be considered as prospective compounds for management of plant diseases. In particular, the extracts from *E. hirta* used in our experiments showed the greatest amounts of phenolic compounds, mainly gallotannins, and hydroxybenzoic and hydroxycinnamic acids (Mekam *et al.*, 2019). These phenolics are known bioactive molecules with antifungal properties (Alves Breda *et al.*, 2016).

Plant-derived alkaloids were extensively studied for their antifungal properties against human pathogenic fungi, such as *Candida albicans* (Mollataghi *et al.*, 2012). Plant alkaloids are also reported to have activity against phytopathogenic fungi. For instance, allosecurinine from *Phyllanthus amarus* (*Euphorbiaceae*) was able to inhibit the growth of *Alternaria* spp. (including *A. solani*), *Fusarium* spp. and other important fungi affecting crop plants (Singh *et al.*, 2008). Liu *et al.*, (2009) found that sanguinaine, an isoquinoline alkaloid from *Macleaya cordata* (*Papaveraceae*) reduced the growth of *R. solani* at a concentration of 0.45 μg mL^{-1}. Nonetheless, the specific roles of alkaloids extracted from the tested plants remain uncertain, as compared to the possible inhibitory role of phenolics present in the same extracts. This role (if any) does not appear to be important since no significant difference in content was found among the three tested plants.

Regarding polysaccharides, one of the best known with confirmed antifungal effects is laminarin (or laminaran), a glucan with different degrees of molecular branching at β-1,3 and β-1,6. For instance, grey mold (*Botrytis cinerea*) and downy mildew (*Plasmopara viticola*) of grapevine were inhibited by spraying laminarin onto vine canopies (Copping, 2004). Laminarin-based products are commercially available as biopesticides (Environmental Protection Agency, 2010). Activity of laminarin in plant tissues is more related to induced resistance than from direct antifungal effects (Aziz *et al.*, 2003). In the present study, polysaccharides were particularly abundant in the WE of *S. cayennensis*, but WE was not more active than EE from this plant. This suggests that polysaccharides may not have inhibitory effects on fungal growth as great as other compounds in the tested extracts – especially the EEs.

The greater antioxidant capability of a plant extract may be correlated with phenolic composition and concentration. The greater the phenolic content the greater is the antioxidant activity. This is due to the phenolic hydroxyl groups, which stop radical chain reactions via radical scavenging (Shahidi and Chandrasekara, 2010).

Principal component analyses for exploration of relationships between plant extracts and biochemical properties assisted description of variance in the set of multivariate data we obtained (polyphenol, tannin, alkaloid contents and associated fungal growth inhibition). Sample splitting showed in the PC bi-plot clearly reflected the main differences due to extraction method, the plant origins, and the phenolic and alkaloid compositions, which influenced antifungal activity of the different extracts.

Under greenhouse condition, foliar application of EE*eu* on tomato before a challenge pathogen inoculation resulted in increased plant height and the reduction of disease severity caused by *R. solani*. Increased plant height was a beneficial effect of foliar application of EE*eu* during the *in planta* experiments. EE*eu* stimulated plant height more than water in treated tomato plants (Figure 7). This growth promoting activity of EE*eu* was possibly attributable to low molecular weight components, such as plant hormones (gibberellic acids), and to major components such as polysaccharides, amino acids and polyphenols that are involved in many aspects of plant physiology and development. These include seed germination, stem and leaf elongation, flower induction, and fruit and seed development (Kamiya and Garcia-Martinez, 1999). This result is similar to those of Nguefack *et al.* (2013), who reported that spraying rice plants with a 2% ethanol extract, followed by a 2% (w/v) aqueous extract of *Callistemon citrinus* or *Cymbopogum citratus*, increased seedling emergence, tillering, panicles/plant and the grain yields by 25–55%. Zakiah *et al.* (2017) reported that low concentrations of a crude extract of *Cassava asiatica* (25 mg L^{-1}) increased height and leaf area of soybean plants. However, in the present study the tomato-EE*eu*-*R. solani* interaction showed less plant height increase, when compared to the tomato-water interaction. Infection by a root/stem pathogen probably interfered with host metabolism and contributed to reduced plant development.

Management of tomato fungal diseases is primarily achieved through breeding resistant cultivars. Many recently developed cultivars are moderately or very resistant to Fusarium wilt, whereas no tomato variety is currently available that shows acceptable tolerance to damping-off/crown rot. Therefore, we chose *R. solani* as the challenging pathogen for the *in vivo* experiments. Disease reduction reached 29 ± 5% in the tomato-EE*eu*-*R. solani* interaction, in comparison to the tomato-water-*R. solani* interaction, at 20 d after inoculation (Table 1). This suggests that EE*eu* contained natural compounds that possibly act alone or in synergy, stimulating plant defense to provide disease control through induced systemic resistance in tomato against damping-off. This is particularly important since damping-off cannot be easily managed, especially in areas where appropriate crop rotations are not followed. Although extracts from tropical plants are reported to provide crop disease control through stimulation of plant defense systems (Baraka *et al.*, 2011; Nashwa and Abo-Elyousr, 2012), *E. hirta* extracts have not been previously examined.

CONCLUSIONS

Natural flora is a source of several biologically active compounds, and some of these have been formulated as botanical agrochemicals and are currently used in the management of agricultural pests (Dubey *et al.*, 2011). The present study established that water and hydro-ethanol extracts obtained from leaves of *O. barrelieri*, *S. cayennensis*, and *E. hirta* are sources of phytochemicals, and these molecules demonstrated inhibitory activity against phytopathogenic fungi. Application of these extracts is, therefore, a promising and environmentally friendly strategy for crop disease control that could contribute to minimizing the risks and hazards of posed by conventional fungicides.

Economic advantages from the use of plant extracts could be particularly relevant in African rural areas, where these pan-tropical plants are common and adapted. Biomolecules may be developed into commercial products by local companies, thus contributing to rural and agro-industrial development, together with increased sustainability for local cropping systems. Current research is devoted to identifying the most effective phenolics and other antifungal compounds (Mekam *et al.*, 2019), and compounds showing effective plant growth promotion, to develop and implement innovative plant disease biocontrol strategies.

ACKNOWLEDGMENTS

This research has been supported in part by the European Commission, under the Erasmus+ KA107 mobility project between the University of Modena and Reggio Emilia (Italy) and the University of Yaoundé I (Cameroon).

LITERATURE CITED

Ademe A., Ayalev A., Woldetsadik K., 2013. Evaluation of antifungal activity of plant extracts against papaya anthracnose (*Colletotrichum gloeosporioides*). *Journal of Plant Pathology and Microbiology* 4: 1-4.

Alves B.C., Marcon G.A., Garcia V.L., Monteiro K.M., Bataglion G.A., Eberlin M.N., Teixeira D.M.C., 2016. Phytochemical analysis and antifungal activity of extracts from leaves and fruit residues of Brazilian savanna plants aiming its use as safe fungicides. *Natural Products and Bioprospecting* 6: 195–204.

Aziz A., Poinssot B., Daire X., Adrian M., Bezier A., … Pugin A., 2003. Laminarin elicits defense responses in grapevine and induces protection against *Botrytis cinerea* and *Plasmopara viticola*. *Molecular Plant-Microbe Interactions* 16: 1118–1128.

Baraka M.A., Fatma M.R., Shaban W.I., Arafat K.H., 2011. Efficiency of some plant extracts, natural oils, biofungicides and fungicides against root rot disease of date palm. *Journal of Biology Chemical and Environment Sciences* 6: 405–429.

Bradford M.M., 1976. A rapid and sensitive method for the quantization of microgram quantities of protein utilizing the principle of protein-dye binding. *Analytical Biochemistry* 72: 248–254.

Copping L.G., 2004. *The Manual of Biocontrol Agents*. 3rd ed. British Crop Protection Council Publications, Alton, Hampshire, UK, 702 pp.

Dakole D.C., Nguefack J., Dongmo L.J.B., Galani Y.J.H., Azah U.R., … Amvam Z.P.H., 2016. Antifungal potential of essential oils, aqueous and ethanol extracts of thirteen plants against *Fusarium oxysporum* f. sp. *lycopersici* and *Phytophtora infestans* (Mont.) de Bary as major tomato pathogens in Cameroon. *International Journal of Current Science* 19: 128–145.

Dubey N.K., Ravindra S., Ashok K., Priyanka S., Bhanu P., 2011. Global scenario on the application of natural products in integrated pest management programmers. In: *Natural Products in Plant Pest Management* (N.K. Dubey, ed.), CAB International, Wallingford, UK, 1–20.

Dubois M., Gille K.A., Hamilton J.K., Rebers P.A., F. Smith, 1956. Colorimetric method for determination of sugars and related substances. *Annals of Chemistry* 28: 350–356.

Eloff J.N., 1998. Which extractant should be used for the screening and isolation of antimicrobial components from plants? *Journal of Ethnopharmacology* 60: 1-8.

Environmental Protection Agency, Office of Pesticide Programs, Biopesticides and Pollution Prevention Division, Biopesticide Registration Action Document: Laminarin - PC Code: 123200. U.S., 2010. Available at: http://www.talunecoproducts.com/wp-content/uploads/2017/01/EPA-Laminarin.pdf. Accessed December 20, 2018.

FAO, 1999. Données statistiques des produits forestiers non-ligneux du Cameroun. Programme de partenariat CE-FAO (1998–2001) - GCP/INT/679/EC. Available at: http://www.fao.org/3/X6699F/X6699F00.htm#TopOfPage. Accessed August 12, 2019.

FAO, 2017. Faostat. Available at: http://faostat3.fao.org/download/Q/QC/E. Accessed August 29, 2018.

Fontem D.A., Gumedzoe M. , Nono-Womdim R., 1999. Biological constraints in tomato production in the western highlands of Cameroon. *Tropicultura* 16: 89–92.

Galani Y.J.H., Nguefack J., Dakole D.C., Fotio D., Petchayo T.S., … Amvam Z.PH., 2013. Antifungal potential and phytochemical analysis of extracts from seven Cameroonian plants against late blight pathogen *Phytophthora infestans*. *International Journal of Current Microbiology and Application Science* 2: 140–154.

Guo M.Q., Hu X., Wang C., Ai L., 2017. Polysaccharides: Structure and Solubility. *IntechOpen*, DOI: 10.5772/intechopen.71570. https://www.intechopen.com/books/solubility-of-polysaccharides/polysaccharides-structure-and-solubility.

Ishii H., Hollomon D.W., 2015. *Fungicide resistance in plant pathogens: Principles and a guide to practical management*. Springer Japan, IX + 490 pp.

Ji P., Momol T.M, Olson S.M., Pradhanang P.M., Jones J.B., 2005. Evaluation of thymol as biofumigant for control of bacterial wilt of tomato under field conditions. *Plant Disease* 89: 497–500.

Jones J.B, Zitter T.A., Momol T.M., Miller S.A., 2014. *Compendium of tomato diseases and pests*. 2nd ed.. APS Press, St. Paul, Minnesota, USA, 168 pp.

Kamiya Y., Garcia-Martinez J.L., 1999. Regulation of gibberellin biosynthesis by light. *Current Opinions in Plant Biology* 2: 398–403.

Karanga Y., Ilboudo O., Bonzi S., Tapsoba I., Somda I., Bonzi-Coulibaly Y.L., 2017. Phytochemical and antifungal properties of *Euphorbia hirta* L. against *Fusarium moliniforme* and *Phoma sorghina*. *Natural Products: an Indian Journal* 13: 105–114.

Kharchoufi S., Parafati L., Licciardello F., Muratore G., Hamdi M., … Restuccia C., 2018. Edible coatings incorporating pomegranate peel extract and biocontrol yeast to reduce *Penicillium digitatum* postharvest decay of oranges. *Food Microbiology*, 74: 107–112.

Khoddami A., Wilkes M.A., Roberts T.H., 2013. Techniques for analysis of plant phenolic compounds. *Molecules* 18: 2328–2375.

Kotze M., Eloff J.N., 2002. Extraction of antibacterial compounds from *Combretum microphyllum* (*Combretaceae*). *South African Journal of Botany* 68: 62–67.

Lapornik B., Prošek M., Wondra A.G., 2005. Comparison of extracts prepared from plant by products using different solvents and extraction time. *Journal of Food Engineering* 71: 214–222.

Liu H., Wang J., Zhao J., Lu S., Wang J., … Zhou L., 2009. Isoquinoline alkaloids from *Macleaya cordata* active against plant microbial pathogens. *Natural Products Communications* 4: 1557–1560.

Logemann J., Jach G., Tommerup H., Mundy J., Schell J., 1992. Expression of a barley ribosome-inactivating protein leads to increased fungal protection in transgenic tobacco plants. *Nature Biotechnology* 10: 305–308.

Lucas J.A., Hawkins N.J., Fraajie B.A., 2015. The evolution of fungicide resistance. *Advances in Applied Microbiology* 90: 29–92.

Mandal S.M., Chakraborty D., Day S., 2010. Phenolic acids act as signaling molecules in plant-microbe symbioses. *Plant Signal Behaviour* 5: 359–368.

Martinez J.A., 2012. Natural Fungicides Obtained from Plants. In: *Fungicides for Plant and Animal Diseases* (Dhanasekeran D., Thajuddin N., Panneerselvam A., ed.). Available at: https://www.intechopen.com/books/fungicides-for-plant-and-animal-diseases/natural-fungicides-obtained-from-plants.

Mekam P.N., Martini S., Nguefack J., Tagliazucchi D., Stefani E., 2019. Phenolic compounds profile of water and ethanol extracts from *Euphorbia hirta* L. leaves showing antioxidant and antifungal properties. *South African Journal of Botany* 127: 319–332.

Mollataghi A., Coudiere E., Hadi A.H.A., Mukhtar M.R., Awang K., … Ata A., 2012. Anti-acetylcholinesterase, anti-α-glucosidase, anti-leishmanial and anti-fungal activities of chemical constituents of *Beilschmiedia* species. *Fitoterapia* 83: 298–302.

Nashwa S.M.A., Abo-Elyousr K.A.M., 2012. Evaluation of various plant extracts against the early blight disease of tomato plants under greenhouse and field conditions. *Plant Protection Sciences* 48: 74–79.

Nguefack J., Wulff E.G., Dongmo L.J.B., Fouelefack F.R., Fotio D., … Torp J., 2013. Effect of plant extracts and an essential oil on the control of brown spot disease, tillering, number of panicles and yield increase in rice. *European Journal of Plant Pathology* 137: 871–882.

Okoye T.C., Akah P.A., Okoli C.O., Ezike A.C., Mbaoji F.M., 2010. Antimicrobial and antispasmodic activity of leaf extracts and fractions of *Stachytarpheta cayennensis*. *Asian Pacific Journal of Tropical Medicine* 3(3): 189–192.

Pusztahelyi T., Holb I.J., Pócsi I., 2015. Secondary metabolites in fungus-plant interactions. *Frontiers in Plant Sciences,* doi: 10.3389/fpls.2015.00573. https://www.frontiersin.org/

Raymond Chia T.W., Dykes G.A., 2010. Antimicrobial activity of crude epicarp and seed extracts from mature avocado fruit (*Persea americana*) of three cultivars. *Pharmaceutical Biology* 48: 753–756.

Re R., Pellegrini N., Proteggente A., Pannala A., Yang M., Evans R.C.A., 1999. Antioxidant activity applying an improved ABTS radical cation decolorization assay. *Free Radical Biology and Medicine* 26: 1231–1237.

Reignault P., Walters D., 2007. Topical induction of inducers for disease control. In: *Induced resistance for plant disease control: A sustainable approach to crop protection* (Walters D., Newton A., Lyon G., ed.) Blackwell Publishing, Oxford, 272 pp.

Ribera A.E., Zuñiga G., 2012. Induced plant secondary metabolites for phytopathogenic fungi control: a review. *Journal of Soil Sciences and Plant Nutrition* 12: 893–911.

Rios J.L., Recio M.C., Villar A., 1988. Screening methods for natural products with antimicrobial activity: a review of the literature. *Journal of Ethnopharmacology* 23: 127–149.

Scavo A., Pandino G., Restuccia C., Parafati C., Cirvilleri G., Mauromicale G., 2019. Antimicrobial activity of cultivated cardoon (*Cynara cardunculus* L. var. *altilis* DC.) leaf extracts against bacterial species of agricultural and food interest. *Industrial Crops and Products*129: 206–211.

Senthikumar S., 2018. A review of pharmacology with medicinal plants. *International Journal of Universal Pharmacology and Bioscience* 7: 45–192.

Sesan T.E., Enache E., Iacomi B.M., Oprea M., Oancea F., Iacomi C., 2015. Antifungal activity of some plant extracts against *Botrytis cinerea* Pers. in the blackcurrant crop (*Ribes nigrum* L.). *Acta Scientiarum Polonorum Hortorum Cultus* 14: 29–43.

Shahidi F., Chandrasekara A., 2010. Hydroxycinnamates and their *in vitro* and *in vivo* antioxidant activities. *Phytochemistry Review* 9: 147–170.

Shuping D.S.S., Eloff J.N., 2017. The use of plants to protect plants and food against fungal pathogens: A review. *African Journal of Traditional, Complementary and Alternative Medicine* 14: 120–127.

Singh A.K., Pandey M.B., Singh S., Singh A.K., Singh U.P., 2008. Antifungal activity of securinine against some plant pathogenic fungi. *Mycobiology* 36: 99–101.

Singleton V.L., Orthofer R., Lamuela-Raventos R.M., 1999. Analysis of total phenols and other oxidation substrates and antioxidants by means of Folin-Ciocalteu reagent. *Methods in Enzymology* 229: 152–178.

Suprapta D.N, 2016. Review of tropical plants with antifungal activities against plant fungal pathogens. *Preprints 2016,* doi: 10.20944/preprints201610.0049.v1. https://www.preprints.org/

Tabasum S., Khare S., Jain K., 2016. Spectrophotometric quantification of total phenolic, flavonoid, and alkaloid contents of *Abrus precatorius* L. seeds. *Asian Journal of Pharmaceutical and Clinical Research* 9(2): 371–374..

Tchiengué B., 2012. *Secondary successions after shifting cultivation in a dense tropical forest of southern Cameroon (Central Africa).* PhD Thesis, Johann Wolfgang Goethe University, Frankfurt/Main, Germany. 274 pp.

Tripathi P., Dubey N.K., 2004. Exploitation of natural products as alternative strategy to control post-harvest fungal rotting of fruits and vegetables. *Postharvest Biology and Technology* 32: 235–245.

Verzelloni E., Tagliazucchi D., Conte A., 2010. Changes in major antioxidant compounds during aging of traditional balsamic vinegar. *Journal of Food Biochemistry* 34: 152–171.

Wink M., 2010. Functions and biotechnology of plant secondary metabolites. In: *Annual Plant Reviews.* 2nd ed. Wiley-Blackwell, Oxford, 424 pp.

Yang B., Jiang Y., Shi J., Chen F., Ashraf M., 2011. Extraction and pharmacological properties of bioactive compounds from longan (*Dimocarpus longan* Lour.) fruit. *Food Research International* 44: 1837–1842.

Zakiah Z., Suliansyah I., Bakhtiar A., Mansyurdin M., 2017. Effect of crude extracts of six plants on vegetative growth of soybean (*Glycine max* Merr.). *International Journal of Advances in Agricultural Science and Technology* 4: 1–12.

Zhishen J., Mengcheng T., Jianming W., 1999. The determination of flavonoid contents in mulberry and their scavenging effects on superoxide radicals. *Food Chemistry* 64: 555–559.

18

Diversity of genes for resistance to stripe rust in wheat elite lines, commercial varieties and landraces from Lebanon and Syria

Rola EL AMIL[1,2,*], Claude DE VALLAVIEILLE-POPE[2], Marc LECONTE[2], Kumarse NAZARI[3]

[1] *Lebanese Agricultural Research Institute (LARI), 287 Zahleh, Bekaa, Lebanon*
[2] *UMR BIOGER INRA, AgroParisTech, Université Paris-Saclay, 78850 Thiverval-Grignon, France*
[3] *Regional Cereal Rust Research Center, P. O. Box 9, 35661, Menemen, Izmir, Turkey*
*Corresponding author: ramil@lari.gov.lb

Summary. Stripe (yellow) rust, caused by *Puccinia striiformis* f. sp. *tritici* (*Pst*), is a major threat to wheat production in Central and West Asia and North Africa (CWANA). Effective fungicides are available, but host resistance remains the most economical, effective and ecologically sustainable method for stripe rust control. Understanding the genetic diversity of resistance to *Pst* is a key element in breeding for durable rust resistance. Multipathotype tests were performed on 87 elite lines of bread wheat from the spring wheat breeding programme at the International Center for Agricultural Research in Dry Areas (ICARDA), 23 Lebanese bread and durum wheat varieties, and 28 Lebanese landraces, with 11 *Pst* pathotypes. Low and high infection types were identified for the resistance genes *Yr1*, *Yr3*, *Yr4*, *Yr6*, *Yr7*, *Yr9*, *Yr17*, *Yr25*, *Yr27*, and *Yr32*. All but one of these genes (*Yr32* being the exception) were postulated. ICARDA elite lines displayed greater diversity for *Yr* genes than the Lebanese varieties and landraces. *Yr27* was the most frequent *Yr* gene postulated singly in the Lebanese varieties. *Yr7*, together with other unidentified *Yr* genes, was the most frequent gene in the ICARDA elite lines. Combinations of two *Yr* genes were common in ICARDA elite lines. These results confirm that the landraces consist of several genotypes. Seventy-five percent of landraces were susceptible to all pathotypes, but they displayed resistance diversity, with different proportions of resistant seedlings. In two landraces, some plants were resistant to the Warrior pathotype, which has recently spread in CWANA regions, and to other pathotypes. This indicates the presence of new resistance genes in these landraces. Some landraces, elite ICARDA lines and Lebanese varieties were completely resistant to all pathotypes, and are therefore potential sources of new resistances.

Keywords. Stripe (yellow) rust, wheat, gene postulation, seedling resistance.

Competing Interests: The Author(s) declare(s) no conflict of interest.

Editor: Diego Rubiales, Institute for Sustainable Agriculture, (CSIC), Cordoba, Spain.

INTRODUCTION

In the context of global warming and the food insecurity, it is likely to cause, feeding the expanding world population through sustainable agricultural practices is a major challenge. With predicted world population of nine billion by 2050, the demand for wheat is expected to increase by 60%. Annual increases in wheat yields will need to increase from the current level of 1% to at least 1.6% to meet this demand (Lucas, 2012).

Stripe rust, caused by the biotrophic fungus *Puccinia striiformis* f. sp. *tritici* (*Pst*), is a common wheat disease of economic importance in all world wheat production regions (de Vallavieille-Pope *et al.*, 2012). In most of these areas stripe rust causes yield losses of 10-70% (Chen, 2005), depending on the time at which initial infections occur, host population density, susceptibility and nutritional status, disease development, and the duration of the epidemics. The last 40 years have seen five major stripe rust epidemics in the Central and West Asia and North Africa region (CWANA), in 1973, 1978, 1995, 2005 and 2010 (Solh *et al.*, 2012). The two most recent epidemics were due to the successive emergence of *Pst* pathotypes with new virulence factors overcoming the widely used *Yr9* and *Yr27* resistance genes (Yahyaoui *et al.*, 2002; Hodson and Nazari, 2010; Sharma-Poudyal *et al.*, 2013). According to Hovmøller *et al.* (2011), PstS2, an aggressive strain with virulence against *Yr* genes *2, 6, 7, 8, 9, 25,* and *27,* was present at high frequency in the Red Sea area, East Africa and in Western and Central Asia between 2003 and 2008. The PstS2 strain was first detected in North America in 2000 (Milus *et al.*, 2009). It was present in 50% of the virulence profiles of rust isolates surveyed in Syria in 2011 (El Amil *et al.*, in press). The recent spread of Warrior pathotypes to this region (Mert *et al.*, 2016) has added an additional dimension to the widespread stripe rust epidemics in wheat-growing areas in CWANA.

This damaging fungus causes losses of wheat grain yield and quality, by reducing tillering and causing grain to shrivel (Roelfs *et al.*, 1992), unless it is controlled by the use of resistant cultivars or timely fungicide applications (Hau and de Vallavieille-Pope, 2006). The deployment and use of resistant cultivars is the most economical and environmentally-friendly measure for controlling this disease (Pathan and Park, 2007). Effective deployment of resistance genes for the management of stripe rust in wheat requires knowledge of the resistance status and diversity of the resistance genes in available cultivars (Nazari *et al.*, 2008). The durability of resistance may be threatened by the frequent emergence of new pathotypes. An understanding of the *Pst* population is therefore crucial in gene deployment strategies (McDonald and Linde, 2002). These strategies involve: i) the deployment of new resistance genes in a controlled manner and over a restricted geographic scale; ii) the combination of several resistance genes within a single cultivar to slow the emergence of new virulent pathotypes; and iii) the combination of race-specific resistance with non race-specific or partial resistance within a single cultivar. These approaches require a knowledge of the resistance genes present in the breeding germplasm and commercial cultivars. It is, therefore, important to identify the resistance genes present in different cultivars, because cultivars may have resistance genes in common, even if they originated from genetically different sources. Knowledge of the resistance genes present makes it possible to prevent the release of mega-cultivars containing the same resistance genes or profiles (Statler, 1984).

Gene postulation is based on the theory of a gene-for-gene relationship (Flor, 1956), according to which it is possible to postulate the existence of race-specific genes for resistance in a cultivar provided with an array of pathotypes bearing diverse combinations of avirulence and virulence genes. This approach can be used for the rapid identification of probable race-specific rust resistance genes (*Yr*) in a large group of wheat lines. This method has traditionally been used for the three rust diseases (Perwaiz and Johnson, 1986; Nazari *et al.*, 2008). Stripe rust resistance genes have been postulated in wild emmer wheat derivatives and advanced wheat lines from Nepal (Sharma *et al.*, 1995), French wheat lines (de Vallavieille-Pope *et al.*, 1990; Robert *et al.*, 2000), Danish wheat cultivars (Hovmøller, 2007), Chinese wheat cultivars and advanced lines (Xia *et al.*, 2007), and Ethiopian bread wheat cultivars (Dawit *et al.*, 2012).

Landraces of cultivated plants were the principal focus of agricultural production until the end of the nineteenth century and the advent of formal plant breeding (Harlan, 1975). According to Camacho Villa *et al.* (2005), "a landrace is a dynamic population of a cultivated plant that has a historical origin, distinct identity and lacks formal crop improvement, as well as often being genetically diverse, locally adapted and associated with traditional farming systems". Modern cereal cultivars are derived from narrow germplasm pools and are mostly adapted to high-input agriculture. A distinction is made between landraces and the modern, so-called "elite" lines generated by formal crop breeding programmes (Newton *et al.*, 2010). Landraces may be good reservoirs of non race-specific or partial resistances capable of conferring durability when combined with major resistance genes commonly exploited in modern cultivars. Given that landraces are adapted to local edaphic and climatic conditions and display tolerance or resistance to many pests and diseases, their use could expand the narrow genetic basis of elite lines (Beharav *et al.*, 1997). Zhang (1995) showed that nine Chinese wheat landraces expressed slow rusting or quantitative resistance to stripe rust. Lebanon is located in the Fertile Crescent, the area in which wheat and its wild relatives are most diverse (Harlan and Zohary, 1966). Lebanese landraces are, therefore, also likely to be promising sources of novel resistance genes with major and partial effects. The identification of seedling rust resistance genes in Lebanese landraces is, therefore, an important first step towards further wheat improvement in the CWANA region.

The genes conferring resistance to wheat stripe rust in ICARDA elite breeding lines, Lebanese cultivars and Lebanese landraces, remain largely unknown. The present study aimed to provide detailed information about specific resistance to wheat stripe rust, detectable at the seedling stage, in 87 elite lines of bread wheat from the spring wheat breeding programme at ICARDA, 23 Lebanese bread and durum wheat varieties and 28 Lebanese landraces. Gene postulation was performed with an array of 11 *Pst* pathotypes that distinguished between low and high infection types for the resistance genes *Yr1*, *Yr3*, *Yr4*, *Yr6*, *Yr7*, *Yr8*, *Yr9*, *Yr17*, *Yr25*, *Yr27*, *Yr32*, *YrSD*, *YrSu* and *YrSP*. Adult plant resistance was also evaluated in some ICARDA lines.

MATERIALS AND METHODS

Pathogen material

The virulence combinations and pathotype codes of the *Pst* isolates used for resistance gene postulation were determined with the European and World sets of 15 differential varieties (Johnson *et al.*, 1972), 13 Avocet lines near-isogenic for *Yr1*, *Yr5*, *Yr6*, *Yr7*, *Yr8*, *Yr9*, *Yr10*, *Yr15*, *Yr24*, *Yr26*, *Yr27*, *Yr32* and *YrSP* (Wellings *et al.*, 2009), and the lines Kalyansona (*Yr2*), Federation 4*/ Kavkaz (*Yr9*), Clement (*Yr9+*), VPM1 (*Yr17+*), TP981 (*Yr25*) and Opata (*Yr27*). Each differential line carries at least one race-specific resistance gene (*Yr*) expressed at the seedling growth stage.

Resistance genes were postulated at the seedling stage at the INRA BIOGER rust facility, on the basis of infection types (IT), with a set of 11 French pathotypes displaying complementary virulences, including the recently propagated Warrior race (Table 1) (de Vallavieille-Pope *et al.*, 2012). As most of these pathotypes present more than one avirulence factor, it was not always possible to infer precise resistance gene combinations.

All isolates from the INRA-Grignon collection were purified from single spores, and were stored in liquid nitrogen. Spore multiplication was performed in a controlled climate chamber. Reference isolate spores were used to inoculate 7-d-old seedlings of the susceptible cultivar 'Victo', which was then incubated in a dew chamber at 8°C for 16 h in the dark to ensure successful infection, before transfer to a controlled climate chamber (day: 16 h, 300 μmol m^{-2} s^{-1}, 17°C; night: 8 h, 14°C). Seedlings were exposed to high-intensity light treatment at 200 μE m^{-2} s^{-1} for at least 8 h before inoculation, to maximize infection efficiency (de Vallavieille-Pope *et al.*, 2002). One week after inoculation, each pot was sealed in a cellophane bag to prevent cross-contamination. Uredospores were collected 18 d post-inoculation, were

Table 1. Pathotype code, name and avirulence/virulence formula of 11 French *Puccinia striiformis* f. sp. *tritici* pathotypes used for the postulation of stripe rust resistance genes in Syrian and Lebanese wheat genotypes.

Pathotype code	Pathotype nomenclature[a]	Avirulence formula[b]	Virulence formula
A	6E16	*1, 3, 4, 5, 9, 10, 15, 17, 24, 25, 26, 27, 32, SD, Su, ND, SP*	*2, 6, 7, 8*
B	6E16*v9v27*	*1, 3, 4, 5, 10, 15, 17, 24, 26, 32, SD, Su, ND, SP*	*2, 6, 7, 8, 9, 25, 27*
C	43E138	*4, 5, 6, 8, 9, 10, 15, 17, 24, 26, 27, 32, Su, SP*	*1, 2, 3, 7, 25, SD, ND*
D	45E140	*4, 5, 7, 8, 9, 10, 15, 17, 24, 26, 27, 32, Su, SP*	*1, 2, 3, 6, 25, SD, ND*
E	106E139	*1, 5, 6, 8, 9, 10, 15, 17, 24, 26, 27, 32, SP*	*2, 3, 4, 7, 25, SD, Su, ND*
F	169E136*v17*	*4, 5, 6, 7, 8, 10, 15, 24, 26, 27, 32, Su, SP*	*1, 2, 3, 9, 17, 25, SD, ND*
G	232E139	*1, 5, 6, 7, 8, 10, 15, 17, 24, 26, 27, 32, SP*	*2, 3, 4, 9, 25, SD, Su, ND*
H	237E140	*5, 7, 8, 10, 15, 17, 24, 26, 27, 32, SP*	*1, 2, 3, 4, 6, 9, 25, SD, Su, ND*
I	237E141*v17*	*5, 7, 8, 10, 15, 24, 26, 27, 32, SP*	*1, 2, 3, 4, 6, 9, 17, 25, SD, Su, ND*
J	237E173*v17 (Oakley/Solstice)*	*5, 7, 8, 10, 15, 24, 26, 27, SP*	*1, 2, 3, 4, 6, 9, 17, 25, 32, SD, Su, ND*
K	239E175*v17 (Warrior)*	*5, 8, 10, 15, 24, 26, 27*	*1, 2, 3, 4, 6, 7, 9, 17, 25, 32, SD, Su, ND, (SP)*[c]

[a] Pathotype nomenclature is based on Johnson *et al.* (1972).
[b] *YrSD*, *YrSu*, *YrND* and *YrSP*, correspond to, respectively, Strubes Dickkopf, Suwon 92 × Omar, Nord Desprez and Spaldings Prolific.
[c] *SP* infection types 5 to 6 were considered to be intermediate reactions, and are shown in parentheses.

dried in a desiccator containing silica gel at 4°C for 3 d, and were then stored in liquid nitrogen (LN). Any cold dormancy of samples stored in LN was broken by subjecting the uredospores to a heat shock (40°C for 10 min) before inoculation in *Yr*-gene postulation tests.

Host material

We tested 138 genotypes at the seedling stage. These included: 87 advanced lines of bread wheat from the spring wheat breeding programme at ICARDA (Table S1), 23 cultivars commonly grown in Lebanon (13 of bread wheat and ten of durum wheat) (Table S2), and 28 Lebanese landraces (21 accessions of bread wheat and seven of durum wheat; Table S2). Seed stocks for elite lines and landraces were obtained, respectively, from ICARDA and the Lebanese Agriculture Research Institute (LARI). Landraces are known to be highly diverse and heterogeneous. Landrace seeds were therefore collected from a number of different sites in Lebanon, to obtain a broad genetic pool. The seeds were then purified at LARI for morphological traits corresponding to the criteria for use in agriculture, before the resistance gene tests (Table 2).

Inoculation and scoring

All seeds were planted in square pots (7 × 7 × 8 cm) filled with standard peat soil. Five seeds of each elite line and variety and 15 seeds of each landrace were planted in two replicated pots. The pots were placed in air-filtered cabinets in a glasshouse, at temperatures between 15 and 25°C, with a 16-h photoperiod extended with sodium vapour lamps, for 10 d. The seedlings were inoculated with spores when they were 2 weeks old and with second leaves fully expanded. The inoculated spores had been stored at −80°C, then taken out of the freezer and immediately heat shocked at 40°C for 10 min before use. Spores (3 mg) were suspended in 600 μL of engineered fluid (NovecTM 7100) for inoculation onto seedlings. Inoculated plants were incubated for 24 h in the dark in a dew chamber at 8°C and 100% relative humidity, after which they were placed in cabinets with the conditions described above. The experiment was performed twice.

Seedling infection types were recorded 15–17 d after inoculation, using a 0–9 scale based on the presence of necrosis, chlorosis, size of sporulation areas and sporulation intensity (McNeal *et al.*, 1971). Infection types (IT) 0 to 4 were considered to indicate various levels of incompatibility (host resistance and pathogen avirulence) between host and pathogen, whereas infection types 7 to 9 were considered to correspond to compatible (host susceptibility and pathogen virulence) interactions. Infection types 5 to 6 were considered to be intermediate reactions (Roelfs *et al.*, 1992). Resistance genes were postulated by comparing the low and high IT patterns obtained with the pathotype array on the tested entities with those of differential lines with known resistance genes. If a wheat variety had a low/high IT pattern similar to that of a differential line with a known resistance profile, the tested genotype was postulated to possess the same resistance genes as the differential. This method was applied successively to all *Yr* genes detectable with the array of 11 pathotypes used for the study.

The diversity of seedling resistance within the Lebanese wheat landrace collections was investigated by assessing the frequency of resistant plants in the overall susceptible landrace population and the frequency of susceptible plants in the overall resistant landrace population, for the 11 *Pst* pathotypes (Table 6).

Assessment of adult plant resistance

The ICARDA elite lines tested at the seedling stage against pathotype 6E16*v9v27* (the most prevalent *Pst* pathotype in Syria and Lebanon during the 2010–2011 cropping season) included 44 lines containing postulated resistance gene and/or gene combinations that were assessed for adult-plant resistance. The experiments on adult plants were performed with this pathotype at the ICARDA research stations in Tal Hadya, Syria in 2009-2011, and Terbol, Lebanon in 2009–2013 (Table 5).

At each site, 30 seeds of each genotype were planted in two 0.5 m rows in a field nursery in November in each year. The highly susceptible 'Morocco' and two cultivars known to carry *Yr9* (Seri-82) and *Yr27* (Cham-8) were planted as spreader rows bordering the trial areas, in all pathways, and at ten-row intervals within the trial. The inoculum used for this study comprised the dominant pathotypes collected separately in Syria or Lebanon from natural infections in the same fields during the previous year. The inoculated pathotypes carried virulence against the genes *Yr2*, *Yr6*, *Yr7*, *Yr8*, *Yr9*, *Yr25*, *Yr27* and *YrSD*. The trail fields were dusted in the evenings with a spore-talc mixtures (1 to 50), at the seedling, tillering, and flag leaf stages. Disease infection types were recorded as described by Roelfs *et al.* (1992), and the modified Cobb scale was used to assess disease severity (Peterson *et al.*, 1948) at the host booting and flag leaf stages. Flag leaf scoring was used to assess the final disease responses of the tested genotypes at the adult plant growth stage.

Table 2. Resistance group, seedling infection types and postulated stripe rust resistance genes in 87 wheat elite lines from ICARDA, tested against 11 *Puccinia striiformis* f. sp. *tritici* pathotypes.

Group	Entry No.	Genotype	Pathotype code [a]											Postulated *Yr* genes
			A	B	C	D	E	F	G	H	I	J	K	
1	**3**	Tabeldi-1	1	1	3	6	1	5	1	3	2	2	5	Resistant[b]
	11	Babaga-3	1	2	3	6	1	5	1	3	2	2	4	Resistant
	20	Sale-6	1	1	2	2	1	2	1	2	2	2	3	Resistant
	32	Hashab-2	1	1	2	2	1	4	1	2	2	2	4	Resistant
	40	Usher-18	1	5	4	3	3	3	2	2	2	3	4	Resistant
	45	Saba/Flag-1	2	6	2	2	2	2	2	2	2	2	4	Resistant
	72	Naji-3	1	2	3	4	1	6	1	2	2	2	5	Resistant
	78	Shuha-8/Ducula	1	1	2	2	1	3	1	2	2	2	4	Resistant
2	**12**	Cham-6	8	7	8	8	9	8	8	8	8	8	8	Susceptible[c]
	75	Nesma*2/14-2//2*Safi-3	8	7	8	8	9	8	8	8	8	8	8	Susceptible
	79	Shuha-8/Ducula	8	6	8	8	8	8	8	8	8	6	8	Susceptible
3	6	Utique 96/Flag-1	4	7	3	8	9	5	9	8	8	8	8	Ni[d]
	7	Hamam-4	1	7	1	1	1	5	7	3	8	8	4	Ni
	16	Durra-8	2	8	3	6	9	5	8	3	8	8	2	Ni
	22	Fow-2/SD8036//SafiI-3/3/NS732/HER//Kauz'S'	3	8	5	8	8	8	8	3	8	5	2	Ni
	30	Sandall-5	2	3	2	2	2	9	2	2	8	8	8	Ni
	38	Temerind-8	1	8	1	8	-[e]	2	3	2	2	2	4	Ni
	39	Gonglase-4	2	2	8	1	2	9	2	8	8	8	8	Ni
	48	Bushraa-3	5	3	3	2	4	8	2	4	3	2	8	Ni
	54	Jasmin-5	2	2	3	7	8	8	8	8	4	3	3	Ni
	57	Qadanfer-5	7	8	4	3	8	5	8	6	2	3	8	Ni
	62	Tevee'S'/3/T.aestivum/SPRW'S'//CA8055/4/Pastor-2/5/Sunbri	1	1	8	8	1	8	1	8	4	8	8	Ni
	69	Manhal-4	2	2	8	8	1	1	6	6	2	8	8	Ni
	21	HD2206/Hork'S'/3/2*NS732/HER//Kauz'S'	4	6	5	8	7	9	8	7	7	8	8	Ni
	85	Girwill-13/2*Pastor-2	1	8	3	2	2	4	8	8	8	8	8	Ni
	14	Jawahir-14	6	1	8	8	1	9	1	8	8	8	8	Ni
		Chinese 166[f]	1	1	9	9	1	8	1	9	9	9	9	*Yr1*
4		**Avocet *Yr1*/6*Avocet S**	2	1	8	8	2	8	1	8	8	8	9	*Yr1*
	60	Crow'S'/Bow'S' -3-1994/95//Tevee'S'/Tadinia	1	1	8	8	1	9	1	8	7	8	7	*Yr1*
	61	Tevee'S'/3/T.aestivum/SPRW'S'//CA8055/4/Pastor-2/5/Sunbri	1	1	8	8	1	9	1	8	8	8	8	*Yr1*
	65	Qafzah-2/Ferroug-2	1	2	8	8	1	9	2	8	8	8	8	*Yr1*
	70	Usher-16	1	2	8	8	1	8	1	8	8	8	8	*Yr1*
	74	Settat-45	2	2	8	8	1	7	1	8	8	8	8	*Yr1*
		Vilmorin 23	2	3	8	8	9	8	8	9	9	9	9	*Yr3*
5	**46**	Bow #1/Fengkang15/3/HYS//DRC*2/7C	5	3	8	8	8	9	7	8	8	8	8	*Yr3*
6		**Hybrid 46**	2	1	1	2	9	2	8	9	9	9	8	*Yr4*
	33	Sanobar-1	2	3	2	2	8	2	8	8	8	8	8	*Yr4*
	18	Sanobar-6	2	2	2	2	8	2	8	6-7	8	8	8	*Yr4*
	82	ESWYT99#18/Arrihane	2	5	4	2	8	3	7	8	8	8	8	*Yr4*
7		**Avocet *Yr6*/6*Avocet S**	8	8	3	8	4	4	4	8	8	8	8	*Yr6*
		Heines Kolben	9	9	2	9	2	2	1	9	9	9	9	*Yr6, Yr2*
	67	Hamam-4/Angi-2	8	8	3	8	3	3	2	8	8	8	8	*Yr6*
	87	Hubara-16/2*Somama-3	7	8	2	8	2	2	3	8	6	8	8	*Yr6*
		Heines Peko	2	4	2	9	2	2	2	9	9	9	8	*Yr6, Yr+*
	2	Zafir-3	2	2	2	9	2	2	2	2	2	7	8	*Yr6, Yr+*

(Continued)

Table 2. (Continued).

Group	Entry No.	Genotype	Pathotype code [a]											Postulated *Yr* genes
			A	B	C	D	E	F	G	H	I	J	K	
	5	Aguilal/Flag-3	2	3	2	8	2	2	2	8	5	8	8	*Yr6, Yr+*
	47	Faisal-1	1	5	4	8	1	3	4	7	8	8	8	*Yr6, Yr+*
	63	Weebill-1/2*Qafzah-21	4	6	3	7	7	2	5	8	8	8	8	*Yr6, Yr+*
	64	Rebwah-12/Zemamra-8	2	1	2	2	1	3	1	8	8	8	8	*Yr6, Yr+*
8		**Avocet *Yr7*/6*Avocet S**	8	8	8	3	8	4	4	2	3	3	8	*Yr7*
		Lee	9	9	9	3	9	3	2	3	3	3	9	*Yr7, Yr+*
		Reichersberg 42	2	4	9	2	9	2	2	2	2	2	8	*Yr7, Yr+*
	43	Soonot-11	1	2	1	1	1	1	2	2	2	2	8	*Yr7, Yr+*
	23	Neem-2	1	2	1	1	1	1	2	2	2	2	8	*Yr7, Yr+*
	34	Cham-10	1	1	1	1	1	1	2	1	2	2	8	*Yr7, Yr+*
	19	Reyna-12	1	1	1	1	1	1	2	1	2	2	8	*Yr7, Yr+*
	36	Reyna-25	1	1	1	1	-	1	2	1	2	2	8	*Yr7, Yr+*
	41	Florkwa-2/Asfoor-5	1	2	1	1	1	1	2	2	2	2	8	*Yr7, Yr+*
	42	Settat-13	1	2	1	1	1	1	3	2	2	2	8	*Yr7, Yr+*
	44	Hubara-15/Zemamra-8	1	1	1	1	1	3	1	2	3	5	8	*Yr7, Yr+*
	37	Reyna-29	1	2	1	1	1	1	2	2	2	2	8	*Yr7, Yr+*
	68	Sisaban-3	1	1	1	1	1	1	2	1	2	2	8	*Yr7, Yr+*
	77	Achta*3//Kanz/KS85-8-4/3/Lakta-8/4/Zemamra-1	2	2	2	1	2	1	2	2	2	2	8	*Yr7, Yr+*
	52	Reyna-24	1	1	1	1	1	1	1	1	1	1	8	*Yr7, Yr+*
	17	Laloub-2	1	1	1	1	1	1	2	1	2	2	8	*Yr7, Yr+*
	81	Achtar/INRA 1764	2	8	2	5	2	3	2	2	4	6	8	*Yr7, Yr+*
	80	Achtar/INRA 1764	2	8	3	3	2	4	3	4	3	6	8	*Yr7, Yr+*
9		**Avocet *Yr9*/6*Avocet S**	2	8	2	2	2	8	8	8	8	8	8	*Yr9*
		Clement	1	5-6	2	1	2	8	8	8	8	8	9	*Yr9*
		Federation 4*/ Kavkaz	1	9	2	1	1	8	8	9	9	9	9	*Yr9*
	66	Haala-50	3	8	5	1	1	8	8	7	8	8	8	*Yr9*
	28	Battell-3	2	8	2	2	2	8	9	-	8	8	8	*Yr9*
	29	Sandall-3	1	8	2	2	2	8	9	-	8	8	8	*Yr9*
	35	ICARDA-SRRL-5	2	8	2	2	2	8	8	-	8	8	8	*Yr9*
10		**VPM1**	2	3	2	2	2	8	2	2	9	9	9	*Yr17*
	13	Ruth-1	1	1	1	2	1	7	1	2	8	8	8	*Yr17*
	49	Nouha-3	1	4	2	2	1	8	2	2	8	8	8	*Yr17*
	31	Samira-2	1	2	5	2	1	5	2	3	8	8	8	*Yr17+*
11		**TP981**	2	8	9	9	9	8	8	9	9	9	9	*Yr25*
	15	Nayzak-3	1	8	8	9	9	9	8	8	8	8	8	*Yr25*
12		**Avocet *Yr27*/6*Avocet S**	5	8	3	2	3	3	2	3	2	2	3	*Yr27*
		Opata	5	8	3	2	3	3	2	3	2	2	3	*Yr27*
	1	Cham-8	1	8	1	1	1	2	2	2	2	2	5	*Yr27*
	8	Inqualab 91/Flag-2	4	8	2	3	2	2	2	2	3	3	4	*Yr27*
	10	Bow#1/Fengkang15/3/HYS//DRC*2/7C	2	8	2	3	2	2	2	2	2	2	2	*Yr27*
	53	NS732/HER//SD8036/3/Saada	2	7	2	2	2	2	5	2	2	2	2	*Yr27*
	56	Loulou-16	1	8	1	1	1	2	2	2	2	2	3	*Yr27*
	73	HIJLEEJ-1	2	8	4	2	5	3	2	2	2	2	3	*Yr27*
	76	NS732/HER*2//Saada	2	8	2	2	3	2	2	2	2	2	2	*Yr27*
	58	Taleh-1	1	8	1	1	1	2	4	2	2	3	4	*Yr27*

(Continued)

Table 2. (Continued).

Group	Entry No.	Genotype	Pathotype code [a]											Postulated *Yr* genes
			A	B	C	D	E	F	G	H	I	J	K	
13	**51**	Sidraa-1	4	6-7	3	4	5	2	5	8	8	8	8	*Yr6* + *Yr9*
	71	Latifa-2	1	8	1	1	1	5	4	8	8	8	8	*Yr6* + *Yr9*
	27	Koukab-2	1	3	2	1	1	4	4	2	8	7	8	*Yr6* + *Yr17*
	86	Hubara-3/Angi-2//Somama-3	1	1	2	1	1	2	1	4	7	8	8	*Yr6* + *Yr17*
	24	Firdous-29	2	1	7	3	2	2	1	2	3	2	8	*Yr7* + *Yr1*
	26	Saamid-2	2	2	2	1	7	1	2	2	2	2	8	*Yr7* + *Yr4*
	4	Soonot-10	2	2	1	1	9	1	2	2	2	2	8	*Yr7* + *Yr4*
	59	Nadia-13	2	2	2	1	8	2	2	2	2	2	8	*Yr7* + *Yr4*
	84	ACSAD 529/Karawan'S'//Somama-3	1	1	3	2	1	9	1	8	8	8	8	*Yr9* + *Yr1*
	83	Hubara-5/3/SHA3/Seri//SHA4/Lira	1	2	2	3	2	8	8	9	8	8	8	*Yr9* + *Yr3*
	9	Qafzah-33/Florkwa-2	1	4	2	2	1	3	8	-	8	8	8	*Yr9* + *Yr4*
	25	Jelmoud-1	1	1	1	1	2	1	8	-	8	8	8	*Yr9* + *Yr4*
	50	Milan/SHA7//Potam*3KS811261-5	1	4	1	1	2	3	8	-	7	8	8	*Yr9* + *Yr4*
	55	Fanoos-14	1	2	1	1	1	3	8	-	8	8	8	*Yr9* + *Yr4*

[a]A = 6E16, B = 6E16*v9v27*, C = 43E138, D = 45E140, E = 106E139, F = 169E136*v17*, G = 232E137, H = 237E141, I = 237E141*v17*, J = 237E173*v17 (Oakley/Solstice)*, K = 239E175*v17 (Warrior)*. Pathotypes are coded according to Johnson *et al.* (1972).
The virulences and avirulences tested were 1, 2, 3, 4, 6, 7, 8, 9, 17, 25, 27, 32, SD, SP, Su. Scoring was performed as described by McNeal *et al.* (1971); Infection types IT0 = No visible uredia, IT1 = Necrotic flecks, IT2 = Necrotic areas without sporulation, IT3-4 = Necrotic and chlorotic areas with restricted sporulation, IT5-6 = Moderate sporulation with necrosis and chlorosis, IT7-8 = Sporulation with chlorosis, IT9 = Abundant sporulation without chlorosis.
[b] Resistant to all *Puccinia striiformis* f. sp. *tritici* pathotypes used.
[c] Susceptible to all *Puccinia striiformis* f. sp. *tritici* pathotypes used.
[d] Ni indicates non-identified resistance genes.
[e] Indicates missing data.
[f] The entries in bold font correspond to the infection type profiles of the tester lines confronted with the array of 11 *Puccinia striiformis* f. sp. *tritici* pathotypes.

Statistical analyses

Principal component analysis (PCA) was performed to illustrate the heterogeneity of the distributions of resistant and susceptible plants for each pathotype tested on landraces. The heterogeneity of resistance in landraces was assessed by determining the percentage of resistant plants in landraces mostly susceptible to one pathotype and the percentage of susceptible plants in landraces mostly resistant to one pathotype. Data were plotted for 11 variables, corresponding to the 11 pathotypes tested. Another PCA was performed on the percentage of plants in each landrace resistant to each of the 11 pathotypes. PCAs were performed with R software (http://www.R-project.org, 2008).

RESULTS

Postulation of seedling resistance genes

The seedling tests conducted on the 138 elite lines, varieties and landraces with 11 French *Pst* pathotypes made it possible to postulate genes for seedling resistance to stripe rust, either singly or in combinations (Tables 2, 3, and 4; Figure 1). The pathotypes used made postulation possible for *Yr1, Yr3, Yr4, Yr6, Yr7, Yr9, Yr17, Yr25, Yr27* and *Yr32*. Based on these postulations, the lines were attributed to 13 stripe rust resistance groups.

Tested lines resistant to all 11 pathotypes were classified into group 1, those susceptible to all 11 pathotypes were classified into group 2, and tested lines that could not be characterized with the set of 11 pathotypes were classified into group 3.

Resistance group 1 corresponded to genotypes resistant to all *Pst* pathotypes. This group included eight ICARDA elite lines of bread wheat (Table 2) and two Lebanese durum wheat varieties (Table 3). These genotypes had low to intermediate ITs for all pathotypes tested. They therefore possessed a *Yr* gene or combination of *Yr* genes without corresponding virulence in the 11 *Pst* pathotypes.

Resistance group 2 corresponded to the genotypes susceptible to all *Pst* pathotypes, which therefore harboured no *Yr* genes, although they may have harboured

Table 3. Resistance group, infection type and postulated seedling-stage stripe rust resistance genes against 11 *Puccinia striiformis* f. sp. *tritici* pathotypes, for 23 Lebanese wheat varieties.

Resistance group	Wheat line	Pathotype code[a]											Postulated *Yr* genes
		A	B	C	D	E	F	G	H	I	J	K	
1	Stork	1	3	2	2	3	4	4	2	2	2	1	Resistant[b]
	Azeghar	3	3	2	2	2	1	2	2	2	2	2	Resistant
2	Super X	8	8	8	8	-[c]	9	-	9	8	8	9	Susceptible[d]
3	Senatore Cappelli	4	4	2	3	8	3	7	3	2	2	1	Ni[e]
	Miki	5	5	4	8	8	5	8	5	3	3	8	Ni
	Tal Amara 2	3	2	3	8	8	5	8	8	8	8	8	Ni
	Icarasha	4	4	2	8	2	2	2	2	4	2	8	Ni
	Tal Amara 1	4	3	1	8	2	2	1	4	2	2	2-4	Ni
	Tal Amara 3	6	5	2	6	2	2	2	4	4	4	8	Ni
	Nab El Jamal	1	8	2	3	2	8	9	9	8	8	8	Ni
5	**Vilmorin 23** [f]	2	3	8	8	9	8	8	9	9	9	9	
	Genessi	3	3	7	8	6-7	9	8	8	8	8	8	*Yr3*
8	**Avocet *Yr7*/6*Avocet S**	8	8	8	3	8	4	4	2	3	3	8	*Yr7*
	Lee	9	9	9	3	9	3	2	3	3	3	9	*Yr7, Yr+*
	Reichersberg 42	2	4	9	2	9	2	2	2	2	2	8	*Yr7, Yr+*
	Haramoun	8	8	8	3	9	3	3	2	2	3	9	*Yr7*
	Tannour	1	8	2	1	2	2	2	2	2	2	8	*Yr7, Yr+*
	885	2	8	2	2	1	2	3	2	2	2	8	*Yr7, Yr+*
11	**TP981**	2	8	9	9	9	8	8	9	9	9	9	
	Florence Aurore	1	8	8	8	9	9	9	9	8	8	8	*Yr25*
12	**Avocet *Yr27*/6*Avocet S**	5	8	3	2	3	3	2	3	2	2	3	*Yr27*
	Opata	5	8	3	2	3	3	2	3	2	2	3	*Yr27*
	MR 1009	3	8	2-3	2	2	3	2	2	2	2	5	*Yr27*
	Katilla	1	8	1	1	1	2	2	2	2	2	2	*Yr27*
	Bouhouth 6	3	8	3	2	3	2	2	2	2	2	4	*Yr27*
	Aammoun	1	8	1	1	1	2	3	2	2	2	5	*Yr27*
	Sham 8	2	8	2	1	2	2	2	3	2	2	5	*Yr27*
13	Lahn	8	7	1	5	2	1	2	6	3	4	8	*Yr6 + Yr7*
	A1103	8	8	3	2	2	3	2	3	2	2	7	*Yr6 + Yr7*
	Naama	1	4	2	2	2	2	2	5	8	8	8	*Yr6 + Yr17*

[a] A = 6E16, B = 6E16*v9v27*, C = 43E138, D = 45E140, E = 106E139, F = 169E136*v17*, G = 232E137, H = 237E141, I = 237E141*v17*, J = 237E173*v17 (Oakley/Solstice)*, K = 239E175*v17 (Warrior)*. Pathotypes are coded according to Johnson *et al.* (1972).
The virulences and avirulences tested were 1, 2, 3, 4, 6, 7, 8, 9, 17, 25, 27, 32, SD, SP, Su. Scoring was performed as described by McNeal *et al.* (1971); Infection types IT0 = No visible uredia, IT1 = Necrotic flecks, IT2 = Necrotic areas without sporulation, IT3-4 = Necrotic and chlorotic areas with restricted sporulation, IT5-6 = Moderate sporulation with necrosis and chlorosis, IT7-8 = Sporulation with chlorosis, IT9 = Abundant sporulation without chlorosis.
[b] Resistant to all *Puccinia striiformis* f. sp. *tritici* pathotypes used.
[c] - indicates missing data.
[d] Susceptible to all *Puccinia striiformis* f. sp. *tritici* pathotypes used.
[e] Ni indicates non-identified resistance genes.
[f] The entries in bold correspond to the infection type profiles of the tester lines confronted with the array of 11 *Puccinia striiformis* f. sp. *tritici* pathotypes.

resistance genes corresponding to the virulence profiles of the 11 tester pathotypes resulting in compatible reactions. This group included three elite lines (Table 2), one Lebanese variety (Table 3) and 21 Lebanese landraces (Table 4; 14 populations of Salamouni wheat, three populations of Abou Shwereb wheat, two populations of Ukranian wheat, and one population each of Bekaii and Haurani wheat).

Table 4. Resistance group, infection type and postulated seedling-stage stripe rust resistance genes against 11 *Puccinia striiformis* f. sp. *tritici* pathotypes, for 28 Lebanese wheat landraces.

Resistance group	Wheat landrace	Pathotype code[a]											Postulated *Yr* genes
		A	B	C	D	E	F	G	H	I	J	K	
1	Waha	1	2	2	2	2	1	1	2	1	2	1	Resistant[b]
2	Abou Shwereb	8	8	8	7	9	9	9	9	8	8	8	Susceptible[c]
	Abou Shwereb	8	8	8	8	9	9	9	9	8	8	9	Susceptible
	Salamouni	8	8	8	8	9	9	9	9	8	8	9	Susceptible
	Salamouni	8	8	8	9	9	8	9	8	8	8	9	Susceptible
	Salamouni	8	8	8	8	9	8	9	8	8	8	8	Susceptible
	Salamouni	8	8	8	8	9	8	9	9	8	8	8	Susceptible
	Salamouni	9	8	8	8	9	8	9	9	8	8	8	Susceptible
	Ukranian	8	8	8	8	9	8	9	9	8	8	9	Susceptible
	Ukranian	8	8	8	8	9	8	9	7	8	8	8	Susceptible
	Salamouni	9	8	8	9	9	9	9	9	7	8	9	Susceptible
	Salamouni	8	8	8	9	9	8	9	9	8	8	8	Susceptible
	Abou Shwereb	9	8	8	9	9	9	9	9	8	8	9	Susceptible
	Salamouni	9	8	8	9	9	9	9	9	8	8	9	Susceptible
	Haurani	8	6-7	8	9	9	9	9	9	8	8	9	Susceptible
	Bekaii	7	8	6-7	7	9	8	9	8	8	8	9	Susceptible
	Salamouni	9	8	8	9	9	8	9	9	8	8	8	Susceptible
	Salamouni	8	8	8	9	9	8	9	9	8	8	8	Susceptible
	Salamouni	8	8	8	8	9	9	9	9	8	8	9	Susceptible
	Salamouni	8	8	8	8	9	9	9	9	8	8	9	Susceptible
	Salamouni	9	8	8	8	9	9	9	9	8	8	9	Susceptible
	Salamouni	8	8	8	8	9	9	9	9	8	8	9	Susceptible
3	Bekaii	8	8	4	8	9	9	9	8	4	3	6	Ni[e]
	Abou Shwereb	3	4	8	8	9	9	9	9	8	8	8	Ni
5	**Vilmorin 23[d]**	2	3	8	8	9	8	8	9	9	9	9	
	Awnless variety	1	3	8	8	9	9	8	9	7	7	8	*Yr3*
6	**Hybrid 46**	2	1	1	2	9	2	8	9	9	9	8	
	Nessr	2	4	2	5	9	5	9	9	8	8	8	*Yr4*
11	**TP981**	2	8	9	9	9	8	8	9	9	9	9	
	Salamouni	1	8	8	8	9	9	9	9	8	8	9	*Yr25*
	Salamouni	1	8	8	8	9	9	9	9	8	8	9	*Yr25*

[a]A = 6E16, B = 6E16*v9v27*, C = 43E138, D = 45E140, E = 106E139, F = 169E136*v17*, G = 232E137, H = 237E141, I = 237E141*v17*, J = 237E173*v17 (Oakley/Solstice)*, K =239E175*v17 (Warrior)*. Pathotypes are coded according to Johnson *et al.* (1972).
The virulences and avirulences tested were 1, 2, 3, 4, 6, 7, 8, 9, 17, 25, 27, 32, SD, SP, Su. Scoring was performed as described by McNeal *et al.*(1971); Infection types IT0 = No visible uredia, IT1 = Necrotic flecks, IT2 = Necrotic areas without sporulation, IT3-4 = Necrotic and chlorotic areas with restricted sporulation, IT5-6 = Moderate sporulation with necrosis and chlorosis, IT7-8 = Sporulation with chlorosis, IT9 = Abundant sporulation without chlorosis.
[b] Resistant to all *Puccinia striiformis* f. sp. *tritici* pathotypes used.
[c] Susceptible to all *Puccinia striiformis* f. sp. *tritici* pathotypes used.
[d] The entries in bold correspond to the infection type profiles of the tester lines inoculated with the array of 11 *Puccinia striiformis* f. sp. *tritici* pathotypes.
[e] Ni indicates non-identified resistance genes.

Resistance group 3 included 15 ICARDA elite lines (Table 2), one bread and six durum wheat varieties from Lebanon (Table 3), and two Lebanese durum wheat landraces that did not display clear differential responses to the 11 pathotypes used in this study. The genotypes in this group, therefore, could not be used for gene postulation (Table 4).

The tested genotypes in resistance group 4 had high ITs (7 to 9) for the seven *Pst* pathotypes virulent against

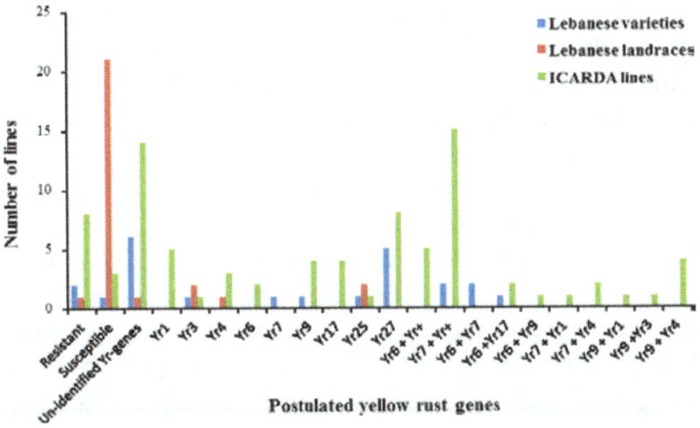

Figure 1. Postulated *Yr* seedling-stage stripe rust resistance in Lebanese wheat varieties, Lebanese wheat landraces and ICARDA wheat lines.

Yr1, and low ITs (1 to 3) for the four pathotypes avirulent against *Yr1*. This led us to postulate the presence of *Yr1* in the five ICARDA elite lines, with resistance profiles similar to those of the tester genotypes Chinese 166 and Avocet *Yr1* (Table 2). This is the first postulation of *Yr1* in Lebanese varieties and landraces.

Genotypes in resistance group 5 had high ITs (7 to 9) for the nine pathotypes virulent against *Yr3*, and low ITs (1 to 4) for pathotypes avirulent against *Yr3*. One ICARDA elite line, one Lebanese variety (Genessi), and one Lebanese bread wheat landrace (awnless variety) were postulated to have *Yr3*. These postulated lines had resistance profiles similar to that of the tester genotype Vilmorin 23 (Tables 2, 3 and 4).

Three elite lines and the bread wheat landrace Nessr were postulated to carry the *Yr4* gene (resistance group 6). The postulated lines had resistance profiles similar to that of the tester genotype Hybrid 46 (Table 1). The low to intermediate ITs (2 to 4, except for one line, ESWYT99#18/Arrihane, which had an IT of 5) for five pathotypes avirulent against *Yr4* and high ITs (7-9) for six pathotypes virulent against *Yr4* led to a postulation of *Yr4* in the tested genotypes in this resistance group (Tables 2 and 4). The IT profile of Hybrid 46 matched that of the tested lines except for the line with an intermediate IT of 5.

Two elite lines had resistance profiles similar to that of the Avocet *Yr6* tester line in tests against the 11 pathotypes (resistance group 7). Based on the high and low infection type patterns of the tested genotypes with the tester line Avocet *Yr6*, the tested genotypes were postulated to carry *Yr6* (Table 2). Five of the ICARDA elite bread wheat lines had high (7–8) and low (1–3) infection type patterns, similar to those of the tester variety Heines Peko when tested against the 11 pathotypes. Heines Peko is known to carry *Yr6* plus additional uncharacterized *Yr*-gene/s (Calonnec *et al*., 1997). These five elite lines (lines 2, 5, 47, 63, and 64), designated *Yr6+* in Table 2, were, therefore, also postulated to carry *Yr6* plus unknown additional gene/s.

Resistance group 8 included one Lebanese bread wheat variety with high ITs (8 to 9) for the pathotypes virulent against *Yr7* and low ITs (2 to 3) for the pathotypes avirulent against *Yr7*. The same infection type pattern was observed when the tester line Avocet *Yr7* was tested against the 11 *Pst* pathotypes (Table 3). Riechersberg 42 is known to carry *Yr7* and additional uncharacterized resistance genes (McIntosh *et al*., 1995). Based on the similarity of the infections patterns of Riechersberg 42, we postulated that *Yr7* and additional uncharacterized genes were present in the 15 elite lines and three Lebanese bread wheat varieties (Table 2).

The genotypes in resistance group 9 were postulated to carry *Yr9*. This group included four bread wheat elite lines. The postulation of *Yr9* in this group was based on similar infection type patterns for the tested lines Avocet *Y9* and Clement, in tests against the seven *Pst* pathotypes virulent against *Yr9*, and the four pathotypes avirulent against this gene (Tables 2 and 3).

Resistance group 10 included three elite lines postulated to carry *Yr17*. These lines had high ITs (IT = 8) against four pathotypes virulent against *Yr17*, and low ITs (2 to 3) for seven pathotypes avirulent against *Yr17*. These lines had resistance profiles similar to that of the tester line VPM1 (Table 2).

Resistance group 11 included one elite line, one Lebanese bread wheat variety (Florence Aurore) and two Lebanese bread wheat landraces (two accessions of Salamouni) which were postulated to carry *Yr25*, with low ITs (IT = 1) for the *Pst* pathotype avirulent against *Yr25*, and high ITs (7 to 9) against the other ten pathotypes virulent against *Yr25*. The resistance profiles of the genotypes postulated to carry *Yr25* were similar to that of the tester line TP981 used as source of *Yr25* in this study (Tables 2, 3 and 4).

Resistance group 12 included eight elite lines (Table 2) and five Lebanese bread wheat varieties (Table 3) postulated to carry *Yr27*. The tested genotypes had high ITs (7 to 8) for the only pathotype virulent against *Yr27*, and low ITs (1 to 5) for the remaining ten pathotypes avirulent against *Yr27*. These postulated lines had resistance profiles similar to that of the tester genotypes Avocet *Yr27* and Opata.

Resistance group 13 included the genotypes for which two *Yr* genes were postulated. Similarities in the infection type patterns of two tester genotypes tested against the 11 pathotypes (Table 1) and those of tested elite lines, Lebanese varieties and landraces, were used to postulate gene combinations in the tested genotypes in this resistance group. In this group, combinations of *Yr6* and *Yr9* were postulated in only two ICARDA elite lines, *Yr6* and *Yr17* were postulated in three of these ICARDA lines, *Yr7* and *Yr1* were postulated in one ICARDA line, *Yr7* and *Yr4* were postulated in three ICARDA lines, *Yr9* and *Yr1*, were postulated in one ICARDA line, *Yr9* and *Yr3* were postulated one ICARDA line, and *Yr9* and *Yr4* were postulated in only four ICARDA elite lines (Table 2). Similarly, the combination of *Yr6* and *Yr7* was postulated in two Lebanese varieties, and the combination of *Yr6* and *Yr17* was postulated in one Lebanese variety (Table 3). With the 11 pathotypes used here, we were able to postulate all the *Yr* genes considered, except for *Yr32,* in the genotypes tested.

Evaluation of adult plant resistance

We evaluated the adult-plant responses of the tested genotypes by performing a field test with 44 ICARDA elite lines at two sites, one in Syria (Tal Hadya) and the other in Lebanon (Terbol). Table 5 shows the seedling and adult plant responses of the tested lines at the two sites, together with the postulated *Yr* genes. Genotypes were considered to carry only adult-plant resistance when the same genotype was susceptible (high ITs of 7 to 9) to pathotype 6E16*v9v27* at the seeding stage, but resistant at the adult-plant stage. Twenty nine out of 44 elites lines that were resistant at the seedling stage, showed low to moderate resistance at adult-plant stage. The elite lines 7, 12, 22, 29, 35, 51, 66, 67, 75, 85 and 87 (Table 5; Figure 2), which were susceptible at the seedling stage, were resistant at adult-plant stage. Elite line 70, which was resistant to the 11 pathotypes used in the gene postulation study, including pathotype 6E16*v9v27,* was also resistant in field tests.

Elite lines postulated to carry *Yr1* (lines 60, 61, and 65), *Yr3* (line 46), and *Yr4* (line 18) were resistant at both the seedling and adult-plant stages. The elite lines postulated to carry *Yr6* (lines 47, 63) and *Yr6+* (64) were resistant in the field. The pathotypes used for inoculation in the field carried virulence against both *Yr6* and *Yr6+*. The field resistance responses of these lines can therefore be considered to indicate the presence of adult-plant resistance in these lines.

Eight lines carrying *Yr7* and additional uncharacterized seedling resistance genes (lines 25, 34, 36, 37, 41, 42, 44, and 68, Table 5) were resistant at both the seedling and adult-plant stages. The source of *Yr7* was susceptible at both the seedling and adult-plant stages and the *Yr7+* source displayed intermediate field responses at the two sites. We therefore considered the strong field responses of the lines postulated to carry *Yr7+* to be due to the combination of an uncharacterized seedling resistance gene and adult-plant resistance genes effective against the field pathotype at both sites.

The line postulated to carry *Yr17* (line 31) displayed an intermediate resistance response in field conditions. The seedling and field responses of the source of *Yr17* showed *Yr17* to be effective against the pathotype used for seedling and adult-plant assessment. Line 39 therefore probably carries *Yr17.* Three lines were postulated to carry *Yr27* (lines 1, 8 and 46), one of which, line 46, displayed moderate resistance at the adult-plant stage suggestive of the presence of adult-plant resistance genes in this line.

Line 51 displayed an adult-plant response of 5R in field conditions, but seedlings of this line had a high infection type with the pathotypes used. As the pathotypes tested were virulent against combinations of the *Yr6* and *Yr9* genes, the adult-plant resistance response of this line was considered to indicate the presence of adult-plant resistance in this line.

The *Yr6+Yr17* combination conferred resistance at both the seedling and adult-plant stages in the Lebanese field for lines 27 and 86, demonstrating the efficacy of *Yr17,* as virulence against *Yr6* was common in both the seedling and adult-plant tests. The line postulated to carry both *Yr7* and *Yr4* (line 4) was resistant at both the seedling and adult-plant stages, demonstrating the efficacy of *Yr4* in field tests. As *Yr1, Yr3,* and *Yr4* were effective against the pathotypes used in both seedling and adult-plant tests, the resistance response of combinations of *Yr9* with *Yr1* (line 84), *Yr3* (line 83), and *Yr4* (line 9) was due to the efficacy of *Yr1, Yr3,* and *Yr4* against the pathotypes used in the seedling and adult-plant tests.

Three of the tested lines for which the seedling resistance could not be postulated by multipathotype testing (lines 7, 22, and 85) were postulated to carry adult-plant resistance, given the high level of infection observed at the seedling stage.

Resistance diversity in Lebanese landraces

The landraces generally displayed considerable diversity in their response to the 11 pathotypes (Table 6; Figure S1). Many landraces displayed variation in their resistance responses to each pathotype, with 0 to 50% susceptible plants in resistant landraces and 0 to 50% resistant plants in susceptible landraces, for any given pathotype. The number of resistant plants was highest with the 6E16, 43E138 and 237E173*v17 (Oakley/ Solstice)* pathotypes, and lowest with 239E175*v17 (War-*

Table 5. Pedigree, postulated seedling-stage stripe rust resistance genes and field responses to stripe rust of 44 advanced bread wheat lines from ICARDA at the Tel Hadya (Syria) and Terbol (Lebanon) research stations.

Entry	Pedigree	Postulated *Yr* genes	Seedling infection type[a]	Adult-plant resistance[b]
11	Tracha'S'//CMH76-252/PVN'S'	Resistant[c]	2	20-30MR
20	Achtar*3//Kanz/KS85-8-4/3/Zemamra-5	Resistant	2	10MR
32	Blass-1/4/CHAT'S'//KVZ/CGN/3/BAU'S'	Resistant	1	50RMR
70	Crow'S'/Bow'S' -1994/95//Asfoor-5	Resistant	5	10R
12	W3918A/JUP	Susceptible[d]	7	10R
75	Nesma*2/14-2//2*Safi-3	Susceptible	8	5R
60	Crow'S'/Bow'S' -3-1994/95//Tevee'S'/Tadinia	*Yr1*	1	10MR
61	Tevee'S'/3/T.aestivum/SPRW'S'//CA8055/4/Pastor-2/5/Sunbri	*Yr1*	1	10MR
65	Qafzah-2/Ferroug-2	*Yr1*	2	10MR
46	Bow #1/Fengkang15/3/HYS//DRC*2/7C	*Yr3*	3	10R
18	SHA3/Seri//Yang87-142/3/2*Towpe	*Yr4*	2	20MR
67	Hamam-4/Angi-2	*Yr6*	8	10R
87	Hubara-16/2*Somama-3	*Yr6*	8	30R
47	MON'S'/ALD'S'//Towpe'S'	*Yr6*	5	10R
63	Weebill-1/2*Qafzah-21	*Yr6*	6	10RMR
64	Rebwah-12/Zemamra-8	*Yr6* +[f]	1	1R
34	Kauz//Kauz/Star	*Yr7* +	1	10MR
36	Cham-4/Shuha'S'/6/2*Saker/5/RBS/Anza/3/KVZ/HYS//YMH/TOB/4/Bow'S'	*Yr7*+	1	5R
37	Cham-4/Shuha'S'/6/2*Saker/5/RBS/Anza/3/KVZ/HYS//YMH/TOB/4/Bow'S'	*Yr7*+	1	5R
41	Florkwa-2/Asfoor-5	*Yr7*+	2	30MR
42	Ferroug-2/Potam*2KS811261-8//Zemamra-8	*Yr7*+	2	10R
44	Hubara-15/Zemamra-8	*Yr7*+	1	10R
19	Cham-4/Shuha'S'/6/2*Saker/5/RBS/Anza/3/KVZ/HYS//YMH/TOB/4/Bow'S'	*Yr7*+	1	5R
68	Shuha-5/Asfoor-1	*Yr7*+	1	10R
66	MON'S'/ALD'S'//Aldan'S'/IAS58/3/Safi-1/4/Zemamra-1	*Yr9*	8	10MR
29	Clement/ALD'S'//Zarzour/5/AU//KAL/BB/3/BON/4/KVZ//CNO/PJ62 (Sandall 3)	*Yr9*	8	5R
35	ICARDA-SRRL-5	*Yr9*	8	20R
31	Shuha-8//Vee'S'/Saker'S'	*Yr17*	2	40MRMS
1	Kauz = JUP/BJY//URES	*Yr27*	8	100S
8	Inqualab 91/Flag-2	*Yr27*	8	10S
10	Bow#1/Fengkang15/3/HYS//DRC*2/7C	*Yr27*	8	30MRMS
51	GV/ALD'S'/5/ALD'S'/4/BB/G11//CNO67/7C/3/KVZ/TI/6/2*Towpe	*Yr6* + *Yr9*	7	5R
27	DVERD-2/*Aegilops squarrosa* (214)//2*ESDA/3/NS732/HER	*Yr6* + *Yr17*	3	5R
86	Hubara-3/Angi-2//Somama-3	*Yr6* + *Yr17*	1	20R
4	Samar-8/Kauz'S'//Cham-4/Shuha'S'	*Yr7* + *Yr4*	2	5R
84	ACSAD 529/Karawan'S'//Somama-3	*Yr9* + *Yr1*	1	5R
83	Hubara-5/3/SHA3/Seri//SHA4/Lira	*Yr9* + *Yr3*	2	5R
9	Qafzah-33/Florkwa-2	*Yr9* + *Yr4*	4	5R
7	*T. aestivum*/SPRW'S'//CA8055/3/Bacanora86	Ni[e]	7	1R
22	Fow-2/SD8036//SafiI-3/3/NS732/HER//Kauz'S'	Ni	8	20MR
38	NS732/HER//Arrihane/3/PGO/Seri//BAU	Ni	8	40S
39	IZAZ-2//Tevee'S'/Shuha'S'	Ni	2	20R
54	Sakha73/5/IAS 58/4/KAL/BB//CJ'S'/3/ALD'S'/6/Goumria-12	Ni	2	20R
85	Girwill-13/2*Pastor-2	Ni	8	10R

[a] Infection type with 6E16*v*9*v*27, the predominant pathotype in the CWANA region for the 2010-2011 season, carrying the *v*2, 6, 7, 8, 9, 25, 27, and SD, based on pathotype surveys conducted by ICARDA for the same year.

[b] R = resistant, MR = moderately resistant, MS = moderately susceptible, and S = susceptible, according to the modified Cobb scale of Peterson *et al.* (1948). The numbers 5-100 are the percentages of the leaf area covered by stripe rust. The score reported is the mean for two seasons in Syria, and three seasons in Lebanon.

[c] Resistant to all *Puccinia striiformis* f. sp. *tritici* pathotypes used.

[d] Susceptible to all *Puccinia striiformis* f. sp. *tritici* pathotypes used.

[e] Ni indicates non-identified resistance genes.

[f] Additional and uncharacterized *Yr* genes.

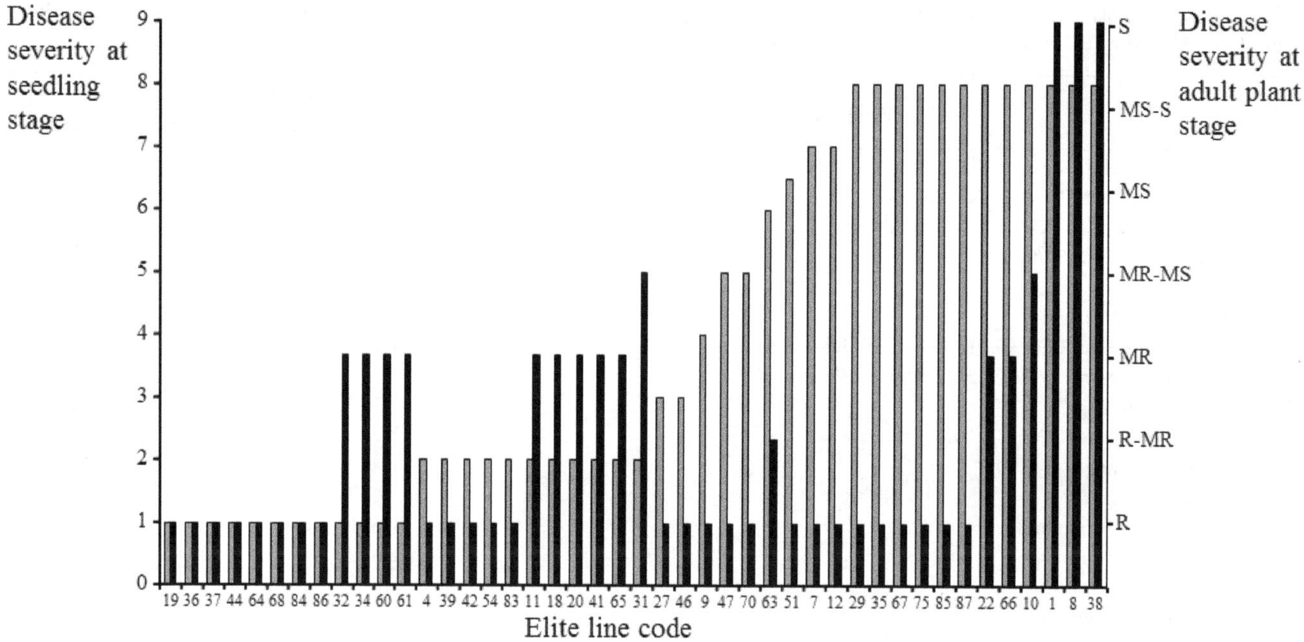

Figure 2. Disease severity at the seedling (gray columns) and adult-plant (black columns) growth stages, for 44 ICARDA bread wheat elite lines inoculated with the predominant pathotype for the 2010-2011 season in Lebanon and Syria. This pathotype carries *v2, 6, 7, 8, 9, 25, 27,* and *SD*, according to the pathotype surveys conducted by ICARDA for the same year. Disease severity at the seedling stage is scored from 0 to 9, where 0 was considered fully resistant and 9 fully susceptible (McNeal *et al.* 1971). At the adult-plant stage, disease severity was recorded as described by Roelfs *et al.* (1992); R = resistant, MR = moderately resistant, MS = moderately susceptible, and S = susceptible. The codes of the elite lines are given in Table 5.

rior), which has the largest number of virulence factors. Interestingly, two durum wheat Bekaii landraces (landraces 24 and 27) from two different sites had 100% and 11% resistant plants, respectively, when tested with the pathotype 239E175*v17 (Warrior)*. These landraces thus carry new resistance genes that could not be detected with the pathotype arrays used here, but the uncharacterized genes in the resistant plants contributed to the resistance of the landraces, particularly in tests against the most virulent race, 239E175*v17 (Warrior)*.

PCA with 11 variables, including the percentage of plants differing from the most frequent reaction with each of the 11 pathotypes, separated the landraces in terms of their heterogeneity of reaction to all 11 pathotypes (Figure 3, Table 7). The two first axes accounted for 54.3% of the variance, with the most heterogeneous landraces to the right of axis 1 (landraces 27, 21, 24, 5, 2, 4, 23) and the most homogeneous ones to the left (landraces 10, 17, 18, 11, 15). Four of the seven most heterogeneous landraces were durum wheat landraces. Axis 2 separated landraces in terms of their heterogeneity of reaction to four pathotypes (three carrying *v7*: 6E16, 43E138, and 106E139). PCA with 11 variables, including the percentage of plants resistant to each of the 11 pathotypes, separated the landraces in terms of their level of resistance for their reaction to all 11 pathotypes (Figure 4; Table 8). The two first axes accounted for 70.1% of total variance, with the most resistant landraces to the right of axis 1 of the PCA (landraces 28, 24, 19, 27) and the most susceptible landraces to the left (landraces 10, 15, 17, 25). Five of the nine most resistant landraces were durum wheat landraces. Axis 2 separated landraces in terms of their resistance to four pathotypes (6E16, 6E16*v9v27*, 45E140 and 169E136*v17*).

DISCUSSION

Host resistance-based approaches remain the most economical and environmentally friendly method of controlling wheat rust diseases. Most of the characterized resistance genes are race-specific and conform to the well-described gene-for-gene model (Flor, 1956). A knowledge of the genetic structure of breeding lines and genetic resources is crucial for the breeding of more durable resistant genotypes and the efficient use of genetic resources. We postulated the *Yr* genes in 138 wheat genotypes from ICARDA, Lebanese varieties and landraces, using an array of 11 *Pst* pathotypes at the seedling stage. Using this set of pathotypes with

Table 6. Assessment of resistance heterogeneity in the wheat landraces, expressed as the percentage of resistant plants (R) among the susceptible landraces, and the percentage of susceptible plants (S) among the resistant landraces, for each of the 11 *Puccinia striiformis* f. sp. *tritici* pathotypes.

Landrace name	Location	Landrace Code	Species[b]	Pathotype[a]																					
				A		B		C		D		E		F		G		H		I		J		K	
				R[c]	S	R	S	R	S	R	S	R	S	R	S	R	S	R	S	R	S	R	S	R	S
Salamouni [d]	Qamouaa, Akkar	1	TA	33[e]	-[f]	0	-	25	-	14	-	8	-	0	-	8	-	4	-	4	-	13	-	0	-
Salamouni	Fneidik, Akkar	2	TA	-	48	17	-	26	-	8	-	17	-	13	-	26	-	0	-	29	-	30	-	0	-
Salamouni	Qamouaa, Akkar	3	TA	0	-	28	-	19	-	-	46	0	-	0	-	0	-	0	-	19	-	26	-	0	-
Salamouni	Laklouk, Jbeil	4	TA	8	-	9	-	19	-	-	41	0	-	29	-	0	-	17	-	-	42	-	36	0	-
Salamouni	Laklouk, Jbeil	5	TA	50	50	32	-	13	-	7	-	9	-	36	-	0	-	0	-	32	-	50	50	0	-
Ukranian Variety	Tel Akhdar, Bekaa	6	TA	-	46	0	-	5	-	3	-	0	-	9	-	0	-	50	-	0	-	0	-	0	-
Ukranian Variety	Tel Akhdar, Bekaa	7	TA	41	-	13	-	16	-	0	-	12	-	18	-	0	-	38	-	4	-	0	-	0	-
Salamouni	Jeb Janine, Bekaa	8	TA	13	-	0	-	-	28	0	-	0	-	0	-	0	-	0	-	32	-	35	-	0	-
Salamouni	Jeb Janine, Bekaa	9	TA	0	-	11	-	22	-	0	-	4	-	16	-	0	-	11	-	48	-	20	-	0	-
Salamouni	Aarida, Akkar	10	TA	0	-	0	-	0	-	0	-	0	-	0	-	0	-	0	-	0	-	0	-	0	-
Salamouni	Northern Bekaa	11	TA	0	-	0	-	-	0	0	-	4	-	5	-	0	-	0	-	0	-	0	-	0	-
Salamouni	Nabha, Bekaa	12	TA	50	50	0	-	40	-	8	-	4	-	10	-	0	-	0	-	0	-	0	-	0	-
Salamouni	Nabha, Bekaa	13	TA	-	0	0	-	4	-	4	-	4	-	0	-	0	-	4	-	4	-	4	-	0	-
Salamouni	Ham, Bekaa	14	TA	7	-	0	-	7	-	0	-	0	-	7	-	4	-	0	-	4	-	4	-	0	-
Salamouni	Ham, Bekaa	15	TA	7	-	0	-	0	-	0	-	0	-	0	-	0	-	0	-	0	-	0	-	0	-
Salamouni	Aarsal, Bekaa	16	TA	0	-	0	-	4	-	0	-	23	-	0	-	4	-	0	-	0	-	0	-	0	-
Salamouni	Northern Bekaa	17	TA	0	-	0	-	0	-	0	-	0	-	0	-	0	-	0	-	0	-	0	-	0	-
Salamouni	Northern Bekaa	18	TA	-	0	0	-	0	-	0	-	0	-	0	-	0	-	0	-	0	-	0	-	0	-
Nessr	Tal Amara, Bekaa	19	TA	-	0	-	0	-	25	-	0	8	-	-	0	0	-	0	-	0	-	0	-	0	-
Awnless Variety	Qamouaa, Akkar	20	TA	-	0	-	0	19	-	10	-	0	-	0	-	9	-	4	-	5	-	4	-	0	-
Abou Shwereb	Qamouaa, Akkar	21	TD	26	-	-	26	34	-	19	-	16	-	13	-	23	-	9	-	30	-	50	50	0	-
Abou Shwereb	Qamouaa, Akkar	22	TD	0	-	0	-	15	-	0	-	8	-	10	-	8	-	4	-	9	-	8	-	0	-
Abou Shwereb	Fneidik, Akkar	23	TD	-	0	-	42	0	-	0	-	0	-	47	-	0	-	0	-	43	-	36	-	0	-
Bekaii	Jeb Janine, Bekaa	24	TD	19	-	38	-	-	31	35	-	4	-	37	-	12	-	50	-	-	0	-	20	-	0
Abou Shwereb	Aarida, Akkar	25	TD	5	-	8	-	0	-	0	-	4	-	0	-	0	-	0	-	0	-	0	-	0	-
Haurani	Tal Amara, Bekaa	26	TD	0	-	-	4	17	-	0	-	0	-	0	-	0	-	0	-	0	-	22	-	0	-
Bekaii	Tal Amara, Bekaa	27	TD	-	33	-	13	-	26	-	43	4	-	36	-	0	-	12	-	-	39	39	-	11	-
Waha	Jeb Janine, Bekaa	28	TD	-	13	-	8	-	0	-	0	-	0	-	8	-	0	-	0	-	0	-	0	-	0

[a] A = 6E16, B = 6E16*v9v27*, C = 43E138, D = 45E140, E = 106E139, F = 169E136*v17*, G = 232E137, H = 237E141, I = 237E141*v17*, J = 237E173*v17 (Oakley/Solstice)*, K = 239E175*v17 (Warrior)*. Pathotypes are coded according to Johnson *et al.* (1972).
The virulences and avirulences tested were 1, 2, 3, 4, 6, 7, 8, 9, 17, 25, 27, 32, SD, SP, Su.
[b] TA = *Triticum aestivum* (L), TD = *T. durum*.
[c] R = seedling resistant (low infection type of 0-6), S = seedling susceptible (high infection type of 7-9).
[d] 30 seedlings per landrace and per pathotype were tested.
[e] Percentage of resistant (R) plants in a mainly susceptible landrace and percentage of susceptible (S) plants in a mainly resistant landrace.
[f] indicates missing data.

complementary virulence spectra, we were able to infer the resistance profiles of most of the lines tested. The 11 pathotypes used here discriminated between the *Yr1*, *Yr3*, *Yr4*, *Yr6*, *Yr7*, *Yr8*, *Yr9*, *Yr17*, *Yr25*, *Yr27*, *Yr32*, *YrSD*, *YrSu* and *YrSP* genes, and we were able to postulate *Yr1*, *Yr3*, *Yr4*, *Yr6*, *Yr7*, *Yr9*, *Yr17*, *Yr25* and *Yr27* singly or in combination in the tested genotypes. These results highlight the utility of this pathotype array for the detection of *Yr* genes. However, a group of genotypes (Group 3) remained for which seedling resistance could not be explained with the pathotypes, and for which the resistance genes present remained unknown. Based on the resistance responses of these genotypes to the wide array of virulence factors of the 11 pathotypes, we postulated that these genotypes might be sources of new stripe rust resistance genes.

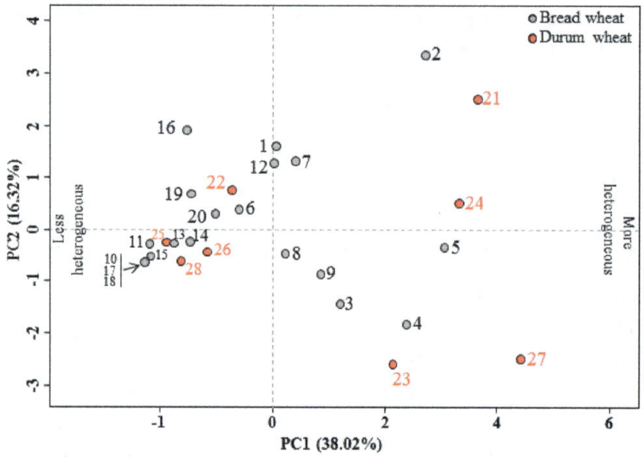

Figure 3. Plot of the first (PC1) and second (PC2) principal component means from an analysis of the 11 variables (percentage of resistant plants in susceptible landraces and percentage of susceptible plants in resistant landraces, for each of the 11 *Puccinia striiformis* f. sp. *tritici* pathotypes, for 28 Lebanese landraces (1-28)), as described in Table 7. Gray symbols = bread wheat, red symbols = durum wheat.

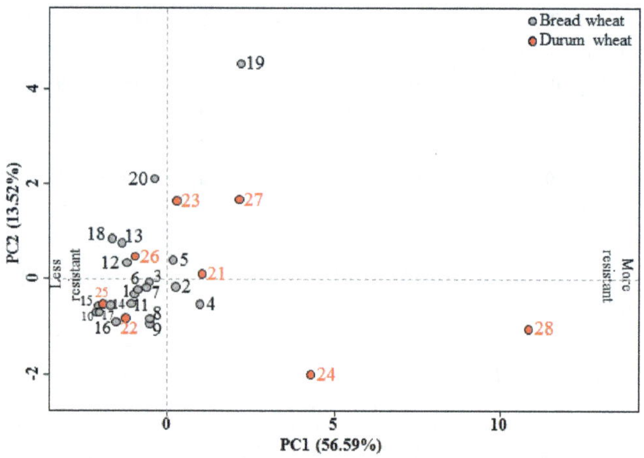

Figure 4. Plot of the first (PC1) and second (PC2) principal component means from an analysis of the 11 variables (percentage of plants resistant to each of the 11 *Puccinia striiformis* f. sp. *tritici* pathotypes, for 28 Lebanese landraces (1-28)), as described in Table 7. Gray symbols = bread wheat, red symbols = durum wheat.

In resistance group 1, 9% of ICARDA elite lines and 9% of Lebanese varieties displayed complete resistance to all pathotypes. Among the resistant Lebanese varieties, only two durum wheat genotypes displayed full resistance. None of the Lebanese bread wheat varieties were completely resistant to all pathotypes. Only one durum wheat landrace was completely resistant to all pathotypes.

With the exception of a few uncharacterized elite lines and landraces, the ICARDA elite lines and Lebanese varieties generally carried at least one effective *Yr* gene providing resistance to at least one of the 11 *Pst* pathotypes, whereas 75% of landraces were susceptible to all

Table 7. Eigen vectors of the two principal components axes (PC1 and 2) for the 11 variables assessed (resistance heterogeneity in the landraces expressed as the percentage of resistant plants in susceptible landraces and the percentage of susceptible plants in resistant landraces, for each of the 11 *Puccinia striiformis* f. sp. *tritici* pathotypes), and their contributions to the variance.

Variables[a]	PC1	PC2
A	0.53	0.42
B	0.76	-0.15
C	0.63	0.37
D	0.67	-0.23
E	0.28	0.75
F	0.80	-0.30
G	0.43	0.71
H	0.33	0.10
I	0.74	-0.31
J	0.86	-0.15
K	0.42	-0.36
Eigen value	4.18	1.80
Percentage of variance (%)	38.02	16.32

[a]A = 6E16, B = 6E16*v9v27*, C = 43E138, D = 45E140, E = 106E139, F = 169E136*v17*, G = 232E137, H = 237E141, I = 237E141*v17*, J = 237E173*v17 (Oakley/Solstice)*, K =239E175*v17 (Warrior)*.

Table 8. Eigen vectors of the two principal components axes (PC1 and 2) for the 11 variables assessed (percentage of plants resistant to each of the 11 *Puccinia striiformis* f. sp. *tritici* pathotypes for Lebanese landraces), and their contributions to the variance.

Variables[a]	PC1	PC2
A	0.53	0.42
B	0.76	-0.15
C	0.63	0.37
D	0.67	-0.23
E	0.28	0.75
F	0.80	-0.30
G	0.43	0.71
H	0.33	0.10
I	0.74	-0.31
J	0.86	-0.15
K	0.42	-0.36
Eigen value	4.18	1.80
Percentage of variance (%)	38.02	16.32

[a]A = 6E16, B = 6E16*v9v27*, C = 43E138, D = 45E140, E = 106E139, F = 169E136*v17*, G = 232E137, H = 237E141, I = 237E141*v17*, J = 237E173*v17 (Oakley/Solstice)*, K = 239E175*v17 (Warrior)*.

pathotypes. Nine of the resistance genes tested (*Yr1*, *Yr3*, *Yr4*, *Yr6*, *Yr7*, *Yr9*, *Yr17*, *Yr25* and *Yr27*) were postulated singly in ICARDA elite lines; *Yr7*, *Yr9*, *Yr25* and *Yr27* were detected in Lebanese varieties and only *Yr3*, *Yr4* and *Yr25* were detected in some landraces. *Yr27* was the most frequent *Yr* gene postulated singly in the Lebanese varieties. *Yr7* in combination with other unidentified *Yr* genes was postulated with the highest frequency in ICARDA elite lines. Combinations of two *Yr* genes were found in 16% of ICARDA elite lines: *Yr6+Yr9*, *Yr6+Yr17*, *Yr7+Yr1*, *Yr7+Yr4*, *Yr9+Yr1*, *Yr9+Yr3*, and *Yr9+Yr4*. The *Yr6+Yr7* combination was found in two Lebanese varieties and *Yr6+Yr17* was found in one landrace only. Only one awnless landrace (BW) contained *Yr3*. Only one Lebanese bread wheat landrace (Nessr) had *Yr4* and two Salamouni bread wheat landraces had *Yr25*. The postulated *Yr* genes, either singly or in combination, were more frequent in Lebanese varieties than in landraces. ICARDA elite lines displayed greater diversity for the postulated *Yr* genes than the Lebanese varieties and landraces (Figure 1).

In general, our study revealed a narrow genetic basis of resistance in the genotypes tested for seedling resistance genes in the absence of effective adult-plant resistance genes, but we did not investigate or characterize such adult-plant resistance genes in this study. With the exception of *Yr3* and *Yr4*, and, to some extent, *Yr1*, which are effective in most of the wheat growing areas in CWANA, the rest of postulated genes were not effective against current pathotypes. The *Yr1*, *Yr3*, and *Yr4* genes will also cease to be effective if the North Western European pathotypes spread to the CWANA region. With the recent incursion of the Warrior pathotype into CWANA, the efficacy of these three *Yr* genes is dwindling, and their use in breeding for rust resistance cannot, therefore, be recommended unless they are used in combination with effective seedling resistance genes and/or adult-plant resistance genes.

Yr1 was postulated in only 6% of ICARDA elite lines. Despite the efficacy of *Yr1* in most of the wheat-growing areas of CWANA, virulence against *Yr1* has been reported in East Asia (Stubbs, 1985), Central Asia and the Caucasus region (Yahyaoui *et al.*, 2002) and Syria (K. Nazari unpublished data), highlighting the race specificity of this gene. Considering the specificity of *Yr1* and the presence of pathotypes virulent against this gene in the *Yr27*-virulent group (Mogens Hovmoller, personal communication), together with the recent spread of the Warrior pathotype to North Africa, Turkey, and Azerbaijan from Europe, the use of elite lines and commercial cultivars bearing only *Yr1* should be restricted in CWANA.

Yr3 and *Yr4* were infrequent in the lines tested. These two resistance genes are very common in winter wheat cultivars and breeding lines in North Western Europe (de Vallavieille-Pope *et al.*, 1990; 2012). However, despite the low frequency of virulence against these two genes in most of the wheat-growing areas of CWANA, sources of *Yr3* and *Yr4* have not been widely used in breeding for stripe rust resistance in spring wheat genotypes. Virulence against these two genes is very common within the *Pst* population in Europe and Australia (de Vallavieille-Pope *et al.*, 2012; Wellings, 2011). *Yr3* and *Yr4* can no longer be recommended as sources of resistance in CWANA, due to the recent spread of the Warrior race to some of the wheat-growing areas of North Africa and West Asia.

Yr6 was postulated singly or in combination with *Yr9* or *Yr17* in 13% of ICARDA elite lines and in combination with *Yr7* in 9% of Lebanese varieties. Varieties carrying *Yr6* were introduced into the CIMMYT wheat breeding program and, hence, into ICARDA germplasm, as sources of leaf rust resistance, including *Lr13* and *Lr34* (Wellings, 1986). However, *Yr6* was not frequent in the ICARDA lines tested and virulence against *Yr6* has been reported to be fixed in all isolates from Asia, Africa and South America tested (GRRC, 2017).

Yr7 was postulated singly in only one Lebanese bread wheat variety (Haramoun) and in combination with additional genes in other two Lebanese bread wheat varieties, Tannour and 885, and 15 elite lines. *Yr7* originated from the durum wheat cv. Iumillo. The gene was transferred to Thatcher wheat, from which the differential variety Lee was derived (McIntosh *et al.*, 2012). *Yr7* is present in a range of winter and spring wheat cultivars (McIntosh *et al.*, 2012). This gene has been defeated in the CWANA region and is no longer effective against the prevalent pathotypes in this region.

Yr9 was postulated singly in four ICARDA elite lines (5%) and in combination with *Yr1*, *Yr3* and *Yr4* in six elite lines (7%). This gene originated from *Secale cereale* and is linked to *Lr26* and *Sr31* in the 1BL.1RS translocation (McIntosh *et al.*, 2012). During the 1990s, most of the adapted wheat germplasm generated and distributed by CIMMYT in spring wheat production areas at low latitudes carried the 1BL.1RS translocation (Bimb and Johnson, 1997). This translocation was also identified in European wheat germplasm by Mettin *et al.* (1973) and Zeller (1973). Virulence against *Yr9* has been common in wheat-growing areas in CWANA and sub-Saharan countries since the 1980s, particularly in countries in which 1B.1R-containing genotypes were distributed, including Ethiopia (Badebo and Bayu, 1992), Syria (Mamluk and El-Naimi, 1992), Turkey (Dusunceli *et al.*, 1996), Iran (Torabi *et al.*, 1995), Pakistan (Bahri *et al.*, 2011), and in Central Asia and Caucasian countries (Yahyaoui, 2005).

The use of this gene in breeding materials should therefore be restricted.

Yr17 was postulated singly in three ICARDA elite lines (3%) and in combination with *Yr6* in one Lebanese bread wheat variety (Naama). The *Yr17*, *Lr37* and *Sr38* gene cluster was transferred to wheat in a translocation from *Aegilops ventricosa* (Dousssinault *et al.*, 1998). It was originally transferred to the VPM1 line (a cross of *Ae. ventricosa*, *Triticum persicum* and *cv.* Marne Desprez) (Bariana and McIntosh, 1993). Virulence against *Yr17* has been detected in the USA (Line *et al.*, 1992), and in North Western Europe (Bayles *et al.*, 2000; Hovmøller *et al.*, 2002), where it remains frequent (de Vallavieille-Pope *et al.*, 2012). The emergence of virulence against *Yr17* with the incursion of the Warrior race into North Africa (Hovmoller *et al.*, 2016) and Turkey (Mert *et al.*, 2016) should restrict the use of *Yr17* sources in isolation. Most of the French isolates studied here carry virulence against *Yr9* and *Yr17*, and it is therefore difficult to postulate these two genes when they are present singly. The use of diagnostic molecular markers for these genes can be very useful.

Yr25, which is common and ineffective in North Western European wheat varieties, was postulated in ICARDA elite line Nayzak-3 (line 15), one Lebanese variety (Florence Aurore), and two accessions of the Salamouni landrace. Virulence against *Yr25* is frequent in CWANA (Yahyaoui *et al.*, 2002, Nazari, unpublished data).

Yr27 was postulated in five Lebanese varieties (22%) and eight ICARDA elite lines (9%). This gene originated from the wheat cultivar Selkirk and derivatives of the cultivar 'McMurachy' (Wellings, 1992), a parent of 'Selkirk'. This gene is also present in many CIMMYT genotypes (Wellings, 2011). Virulence against *Yr27* has been detected in New Zealand (Wellings and Burdon, 1992), Pakistan (Bahri *et al.*, 2011; Ali *et al.*, 2014), India (Prashar *et al.*, 2007), Tajikistan, Kyrgyzstan (Singh *et al.*, 2004), Iran (Nazari and Torabi, 2000) and Syria (Nazari *et al.*, 2011). In the last few years, major wheat stripe rust epidemics have occurred in CWANA, sub-Saharan Africa, the Caucasus region and the Indian subcontinents, mostly due to the widespread cultivation of *Yr27* genotypes (Atilla and Kauz derivatives). Interestingly, that the Warrior pathotype does not carry virulence against *Yr27* and *Yr27* wheat cultivars may therefore once again come to predominate in the region.

In Lebanon and Syria, in 2010/2011, the *Yr1*, *Yr3* and *Yr4* genes remained effective, but virulence against *Yr2*, *Yr6*, *Yr7*, *Yr9*, *Yr25*, and *Yr27* predominated, with virulence against *Yr8* and *Yr17* occurring at only low to moderate frequencies (El Amil *et al.*, in press).

Adult-plant resistance is often race non-specific and more durable than race-specific seedling resistance. Adult-plant resistance is generally controlled by temperature-sensitive, minor or additive genes. The presence of adult-plant resistance genes has been reported in various winter and spring wheats (Johnson, 1980; Singh and Rajaram, 1994; Chen *et al.*, 2014).

We therefore tested adult-plant resistance in Lebanese and Syrian fields, for a subset of elite lines. The three lines with *Yr1* were moderately resistant in the field and virulence against this gene (*v*1) was not detected in the survey conducted in 2010-2011; *v*1 may therefore have been present, at low frequency, at Terbol (LB). Lines 63 and 64, which carry *Yr6*, were resistant and moderately resistant, respectively, in the field, despite the presence of *v6* in Syria and Lebanon, indicating the presence of additional adult-plant resistance in these lines. Line 34, which carries only *Yr7*, was less resistant in the field than line 4, which carries the *Yr7+Yr4* combination. Given that the *Yr4* elite line 18 tested was moderately resistant when tested at Terbol (LB), the *Yr7+Yr4* combination seems to be responsible for conferring full resistance. The combination of two seedling resistance genes, *Yr7+Yr4*, was effective at both sites. Line 1, in which *Yr27* was postulated singly, was susceptible in Syria. Elite line 8 was susceptible in Lebanon and line 10 was moderately resistant to moderately susceptible, suggesting that line 10 displayed adult-plant resistance. The adult-plant resistance test confirmed the presence of *v27* in Syria and Lebanon in 2010 and 2012.

Seedling resistance is of short duration and is rapidly overcome by the pathogen population in the absence of adult-plant resistance. Combinations of seedling resistances prolong the efficacy of the genes, but are rarely durable. Quantitative trait loci (QTLs) for adult-plant resistance provide partial resistance, but seldom protect the plant early in its life. A combination of both types of resistance is, therefore, crucial for the protection of the plant throughout the entire growing season. Durable stripe rust resistance has been observed in four French cultivars and one English cultivar combining both seedling resistance genes and QTLs active at the adult plant stage: *cv.* Renan (Dedryver *et al.*, 2009), *cv.* Camp Rémy (Mallard *et al.*, 2005), *cv.* Apache (Paillard *et al.*, 2012), *cv.* Soissons (de Vallavieille-Pope *et al.*, 2012) and *cv.* Claire (Powell *et al.*, 2013).

Landraces are considered to be potential sources of disease resistance and agronomic traits. Considerable heterogeneity has already been reported, for plant height and days to heading, in Israeli bread and durum wheat landrace populations (Beharav *et al.*, 1997), and has been advocated as a potential source of stripe and leaf rust resistance in nine Chinese landraces (Zhang, 1995). Our study confirms that landraces are composed of several genotypes, and it will be of particular interest to investigate resistance genes in the genotypes resistant to 239E175*v17 (Warrior)*, the most frequent pathotype in North Western Europe. Further studies may determine whether the resistance gene found in the landraces differs from the genes already identified, and whether resistant landraces could be exploited for rust resistance and other agronomic traits.

ACKNOWLEDGEMENTS

We thank Mr Laurent Gérard and Mrs Nathalie Retout for technical assistance with the sowing of all plant material, and Mr Simon Khairallah for delivering the landrace seeds from the collection that he maintains at the Gene Bank in LARI, Tal Amara, Lebanon. We also thank Julie Sappa, from Alex Edelman & Associates, for assisting with English language presentation of this paper.

LITERATURE CITED

Ali S., Leconte M., Rahman H., Saqib M.S., Gladieux P., ... de Vallavieille-Pope C., 2014. A high virulence and pathotype diversity of *Puccinia striiformis* f. sp. *tritici* at its centre of diversity, the Himalayan region of Pakistan. *European Journal of Plant Pathology* 140: 275–290.

Badebo A., Bayu W., 1992. The importance of stripe rust in the major bread wheat-producing region of Ethiopia 1988–90. In: *Proceedings of the 7th Regional Wheat Workshop for Eastern, Central and Southern Africa, Nakuru, Kenya* (D.G. Tanner, W. Mwangi ed), CIMMYT, Mexico, D.F., Mexico, 96–202.

Bahri B., Shah S.J.A., Hussain S., Leconte M., Enjalbert J., de Vallavieille-Pope C., 2011. Genetic diversity of wheat yellow rust population in Pakistan and its relationship with host resistance. *Plant Pathology* 60: 649–660.

Bariana H.S., McIntosh R.A., 1993. Cytogenetic studies in wheat location of rust resistance genes in VPM1 and their genetic linkage with other disease resistance genes in chromosome 2A. *Genome* 36: 476–482.

Bayles R.A., Flath K., Hovmøller M.S., de Vallavieille-Pope C., 2000. Breakdown of the *Yr17* resistance to yellow rust of wheat in northern Europe. *Agronomie* 20: 805–811.

Beharav A., Golan G., Levy A., 1997. Evaluation and variation in response to infection with *Puccinia striiformis* and *Puccinia recondita* of local wheat landraces. *Euphytica* 94: 287–293.

Bimb H.M., Johnson R., 1997. Breeding resistance to yellow (stripe) in wheat. Wheat Special Report No 41. CIMMYT, Mexico, D.F., Mexico, 20 pp.

Calonnec A., Johnson R., de Vallavieille-Pope C., 1997. Genetic analysis of resistance to *Puccinia striiformis* in the wheat differential varieties Heines VII, Heines Peko and Strubes Dickkopf. *Plant Pathology* 46: 373–386.

Camacho Villa T.C., Maxted N., Scholten M.A., Ford-Lloyd B.V., 2005. Defining and identifying crop landraces. *Plant Genetic Resources* 3: 373–384.

Chen X.M., 2005. Epidemiology and control of stripe rust (*Puccinia striiformis* f. sp. *tritici*) on wheat. *Canadian Journal of Plant Pathology* 27: 314–337.

Chen W., Wellings C., Chen X., Kang Z., Liu T., 2014. Wheat stripe (yellow) rust caused by *Puccinia striiformis* f. sp. *tritici*. *Molecular Plant Pathology* 15: 433–446.

Dawit W., Flath K., Weber W.E., Schumann E., Röder M.S., Chen X.M., 2012. Postulation and mapping of seedling stripe rust resistance genes in Ethiopian bread wheat cultivars. *Plant Pathology* 94: 403–409.

de Vallavieille-Pope C., Picard-Formery H., Radulovic S., Johnson R., 1990. Specific resistance factors to yellow rust in seedlings of some French wheat varieties and races of *Puccinia striiformis* Westend. in France. *Agronomie* 10: 103–113.

de Vallavieille-Pope C., Huber L., Leconte M., Bethenod O., 2002. Preinoculation effects of light quantity on infection efficiency of *Puccinia striiformis* and *P. triticina* on wheat seedlings. *Phytopathology* 92: 1308–1314.

de Vallavieille-Pope C., Ali S., Leconte M., Enjalbert J., Delos M., Rouzet J., 2012. Virulence dynamics and regional structuring of *Puccinia striiformis* f. sp. *tritici* in France between 1984 and 2009. *Plant Disease* 96: 131–140.

Dedryver F., Paillard S., Mallard S., Robert O., Trottet M., ... Jahier J., 2009. Characterization of genetic components involved in durable resistance to stripe rust in the bread wheat 'Renan'. *Phytopathology* 99: 968–973.

Doussinault G., Dosba F., Jahier J., 1998. Use of a hybrid between *Triticum aestivum* L. and *Aegilops ventricosa* Tausch in wheat breeding. In: *Proceedings of the Seventh International Wheat Genetics Symposium* (T.E. Miller, R.M.D. Koebner, ed.), Institute of Plant Science Research, Cambridge, UK, 253–258.

Dusunceli F., Cetin L., Albustan S. Beniwal S.P.S., 1996. Occurrence and impact of wheat stripe rust (*Puccinia striiformis*) in Turkey in 1994/95-crop season. In:

Proceedings of the 9th European and Mediterranean Cereal Rusts and Powdery Mildews Conference (G.H.J. Kema, R.E. Niks, R.A. Daamen, ed.), Lunteren, The Netherlands, p 309.

El Amil R., Ali S., Bahri B, Leconte M., de Vallavieille-Pope C., Nazari K, 2020. Pathotype diversification in the invasive PstS2 clonal lineage of *Puccinia striiformis* f. sp. *tritici* causing yellow rust on durum and bread wheat in Lebanon and Syria in 2010-2011. *Plant Pathology* in press

Flor H.H., 1956. The complementary gene systems in flax and flax rust. *Advances in Genetics* 8: 29–54.

GRRC, 2017. Summary of *Puccinia striiformis* race analysis 2017. Available at: https://agro.au.dk/en/research/research-areas/global-rust-reference-center/.

Harlan J.R., Zohary D., 1966. Distribution of wild wheats and barley. *Science* 153: 1074–1080.

Harlan J.R., 1975. Our vanishing genetic resources. *Science* 188: 618–621.

Hau B., de Vallavieille-Pope C., 2006. Wind-dispersed diseases. In: The Epidemiology of Plant Diseases (B. M. Cooke, D. Gareth Jones, B. Kaye, ed.), Second Edition, Springer, Dordrecht, The Netherlands. 387–416.

Hodson D., Nazari K., 2010. Serious outbreaks of wheat stripe rust or yellow rust in Central and West Asia and North Africa, March/April 2010. Available at: http://globalrust.org/traction/permalink/Pathogen206.

Hovmøller M.S., Justesen A.F., Brown J.K.M, 2002. Clonality and long-distance migration of *Puccinia striiformis* f. sp. *tritici* in north-west Europe. *Plant Pathology* 51: 24–32.

Hovmøller M.S., 2007. Sources of seedling and adult plant resistance to *Puccinia striiformis* f. sp. *tritici* in European wheats. *Plant Breeding* 126: 225–233.

Hovmøller M.S., Sørensen C.K., Walter S., Justesen A.F., 2011. Diversity of *Puccinia striiformis* on cereals and grasses. *Annual Review of Phytopathology* 49: 197–217.

Hovmøller M.S., Walter S., Bayles R.A., Hubbard A., Flath K., ... de Vallavieille-Pope C., 2016. Replacement of the European wheat yellow rust population by new races from the centre of diversity in the near-Himalayan region. *Plant Pathology* 65: 402–411.

Johnson D.A., 1980. Effect of low temperature on the latent period of slow and fast rusting winter wheat genotypes. *Plant Disease* 64:1006–1008.

Johnson R., Stubbs W., Fuchs E., Chamberlain N.H., 1972. Nomenclature for physiologic races of *Puccinia striiformis* infecting wheat. *Transactions of the British Mycological Society* 58: 475–480.

Line R.F., Chen X.M., Qayoum A., 1992. Races of *Puccinia striiformis* in North America, identification of resistance genes and durability of resistance. *Vorträge für Pflanzenzüchtung* 24: 280–282.

Lucas H., 2012. Briefing paper- Wheat Initiative. GCARD 2012. Second Global Conference on Agricultural Research for Development, 19 October – 1 November 2012, Punta del Este, Uruguay.

Mallard S., Gaudet D., Aldeia A., Abelard C., Besnard A.L., ... Dedryver F., 2005. Genetic analysis of durable resistance to yellow rust in bread wheat. *Theoretical and Applied Genetics* 110: 1401–1409.

Mamluk O.F., El-Naimi M., 1992. Occurrence and virulence of wheat yellow rust in Syria. In: *Proceedings of the 8th European and Mediterranean Cereal Rusts and Mildews Conference* (F.J. Zeller, G. Fischbeck, ed.), Weihenstephan, Germany, 115–117.

McDonald B.A., Linde C., 2002. Pathogen population genetics, evolutionary potential, and durable resistance. *Annual Review of Phytopathology* 40: 349–379.

McIntosh R.A., Wellings C.R., Park R.F., 1995. Wheat rusts: An atlas of resistance genes. CSIRO Australia, Kluwer Acad. Publ., Dordrecht, the Netherlands, 200 pp.

McIntosh R.A., Dubcovsky J., Rogers W.J., Morris C., Appels R., Xia X.C., 2017. Catalogue of Gene Symbols for Wheat: 2017 Supplement. Available at: http://www.wheat.pw.usda.gov/GG2/pubs.

McNeal F.H., Konzak C.F., Smith E.P., Tate W.S. Russel T.S., 1971. A uniform system for recording and processing cereal research data. In: Agricultural Research Service Bulletin, United States Department of Agriculture, Washington, DC, 34–121.

Mert Z., Nazari K., Karagöz E., Akan K., Öztürk İ. Tülek A., 2016. First incursion of the Warrior race of wheat stripe rust (*Puccinia striiformis* f. sp. *tritici*) to Turkey in 2014. *Plant Disease* 100: 528.

Mettin D., Bluthner W.D., Schlegel G., 1973. Additional evidence on spontaneous 1B.1R wheat-rye substitutions and translocations. In: *Proceedings 4th International Wheat Genetics Symposium* (E.R. Sears, L.M.S. Sears, ed.), Agricultural Experiment Station, University of Missouri: Columbia, Missouri, USA, 179–184.

Milus E.A., Kristensen K., Hovmøller M.S., 2009. Evidence for increased aggressiveness in a recent widespread strain of *Puccinia striiformis* f. sp. *tritici* causing stripe rust of wheat. *Phytopathology* 99: 89–94.

Nazari K., Torabi M., 2000. Distribution of yellow rust (*Yr*) resistance genes in Iran. *Acta Phytopathologica et Entomologica Hungarica* 35: 121–131.

Nazari K., Wellings C.R., Park R.F., 2008. Characterisation of seedling resistance to rust diseases in wheat cultivars from Central Asia and the Caucasus. *International Journal of Plant Breeding* 2: 52–63.

Nazari K., Hodson D., Hovmøller M.S., 2011. Yellow rust in CWANA in 2010 and 2011: The situation and measures taken and how to control it. Borlaug Global Rust Initiative workshop, June 13–16, St. Paul, Minnesota, USA.

Newton A.C., Akar T., Baresel J.P., Bebeli P.J., Bettencourt E., ... Vaz Patto M.C., 2010. Cereal landraces for sustainable agriculture. A review. *Agronomy for Sustainable Development* 30: 237–269.

Paillard S., Trotoux-Verplancke G., Perretant M.R., Mohamadi F., Leconte M., ... Dedryver F., 2012. Durable resistance to stripe rust is due to three specific resistance genes in French bread wheat cultivar Apache. *Theoretical and Applied Genetics* 125: 955–965.

Pathan A.K., Park R.F., 2007. Evaluation of seedling and adult plant resistance to stem rust in European wheat cultivars. *Euphytica* 155: 87–105.

Perwaiz M.S., Johnson R., 1986. Genes for resistance to yellow rust in seedlings of wheat cultivars from Pakistan tested with British isolates of *Puccinia striiformis*. *Plant Breeding* 97: 289–296.

Peterson R.F., Campbell A.B, Hannah A.E., 1948. A diagrammatic scale for estimating rust intensity on leaves and stems of cereals. *Canadian Journal* of *Research* 26: 496–500.

Powell N.M., Lewis C.M., Berry S.T., MacCormack R., Boyd L.A., 2013. Stripe rust resistance genes in the UK winter wheat cultivar Claire. *Theoretical and Applied Genetics* 126: 1599–1612.

Prashar M., Bhardwaj S.C., Jain S.K., Datta D., 2007. Pathotypic evolution in *Puccinia striiformis* in India during 1995–2004. *Australian Journal of Agricultural Research* 58: 602–604.

R Development Core Team, 2008. R: A language and environment for statistical computing. R Foundation for Statistical Computing, Vienna, Austria. ISBN 3-900051-07-0, Available at: http://www.R-project.org.

Robert O., Dedryver F., Leconte M., Rolland B., de Vallavieille-Pope C., 2000. Combination of resistance tests and molecular tests to postulate the yellow rust resistance gene *Yr*17 in bread wheat lines. *Plant Breeding* 119: 467–472.

Roelfs A.P., Huerta-Espino J., Singh R., Saari E.E., 1992. Rust diseases of wheat: concepts and methods of disease management. In: *Concepts and methods of disease management* (G.P. Hettel, ed.), CIMMYT, Mexico, D.F., Mexico, 81 pp.

Sharma S., Louwers J.M., Karki C.B., Snijders C.H.A., 1995. Postulation of resistance genes to yellow rust in wild emmer wheat derivatives and advanced wheat lines from Nepal. *Euphytica* 81:271–277.

Sharma-Poudyal D., Chen X.M., Wan A.M., Zhan G.M., Kang Z.S., ... Patzek L.J., 2013. Virulence characterization of international collections of the wheat stripe rust pathogen, *Puccinia striiformis* f. sp. *tritici*. *Plant Disease* 97: 379–386.

Singh R.P., Rajaram S., 1994. Genetics of adult plant resistance to stripe rust in ten spring bread wheats. *Euphytica* 72: 1–7.

Singh R.P., William H.M., Huerta-Espino J., Rosewarne G., 2004. Wheat rust in Asia: meeting the challenges with old and new technologies. In: *Proceedings of the 4th International Crop Science Conference* (T. Fischer, N. Turner, J. Angus, L. McIntyre, M. Robertson, A. Borrell, D. Lloyd, ed.). 26 September – 1 October, 2004, Brisbane, Australia.

Solh M., Nazari K., Tadesse W., Wellings C.R., 2012. The growing threat of stripe rust worldwide. BGRI 2012 Technical Workshop. 1–4 September 2012, Beijing, China.

Statler G.D., 1984. Probable genes for leaf rust resistance in several hard red spring wheats. Crop Science 24: 883–886.

Stubbs R.W., 1985. Stripe rust. In: *The Cereal Rusts, Diseases, Distribution, Epidemiology and Control* (A. P. Roelfs, W. R. Bushnell, ed.), Academic Press, London, 61–101.

Torabi M., Mardoukhi V., Nazari K., Afshari F., Forootan ... Kashani A.S., 1995. Effectiveness of wheat yellow rust resistance genes in different parts of Iran. *Cereal Rusts and Powdery Mildew Bulletin* 23: 9–12.

Wellings C.R., 1992. Resistance to stripe (yellow) rust in selected spring wheats. *Vorträge für Pflanzenzüchtung* 24: 273–275.

Wellings, C.R., 1986. *Host: pathogen studies of wheat stripe rust in Australia*. PhD thesis, University of Sydney, Sydney, Australia.

Wellings C.R., Burdon J.J., 1992. Variability in *Puccinia striiformis* f. sp. *tritici* in Australasia. *Vorträge für Pflanzenzüchtung* 24: 114.

Wellings C.R., 2011. Global status of stripe rust: a review of historical and current threats *Euphytica* 179: 129–141.

Wellings C.R., Singh R.P., Yahyaoui A.H., Nazari K., McIntosh R.A., 2009. The development and application of near-isogenic lines for monitoring cereal rust pathogens. BGR2 TechnicalWorkshop, Obregon Mexico 2009. Available at: https://globalrust.org/all-bgri-abstracts.

Xia X.C., Li Z.F., Li G.Q., Singh R.P., 2007. Stripe rust resistance in Chinese bread wheat cultivars and lines. In: *Wheat Production in Stressed Environments* (H.T.

Buck, J.E.Nisi, N. Salomon, ed.). Springer, Amsterdam, The Netherlands, 77–82.

Yahyaoui A., Hakim M.S., El Naimi M., and Rbeiz N., 2002. Evolution of physiologic races and virulence of *Puccinia striiformis* on wheat in Syria and Lebanon. *Plant Disease* 86: 499–505.

Yahyaoui A., 2005. Cereal rust monitoring in Central, West Asia and North Africa: current status and future challenges. In: *Global Landscapes in Cereal Rust Control*, 20–21 September, 2005, Katoomba, Australia, p 40.

Zeller F.J., 1973. 1B/1R wheat-rye chromosome substitutions and translocations. In: *Proceedings of the 4th International Wheat Genetics Symposium*. (E.R. Searsm L.M.S. Searsm ed.), Agricultural Experiment Station, University of Missouri: Columbia, Missouri, USA, 209–221.

Zhang Z.J., 1995. Evidence of durable resistance in nine Chinese landraces and one Italian cultivar of *Triticum aestivum* to *Puccinia striiformis*. *European Journal of Plant Pathology* 101: 405–409.

19

Development and validation of a severity scale for assessment of fig rust

Gabriella Cristina Botelho Mageste da SILVA[1], Rafael PIO[1], Renata Cristina Martins PEREIRA[2], Pedro Maranha PECHE[1], Edson Ampélio POZZA[2,*]

[1] *Departament of Agriculture, Federal University of Lavras (UFLA) - Lavras – MG - 37200-000, Brazil*
[2] *Departament of Phytopathology, Federal University of Lavras (UFLA) - Lavras – MG - 37200-000, Brazil*
*Corresponding author: edsonpozza@gmail.com

Competing Interests: The Author(s) declare(s) no conflict of interest.

Editor: Tito Caffi, DI.PRO.VE.S., Università Cattolica del Sacro Cuore, Piacenza, Italy.

Summary. Standardized methods for quantifying rust severity (*Cerotelium fici*) on fig leaves (*Ficus carica* L.) are required, so this study aimed to develop and validate a diagrammatic scale to assessment the severity of this disease. Fig leaves that exhibited varying severities of rust symptoms were collected in the field. The actual severity, maximum and minimum limits, and intermediate levels of the scale were determined based on the frequency distribution of the severity values found in the field. In validation of the scale, eight evaluators estimated the severity in 50 leaves with different levels of symptoms with and without the use of the diagrammatic scale. Accuracy and precision of the data were evaluated, and linear regression was used to assess the repeatability and reproducibility of the estimates. The use of the diagrammatic scale provided adequate results for the parameters analysed when compared assessments made without use of the scale, confirming reliability of the estimates to evaluate rust severity on fig leaves.

Keywords. *Ficus carica*, *Cerotelium fici*, Lin's method.

INTRODUCTION

Fig (*Ficus carica* L.) is among the most important cultivated world fruit species. The use of fig fruit as food, and of fig plants for ornamental purposes, have been recorded for thousands of years including in the Bible (Eisen, 1901). Turkey is the largest producer of figs in the world, producing 305,450 tonnes per year, followed by Egypt, Morocco, Algeria and Iran (FAO, 2018). European countries, including Portugal, Spain and Italy, are also major producers and exporters of figs (Khemira and Mars, 2017). In subtropical regions, fig crops are grown to produce ripe figs to supply fresh fruit markets, or unripe fruit for the production of sweets, compotes and crystallized figs (Dalastra *et al.*, 2009).

Although fig originated in temperate regions (Pio *et al.*, 2019), it can adapt to different climates and soil conditions, which has boosted expansion of fig production to Brazilian tropical and subtropical regions (Chalfun *et al.*, 2012). In the last 10 years, Brazil's cultivated fig area has remained steady at approx. 2,591 ha (IBGE. 2019). The primary fig-producing states in Brazil include Rio Grande do Sul (11,918 tonnes), São Paulo (10,903 tonnes) and Minas Gerais (1,698 tonnes). Recent decades have seen increased exploitation of fig crops in Brazil and Chile, the produce from which is destined for export to North African and European countries during the production off-season in those regions (Pio *et al.*, 2017). However, when grown in subtropical regions, some diseases affect fig crops.

Fig rust (*Cerotelium fici* (Cast.) Arth.) is the principal disease that affects fig crops (Galleti and Rezende, 2005). Symptoms of the disease on adaxial surfaces of fig leaves appear as angular yellow-green spots that progress to brown. On abaxial leaf surfaces orange-red pustules develop that contain powdery masses of spores. In severe infections, the leaves fall, and growth and ripening of the figs are halted. With the premature fall of leaves, there is a reduction in the accumulation of carbohydrates, which compromises the next fruit production cycle (Galleti and Rezende, 2005; Solano-Báez *et al.*, 2017).

Due to the losses associated with fig rust, appropriate disease management methods are required. These include development of resistant cultivars, fungicide applications, development of biological control agents, resistance inducers, appropriate pruning techniques or crop management. However, to measure the effectiveness of these techniques and to identify which can be integrated into crop management, it is necessary to quantify the disease (Gomes *et al.*, 2004).

Quantifying disease enables control measures to be evaluated for whether they will be effective and therefore recommended for application in the field. For producers, the benefits of disease quantification include the assistance for efficient crop management decisions and prioritization of resources to enable low environmental impacts in sustainable disease management (Bergamin Filho and Amorim, 1996).

Among the methods for assessment of plant diseases, the most commonly implemented are those that are visual. These are simple because they do not require the use of sophisticated equipment, and they are accurate and precise (Campbell and Madden. 1990). Key tools for employing these techniques are diagrammatic disease severity scales. This method helps to define disease severity using photographs or diagrams of symptomatic plants or their organs. However, although this approach is simple, development must meet criteria to ensure the correct quantification of disease severity (Bergamin Filho and Amorim, 1996).

The primary aspects to be evaluated in the development of a diagrammatic disease severity scale are the minimum and maximum limits of the scale corresponding to the disease levels found in the field, and use of images that display a pattern compatible with the symptoms representing the levels of disease. A further important consideration is the limits of visual acuity of the human eye, according to Weber-Fechner's Law, assigning scores with respective severity intervals, as the human eye has difficulty seeing points or precise percentage values (Horsfall and Barratt, 1945; Nutter and Schultz. 1995).

The present study aimed to develop and validate a diagrammatic scale for accurate and precise assessment of fig rust, because no standardized methods were available for quantifying severity of the disease, which is the most important disease affecting fig orchards in subtropical conditions.

MATERIALS AND METHODS

Diagrammatic scale development

To develop the diagrammatic scale, 190 fig leaves from the field were randomly collected, that displayed different levels of disease severity. The leaves, naturally infected, were from several fig trees in an experimental orchard at the Federal University of Lavras, Brazil. The municipality is located at 21° 13' 40" south latitude and 44° 57' 42" west longitude, at an average altitude of 970 m above sea level. According to the Köppen climate classification, Lavras has a tropical climate of the Cwa type, characterized by dry winters and hot, humid summers (Alvares *et al.*, 2014). To confirm the causal agent of the disease on the leaves, anatomical sections were prepared from a diseased leaf and analysed for the pathogen morphology.

All plant material was photographed on a white background, using a Nikon d3100 digital camera, in automatic mode, with 18–55 mm lens focal length. Subsequently, the diseased and total leaf area were determined for each leaf using the Assess® software (American Phytopathological Society). Pustules and the areas with necrotic and chlorotic tissue caused by the disease were considered as diseased areas.

According to the minimum and maximum levels found, a frequency plot was constructed, plotting the percentage of damaged leaf area (x-axis), in severity intervals of 5% (y-axis). These values were then fitted to a simple linear model and to non-linear exponential and logarithmic models (Campbell and Madden. 1990). The model that best fitted the frequency plot was chosen as indicated from the largest R^2 and the significance of the parameters of the equations in the t-test. The disease severity scale was created according to the intervals with the greatest concentration of leaves having the same percentage of damaged area. The severity intervals for each score were established according to Weber-Fechner's visual acuity law (Horsfall and Barrat, 1945; Nutter and Schultz, 1995) and according to the shape and distribution of the lesions. Photographs of leaves with disease lesions were then used to develop the scale.

Diagrammatic scale validation

To validate the diagrammatic scale, 50 leaves of fig showing symptoms of rust were used, representing all variation levels of dis-ease severity. In three evaluations, 8 evaluators without experience in quantification of plant disease observed images of diseased leaves using Microsoft PowerPoint 2010. The first evaluation was performed without using the scale. After an interval of 7 days, a second evaluation was performed aided by the diagrammatic scale. To assess the repeatability of the observed values, a third evaluation was performed after 7 days, also using the proposed scale.

Based on the data obtained from each evaluator the accuracy and precision the developed scale were determined using Lin's method. Lin's concordance correlation coefficient (Pc) (Lin. 1989), to assess agreement between pairs of observations, was used to measure adjustment between the actual values and estimated disease severities. The method also includes other variables to aid in validation. The scale shift factor, where 1 = perfect agreement between x and y, measures the difference between actual and estimated values, and is calculated as the difference between the slope of the fitted regression lines and the concordant line. The location shift factor, where 0 = perfect agreement between x and y, estimates the change of the fitted regression line relative to the concordant line, by measuring the difference in height between the two lines. The BIAS correction factor, which measures how far the fitted line deviates from the concordant line, was calculated from the location shift factor and the scale shift factor, derived from the means and standard deviations of x and y. In addition to these factors, Pearson's correlation was used to evaluate the precision of the assessments. The confidence interval (CI) ($P < 0.05$) between the groups of evaluators, with and without the use of the scale, was calculated to determine if there were significant differences between the evaluations.

The repeatability of the estimates from each evaluator was determined by R^2 values of the linear regression between two assessments using the scale (Nutter *et al.*, 1993). The reproducibility of the estimates was evaluated by R^2 values obtained from linear regressions between the estimated severities of the same sample unit using different evaluators in pairs (Kranz. 1988; Campbell and Madden. 1990; Nutter and Schultz. 1995).

The data were tabulated and the statistical analyses performed using the RStudio software (R Core Team, 2018), and the epi.ccc function of the epiR package (Stevenson *et al.*, 2018) to determine the Lin's concordance correlation coefficient.

RESULTS

Scale development

The minimum and maximum severity of fig rust was 0% and the maximum severity was 89.3%. A high proportion (43%) of leaves were in the frequency intervals up to 5% severity (Table 1). Based on the disease severity found in natural infections, the scale had a maximum level of 89.3%, with clorotic and necrotic areas.

The best model adjusted for the frequency values in the severity intervals was logarithmic, in this case in according of Weber-Fechner's law, with the greatest R^2 (87%) and significance of the parameters of the equations in the t-test (Table 2).

The severity scale was developed using six scores or percentage intervals (Figure 1), three of which were distributed into intervals ranging up to 15.0% of diseased leaf area. The interval up to 1% included 11.6% of the total leaves, constituting the greatest frequency unit interval. The six percentage severity intervals of the scale were 0, 0.1–5.0%, 5.1–15.0%, 15.1–25%, 25.1–50.0% and >50%.

Scale validation

According to Lin's method, estimates of disease severity assessments improved with the use of the proposed scale (Table 3). According to the concordance coefficient and correlations between the actual and estimated values, greater estimation efficiency was obtained with use of the scale (a = 0.80) compared to evaluations without use of the scale (a = 0.71). The evaluators overestimated disease severity when not using the scale (c = 0.33), and underestimated severity when they used the scale (c = -0.28). The confidence interval between the two evaluations did not differ significantly, however, proving that there was no significant improvement in the variable under analysis. The Pearson's correlation

Table 1. Frequency distribution, in unit intervals, of disease severity values (%) of rust on fig leaves.

Interval (Severity %)	Frequency	Percentage (%)	Cumulative frequency	Cumulative Percentage (%)	Interval (Severity %)	Frequency	Percentage (%)	Cumulative frequency	Cumulative Percentage (%)
0-1	22	11.6	22	11.6	26-27	0	0.0	108	56.8
1-2	2	1.1	24	12.6	27-28	3	1.6	111	58.4
2-3	7	3.7	31	16.3	28-29	2	1.1	113	59.5
3-4	5	2.6	36	19.0	29-30	1	0.5	114	60.0
4-5	7	3.7	43	22.7	30-31	1	0.5	115	60.5
5-6	2	1.1	45	23.7	31-32	4	2.1	119	62.6
6-7	4	2.1	49	25.8	32-33	1	0.5	120	63.1
7-8	4	2.1	53	27.9	33-34	1	0.5	121	63.7
8-9	4	2.1	57	30.0	34-35	4	2.1	125	65.8
9-10	3	1.6	60	31.6	35-36	1	0.5	126	66.3
10-11	1	0.5	61	32.1	36-37	2	1.1	128	67.4
11-12	1	0.5	62	32.6	37-38	1	0.5	129	67.9
12-13	2	1.1	64	33.7	38-39	1	0.5	130	68.4
13-14	3	1.56	67	35.3	39-40	0	0.0	130	68.4
14-15	4	2.1	71	37.4	40-41	3	1.6	133	70.0
15-16	6	3.2	77	40.5	41-42	2	1.1	135	71.0
16-17	2	1.1	79	41.6	42-43	3	1.6	138	72.6
17-18	5	2.6	84	44.2	43-44	3	1.6	141	74.2
18-19	3	1.6	87	45.8	44-45	0	0.0	141	74.2
19-20	1	0.5	88	46.3	45-46	0	0.0	141	74.2
20-21	4	2.1	92	48.4	46-47	2	1.1	143	75.3
21-22	5	2.6	97	51.0	47-48	1	0.5	144	75.8
22-23	3	1.6	100	52.6	48-49	0	0.0	144	75.8
23-24	1	0.5	101	53.3	49-50	1	0.5	145	76.3
24-25	5	2.6	106	55.8	>50	45	23.7	190	100.00
25-26	2	1.1	108	56.8					

Table 2. Parameters of the linear and non-linear models for the frequency of severity of fig rust, in severity intervals.

Model	R^2 [a]	r [b]	y_0 [c]
Exponential	0.75	0.04***	39.42***
Logarithmic	0.87	-0.0004***	0.02***
Linear	0.57	-0.28***	22.95***

[a] Coefficient of determination (R^2).
[b] Progress rate (r).
[c] Initial inoculum (y_0).
*** Significant according to the t-tests ($P = 0.001$).

coefficient indicated increased precision of the evaluators when using the scale (e = 0.86), compared to the evaluations without the scale (e = 0.76). However, the value of the BIAS correction factor without the use of the scale (d = 0.94) was greater than that of the estimates obtained using the scale (d = 0.93). This indicated that there was no increase in the accuracy of the evaluators. Considering the confidence intervals, the assessments for fig rust with and without the use of the diagrammatic scale differed significantly at the 95% confidence interval, except for the location shift factor.

For reproducibility, without using the diagrammatic scale the value of the determination coefficient (R^2) ranged from 64 to 88%, with a mean of 81.1% (Table 4). With use of the scale, R^2 values ranged from 71 to 91% (mean = 80.7%) for the first evaluation, and from 61 to 81% (mean = 72.2%) in the second evaluation, with $R^2 \geq 70\%$ in approximately 82% of the combinations of evaluators.

There was good repeatability between the estimates of the same evaluator (Table 5). Between the two evaluations with the use of the scale, only one evaluator (A) exhibited a slope significantly different from 1, with good precision of the estimates of 87.5% of the evaluators. The evaluators all presented good repeatability in the estimates of leaf rust severity, as the mean variation between the first evaluation and the second evaluation was approx. 70%.

Figure 1. Diagrammatic scale for assessment of rust severity on fig leaves. The numbers represent percentages of leaf area diseased. leaf area.

Table 3. Lin's concordance correlation coefficients for eight evaluators without or with the diagrammatic disease severity scale, used to estimate rust severity on fig leaves.

Lin´s statistic	Without scale	With scale	95% CI[f]
Lin's concordance correlation coefficient [a]	0.71	0.80	**0.6872*; 0.8123***
Scale shift factor [b]	1.16	0.76	**0.8738*; 1.4020***
Location shift factor [c]	0.33	-0.28	-0.1705; 0.3112
Bias correction factor [d]	0.94	0.93	**0.7948*; 0.9328***
Pearson's correlation [e]	0.76	0.86	**0.8603*; 0.8837***

[a] Lin's concordance correlation coefficient.
[b] Scale shift factor relative to perfect agreement.
[c] Location shift factor relative to perfect agreement.
[d] Bias correction factor.
[e] Pearson's correlation.
[f] Upper and lower limits of the 95% confidence intervals.
Bold* represents a significant difference ($P \leq 0.05$) between the two evaluations, according to the t-tests.

Most evaluators presented good precision, regardless of whether the scale was used. Sixty-three percent of the participants presented R^2 values in the second evaluation that were greater or equal to those for the first evaluation, suggesting equal or greater precision with the second evaluation. Absolute errors were reduced when the scale was used, decreasing the range of values between the first and second evaluations (Figure 2). However, in the second evaluation, using the scale, the minimum and maximum values observed for the residuals of all the evaluators were, respectively, -49.67 and 70.90, increasing the range of the determined values.

DISCUSSION

The diagrammatic scale developed in here allowed the evaluators to obtain accurate and precise estimates of fig rust severity, according to the validation analyses.

Linear and non-linear models were fitted to the data to determine if the scale levels should increase logarithmically or linearly. The particulars of each pathosystem are considered for determination of scale intervals, and this fitting is required to assess the accuracy of the assessments. The model with the best fit was the logarithmic model, and with this the intermediate levels of the scale were determined based on the highest frequency intervals of disease levels on the leaves, combined with the logarithmic increase in severity, in accordance with Weber Fechner law (Campbell and Madden, 1990).

Each level of the scale was defined according to the frequency distribution of the number of leaves with a

Table 4. Coefficients of determination (R^2) of the linear regression equation between pairs of different evaluators, with or without the use of the disease severity assessment scale in two evaluations, estimating rust severity on fig leaves.

Evaluator	Without scale						
	B	C	D	E	F	G	H
A	0.73	0.76	0.75	0.68	0.64	0.77	0.73
B		0.83	0.78	0.88	0.75	0.85	0.71
C			0.82	0.80	0.81	0.86	0.77
D				0.80	0.74	0.81	0.74
E					0.82	0.83	0.66
F						0.87	0.84
G							0.84
	With scale – 1st assesment						
	B	C	D	E	F	G	H
A	0.74	0.79	0.71	0.86	0.87	0.78	0.81
B		0.78	0.79	0.82	0.79	0.75	0.76
C			0.91	0.78	0.85	0.80	0.82
D				0.79	0.79	0.77	0.78
E					0.83	0.81	0.85
F						0.86	0.89
G							0.83
	With scale – 2nd assesment						
	B	C	D	E	F	G	H
A	0.70	0.72	0.65	0.79	0.74	0.65	0.78
B		0.61	0.63	0.80	0.81	0.77	0.72
C			0.69	0.64	0.74	0.65	0.72
D				0.78	0.73	0.75	0.69
E					0.75	0.79	0.68
F						0.78	0.77
G							0.69

Table 5. Intercept (β_0), slope (β_1) and coefficient of determination (R^2) of the linear regression equations relating the first to second estimates of rust severity on fig leaves, for estimates performed by eight evaluators using the disease severity scale.

Evaluator	Coefficients		
	β_0	β_1	R^2
A	10.55*	0.80 ns	0.62
B	8.67 ns	0.69 ns	0.64
C	2.63 ns	0.71 ns	0.51
D	0.56 ns	1.04 ns	0.73
E	2.82 ns	0.90 ns	0.80
F	1.47 ns	0.85 ns	0.79
G	2.61 ns	0.80 ns	0.74
H	6.22 ns	0.83 ns	0.75

* ns represent situations where the null hypothesis ($\beta_0 = 0$ or $\beta_1 = 1$) was, respectively, rejected and not rejected according to t-tests ($P = 0.05$).

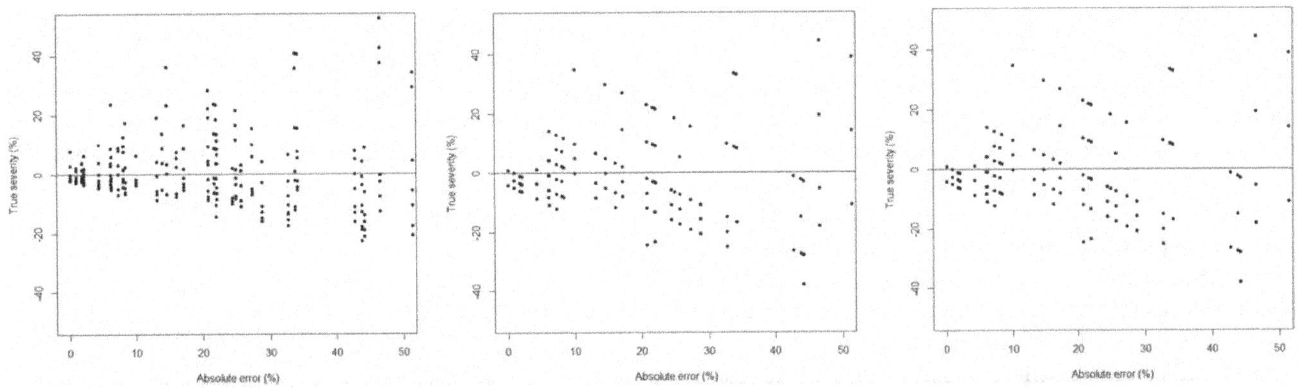

Figure 2. Distributions of residuals (estimated severity – actual severity) of estimates of rust severity on fig leaves, with or without the use of the disease severity scale, in two assessments by eight evaluators.

specific disease leaf area found in the field. Although the greatest unit interval of diseased leaves was between 0 and 1%, this interval was not used in the scale due to the difficulty in visually locating a lesion of that size on a fig leaf. Lorenzetti (2008) proposed a diagrammatic scale to quantify the severity of the same rust disease. However, their scale was expressed in percentages, with no values less than 4% of disease severity, and with large intervals between the percentages, factors that increase the subjectivity of the estimates of actual disease severity.

Based on the severity of fig rust found in natural infections, the scale had a maximum level of 89.3%, with chlorotic and necrotic areas. The absence of leaves with severities greater than 89% in the current study could be characteristic of the pathogen–host interaction of this disease. At greater disease intensities, leaf tissue necrosis was present that led to early leaf abscission.

Severity values greater than 50% were combined in the proposed scale because the human eye has difficulty distinguishing disease severity greater than this percentage (Campbell and Madden. 1990), and few leaves have been found in the unit intervals above this value due to defoliation caused by the disease (Pastore *et al.* 2016). The scale developed by Angeloti *et al.* (2011) for assessment of grapevine rust found a maximum level of severity in leaves of 75% for a similar reason. Dolinski *et al.* (2017), when developing a scale to quantify severity of peach rust severity, defined a maximum level of 30%. Although these authors found leaves with greater severity levels, as the variation in the disease severity can be due to cultivar susceptibility differences, cultivation practices and climate variations.

For the construction of the fig rust scale, photographs were used instead of graphical representations, which is a common practice. Belan *et al.* (2014) noted that this method increases the precision and accuracy of disease assessments. Using the real images or photographs, rather than black and white or colour diagrams, draws evaluators to the reality, facilitating the disease assessments.

In most studies involving validation of diagrammatic scales to determine disease severity in plant leaves, evaluators have exhibited tendency to overestimate the severity of particular diseases (Capucho *et al.*, 2011; Belan *et al.*, 2014; Freitas *et al.*, 2015). In some cases, such as early blight in potato, leaf disease severity was underestimated (Michereff *et al.*, 2000; Gomes *et al.*, 2004). In the present study, it is not possible to make such an inference because there was no significant difference between the evaluations.

In the validation of other rust severity evaluation scales, increased accuracy and precision by evaluators has been observed with their use. Capucho *et al.* (2011), using a diagrammatic scale for coffee leaf rust, validated the results using Lin's method, and the mean Pearson's correlation coefficient increased from 0.77 to 0.87 when using the proposed scale. In a study of diagrammatic scale validation for sugarcane orange rust, Klosowski *et al.* (2013) determined the indices by simple linear regression and obtained satisfactory results, with 100% of the evaluators obtaining intercepts statistically equivalent to zero and slope values equal to 1, indicating the absence of systematic deviations.

Although it is the most commonly used method for validating scales, linear regression does not detect the values of intercept 0 (β0) and slope (β1), when the data are scattered (Bock *et al.*, 2010), and this may lead to erroneous conclusions. Lin's method provides a single index ("Lin's concordance correlation coefficient"), and the accuracy and precision of severity estimates. This method has been used to analyse how disease severity data behave, and how they relate to actual estimates and with evaluations of the repeatability of estimates (Nita *et al.*, 2003; Bock *et al.*, 2008).

The reproducibility of the disease severity estimates among the evaluators was analysed using paired linear regression (Nutter and Schultz, 1995), and greater standardization was observed in the estimations with the use of the described here. However, in some pairs the coefficient of determination reached values between 61 and 69% in the second assessment using the scale. It is possible that these results were due to inexperience among some evaluators with disease quantification.

More than 75% of the evaluator pairs presented R^2 values greater than 70% using the diagrammatic scale, similar to that found in the validation of scales for other pathosystems, such as rust (Capucho *et al.*, 2011) and bacterial blight (Belan *et al.*, 2014) of coffee leaves, with mean values, respectively, of 87% and 99%.

The range of the residuals in the assessments using the scale described here were -57.50 to 72.03 for the first evaluation, and -49.67 to 70.90 for the second. The high values are explained by the difficulty in evaluating the disease. Due to the characteristics of the lesions, which are individual and small, and scattered on fig leaf surfaces, this causes evaluators to underestimate or overestimate the diseased leaf areas. This can influence the quality of disease estimation through psychological stimuli and responses, including the complexity of the sample units, size and shape of the lesions, colour and number of lesions, evaluator fatigue or difficulty to concentrate on the task (Sherwood *et al.*, 1983; Kranz, 1988).

Disease severity evaluation results are considered satisfactory when the means of the absolute errors are between 10 and 15%. This was described by De Paula *et al.* (2016), proposing and validating diagrammatic scales to assess brown eye spot in red and yellow coffee cherries, and also by Godoy *et al.* (2006) validating a scale for quantification of soybean rust. Belan *et al.* (2014) reported mean absolute errors between -20.95 and 20.01 in two evaluations, with a scale for assessment of bacterial blight in coffee leaves, especially at high severity levels, which is contrary to the observations in the present study.

In conclusion, we have developed and validated a diagrammatic scale for assessment of rust severity on fig leaves. The disease severity scale outlined here provides good accuracy, precision, repeatability and reproducibility, for evaluation of this disease.

ACKNOWLEDGEMENTS

Financial support for this study was provided by the Brazilian Federal Agency for the Support and Evaluation of Graduate Education (CAPES) – Financing Code 001, for a scholarship awarded to the first author. The Fruticulture sector of the Department of Agriculture and the post-graduate students of the Department of Phytopathology of the Federal University of Lavras/MG provided assistance to develop and validate the diagrammatic scale described here, and with interpreting the results obtained.

LITERATURE CITED

Alvares C.A., Stape J.L., Sentelhas P.C., Gonçalves J.L.M., Sparovek, G., 2014. Koppen's climate classification map for Brazil. *Meteorologische Zeitschrift*, 22: 711–728.

Angelotti F., Scapin C.R.; Tessman D.J., Vida, J.B. Oliveira R.R., Canteri, M.G. (2008). Diagrammatic scale for assessment of grapevine rust. *Tropical Plant Pathology*, 33, 439–443.

Belan L.L., Pozza E.A., Freitas M.L.O., Souza R.M., Jesus Junior W.C., Oliveira J.M., 2014. Diagrammatic scale for assessment of bacterial blight in coffee leaves. *Journal of Phytopathology*, 162: 801–810.

Bergamin Filho A., Amorim, L., 1996. Doenças de Plantas Tropicais: Epidemiologia e Controle Econômico. *Agronômica Ceres*, São Paulo, São Paulo, Brazil, XXX pp.

Bock C.H., Parker P.E., Cook A.Z., Gottwald T.T., 2008. Visual rating and the use of image analysis for assessing different symptoms of citrus canker and the grapefruit leaves. *Plant Disease*, 92: 530–541.

Bock, C.H., Poole, G.H., Parker, P.E., & Gottwald, T.T., 2010. Plant disease severity estimated visually, by digital photography and image analysis, and by hyperpectral imaging. *Critical Reviews in Plant Science*, 29: 59–107.

Campbell C.L., Madden L.V., 1990. *Introduction to plant disease epidemiology*. John Wiley, New York, USA, XXX pp.

Capucho, A. S., Zambolim, L., Duarte, H. S. S., & Vaz, G. R. O., 2011. Development and validation of a standard area diagram set to estimate severity of leaf rust in *Coffea arabica* and *C. canephora*. *Plant Pathology*, 60: 1144–1150.

De Paula P.V.A.A., Pozza E.A., Santos L.A., Chaves E., Maciel M.P., Paula J.C.A. (2016). Diagrammatic scales for assessing Brown Eye Spot (*Cercospora coffeicola*) in red and yellow coffee cherries. *Journal of Phytopathology*, 164: 791–800.

Dolinski M.A., Duarte H.S.S., Silva J.B., De Mio L.M.M., 2017. Development and validation of a standard area diagram set for assessment of peach rust. *European Journal of Plant Pathology*, 148: 817–824.

Eisen G. A. 1901. *The Fig: its History, Culture, and Curing*. Washington: Governament printing Office.

FAO: Food And Agriculture Organization, 2018. *Figs*. http://www.fao.org.

Freitas M.L., Pozza E.A., Belan L.L., Silva J.L., Abreu M.S., 2015. Diagrammatic scale for blister spot in leaves of coffee treee. *African Journal of Agricultural Research*, 2068–2070.

Galleti S.R., Rezende J.A.M., 2005. Doenças da figueira. In H. Kimathi *et al.* Eds., *Manual de Fitopatologia: doenças de plantas cultivadas*, Agronômica Ceres, São Paulo, Brazil, 351–354 pp.

Godoy C.V., Koga L.J., Canteri, M.G., 2006. Diagrammatic scale for assessment of soybean rust severity. *Fitopatologia Brasileira*, 31: 63–68.

Gomes A. M. A., Michereff S. J., Mariano R. L. R., 2004. Elaboração e validação de escala diagramática para cercosporiose da alface. *Summa Phytopathologica*, 30: 38–42.

Horsfall J.C., Barratt R.W., 1945. An improved grading system for measuring plant diseases. *American Phytopathology Society*, 35: 655–655.

IBGE: Instituto Brasileiro de Geografia e Estatística. SIDRA: Banco de dados agregados do IBGE. http://www.sidra.ibge.gov.br/.

Khemira H., Mars M., 2017. Fig production in subtropical south-western Saudi Arabia. *Acta Horticulturae*, Iran, 169–172 pp.

Klosowski A.C., Ruaro L., Bepalhock Filho J.C., De Mio, L.L.M., 2013. Proposta e validação de escala para a ferrugem alaranjada da cana-de-açúcar. *Tropical Plant Pathology*, 38:166–171.

Kranz J., 1988. Measuring plant disease. In: J. Kranz, & J. Rotem (Eds.), *Experimental Techniques in Plant Disease Epidemiology*. Springer-Verlag, Berlin, Heidelberg 35–50 pp.

Lin L., 1989. A concordance correlation coefficient to evaluate reproducibility. *Biometrics*, 45: 255–268.

Lorenzetti E.R., 2008. Escala diagramática para avaliação da severidade de ferrugem do figo causada por *Cerotelium fici* (Cast.). In XX Congresso Brasileiro De Fruticultura, 2008, Vitória, ES, Brazil.

Michereff S. J., Maffia L. A., Noronha M. A., 2000. Escala diagramática para avaliação da severidade da queima das folhas do inhame. *Fitopatologia Brasileira*, 25: 612–629.

Nita M., Ellis M.A. Madden,L.V.,2003. Reliability and accuracy of visual estimation of Phomopsis leaf blight of strawberry. *Phytopathology*, 93: 995–1005.

Nutter F.W., Gleason M.L., Jenco J.H., Christians N.C., 1993. Assessing the accuracy, intra-rater repeatability, and inter-rater reliability of disease assessment systems. *Phytopathology*, 83: 806–812.

Nutter F.W., Schultz P.M., 1995. Improving the accuracy and precision of disease assessments: selection of methods and use of computer-aided training programs. *Canadian Journal of Plant Pathology*, 17: 174–184.

Pastore R.L., Boff P., Golinski N.G., Boff M.I.C., 2016. Rust resistance of fg landraces in an organic cropping system in Santa Catarina. *Biological agriculture & Horticulture*, Brazil, 63–71 pp.

Pio R., Souza F. B. M., Kalcsits L., Bisi R. B., Farias D. H., 2019. Advances in the production of temperate fruits in the tropics. *Acta Scientiarum. Agronomy*, 41: 1-10.

Pio R., Oliveira A.C.L., Pasqual M., Pio L.A.S., Curi P.N., Bisi R.B., 2017. Protocol for flow cytometric estimation of nuclear DNA amount in fig (*Ficus carica* L.) cultivars. *Acta Horticulturae*, 1: 99–104.

R Core Team, 2018. R: a language and environment for statistical computing. https://www.R-project.org/.

Solano-Báez A.R., Jiménez-Jiménez B., Camacho-Tapia M., Leyva-Mir S.G., Nieto-López E.H, Tovar-Pedraza J.M., 2017. First confirmed report of *Cerotelium fici* causing leaf rust on *Ficus carica* in Mexico. *Plant Pathology & Quarantine*, 7: 160–163.

Stevenson M., Nunes T., Sanchez J., Thornton, R., Reiczigel J. … Sebastiani P., 2018. *An R package for the analysis of epidemiological data*. http://CRAN.R-project.org/package=epiR.

Sherwood R.T., Berg C.C., Hoover M.R., Zeiders K.E., 1983. Illusions in visual assessment of stagonospora leaf spot of orchardgrass. *Phytopathology*, 73: 173–177.

Permissions

All chapters in this book were first published in PM, by Firenze University Press; hereby published with permission under the Creative Commons Attribution License or equivalent. Every chapter published in this book has been scrutinized by our experts. Their significance has been extensively debated. The topics covered herein carry significant findings which will fuel the growth of the discipline. They may even be implemented as practical applications or may be referred to as a beginning point for another development.

The contributors of this book come from diverse backgrounds, making this book a truly international effort. This book will bring forth new frontiers with its revolutionizing research information and detailed analysis of the nascent developments around the world.

We would like to thank all the contributing authors for lending their expertise to make the book truly unique. They have played a crucial role in the development of this book. Without their invaluable contributions this book wouldn't have been possible. They have made vital efforts to compile up to date information on the varied aspects of this subject to make this book a valuable addition to the collection of many professionals and students.

This book was conceptualized with the vision of imparting up-to-date information and advanced data in this field. To ensure the same, a matchless editorial board was set up. Every individual on the board went through rigorous rounds of assessment to prove their worth. After which they invested a large part of their time researching and compiling the most relevant data for our readers.

The editorial board has been involved in producing this book since its inception. They have spent rigorous hours researching and exploring the diverse topics which have resulted in the successful publishing of this book. They have passed on their knowledge of decades through this book. To expedite this challenging task, the publisher supported the team at every step. A small team of assistant editors was also appointed to further simplify the editing procedure and attain best results for the readers.

Apart from the editorial board, the designing team has also invested a significant amount of their time in understanding the subject and creating the most relevant covers. They scrutinized every image to scout for the most suitable representation of the subject and create an appropriate cover for the book.

The publishing team has been an ardent support to the editorial, designing and production team. Their endless efforts to recruit the best for this project, has resulted in the accomplishment of this book. They are a veteran in the field of academics and their pool of knowledge is as vast as their experience in printing. Their expertise and guidance has proved useful at every step. Their uncompromising quality standards have made this book an exceptional effort. Their encouragement from time to time has been an inspiration for everyone.

The publisher and the editorial board hope that this book will prove to be a valuable piece of knowledge for researchers, students, practitioners and scholars across the globe.

List of Contributors

Fernando Martínez-Moreno and Ignacio Solís
ETSIA, Dep. Ciencias Agroforestales, University of Seville, Ctra de Utrera km1, 41013 Seville (Spain)

Nilwala S. Abeysekara
Department of Microbiology and Plant Pathology, University of California- Riverside, Riverside, CA, 92507, USA

Heather Hickman, Sara Westhafer, Thomas A. Evans, Nancy F. Gregory and Nicole M. Donofrio
Department of Plant and Soil Sciences, University of Delaware, Newark, DE, 19716, USA

Gordon C. Johnson
Department of Plant and Soil Sciences, University of Delaware, Georgetown, DE, 19947, USA

Saeedeh Dehghanpour-Farashah, Parissa Taheri and Mahrokh Falahati-Rastegar
Department of Plant Protection, Faculty of Agriculture, Ferdowsi University of Mashhad, Mashhad, Iran

Joana Horta, Isabel Conceicao and Isabel Abrantes
Centre for Functional Ecology, Department of Life Sciences, University of Coimbra, Calçada Martim de Freitas, P 3000 456 Coimbra, Portugal

Maria Clara Vieira Dos Santos
Centre for Functional Ecology, Department of Life Sciences, University of Coimbra, Calçada Martim de Freitas, P 3000 456 Coimbra, Portugal
CBMA- Centre of Molecular and Environmental Biology, Department of Biology, University of Minho, Campus de Gualtar, 4710-057 Braga, Portugal

Luisa Moura and David V. Pires
Mountain Research Centre (CIMO)/Escola Superior Agrária (ESA), Instituto Politécnico de Viana do Castelo, Refóios, 4990-706 Ponte de Lima, Portugal

Sofia R. Costa
Mountain Research Centre (CIMO)/Escola Superior Agrária (ESA), Instituto Politécnico de Viana do Castelo, Refóios, 4990-706 Ponte de Lima, Portugal
CBMA- Centre of Molecular and Environmental Biology, Department of Biology, University of Minho, Campus de Gualtar, 4710-057 Braga, Portugal

Supriya Sachdeva, Basavanagouda Siddanagouda Patil, Sanjeev K. Chauhan and Chellapilla Bharadwaj
Division of Genetics, Indian Agricultural Research Institute (ICAR), Pusa, New Delhi 110012, India

Shruti Dawar, Upasana Rani and Sarvjeet Singh
Principal Pulse Breeder, Division of Plant Breeding and Genetics, Punjab Agricultural University, Ludhiana, Punjab, India

Khela Ram Soren
ICAR-Indian Institute of Pulse Research, Kanpur, UP, India

SatishKumar Sanwal
ICAR-CSSRI, Karnal, Haryana, India

Noura Omri Ben Youssef, Imen Halila, Noura Jammezi, Ahlem Mbazia and Mohamed Kharrat
Laboratoire des Grandes Cultures, Institut National de la Recherche Agronomique de Tunisie, Université de Carthage, Rue Hédi Karray, EL Manzah, 1004, Tunis, Tunisie

Lydie Kerdraon
INRA, UMR, UMR BIOGER, INRA-AgroParisTech, Avenue Lucien Brétignières BP 01 78850 Thiverval-Grignon, France

Lucie Mieuzet
INRA, UMR 1349 IGEPP, Domaine de la Motte, 35653 Le Rheu, France

Christophe Le May
INRA, UMR 1349 IGEPP, Domaine de la Motte, 35653 Le Rheu, France
Agrocampus Ouest, UP ESP, 65 rue de Saint Brieuc, 35043 Rennes, France

Michael Chatzidimopoulos
Department of Agriculture, Crop Production and Rural Environment, Laboratory of Plant Pathology, University of Thessaly, Fitokou Street, 384 46, N. Ionia, Volos, Greece

Athanassios C. Pappas
17 Bizaniou, 1156 69 Papagos, Greece

Marco Fierro, Davide Palmieri, Filippo De Curtis, Domenico Vitullo and Giuseppe Lima
Dipartimento Agricoltura, Ambiente e Alimenti, Università del Molise, via De Sanctis, 86100 Campobasso, Italy

Juan Gil and Teresa Millan
Universidad de Córdoba, Departamento de Genética-ETSIAM, Campus de Rabanales, 14071 Córdoba, Spain

Josefa Rubio
Instituto de Investigación y Formación Agraria y Pesquera de Andalucía (IFAPA), Area de Genómica y Biotecnología, Alameda del Obispo, 3092 Córdoba, Spain

Bahria Khenfous-Djebari and Zouaoui Bouznad
Laboratoire de Phytopathologie et de Biologie Moléculaire, Département de Botanique, ENSA El Harrach, Alger, Algérie

Mohamed Kerkoud
DiagGene, 8, Rue le Nôtre 49066 Angers, France

Marion Fischer-Le Saux
IRHS, INRA, Agrocampus-Ouest, Université d'Angers, SFR 4207 QuaSaV, 49071, Beaucouzé, France

Traki Benhassine
Département de Biologie Cellulaire et de Biologie Moléculaire, USTHB, Bab Ezzouar, Alger, Algerie

Ralf Koebnik
IRD, Cirad, Université de Montpellier, IPME, Montpellier, France

Claude Bragard
Applied Microbiology-Phytopathology, Earth & Life Institute, UC Louvain, Louvain-la- Neuve, Belgium

Djihad Aribi, Zouaoui Bouznad and Abdelaziz Kedad
ENSA-El Harrach, Laboratoire de Phytopathologie et Biologie Moléculaire. Algiers, Algeria

Djida Ayad
ENSA-El Harrach, Laboratoire de Phytopathologie et Biologie Moléculaire. Algiers, Algeria
UMR 1345 IRHS, Université d'Angers-INRA-Agrocampus Ouest, SFR QUASAV, 42 Rue G. Morel, F-49071 Beaucouzé, France

Bruno Hamon and Philippe Simoneau
UMR 1345 IRHS, Université d'Angers-INRA-Agrocampus Ouest, SFR QUASAV, 42 Rue G. Morel, F-49071 Beaucouzé, France

Lucia Guerin-Dubrana
Université de Bordeaux, ISVV, UMR1065 Santé et Agroécologie du Vignoble, Bordeaux Sciences Agro, F-33175 Gradignan, France

Florence Fontaine
Structure Fédérative de Recherche (SFR) Condorcet-FR CNRS 3417, Université de Reims Champagne-Ardenne, Unité de Recherche *Résistance* Induite et Bioprotection des Plantes EA 4707, BP 1039, (Cedex 2), F-51687 Reims, France

Laura Mugnai
Dipartimento di Scienze e Tecnologie Agrarie, Alimentari, Ambientali e Forestali (DAGRI) - Sez. Patologia vegetale ed Entomologia, University of Florence, Piazzale delle Cascine 28, 50144 Florence, Italy

Arzu Çelik Oğu
Ankara University, Faculty of Agriculture, Department of Plant Protection, 06110, Ankara, Turkey

Leonardo Varvaro
Department of Agriculture and Forestry Science, University of Tuscia, Via S. Camillo de Lellis s.n.c., 01100 Viterbo (VT), Italy

Giulia Casini
Department of Agriculture and Forestry Science, University of Tuscia, Via S. Camillo de Lellis s.n.c., 01100 Viterbo (VT), Italy
Enbiotech S.r.l., via Quarto dei mille, 6, 90144 Palermo (PA), Italy

Sandro Drago
Enbiotech S.r.l., via Quarto dei mille, 6, 90144 Palermo (PA), Italy

Thaer Yaseen
Regional Office for the Near East and North Africa Region (RNE), Food and Agriculture Organization of the United Nations (FAO), Cairo, Egypt

Leonardo Schena
Dipartimento di Agraria, Università Mediterranea di Reggio Calabria, Località Feo di Vito, 89122 Reggio Calabria (RC), Italy

Ahmed Abdelfattah
Dipartimento di Agraria, Università Mediterranea di Reggio Calabria, Località Feo di Vito, 89122 Reggio Calabria (RC), Italy
Department of Ecology, Environment and Plant Sciences, Stockholm University, SE-106 91, Stockholm, Sweden

Franco Santoro
CIHEAM - Istituto Agronomico Mediterraneo, Via Ceglie 9, 70010 Valenzano (BA), Italy

Samira Chekali
Laboratoire de Protection des Végétaux, Institut National de Recherche Agronomique de Tunisie, Pôle Régional de Recherche-Développement Agricole du Nord-Ouest Semi-Aride au Kef, Université de Carthage, BP 221, 7100, Le Kef, Tunisia

Samia Gargouri
Laboratoire de Protection des Végétaux, Institut National de la Recherche Agronomique de Tunisie, Université de Carthage, Rue Hédi Karray, 2049, Ariana, Tunisia

Moncef Ben Hammouda
Laboratoire de Physiologie de la Production des Céréales, Ecole Supérieure d'Agriculture du Kef, Université de Jandouba, 7119 Le Kef, Tunisia

Hatem Cheikh M'hamed
Laboratoire de Recherche de Science et Techniques Agronomiques, Institut National de Recherche Agronomique de Tunisie, Université de Carthage, Rue Hédi Karray, 2049, Ariana, Tunisia

Bouzid Nasraoui
Laboratoire de Recherche des Bioagresseurs et Protection Intégrée en Agriculture, Institut National Agronomique de Tunisie, Université de Carthage, 43 Avenue Charles Nicolle, 1082, Tunis-Mahrajène, Tunisia

Mohamed Zouaoui, Badiaa Essghaier, Maroua Weslati, Marwa Smiri and Najla Sadfi-Zouaoui
Laboratoire de Mycologie, Pathologies et Biomarqueurs LR16ES05, Faculté des Sciences de Tunis, Université de Tunis El Manar 2092, Tunis, Tunisie

Mohamed Rabeh Hajlaoui
Laboratoire de Biotechnologie Appliquée à l'Agriculture, INRA Tunisie, Université de Carthage, 2049 Ariana, Tunisie

Virgilio Balmas, Barbara Scherm, Domenico Rau, Salvatore Camiolo and Vanda Assunta Prota
Dipartimento di Agraria, Universita degli Studi di Sassari, Via E. De Nicola 9, 07100 Sassari, Italy

Safa Oufensou
Dipartimento di Agraria, Universita degli Studi di Sassari, Via E. De Nicola 9, 07100 Sassari, Italy
Laboratoire de Bio-surveillance de l'Environnement, Faculte des Sciences de Bizerte, Route de Tunis, 7021Zarzouna, Universite de Carthage, Tunisia

Quirico Migheli
Dipartimento di Agraria, Universita degli Studi di Sassari, Via E. De Nicola 9, 07100 Sassari, Italy
Unita di ricerca-Istituto Nazionale di Biostrutture e Biosistemi, Universita degli Studi di Sassari, 07100 Sassari, Italy

Mossadok Benattia
Laboratoire de Bio-surveillance de l'Environnement, Faculte des Sciences de Bizerte, Route de Tunis, 7021Zarzouna, Universite de Carthage, Tunisia

Samia Gargouri
Laboratoire de Protection des Vegetaux, Institut National de Recherche Agronomique de Tunisie, Rue Hedi Karray, 2049 Ariana, Tunisia

Matias Pasquali
DeFENS-Department of Food Environmental and Nutritional Sciences, Universita degli Studi di Milano, Via Celoria, 2, 20133 Milano, Italy

Safia El-Bok
Laboratoire de Biodiversite, Biotechnologies et Changements Climatiques, Faculte des Sciences de Tunis, Universite Tunis-El Manar, 2092 Tunis, Tunisia

Julienne Nguefack and Ghislaine Ndonkeu Mangoumou
Department of Biochemistry, Faculty of Science, University of Yaounde 1, Cameroon

Pascal Noel Mekam
Department of Biochemistry, Faculty of Science, University of Yaounde 1, Cameroon
Department of Life Sciences, University of Modena and Reggio Emilia, Via Amendola 2, 42122 Reggio Emilia, Italy

Serena Martini and Davide Tagliazucchi
Department of Life Sciences, University of Modena and Reggio Emilia, Via Amendola 2, 42122 Reggio Emilia, Italy

Emilio Stefani
Department of Life Sciences, University of Modena and Reggio Emilia, Via Amendola 2, 42122 Reggio Emilia, Italy
University Centre for International Cooperation and Development (CUSCOS), via Universita 4, 41121 Modena, Italy

Rola El Amil
Lebanese Agricultural Research Institute (LARI), 287 Zahleh, Bekaa, Lebanon
UMR BIOGER INRA, AgroParisTech, Université Paris-Saclay, 78850 Thiverval- Grignon, France

Claude De Vallavieille-Pope and Marc Leconte
UMR BIOGER INRA, AgroParisTech, Université Paris-Saclay, 78850 Thiverval- Grignon, France

Kumarse Nazari
Regional Cereal Rust Research Center, Menemen, Izmir, Turkey

Gabriella Cristina Botelho Mageste da Silva, Rafael Pio and Pedro Maranha Peche
Departament of Agriculture, Federal University of Lavras (UFLA) - Lavras – MG - 37200-000, Brazil

Renata Cristina Martins Pereira and Edson Ampelio Pozza
Departament of Phytopathology, Federal University of Lavras (UFLA) - Lavras – MG - 37200-000, Brazil

Index

www.ingramcontent.com/pod-product-compliance
Lightning Source LLC
LaVergne TN
LVHW080920210926
848485LV00004B/689